Solitons in Crystalline Processes (2nd Edition)

Irreversible thermodynamics of structural phase transitions and superconductivity

Solitons in Crystalline Processes (2nd Edition)

Irreversible thermodynamics of structural phase transitions and superconductivity

Minoru Fujimoto
University of Guelph, Ontario, Canada

IOP Publishing, Bristol, UK

ISBN 978-0-7503-2572-1 (ebook)
ISBN 978-0-7503-2570-7 (print)
ISBN 978-0-7503-2571-4 (myPrint)
ISBN 978-0-7503-2573-8 (mobi)

DOI 10.1088/978-0-7503-2572-1

Version: 20191101

IOP ebooks

British Library Cataloguing-in-Publication Data: A catalogue record for this book is available from the British Library.

Published by IOP Publishing, wholly owned by The Institute of Physics, London

IOP Publishing, Temple Circus, Temple Way, Bristol, BS1 6HG, UK

US Office: IOP Publishing, Inc., 190 North Independence Mall West, Suite 601, Philadelphia, PA 19106, USA

To the memory of Jan Stankowski

Contents

Notes on the second edition

Published more than a year ago, the soliton theory in the first edition is acceptable with a discipline of thermodynamics, revising in a more readable form to explore this new area of research. It is the author's responsibility to publish an adequately revised edition for concerned readers. The second edition is thus prepared for comprehensive materials in a textbook to teach readers at graduate level. There are a number of untraditional discussions in this edition, however, it is important to follow them as discussed in sequence for efficient reading.

In the first edition, the superconductivity was logically discussed in terms of the soliton theory for a variety of crystals, restricted to the critical phenomena. In this edition, however, brief discussions for modulated lattice were added to emphasize soliton concepts to be essential to understand the role played by deformed lattice. Two groups of synthetic layer compounds of 3d transition-ions are dominated in the present literature on high T_c superconductors, which are briefly discussed in this edition.

Subjects on polymers and liquid crystals should be discussed extensively with respect to practical observations; these should, however, be studied under a different discipline, as, for example, in *Physics of Liquid Crystals* by de Gennes and Prost. In contrast, the soliton dynamics in the present form offers a logical approach to irreversible superconducting processes, as evidenced for example by the common mechanism of distorted lattice for superconductivity.

Minoru Fujimoto
May 2019

Preface to the first edition

This monograph is intended to serve as an introductory book for structural phase transitions and related mesoscopic disorder, presenting the soliton theory applied to modulated crystalline states. In past decades, modulated crystals have not been much discussed systematically using soliton theory, in spite of the significance not only for phase transitions but also for superconductivity, liquid crystals and other properties of modern materials, as stimulated particularly by recent high-pressure experiments. That motivated me to write about modern thermodynamics of crystals, which should be helpful in searching for new information from diversified literature.

In thermodynamics of condensed matters, the *internal energy* needs to be redefined more precisely than originally, in the absence of reliable knowledge of cohesive force in today's physics. Moreover, microscopic variables in collective motion play a basic role in distorted crystals, while presently discussed only with microscopic order variables that are primarily assumed as independent from each other. For instance, collective motion in isothermal processes cannot be dealt with in the mean-field approximation, unless the system is dominantly conservative in nature. In traditional theories, the internal Weiss field or equivalent potential are assumed as if applied uniformly from outside, so that recent experimental studies demand updating existing theories with the *soliton field* to deal with internal correlations in whole crystals. Soliton theory is therefore essential for collective motion to be logically discussed, constituting the lattice dynamics of irreversible processes in crystalline states.

In traditional theories, order variables in crystals are discussed with linear differential equations for symmetrical lattice structure as granted. However, interactions in modulated structure are essentially nonlinear in character, for which the soliton theory is required for a logical pathway to the intrinsic Weiss field, which is necessary to deal with spontaneous changes in crystals. For example, at extreme-low temperatures in metals, Cooper's electron pairs bound via *soliton mechanism* derive the corresponding *persistent current* in superconducting state, otherwise their charge–current continuity cannot be confirmed.

Phase transitions followed by mesoscopic structural disorder constitute basic subjects of irreversible thermodynamics of crystalline states, which can be discussed logically with the soliton concept, as substantiated by experimental results. In this book, transition anomalies due to lattice fluctuations are fully discussed in chapters 1–7 with solitons in mind, which are then extended with the soliton concept in chapters 8–17 to mesoscopic states signified by nonlinearity.

The statistical *scaling theory* is not particularly included, while critical exponents are believed to be derivable from the soliton theory for experimental data that are characterized in idealized conditions. Pedagogically however, it is more fundamental to learn about the soliton theory for irreversible thermodynamics; besides, the concept can naturally be involved in collective motion in modulated crystals. Instead, *Toda's theory* of the soliton lattice is fully introduced in chapter 11 for convenience of discussing binary systems.

Written as an introductory treatise, this book should be a useful reference, readable with standard knowledge of classical thermodynamics, electromagnetic theory, quantum theory and statistical mechanics of isotropic systems. While mathematical discussions are kept to minimum necessities, those interested in detail are referred, for example, to my previous book *Introduction to the Mathematical Physics of Nonlinear Waves* (IOP Publishing, 2014). Exercise problems are listed at most chapter ends for reviewing convenience of key issues. Discussed at the level of undergraduate mathematics, this book should not be hard to read by students and researchers in physics and material sciences. Required formulae of hyperbolic and elliptic functions are listed in the appendix for convenience for unfamiliar readers. It is the author's hope to provide stimulating readable materials in modern solid-state physics for those interested in engineering applications as well.

Acknowledgments

Engaged mostly in experimental work myself, I was guided profoundly towards practical crystals by a number of pioneering books in this area of irreversible thermodynamics written by Professors M Born, K Huang, I Prigogine, G L Lamb Jr and M Toda, to whom I express my heartfelt gratitude. I should also mention with sincere appreciation that the writing benefited a great deal from numerous discussions and valuable comments with my colleagues and students, and that the draft manuscript was read by Professor N Akhmediev prior to publication. Finally, my thanks also go to my wife Haruko for her continuous encouragement.

Minoru Fujimoto
April 2016

Author biography

Minoru Fujimoto

Minoru Fujimoto is a retired professor from the University of Guelph, Ontario, Canada. During his association with the university, his research area was in the field of magnetic resonance studies on structural phase transitions in crystals, which has currently been extended to theoretical work with soliton dynamics. He is the author of *Physics of Classical Electromagnetism and Thermodynamics of Crystalline State* (Springer), and *Introduction to Mathematical Physics of Nonlinear Waves* (IOP Publishing). He lives in Mississauga, Ontario; mfujimotp@gmail.com.

Introduction

Classical thermodynamics is a well-established discipline of physics today. Applying to crystals however, irreversible processes of collective order variables below transition temperatures are essentially nonlinear, responding to the hosting lattice, so that the traditional approach should be revised for the *soliton theory* to deal with *mesoscopic phenomena*. Signified by the *entropy production* with varying volume and temperature, the *internal correlation energies* among constituent molecules are responsible for the *irreversible processes* in crystalline states.

In this introduction, revising the traditional statistical approach, the basic concepts are redefined for modulated crystals in finite size in the thermal environment. In the following chapters, nonlinear dynamics constitutes the basic objective for irreversible processes in crystals, extending the traditional theory in isotropic systems [1] to irreversible processes in modulated crystals.

0.1 The internal energy of equilibrium crystals

In equilibrium with surroundings, crystalline states of chemically pure materials are characterized by the uniform lattice structure, as determined by the crystallographic analysis. In the absence of reliable knowledge on the molecular coagulation in today's physics however, the equilibrium lattice structure defined by *symmetry groups* is assumed to be a fact of nature. In this view, the modulated lattice is not always in equilibrium with surroundings, where the *elastic strain energy* in the lattice is distributed within structure, hence the non-equilibrium crystalline state at given pressure from the surroundings is generally temperature-dependent. On the other hand, the internal dynamics in equilibrium with the lattice should be *conservative* as determined by canonical equations for steady states; otherwise it shoulf be *dispersive* and *dissipative* in non-equilibrium processes.

Referring to *space* and *point groups*, the equilibrium structure of crystals is invariant under symmetry operations of the lattice. The point group specifies for all

doi:10.1088/978-0-7503-2572-1ch0

identical constituents to remain static at fixed lattice sites, whereas the space group confirms the internal structural invariance of crystals against spatial translations.

Denoting the *internal energy* of an equilibrium crystal in the surroundings by a constant energy U_0 of the lattice structure, the conservation law for the internal energy U determined by external pressure p and temperature T can be expressed in general as

$$U - U_0 = \Delta U = Q - p\Delta V, \tag{0.1}$$

where Q and $-p\Delta V$ represent respectively heat change with the lattice at T and the energy transfer for a volume change ΔV at constant p of the surroundings; here equation (0.1) represents the *first law of thermodynamics*. Invariant geometrical symmetry itself does not represent properties of crystals in the thermodynamic environment, however, it can be violated during irreversible processes, which is therefore considered as responsible for *modulated structure*. Moreover, for crystals specified by finite volume, the volume change ΔV cannot be ignored in principle, while the condition $\Delta V = 0$ is technically difficult to maintain in practical crystals.

Heat transfer processes are always *irreversible*, where the transferred heat quantity Q' is in fact always *less than* a given Q, so that equation (0.1) should be replaced by an inequality

$$\Delta U \geqslant Q' - p\Delta V,$$

as described by the *second law of thermodynamics*. Mathematically, finding such a temperature T as to keep $Q'/T = \Delta S$ a *total differential of the entropy* S, and the above inequality can be re-expressed by $\Delta U \geqslant T\Delta S - p\Delta V$; hence we write

$$\Delta F \geqslant -p\Delta V, \tag{0.2}$$

where the *Helmholtz function* $F = U - TS$ defined from (0.2) can be used for equilibrium conditions at constant volume with a given pressure p; the free energy F should take a minimum value at $\Delta F = 0$, which can be regarded to be at a constant volume condition.

Related to the entropy S, the temperature T is associated with the heat Q that can be attributed microscopically to *random collisions* of air particles in the surroundings. Thereby, the *thermodynamic probability* $g(T)$ can be defined statistically by the equation $S = k_B \ln g(T)$, where k_B is the Boltzmann constant.

0.2 Microscopic order variables and their fluctuations

Spontaneous structural changes are significant phenomena in crystals, which can be attributed to *internal order variables* $\boldsymbol{\sigma}_n$ associated with constituents at all lattice sites n, characterized by *space symmetry or internal degree of freedom* of the constituent. Thermodynamically, $\boldsymbol{\sigma}_n$ should be regarded as continuous variables of space-time average at low energies, as defined by Kirkwood and Oppenheim [2] for internal variables in chemical thermodynamics. However, such internal variables $\boldsymbol{\sigma}_n$ called *order variables* in crystals are mutually correlated among lattice sites. Assuming *binary correlations* that are predominant at short distances, the correlation energy can be written as $-\sum_m J_{mn}\boldsymbol{\sigma}_m\boldsymbol{\sigma}_n = -\boldsymbol{\sigma}_n\sum_m J_{mn}\boldsymbol{\sigma}_m$, where the quantity $X_n = \sum_m J_{mn}\boldsymbol{\sigma}_m$ can be defined as a *field* acting on $\boldsymbol{\sigma}_n$ by analogy of *Weiss' molecular field* in magnetic crystals. Referring to it as the *Weiss field* in general, such local fields X_n can be defined as a continuous field at low energies in crystals. Assuming these $\boldsymbol{\sigma}_n$ and X_n as continuous variables, equation (0.1) can be generalized as $\Delta U = T\Delta S - p\Delta V - \sum_n \sigma_n \Delta X_n$, including an *inevitable volume change* ΔV, thereby expressing the equilibrium condition as

$$\Delta G \geqslant 0, \tag{0.3}$$

where $G = U - TS + pV + \Sigma_n \boldsymbol{\sigma}_n X_n$ is called the *Gibbs free energy*, indicating that the equilibrium at given T and p can be obtained by minimizing G with respect to σ_n. Born and Huang [3] showed in their theory of lattice dynamics that the Weiss field is a valid concept in the *adiabatic approximation*. It is noted that in the presence of internal $\sigma_n X_n$ or *correlations* $J_{mn}\sigma_m\sigma_n$, the volume is not usually constant, i.e. $\Delta V \neq 0$, hence the Gibbs function should be minimized to determine equilibrium conditions. In fact, in order for (0.2) to express the thermodynamic equilibrium, we must consider the relation

$$-p\Delta V = \sum_n \sigma_n \Delta X_n,$$

attributing all internal variations to an effective volume change ΔV, which is particularly significant for *relatively larger molecular constituents*. Therefore, it is more convenient to use the Gibbs function than the Helmholtz function, if we need to consider the internal variable σ_n in finite size.

Microscopic order variables in crystals are considered to *fluctuate* among lattice points, which are in fact unavoidable during transitions between different crystalline phases in particular. Crystalline states in transition are *inhomogeneous* in density, so we consider mean-field averages $g_1 = G_1/V_1$ and $g_2 = G_2/V_2$ for the phases 1 and 2, respectively. Their equilibrium can then be determined by minimizing the difference $g_1 - g_2 = \Delta g$, namely

$$\Delta g = \frac{1}{2}\left\langle \frac{\partial^2 \Delta g}{\partial\sigma_1\partial\sigma_2} \right\rangle_{p,T} \langle \Delta\sigma_1\Delta\sigma_2 \rangle_{p,T} + \cdots, \tag{0.4}$$

where the brackets $\langle\cdots\rangle_{p,T}$ indicate the mean-field average of fluctuations. In deriving (0.4), we consider that averages of order variables themselves should vanish, so that $\langle\sigma_1\rangle_{p,T} = \langle\sigma_2\rangle_{p,T} = 0$ are utilized. In this case, from (0.4) the average of binary correlations should be non-zero, as expressed by the inequality $\langle\Delta\sigma_1\Delta\sigma_2\rangle_{p,T} \neq 0$, which is *necessary* for the transition to be *second-order* in the *Ehrenfest classification* of phase transitions. Nonetheless, observed critical anomalies indicate evidence for such binary correlations to describe thermodynamically *adiabatic fluctuations* in crystals.

The transition from uncorrelated to correlated variables is associated with a dynamical *bifurcation* for the nonlinear process; on the other hand, observed anomalies can be attributed to unavoidable *quantum-mechanical space-time uncertainties* during the transition.

0.3 Collective order variables in propagation

Most order variables $\boldsymbol{\sigma}_n$ at lattice sites n are related to *partial displacements* inside constituents, representing *relative displacement* with respect to the lattice. In crystals, such σ_n are in collective fluctuations in a mesoscopic phase that should be accompanied with *counter-fluctuations* in the periodic structure, resulting in collective displacements of lattice sites. Mathematically, such collective fluctuations *in periodic structure in finite size* can be described by *Fourier series* [4] as expressed by

$$\boldsymbol{\sigma}_n = \sum_{\boldsymbol{k}} \boldsymbol{\sigma}_k \exp i(\boldsymbol{k} \cdot \boldsymbol{r}_n - \omega t_n) \tag{0.5}$$

which is convenient to use in crystals as characterized by the *Bloch theorem*, where $\omega = v|\boldsymbol{k}|$ and v are the frequency and speed of propagation along a specified direction in periodic structure.

In sufficiently large crystals, where surfaces are neglected, the local phase

$$\phi_n = \boldsymbol{k} \cdot \boldsymbol{r}_n - \omega\, t_n$$

at a site n varies in range $-\pi/2 \leqslant \phi_n \leqslant \pi/2$, but can effectively be redefined by a continuous phase $\phi = \boldsymbol{k} \cdot \boldsymbol{r} - \omega\, t$, if $|\boldsymbol{k}|$ is sufficiently small in crystals in large size. Accordingly, the continuous phase ϕ in range $-\pi/2 \leqslant \phi \leqslant \pi/2$ can be employed, instead of ϕ_n, to express the collective $\sigma(\boldsymbol{r}, t)$ as a thermodynamic variable.

Expressing the collective order variable by $\boldsymbol{\sigma}(\boldsymbol{r}, t) = \boldsymbol{\sigma}_0 \exp i\phi$ for small values of $|\boldsymbol{k}|$, we can use the Fourier transform

$$\boldsymbol{\sigma}_k = \frac{1}{\sqrt{N}} \sum_n \boldsymbol{\sigma}_n \exp\{-i(\boldsymbol{k} \cdot \boldsymbol{r}_n - \omega t_n)\},$$

which can be simplified for a single k-vector as

$$\boldsymbol{\sigma}_k(\phi) = \boldsymbol{\sigma}_0 \exp(-i\phi) \quad \text{where} \quad \phi = \boldsymbol{k} \cdot \boldsymbol{r} - \omega\, t, \tag{0.6}$$

and inversion symmetry in finite crystals can thus be signified by the *coherent phase inversion* $\phi \to -\phi$.

Theorem: For thermodynamics, we consider *spatial inversion* $\boldsymbol{r} \to -\boldsymbol{r}$ *only, disregarding the time inversion.* That is the basic field-theoretical approach to thermodynamics of crystals, where the surface can be ignored by considering volumes in sufficiently large size, while important for thermodynamics of heat exchange with the surroundings.

We therefore define renormalized phase variables as $-\pi/2 \leqslant \pm\phi \leqslant +\pi/2$ for order variables, which are employed to express *collective order variables* in crystals. On the other hand, phases ϕ_n are referred to as microscopic space-time $(\boldsymbol{r}_n, t_n)$. In fact, it is particularly convenient to define such a range as $-\pi/2 \leqslant \pm\phi \leqslant +\pi/2$ to cover phase inversion in the whole crystal from one side to the other of *finite periodic structure*. Assuming that one-dimensional length of a crystal is composed of integral multiples of a short renormalized unit, the *terminal surfaces* on the right and left can be determined by boundary conditions

$$-\pi/2 \leqslant \pm\phi \quad \text{and} \quad \pm\phi \leqslant +\pi/2, \tag{0.7}$$

respectively, offering a simple but realistic theoretical model for crystal surfaces to interact with thermodynamic surroundings.

Theorem: In field-theoretical approximation valid for small values of $|\boldsymbol{k}|$, the motion of $\boldsymbol{\sigma}_k(\phi)$ is internally driven by a *force* $-\partial\Delta U_k/\partial\phi$, where U_k is a function of ϕ, and ΔU_k should be responsible for wave motion of $\sigma_k(\phi)$. For dynamics of $\boldsymbol{\sigma}_k(\phi)$ in general, we should consider the corresponding *kinetic energy* $K(\dot{\boldsymbol{\sigma}}_k)$ as well, where $\dot{\boldsymbol{\sigma}}_k$ is a momentum variable conjugate to $\boldsymbol{\sigma}_k$. By the least-action principle, the integral of *Lagrangian* $L(\dot{\boldsymbol{\sigma}}_k, \boldsymbol{\sigma}_k) = K(\dot{\boldsymbol{\sigma}}_k) - U(\boldsymbol{\sigma}_k)$, namely the action variable $\int_V L(\dot{\boldsymbol{\sigma}}_k, \boldsymbol{\sigma}_k)dV$ should be minimized for a *conservative* system to determine for the Hamiltonian $\mathcal{H} = K(\dot{\boldsymbol{\sigma}}_k) + U(\boldsymbol{\sigma}_k)$ to be a constant of time, characterizing the equilibrium system by its eigenvalue. Experimentally, the potential $\Delta U(\sigma_k)$ or ΔU_k has been confirmed to exist in equilibrium crystals.

0.4 Crystal surfaces and entropy production

In thermodynamics, crystal surfaces play a significant role for heat exchange with the surroundings. On the other hand, we consider idealized crystals of sufficiently large volume, whose properties are determined by the whole crystal, thereby dynamically ignoring surfaces from the bulk crystals. However, the surfaces cannot be ignored thermodynamically, because of their role for order variables to be specified by a finite value on surfaces; nevertheless, we minimize surface contributions to properties of periodic structure for mathematical convenience, assuming a crystal to consist of a large number of unit structures in repetition. Disregarding surfaces, we can consider that surfaces are represented by *crystal planes*, which actually are convenient for discussing *domain structure* as well. Nonetheless, for heat exchange with surroundings, we do not have to specify real surfaces, but mathematically require *nodal planes* for $\boldsymbol{\sigma}_k(\phi)$ at $\phi = \pm(\pi/2) \times$ integer.

Singularities of the function $\boldsymbol{\sigma}_k(\phi)$ can be responsible for energy transfer to the crystalline media and to surroundings, where the process is naturally irreversible, as manifested by the second law of thermodynamics.

In the *soliton theory*, the energy transfer can be described conveniently by *quantized soliton particles*[note1] for thermodynamic description of *entropy production*.

As defined in chemical thermodynamics [1], a continuous parameter for reacting species is a convenient measure for *isotropic reactions* to be described as an irreversible energy transfer process from $\boldsymbol{\sigma}_k(\phi)$ to the correlation energy $U_k(\phi)$,

[note1] Here, the word 'quantized' refers to discrete soliton numbers in crystals, rather than quantum-mechanical quantization with respect to the Planck's constant $h = 6.62618 \times 10^{-34}$ *J-s*. Nevertheless, solitons can be regarded as quantized particles in crystalline media, regarding discrete lattice energies.

which is usually expressed in terms of a *chemical potential* μ and *number n* of soliton particles.

However, it is important to realize that such a dynamical argument as above should be limited to *macroscopically uniform crystals.* In order to analyze observed results, sample crystals should be in *ellipsoidal shape* in principle.

At this point, we propose considering another quantity called *solitons* to crystals, thereby making the structure *inhomogeneous,* as in flowing reactions in chemical systems, as will be discussed in later chapters. Writing the *soliton density n* in the Gibbs function as $G(p, T; n)$ for inhomogeneous crystals, we have

$$G(p, T; n) = \Delta U - T\,\Delta S + p\Delta V + \mu\,\Delta n, \tag{0.8}$$

where Δn is a variation of *soliton number of order variables* that is a function of T and p of the surroundings, and μ is the *chemical potential*; such Δn can be written as $\Delta n(T)$ and $\Delta n(p)$ for isothermal and isobaric processes, respectively. Combining the second and fourth terms on the right of (0.8), entropy production can be expressed as

$$T\left(\Delta S - \frac{\mu\,\Delta n(T, p)}{T}\right),$$

where the second term $\Delta S' = -\frac{\mu\,\Delta n(T,p)}{T}$ represents effectively additional entropy production to the conventional heat $Q = T\Delta S$. In this case, as in chemical systems, $Q' = T\Delta S'$ represents effectively either $-\mu\,\Delta n(T)$ or by $-\mu\,\Delta n(p)$ for entropy production $\Delta S'$ in isothermal or isobaric processes, respectively.

While valid in chemical systems in liquid phase, it is also significant in crystals that such entropy exchanges ΔS and $\Delta S'$ take place not only between the system and surroundings, but also between order variables and lattice internally, as theoretically described by *Onsager's reciprocity theorem.*

Quantized lattice vibrations represented by phonons are also signified for their energies to be determined by the *equipartition theorem* as proportional to thermal energy k_BT, so that $\Delta n \propto T$, hence $\Delta S' = 0$ *for phonons.* On the other hand, in crystals such a quantization should be applied to modulated structure, for which $\Delta n(T)$ is not simply proportional to T, to make adiabatic entropy production $\Delta S' \neq 0$, but accompanied by finite temperature change $\Delta T \neq 0$. In this sense of thermodynamics, the *time scale can conveniently be converted to the temperature scale to deal with entropy production.*

Furthermore, in modulated crystalline states $\mu\,\Delta\,n$ expresses internal work equivalent to mechanical work $-p_{\text{int}}\Delta V'$ at a temperature T, where $\Delta V'$ is an *adiabatic volume change* and p_{int} is an effective internal pressure, which is more appropriately expressed as *external work* $+V'\,\Delta p_{\text{ext}}$, if the crystal is under external pressure p_{ext}[note2]. In this case, $\mu\Delta n$ can be observed with varying external work on degrading lattice symmetry against surroundings. The soliton theory is primarily

[note2] This relation is consequent on the soliton gas that obeys the law of ideal gas, i.e. $pV' = RT$ at a constant T, applied to the equilibrium condition $\Delta G_{\text{int}} + \Delta G_{\text{ext}} = 0$.

established for isotropic media at constant V', however, in crystalline states $\Delta V' \neq 0$ occurs normally as determined by changing lattice symmetry.

0.5 Lattice symmetry and the internal energy in crystals

Crystals constitute a specific group of condensed matter, characterized by periodic arrangement of constituent ions and molecules, specified by invariant structure by geometrical operations such as translation, reflection, rotation, etc, to maintain lattice symmetry. In this book, we shall not discuss the mathematical detail of finite group theory, referring the reader instead to excellent references available in the literature [5].

However, some important aspects of the internal energy of crystals require specific attention in thermodynamics, which is significant particularly for the structure in thermodynamic processes.

We have to consider a variation in thermodynamic processes due to a variety of changes in surrounding conditions. Considering a change in the vibrational energy $\Delta\mathcal{H}_{\text{vib}}$ due to the internal potential ΔU_{o}, according to least-action principle in section 0.4, we have the relation

$$\left\langle Q_i|\Delta\mathcal{H}_{\text{vib}} - \Delta U_{\text{o}}(p, T)|Q_j\right\rangle = \Delta E_i\, \delta_{ij}, \tag{0.9a}$$

where the operator $\langle Q_i|$ represents structural transformation of the lattice, and $\delta_{ii} = 1$, $\delta_{ij} = 0$ for $i \neq j$ (Kronecker's delta). Then, (0.9*a*) signifies

$$\left\langle Q_i|\mathcal{H}_{\text{vib}}|Q_i\right\rangle_{p,T} = \left\langle Q_i|U_{\text{o}}|Q_i\right\rangle_{p,T} = E_i, \tag{0.9b}$$

indicating that both $|\mathcal{H}_{\text{vib},i}|_{p,T}$ and $|U_{\text{o},i}|_{p,T}$ have the common eigenvalue E_i, allowing for the former to represent the latter. That is important for both to represent the *thermal equilibrium at E_i characterized by the corresponding T_i at constant p*, i.e. $E_i \propto T_i$ that is consistent with principles of statistical mechanics.

0.6 Timescales for sampling modulated structure and thermodynamic measurements

In the thermodynamics of an irreversible process the collective motion of order variables is vital in the modulated structure in mesoscopic scale. The timescale for modulation is not determined entirely from the quantization process, unless the minimum modulation energy is specified. Paying attention to the discreteness of the modulated lattice, however, we need to consider observing frequencies under equilibrium conditions. To sample modulated lattices with a known frequency, the timescale of the measurement is significant for correct analysis of experimental results.

Structural changes in crystals are detected from observed anomalies due to fluctuations of $\boldsymbol{\sigma}_k(\phi)$, where phase variations $\Delta\phi$ dominate the critical region. Such fluctuations observed as a *mesoscopic modulation* of $\sigma_k(\phi)$ can be visualized by an appropriate sampling experiment in timescale t_o [6]; in contrast, conventional thermal measurements are *macroscopic* and performed on very long timescales. Although theoretically insignificant, it is important for practical observations, if the timescale t_o of experiments is comparable with the timescale t of fluctuating $\sigma_k(\phi)$. In the mesoscopic observation, measured quantities are determined by the time average between $+t_o$ and $-t_o$.

Sampling a fluctuating Gibbs function by an experiment in timescale $t = 2\pi/\omega$, the observed quantity is determined by such a time average as determined by $\pm t_o$. Accordingly, normal sampling results are related to the thermodynamic average

$$\langle G(\phi)\rangle_{p,T} = \frac{1}{2t_o}\int_{-t_o}^{+t_o} \langle G(\boldsymbol{r}, t)\rangle_{\text{space av.}} \cos \omega t \, dt = \langle G(\boldsymbol{r})\rangle_{\text{space av.}} \frac{\sin \omega t_o}{\omega t_o}, \qquad (0.10)$$

which is approximately equal to $\langle G(\boldsymbol{r})\rangle_{\text{space av}}$. if $\omega t_o \approx 1$, otherwise (0.10) is averaged out. Nevertheless, $\langle G(\boldsymbol{r})\rangle_{\text{space av.}}$ is *measurable* in timescale t_o, provided that $t_o < t = 2\pi/\omega$. We discuss sampling practices in detail in Part II.

In contrast to sampling experiments, thermodynamic quantities such as $G(\phi)$ are determined by the surroundings in equilibrium at p and T, for which the timescale of observation is regarded as infinity. For example, the specific heat of a crystal or any other thermodynamic quantity is measured as a function of T under $p = \text{const.}$, where a temperature-dependent frequency $\omega = \omega(T)$ is detected, referred to as *soft mode.*

0.7 Statistical theories and the mean-field approximation

The traditional statistical theory of solids is based on the fact that the *lattice vibration* is random in character, permitting statistical averages of dynamical variables in crystals determined by the temperature of the surroundings. Ordering processes in crystalline media are therefore discussed with statistical theories, considering the lattice as the heat reservoir. The Weiss field was defined originally in mean-field approximation, dominating thermodynamic discussions of solid-state phenomena. In mean-field accuracy, however, lattice symmetry is only an implicit variable for thermodynamic analysis of equilibrium crystals at constant volume. As has emerged in statistical arguments, important concepts of *order parameter* and adiabatic potential should, however, be redefined for *mesoscopic collective variables* in soliton theory. In this section, these statistical concepts are reviewed as important prerequisites for revising existing theories.

Although somewhat specific, we consider *binary order* as fundamental for structural changes of crystals, while it is mathematically simple among others. Binary processes are the subject of detailed discussion in Parts I and II; based on which we proceed to nonlinear soliton theories in crystals in Parts III and IV.

0.7.1 Probabilities and the domain structure

In order–disorder phenomena of binary alloys of two atoms A and B, we consider probabilities $p_n(\text{A})$ and $p_n(\text{B})$ at lattice sites n that are occupied by either A or B atoms. We assume that there are no vacant lattice sites; each site must be occupied by either one of these atoms. Denoting number of sites by N_A and N_B, we have then $N_A = N_B = N/2$, where $N = N_A + N_B$.

Physical properties of alloys, such as β–brass CuZn, are well documented, exhibiting order–disorder transitions at critical temperatures T_c. In their original statistical theory, Bragg and Williams [7, 8] considered probabilities for pairs of like atoms, A–A and B–B, and for unlike pairs, A–B and B–A, at nearest-neighbor sites, thereby introducing the concept of short-range order. Judging from the order of the transition temperature $T_c \approx 450°C$, β–brasscannot be so rigid that the constituent atoms should be in diffusive motion, and the lattice may be unsteady in the timescale of observation; nevertheless, these aspects were ignored in early physics.

By definition, we have the relation

$$p_n(\mathrm{A}) + p_n(\mathrm{B}) = 1 \quad 0 \leqslant p_n(\mathrm{A}),\ p_n(\mathrm{B}) \leqslant 1 \tag{0.11a}$$

from which we can define the order variable as

$$\sigma_n = p_n(\mathrm{A}) - p_n(\mathrm{B}) \quad -1 \leqslant \sigma_n \leqslant +1 \tag{0.11b}$$

The complete order can be specified by $\sigma_n = \pm 1$, whereas $\sigma_n = 0$ signifies complete disorder, attributing signs of σ_n to *opposite domains*. In addition, the former corresponds to two domains related by inversion $\sigma_n \to -\sigma_n$, and the latter signifies the relations

$$p_n(\mathrm{A}) = p_n(\mathrm{B}) = 1/2. \tag{0.11c}$$

Thus, domains are independent of each other in existing theories, which, however, must be revised for their correlations, if collective motion is considered.

0.7.2 Short-range correlations and the mean-field approximation

The idea of mean-field averages can be applied to order–disorder phenomena to deal with long-range correlations. We use simply mathematical averages $\langle \ldots \rangle = \frac{1}{N}\sum_1^N n$ to obtain averages $\langle p_n(\mathrm{A})\rangle = p(\mathrm{A})$ and $\langle p_n(\mathrm{B})\rangle = p(\mathrm{B})$, and $\langle \sigma_n \rangle = \eta$ is defined as the order parameter.

It is logical to consider the correlations between adjacent atoms in crystals, as the first approximation. We calculate such short-range interactions between nearest-neighbors, which primarily do not violate space symmetry. Denoting interatomic energies by $\varepsilon_{mn}^{\mathrm{AB}}$ etc, to indicate the interaction between A and B at sites m and n, respectively, the short-range interaction energy E_n at site n can be expressed as

$$E_n = \sum_m \left\{ \varepsilon_{mn}^{\mathrm{AA}} p_m(\mathrm{A}) p_n(\mathrm{A}) + \varepsilon_{mn}^{\mathrm{BB}} p_m(\mathrm{B}) p_n(\mathrm{B}) + \varepsilon_{mn}^{\mathrm{AB}} p_m(\mathrm{A}) p_n(\mathrm{B}) + \varepsilon_{mn}^{\mathrm{BA}} p_m(\mathrm{B}) p_n(\mathrm{A}) \right\}.$$

Substituting probabilities by order variables defined in (0.2),

$$E_n = \sum_m E_{mn},$$

where

$$E_{mn} = \frac{1}{2}\left(2\varepsilon_{mn}^{AB} + \varepsilon_{mn}^{AA} + \varepsilon_{mn}^{BB}\right) + \frac{\varepsilon_{mn}^{AA} - \varepsilon_{mn}^{BB}}{4}(\sigma_m + \sigma_n) + \frac{2\varepsilon_{mn}^{AB} - \varepsilon_{mn}^{AA} - \varepsilon_{mn}^{BB}}{4}\sigma_m\sigma_n,$$
$$= \text{const.} - K_{mn}(\sigma_m + \sigma_n) - J_{mn}\sigma_m\sigma_n$$

where $K_{mn} = -(\varepsilon_{mn}^{AA} - \varepsilon_{mn}^{BB})/4$ and $J_{mn} = -(2\varepsilon_{mn}^{AB} - \varepsilon_{mn}^{AA} - \varepsilon_{mn}^{BB})/4$. For like-pairs A–A and B–B, we assume that $\varepsilon_{mn}^{AA} \approx \varepsilon_{mn}^{BB}$ in binary system, hence $K_{mn} \approx 0$ in this case, and we have

$$E_{mn} = \text{const.} - J_{mn}\sigma_m\sigma_n.$$

If σ_n are uncorrelated, we consider that $\langle\sigma_m\sigma_n\rangle = 0$, hence $\langle\text{const.}\rangle = 0$ should be held for the average $\langle E_{mn}\rangle = 0$. Accordingly, for a correlated case, the correlation energy can be expressed as

$$\langle E_{mn}\rangle = -J\langle\sigma_m\sigma_n\rangle, \quad \text{where} \quad J = \langle J_{mn}\rangle, \tag{0.12}$$

which is the binary correlation energy in general form, consistent with the average $\langle\sigma_m\sigma_n\rangle \neq 0$.

0.7.3 The Bragg–Williams theory

Bragg and Williams [6] assumed that the number of unlike pairs A–B determines an ordered arrangement specified by the order parameter η. Using $p(\text{A}) = \frac{1}{2}(1 + \eta)$ and $p(\text{B}) = \frac{1}{2}(1 - \eta)$ for a crystal where the number of nearest neighbors is z, the total number of such pairs can be expressed by

$$N_{\text{AB}} = 2Nzp(\text{A})p(\text{B}) = \frac{1}{2}Nz(1 - \eta^2).$$

Hence the ordering energy can be written as

$$E(\eta) = \frac{1}{2}NzJ(1 - \eta^2) \quad \text{and} \quad E(0) = \frac{1}{2}NzJ,$$

so that

$$\Delta E(\eta) = E(\eta) - E(0) = -\frac{1}{2}NzJ\eta^2 \tag{0.13}$$

is the amount of lowered energy from the disordered state.

The energy $\Delta E(\eta)$ is highly degenerate, since there are a large number of combinations for choosing an A–B pair in the crystal, which need to be calculated as thermodynamic quantities. The statistical weight of the ordering energy $\Delta E(\eta)$ is given by

$$g(\eta) = \binom{N}{Np(\text{A})}\binom{N}{Np(\text{B})} = N^2\binom{1}{\frac{1+\eta}{2}}\binom{1}{\frac{1-\eta}{2}},$$

with which the partition function $Z = Z_L g(\eta)\exp\{-\frac{\Delta E(\eta)}{k_B T}\}$ is calculated, where Z_L is the partition function of the lattice. Then, the equilibrium can be obtained by minimizing Helmholtz' free energy $F = -k_B T \ln Z$. Namely, setting the equation $(\partial F/\partial \eta)_V = 0$, we obtain

$$\frac{\partial}{\partial \eta}\left\{\ln Z_L + \ln g(\eta) + \frac{NzJ\eta^2}{k_B T}\right\} = 0,$$

which is to be solved for η. We evaluate the second term in the brackets by *Stirling' formula* for a large N, i.e. $\frac{\partial \ln \eta}{\partial \eta} = -\frac{N}{2}\ln\frac{1+\eta}{1-\eta}$, resulting in $\frac{zJ}{k_B T} = \ln\frac{1+\eta}{1-\eta}$. Consequently, we arrive at the expression

$$\eta = \tanh\frac{zJ\eta}{2k_B T}. \tag{0.14}$$

To solve equation (0.14) graphically, we set $zJ\eta/2k_B T = y$ and $\eta = \tanh y$ to find the crossing point of the line and the curve in the y-η plane, as illustrated in figure 0.1. It is noted that the tangent at the origin $\eta = y = 0$ corresponds to a critical temperature $T = T_c$; there is one crossing point P if $T < T_c$ but no crossing for all temperatures for $T > T_c$. The critical point is determined by T_c.

For $T < T_c$, if $T_c - T$ is small, we can derive an approximate expression $\eta^2 \simeq 3(T_c - T)/T_c$, yielding a parabolic temperature dependence of the order parameter, $\eta \propto \sqrt{T_c - T}$, in the vicinity of T_c, as shown in figure 0.2(a). In addition, the specific heat $C_V = (\partial \Delta E/\partial T)_V$ calculated with (0.4) shows a discontinuity, $\Delta C_V = \frac{NzJ}{2}\left(\frac{d\eta^2}{dT}\right)_{T_c} = \frac{1}{2}Nk_B$ at $T = T_c$, as illustrated in figure 0.2(b), which is a consequence of the mean-field approximation.

It is notable that equation (0.14) should be consistent with the definition of probabilities (0.10) and order parameter (0.11). To confirm this, we write

$$p(\mathrm{A}) = \frac{1}{Z}\exp\frac{-zJ\eta}{2k_B T}$$

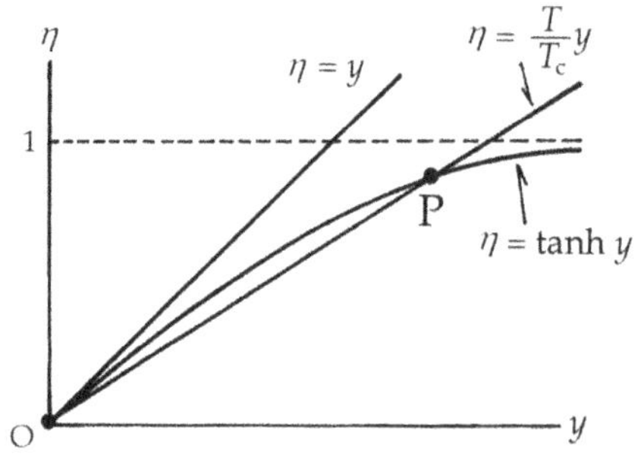

Figure 0.1. The η-y curve of Bragg–Williams' theory. There is one solution at $P(\eta)$ for $T < T_c$, whereas $\eta = 0$ is the solution for $T > T_c$. $T = T_c$ is the critical temperature.

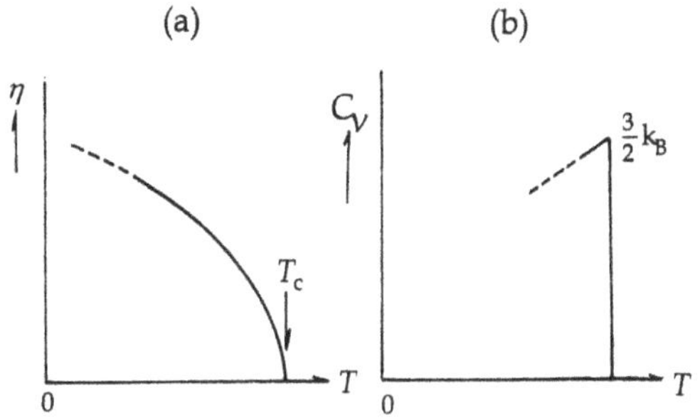

Figure 0.2. (a) Parabolic temperature dependence of $\eta \propto \sqrt{T_c - T}$ for $T < T_c$. (b) Critical discontinuity of ΔC_V at $T = T_c$ in the mean-field approximation.

and

$$p(\mathrm{B}) = \frac{1}{Z} \exp \frac{zJ\eta}{2k_BT},$$

where

$$Z = \exp \frac{-zJ\eta}{2k_BT} + \exp \frac{zJ\eta}{2k_BT},$$

for the energy gap $2zJ\eta = \Delta E(\eta) - \Delta E(-\eta)$, we obtain

$$\eta_+ = p(\mathrm{A}) - p(\mathrm{B}) \quad \text{and} \quad \eta_- = p(\mathrm{B}) - p(\mathrm{A}), \tag{0.15}$$

which are order parameters in *domains* + *and* −, respectively. That is an obvious consequence of the Boltzmann statistics, although the basic result (0.14) is quoted in literature following the original article. Needless to say, it is important to realize that the Boltzmann statistics is a valid concept, as supported by the randomness of phonon collisions in equilibrium lattice, as will be confirmed in chapter 1.

However, it is a notable conflict that the Bragg–William theory is not compatible with the lattice dynamics in *finite crystals*, as will be discussed in chapter 8; so, traditional statistical results should be re-evaluated for finite crystals.

0.7.4 Ferromagnetic order and the Weiss field

Composed of magnetic ions, a ferromagnetic crystal is magnetized along a specific crystallographic axis. In the case of iron metals, we can consider a chain of Fe^{3+} ions in periodic arrangement, for which Heisenberg (1929) proposed a model of spin–spin exchange interaction $-2J_{mn}s_m \cdot s_n$, where J_{mn} is the exchange integral. Writing $2J_{mn} = J_{mn}$ and considering spins as order variables, the exchange interaction is in the same form as $-J_{mn}\boldsymbol{\sigma}_m \cdot \boldsymbol{\sigma}_n$, which is convenient to use in this section.

For a chain of magnetic spins $\boldsymbol{\sigma}_n$ arranged in the direction of magnetization, we apply a uniform magnetic field $\boldsymbol{B}_0$ in parallel, and write the Hamiltonian for the spin $\boldsymbol{\sigma}_n$ at site n as

$$\mathcal{H}_n = -\boldsymbol{\sigma}_n \cdot \boldsymbol{B}_0 - \sum_m J_{mn}\boldsymbol{\sigma}_m \cdot \boldsymbol{\sigma}_n.$$

Here, we can consider that the quantity $\langle \sum_m J_{mn}\boldsymbol{\sigma}_m \rangle = \boldsymbol{B}_n$ represents a local magnetic field acting on $\boldsymbol{\sigma}_n$, expressed in the mean-field approximation. Accordingly, in the mean-field approximation we write

$$\langle \mathcal{H}_n \rangle = -\langle \boldsymbol{\sigma}_n \rangle \cdot (\boldsymbol{B}_\text{o} + \langle \boldsymbol{B}_n \rangle)$$

for discussing magnetic ordering. The average $\langle \boldsymbol{\sigma}_n \rangle$ represents the macroscopic magnetization $\boldsymbol{M}$ of a crystal, for which Weiss postulated the relation $\langle \boldsymbol{B}_n \rangle = \lambda \boldsymbol{M}$ with the proportionality constant λ, writing a susceptibility formula for M in the effective field $B_\text{o} + \lambda M$, i.e.

$$M = \chi_\text{o}(B_\text{o} + \lambda M), \tag{0.16}$$

where $\chi_\text{o} = C/T$ is the Curie's law for paramagnetic susceptibility; C is Curie's constant. Solving (0.16) for M, a ferromagnetic susceptibility formula can be derived as

$$\chi = \frac{M}{B_\text{o}} = \frac{C}{T - C\lambda} = \frac{C}{T - T_c} \quad \text{for } T > T_c, \tag{0.17}$$

where $T_c = C\lambda$ is the critical temperature. Equation (0.17) is known as the Curie–Weiss law, applicable for $T > T_c$.

At temperatures below T_c, on the other hand, the spin order should be discussed with respect to the order parameter $\eta = (N_+ - N_-)/N$, and the domains $\pm$ are separated by the ordering energy in the mean-field approximation. In this case, the ordering energies of two domains can be expressed as

$$\Delta E_\pm = \pm M B_\text{o} + \lambda M^2 = \pm N_\pm \beta B_\text{o} + \frac{1}{2} N \lambda \beta^2 \eta^2,$$

where β is Bohr's *magneton.* Similar to Bragg–Williams' theory, the partition function can be written as

$$Z = Z_+ Z_- = \binom{N}{N_+}\binom{N}{N_-} \exp \frac{-(\Delta E_+ - \Delta E_-)}{\mathrm{k_B} T},$$

thereby minimizing the free energy $F = -\mathrm{k_B} T \ln Z$ to obtain N_+ and N_- in equilibrium.

Using the relation $\frac{d}{dN_\pm} \ln \binom{N}{N_\pm} = \frac{d}{dN_\pm}(\mp N_+ \ln N_+ \pm N_- \ln N_-) \pm \frac{1}{\mathrm{k_B} T} \frac{d\Delta E_\pm}{dN}$, we drive

$$\ln \frac{N_-}{N_+} = -\frac{2}{N \mathrm{k_B} T} \frac{d\Delta E_\pm}{d\eta}, \quad \text{hence} \quad \ln \frac{1 - \eta}{1 + \eta} = -\frac{2}{\mathrm{k_B} T}(\lambda \beta^2 \eta \pm \beta B_\text{o}).$$

The last equation can be solved for the two domains $\pm$ separately, and

$$\eta_\pm = \tanh\left(\frac{\lambda \beta^2 \eta_\pm}{2 \mathrm{k_B} T} \mp \frac{\beta B_\text{o}}{\mathrm{k_B} T} \right). \tag{0.18}$$

Nevertheless, as in the case of $B_\text{o} = 0$, equation (0.18) can be solved graphically in the same way, as illustrated for $B_\text{o} \neq 0$ in figure 0.3(a). Defining $y' = \frac{\lambda \beta^2}{2 \mathrm{k_B} T}(\eta_\pm \mp \frac{2 B_\text{o}}{\lambda \beta})$

and $\eta_\pm = \tanh y'$, we discuss the crossing point between these. As seen from the figure, the horizontal coordinate shifts as $y' \to y \pm (\beta B_o/k_B T)$ by increasing B_o. Hence the critical temperature changes as $T_c' \to T_c - (\beta B_o/k_B)$ in two domains, which is however very small in a practical field of $B_o \sim 10^4$gauss, i.e. $\beta B_o/k_B \sim 1$ K.

Assuming $\Delta T_c \sim 0$, we can write $y_\pm' = \frac{T_c}{T}\eta_\pm \mp \frac{\beta B_o}{k_B T}$ to describe the volume variation of domains, which is shown in figure 0.3(b). Here, the difference $y_+' - y_-' = 2\beta B_o/k_B T$ represents the volume change signified by

$$V_+' - V_-' \propto \frac{B_o}{T}, \tag{0.19}$$

resulting in *domain structure*. Domains are thus convertible, however, it is possible to fabricate a single-domain crystal by either applying B_o externally, or separating by force.

The probability and Weiss's adiabatic field defined above are basic concepts in the statistical theory applied to crystalline transitions, which, however, need to be re-evaluated for phenomena in mesoscopic timescale. Further, a remark can be made at this point that is about the homogeneity of the internal field, which is important for practical experiments.

In the foregoing ferromagnetic crystals, we considered a one-dimensional magnet to simplify complex interactions in a three-dimensional structure. This assumption can be justified in a macroscopic crystal in *ellipsoidal shape* that can be magnetized uniformly by a constant applied field [7].

Thermodynamically, applying $\boldsymbol{B}_o$ is regarded for an external adiabatic potential energy $-\boldsymbol{M}.\ \boldsymbol{B}_o$ as adequate for dielectric crystals as well. However, *isotropic* $-p\Delta V$ must always be considered for all materials including polarizable crystals. A single crystal in particular symmetry at a given temperature is characterized as macroscopically *uniaxial*, so that $p\Delta V = -pA\Delta x$ is the adiabatic potential, where A and Δx are the surface area and its displacement, respectively, as equivalent geometrically to an ellipsoidal crystal. Nevertheless, such an adiabatic work as $-p\Delta V$ for

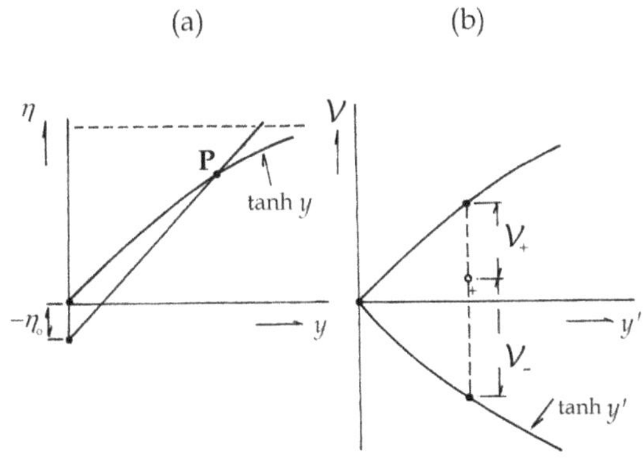

Figure 0.3. (a) A η-y plot of magnetic order in the presence of an applied magnetic field $B_o \propto \eta_o$ to be compared with figure 0.1. (b) Domain-volumes V_+ and V_-, varying with an applied field B.

isothermal process is not necessary to consider, if holding the relation $\Delta V = 0$. In contrast, the condition $\Delta V \neq 0$ is unavoidable or technically difficult to maintain, but significant under external pressure. Accordingly, statistical quantities of the probability and Weiss field need to be redefined as valid in *finite crystals.*

0.8 Remarks on notations in mesoscopic states

For mesoscopic states, it is significant that the order variables need to be expressed in collective motion in finite crystals, where their space-time dependence is essentially described in terms of a *phase function.* Accordingly, specific notations are employed in the following chapters, helping to avoid confusion from the mathematical form. Therefore, we remark on these notations as listed next.

1. Mesoscopic order variables are usually expressed in complex form as $\sigma = \sigma_0 \exp i\phi$, which is expressed as $\sigma(\phi)$, if dominated by the phase ϕ. In this case, the phase function is always written in italic ϕ. In contrast, $\mathrm{\phi}$ in roman style is reserved for potentials, which can be used for the correlation potential as $\boldsymbol{\phi}$.
2. Lax' Hamiltonian $\boldsymbol{L}$ is particularly distinct from conventional $\mathcal{H}$ in the relative coordinate system, signifying kinetic energies.
3. $\boldsymbol{B}$ and B express the developing operator and magnetic induction vector, respectively.
4. All other notations are conventional, while electromagnetic units are in MKS with respect to ε_0 and μ_0 [9].

Exercises

1. Discuss the Ehrenfest classification of phase transitions with respect to microscopic order variables, and show that their fluctuations should exist during transition.
2. Arfgue that the dynamical system representing crystals in equilibrium with surroundings must be conservative in nature.
3. Discuss the Weiss field. Is Weiss' mean-field definition an acceptable idea? If not, why?
4. Why do we have to use the Gibbs function if order variables are active? Can we not assume a constant volume in this case? In fact, the Bragg–William theory assumed $dV = 0$, which seems to be conflicting. Discuss the issue.
5. Why do we need the field-theoretical approximation for thermodynamics of solid states?

References

[1] Prigogine I 1955 *Introduction to Thermodynamics of Irreversible Processes* (New York: Wiley)
[2] Kirkwood J G and Oppenheim I 1961 *Chemical Thermodynamics* (New York: McGraw-Hill)
[3] Born M and Huang K 1968 *Dynamical Theory of Crystal Lattices* (London: Oxford University Press)
[4] Haken H 1973 *Quantenfeldtheorie des Festkörpers* (Stuttgart: Teubner)

[5] Tinkham M 1964 *Group Theory and Quantum Mechanics* (New York: McGraw-Hill)
Bradley C J and Cracknell A P 1972 *The Mathematical Theory of Symmetry in Solids* (Oxford: Clarendon)
[6] Fujimoto M 2005 *The Physics of Structural Phase Transitions* (New York: Springer)
[7] Ziman J M 1979 *Models of Disorder* (London: Cambridge University Press)
[8] Becker R 1967 *Theory of Heat* 2nd edn (New York: Springer)
[9] Fujimoto M 2007 *Physics of Classical Electromagnetism* (New York: Springer)

Part I

Binary transitions

In crystalline processes, the collective mode of order variable $\sigma(\phi)$ plays the fundamental role in structural changes, for which the Bloch theory for periodic system is utilized to form collective motion, where the phase ϕ of propagation specifies the conventional space-time.

IOP Publishing

Chapter 1

Phonons and lattice stability

It is a fundamental postulate that correlated ions or molecular complexes in stable crystals are in thermal equilibrium with their surroundings as signified by the space group. Attributed to mutual correlations, stable crystal structures are taken for granted, where the correlation energy U is assumed to be constant as determined by the external temperature T and pressure p. Accordingly, any change $\Delta U = U - U_\text{o}$ related to distorted structure is a function of temperature, i.e. $\Delta U = \Delta U(T)$ at an external pressure p. Besides, in equilibrium, the volume is finite at constant p, where the surfaces and imperfections disrupt the lattice periodicity locally. An idealized crystal is characterized by minimal imperfections to keep the whole periodicity in sizable volume. Ignoring surfaces, near perfect periodicity can be assumed with an ignorable number of imperfections for calculating thermodynamic properties of sufficiently large crystals.

Setting *point symmetry* aside, lattice vibrations are discussed in this chapter for an idealized crystal signified by *space group*, which determines thermal properties at any temperature evaluated by the phonon spectra. Nevertheless, such idealized crystals become *unstable* against any *anharmonic* disturbances from the inside and the surroundings. Characterized by a large heat capacity of phonons in sizable crystals, the harmonic lattice is considered for the heat reservoir of order variables.

1.1 The space symmetry group and the internal energy in crystals

Single crystals are normally simulated by their geometrical structures, as characterized by symmetry operations such as rotation, translation, reflection and others to reproduce structure. The Hamiltonian $\mathcal{H}$ of lattice vibrations can then be *diagonalized* with respect to symmetry axes, thereby the stable structure is characterized with respect to its eigenvalues.

doi:10.1088/978-0-7503-2572-1ch1

Theorem: Writing a symmetry operator as $\boldsymbol{Q}$, the Hamiltonian of a crystal should be diagonalized as $(\boldsymbol{Q}^{-1}\mathcal{H}\boldsymbol{Q})_{ij} = E_i\delta_{ij}$, where δ_{ij} is *Kronecker's delta*, i.e. $\delta_{ij} = 0$ for $i \neq j$, and eigenvalues E_i for $\delta_{ii} = 1$ are invariant of $\boldsymbol{Q}$ with respect to lattice correlations.

A stable structure at a given temperature and pressure can therefore be specified by one of these E_i in discrete spectrum at given temperature and pressure.

Specified by *space and point groups*, practical crystals are more complex than monatomic lattices, as the constituents usually have additional structures determined by the *point group*. In practical crystals, there may be additional symmetry operations such as *screw axis* for cyclic rotation; crystals can often be characterized further by extra special operations, however, we shall not discuss the mathematical detail here, leaving it to excellent articles of group theory on solid states available in the literature [1, 2].

Structural problems in crystals can be solved primarily with the space group, considering next the point group, when necessary. Here, it is important to realize that *harmonic lattice vibrations* are compatible with stable structure, as remarked in the Introduction, playing the responsible role for stability. On the other hand, any *anharmonic* motion makes lattice structure unstable.

1.2 Normal modes in a monatomic lattice

A crystal of chemically identical constituents has a steady structure, if in equilibrium with the surroundings characterized by *invariance* in translation. Referring to symmetry axes, the physical properties can be determined as related to translational symmetry, in consequence of energy-momentum conservations among the constituents.

Constituent ions and molecules are bound together by restoring forces in the structure of stable crystals. Assuming a cubic lattice of N^3 identical mass-particles in a cubic crystal in sufficiently large size, we first discuss the classical equations of motion for a lattice with nearest-neighbor interactions. Although such a model should be discussed quantum-mechanically, classical arguments are also instructive. Lattice symmetry is unchanged by harmonic interactions with nearest neighbors, assuring structural stability in equilibrium crystals. In this case, we have a set of linear differential equations all in one dimension, which has $3N$ independent waves, namely $3N$ normal modes of N constituents along the symmetry axes x, y and z [3]. Denoting the displacement vector of a constituent by a vector $\boldsymbol{q}_n$ from a site n, we write equations of motion for the components q_{nx}, q_{ny} and q_{nz}, as

$$\ddot{q}_{x,n} = \omega^2(q_{x,n+1} + q_{x,n-1} - 2q_{x,n}),$$

$$\ddot{q}_{y,n} = \omega^2(q_{y,n+1} + q_{y,n-1} - 2q_{y,n})$$

and

$$\ddot{q}_{z,n} = \omega^2(q_{z,n+1} + q_{z,n-1} - 2q_{z,n}),$$

where $\omega^2 = \kappa/m$; κ and m are the mass of a particle and the spring constant, respectively. As these equations are in identical form, a representative one can be considered by dropping x, y and z. i.e.

$$\ddot{q}_n = \omega^2(q_{n+1} + q_{n-1} - 2q_n) \tag{1.1}$$

in one dimension.

Considering the conjugate momentum $p_n = m\dot{q}_n$, the Hamiltonian of a harmonic lattice can be expressed as

$$\mathcal{H} = \sum_{n=0}^{N}\left\{\frac{p_n^2}{2m} + \frac{m\omega^2}{2}(q_{n+1} - q_n)^2 + \frac{m\omega^2}{2}(q_n - q_{n-1})^2\right\} \tag{1.2}$$

for one-dimensional chain of N identical masses, illustrated in figure 1.1(a).

Normal coordinates and conjugate momenta, Q_k and P_k, are defined with the *Fourier amplitudes* in expansions of q_n and p_n

$$q_n = \frac{1}{\sqrt{N}}\sum_{k=0}^{k_N} Q_k \exp(ikna) \quad \text{and} \quad p_n = \frac{1}{\sqrt{N}}\sum_{k=0}^{k_N} P_k \exp(ikna), \tag{1.3}$$

where a is the lattice constant. By inversion of $k \to -k$, these amplitudes are related as

$$Q_{-k} = Q_k^*,\ P_{-k} = P_k^* \quad \text{and} \quad \sum_{n=0}^{N} \exp i(k - k')na = N\delta_{kk'}, \tag{1.4}$$

where $\delta_{kk'} = 1$ if $k = k'$.

Using normal coordinates and conjugate momenta, the Hamiltonian can be expressed by

$$\mathcal{H} = \frac{1}{2m}\sum_{k=0}^{2\pi/a}\left\{P_kP_k^* + Q_kQ_k^* m\omega^2\left(\sin^2\frac{ka}{2}\right)\right\}, \tag{1.5}$$

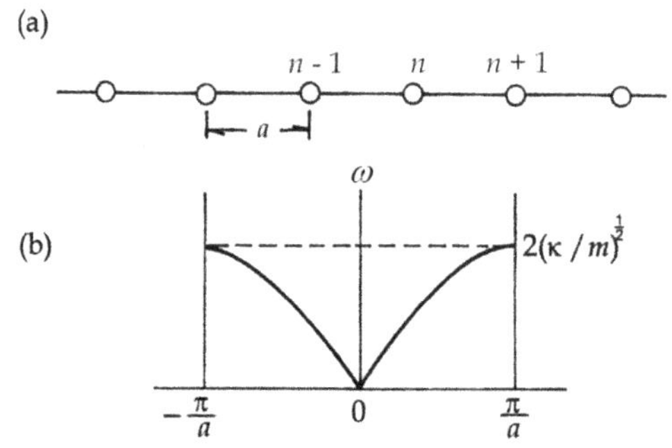

Figure 1.1. (a) One-dimensional monatomic chain; a is the lattice constant. (b) Dispersion curve ω versus k of the chain lattice.

from which the canonical equation of motion can be derived as

$$\ddot{Q}_k = -m^2\omega_k^2\, Q_k \quad \text{and} \quad \omega_k = 2\omega \sin\frac{ka}{2} = 2\sqrt{\frac{\kappa}{m}}\, \sin\frac{ka}{2}. \tag{1.6}$$

As indicated by (1.6), the mode of coupled oscillators specified by k is *dispersive*, and linearly independent from other k'-modes. Accordingly, $\mathcal{H}$ is composed of N independent oscillators, each of which is determined by Q_k and P_k.

Applying the *Born–von Kármán boundary conditions* in periodic lattice of constant a, the values of k can be determined as $k = \frac{2\pi n}{Na}$, where $n = 0, 1, 2, \ldots, N$. Figure 1.1(b) shows the dispersion curve ω_k versus k of (1.6).

With initial values $Q_k(0)$ and $\dot{Q}_k(0)$ at $t = 0$, the solution of (1.6) can be given by

$$Q_k(t) = Q_k(0)\cos \omega_k t + \frac{\dot{Q}_k(0)}{\omega_k} \sin \omega_k t \tag{1.7}$$

Accordingly, we can write

$$q_n(t) = \frac{1}{\sqrt{N}} \sum_{k=0}^{k_N} \sum_{n'=n,n\pm 1} \times \left[q_{n'} \cos\{ka(n-n') - \omega_k t\} + \frac{\dot{q}_{n'}(0)}{\omega_k} \sin\{ka(n-n') - \omega_k t\} \right], \tag{1.8}$$

where $a(n - n')$ is the distance between sites n and n', so that we can write $x = a(n - n')$.

Considering a crystal of a large number of cubic units of volume $V = L^3$, where $L = Na$, disregarding surfaces, the periodic boundary conditions can be set as $q_{n=0}(t) = q_{n=N}(t)$ at arbitrary time t. Lattice points between boundaries are then given by $x = na$, where $0 \leqslant x \leqslant Na$, and (1.8) can be expressed as

$$q(x, t) = \sum_k \{A_k \cos(\pm kx - \omega_k t) + B_k \sin(\pm kx - \omega_k t)\}, \tag{1.9a}$$

where $A_k = \frac{q_k(0)}{\sqrt{N}}$ and $B_k = \frac{\dot{q}_k(0)}{\omega_k\sqrt{N}}$, and we can assume that x is virtually continuous in the range $0 \leqslant x \leqslant L$, if $L \gg a$. In this case, (1.9*a*) can be re-expressed equivalently as

$$q\,(x, t) = \sum_k C_k \exp i(\pm kx - \omega_k t + \varphi_k), \tag{1.9b}$$

where $C_k^2 = A_k^2 + B_k^2$ and $\tan\varphi_k = B_k/A_k$. These expressions can be easily transferred to three-dimensional crystals by writing for three symmetry directions.

1.3 Quantized normal modes

Considering the normal coordinate Q_k and conjugate momentum P_k as *quantum-mechanical operators*, the one-dimensional harmonic oscillator can be quantized by applying quantization conditions:

$$[Q_k, Q_{k'}] = 0, \quad [P_k, P_{k'}] = 0 \qquad \text{and} \qquad [P_k, Q_{k'}] = i\hbar\delta_{k,k'}, \tag{1.10}$$

where the square brackets $[\cdots]$ are a *commutator*, indicating, e.g. $[P_k, Q_k] = P_kQ_k - Q_kP_k = i\hbar$, hence we have $P_k = -i\hbar\frac{\partial}{\partial Q_k}$ [3]. With those operators, the Hamiltonian in (1.5) is an operator

$$\mathcal{H}_k = \frac{1}{2m}\left(P_kP_k^\dagger + m^2\omega_k^2\, Q_kQ_k^\dagger\right),$$

where $Q_k^\dagger$ and $P_k^\dagger$ are transposed matrix operators of Q_k and P_k matrices, respectively.

Denoting the eigenvalue of $\mathcal{H}$ by ε_k, the wavefunction ψ_k should satisfy the equation

$$\mathcal{H}_k\psi_k = \varepsilon_k\psi_k, \tag{1.11}$$

where the energy eigenvalue ε_k must be a *real* number. Therefore, Q_k and P_k should be *Hermitian operators*, signified by the relations $Q_k^\dagger = Q_{-k}$ and $P_k^\dagger = P_{-k}$, respectively. Defining a set of operators,

$$b_k = \frac{m\omega_kQ_k + iP_k^\dagger}{\sqrt{2m\varepsilon_k}} \qquad \text{and} \qquad b_k^\dagger = \frac{m\omega_kQ_k^\dagger - iP_k}{\sqrt{2m\varepsilon_k}}, \tag{1.12}$$

we can calculate that

$$\begin{aligned} b_kb_k^\dagger &= \tfrac{1}{2m\varepsilon_k}\left(m^2\omega_k^2Q_k^\dagger Q_k + P_k^\dagger P_k\right) + \tfrac{i\omega_k}{2\varepsilon_k}\left(Q_k^\dagger P_k^\dagger - P_kQ_k\right) \\ &= \tfrac{\mathcal{H}_k}{\varepsilon_k} + \tfrac{i\omega_k}{2\varepsilon_k}(Q_{-k}P_{-k} - P_kQ_k). \end{aligned}$$

From the last expression, we can obtain that

$$\mathcal{H}_k = \hbar\omega_k\left(b_k^\dagger b_k + \frac{1}{2}\right) \qquad \text{if } \varepsilon_k = \frac{\hbar\omega_k}{2}, \tag{1.13a}$$

which is the *minimum vibrational energy* at $T = 0$ K, known as the *zero-point energy*.

Further, we can derive a significant feature of operators b_k and $b_k^\dagger$ defined above, considering that $\frac{1}{2}\hbar\omega_k$ is the minimum lattice energy of vibration. From (1.13*a*), we notice that the product operator $b_k^\dagger b_k$ is commutable with the Hamiltonian $\mathcal{H}_k$, that is

$$[\mathcal{H}_k, b_k^\dagger b_k] = 0. \tag{1.13b}$$

Also, from the definition (1.12),

$$[b_{k'}, b_k^\dagger] = \delta_{k',k}, \ [b_{k'}, b_k] = 0 \quad \text{and} \quad [b_{k'}^\dagger, b_k^\dagger] = 0. \tag{1.13c}$$

Accordingly, we have

$$[\mathcal{H}_k, b_k^\dagger] = \hbar\omega_k b_k^\dagger \quad \text{and} \quad [b_k, \mathcal{H}_k] = \hbar\omega_k b_k. \tag{1.13d}$$

Using the wave equation (1.11) with these commutator relations, we can derive

$$\mathcal{H}_k\left(b_k^\dagger\psi_k\right) = (\varepsilon_k + \hbar\omega_k)\left(b_k^\dagger\psi_k\right) \text{ and } \mathcal{H}_k(b_k\psi_k) = (\varepsilon_k - \hbar\omega_k)(b_k\psi_k), \tag{1.13e}$$

which indicate that $b_k^\dagger\psi_k$ and $b_k\psi_k$ are eigenfunctions for eigenvalues $\varepsilon_k + \hbar\omega_k$ and $\varepsilon_k - \hbar\omega_k$, respectively. In this context, $b_k^\dagger$ and b_k are called *creation* and *annihilation* operators of a quantum $\hbar\omega_k$, for which we have the relation

$$b_k^\dagger b_k = 1 \tag{1.14}$$

The *boson* properties of phonons are signified by $b_k^\dagger$ and b_k, which are listed by the above equations (1.13*a*–*e*) and (1.14).

Applying the creation operator n_k times on the ground state ψ_k, we obtain an energy state $\left(n_k + \frac{1}{2}\right)\hbar\omega_k$ for the wavefunction $(b_k^\dagger)^{n_k}\psi_k$, where n_k quanta of $\hbar\omega_k$ has been created on the ground state ψ_k of energy $\frac{1}{2}\hbar\omega_k$. Considering $\hbar\omega_k$ like a particle that is unidentifiable among n_k quanta, the wavefunction should be normalized as $(b_k^\dagger)^{n_k}\psi_k/\sqrt{n_k!}$, where $n_k!$ is the permutation number of identical quanta. Here for the energy $\hbar\omega_k$, we have to define the momentum, in order to take it as a dynamical object *photon*. Nevertheless, leaving it to the next section 3.4, for the lattice as a whole we need to express the *global wavefunction* of multiple states ψ_k, where $k = 1, 2, \ldots,$ when $n_1, n_2, \ldots$ photons are created. The global state should be expressed by the internal energy

$$U(n_1, n_2, \ldots) = U_o + \sum_k n_k\hbar\omega_k, \quad \text{where } U_o = \sum_k \frac{\hbar\omega_k}{2}, \tag{1.15a}$$

and the wavefunction

$$\Psi(n_1, n_2, \ldots) = \frac{(b_1^\dagger)^{n_1}(b_2^\dagger)^{n_2}\cdots}{\sqrt{n_1!n_2!\ldots}}(\psi_1\psi_2\ldots); \tag{1.15b}$$

U_o is called the zero-point energy of the vibrational field. Here the total number of phonons is given by $N = n_1 + n_2 + \cdots$, which cannot be a fixed number dynamically, but evaluated thermodynamically as related to thermal excitation levels of the field if the temperature is specified.

1.4 Phonon field and momentum

In the classical normal-mode theory, we consider the mode q_k represents linearly independent displacements, however, there are product terms $q_{ki}q_{kj}$, unless $\sum_k \mathcal{H}_k$ is diagonalized properly. Therefore, it is more appropriate to employ a view of the whole of the collective vibrations as a *field*, which is an adequate approach particularly for a small k at low level of lattice excitation.

Considering an orthorhombic crystal for simplicity, classical equations of normal vibration can be written with respect to symmetry axes x, y and z, as

$$\frac{p_{x,n_x}^2}{2m} + \frac{\kappa}{2}\left\{(q_{x,n_x} - q_{x,n_x+1})^2 + (q_{x,n_x} - q_{x,n_x-1})^2\right\} = \varepsilon_{x,n_x}$$

and similar equations for y and z components, where

$$\varepsilon_{x,n_x} + \varepsilon_{y,n_y} + \varepsilon_{z,n_z} = \varepsilon_{n_x,n_y,n_z}$$

and κ is the force constant. In these expressions, the direction of displacement and the energy of propagation are specified by the vector $\boldsymbol{q} = (q_{x,n_x}, q_{y,n_y}, q_{z,n_z})$ and $\varepsilon_{n_x,n_y,n_z}$, respectively.

However, in view of the dispersive nature, these component waves may not be independent, which is desirable if the theory predicts otherwise. The classical vector $\boldsymbol{q}$ can be replaced by a *product* wavefunction $q(x, t)q(y, t)q(z, t)$ in quantum theory, thereby allowing us to consider possible interference of component functions, which is a logical approach to the vibrational field.

In the one-dimensional classical theory, we obtained that $q(x, t)$ can be expressed as (1.9*b*) in exponential function. Writing $q(y, t)$ and $q(z, t)$ similar to (1.9*b*), the wavefunction in the vibration field can be expressed as

$$\Psi(n_x, n_y, n_z) = \sum_{\boldsymbol{k}} A_k \exp i\left(\pm\boldsymbol{k}\cdot\boldsymbol{r} - \frac{n_x\varepsilon_{x,n_x} + n_y\varepsilon_{y,n_y} + n_z\varepsilon_{z,n_z}}{\hbar}t + \varphi_{\boldsymbol{k}}\right),$$

where $A_k = C_{k_x}C_{k_y}C_{k_z}$, $\boldsymbol{k} = (k_x, k_y, k_z)$, and $\varphi_{\boldsymbol{k}} = (\varphi_{k_x}, \varphi_{k_y}, \varphi_{k_z})$ are the amplitude, wavevector and phase constant, respectively. Further, defining a short-hand notation

$$n_x\varepsilon_{x,n_x} + n_y\varepsilon_{y.n_y} + n_z\varepsilon_{z.n_z} = \hbar\omega_{\boldsymbol{k}}(n_x, n_y, n_z),$$

the wave $\Psi(n_x, n_y, n_z)$ can be reduced to a simplified component wave

$$\Psi(\boldsymbol{k}, \omega_{\boldsymbol{k}}) = A_{\boldsymbol{k}} \exp i(\pm\boldsymbol{k}\cdot\boldsymbol{r} - \omega_{\boldsymbol{k}}t + \varphi_{\boldsymbol{k}}), \tag{1.16a}$$

propagating with the energy $\hbar\omega_{\boldsymbol{k}}$ and momentum $\pm\hbar\boldsymbol{k}$ in the direction (n_x, n_y, n_z). Noted clearly from (1.16*a*), we have the relation $\omega_{\boldsymbol{k}} = v|\boldsymbol{k}|$, where the speed of propagation v is constant of the crystalline medium. Accordingly, phonon particles $(\hbar\omega_{\boldsymbol{k}}, \pm\hbar\boldsymbol{k})$ can move in momentum space in all directions, as if they are free classical gas particles. Nonetheless, (1.16) is only approximate valid to a continuous crystal, and the direction of propagation should be determined in the discrete $\boldsymbol{k}$ space. In a cubic crystal for example,

$$k_x = \frac{2\pi n_x}{L},\ k_y = \frac{2\pi n_y}{L} \text{ and } k_z = \frac{2\pi n_z}{L}, \tag{1.16b}$$

where $L = N \times$ (lattice constant), and $|\boldsymbol{k}| = \frac{2\pi}{L}\sqrt{n_x^2 + n_y^2 + n_z^2}$. Figure 1.2 shows a two-dimensional view in the $k_x - k_y$ plane, where dots indicate $\boldsymbol{k}$ states permitted by (1.16*b*).

Quasi-particles of phonons are important for thermodynamics, because they are responsible for heat transfer processes between crystals and their surroundings. Accordingly, phonon scattering processes play essential roles in crystals, as will be discussed in the following chapters.

1.5 Specific heat of monatomic crystals

The specific heat at a constant volume $C_V = \left(\frac{\partial U}{\partial T}\right)_V$ is a measurable quantity with varying temperature under a constant pressure. The *phonon gas* is an appropriate model for evaluating thermal results from simple monatomic crystals, as characterized with no structural changes [4].

For a monatomic cubic crystal, the internal energy and the specific heat can be calculated by statistical mechanics of quantized phonon energies $\varepsilon_k = \left(n_k + \frac{1}{2}\right)\hbar\omega_k$, assuming that the $\boldsymbol{k}$ vectors are equally distributed in all directions in the reciprocal lattice space. Considering that we have $3N$ phonons in total, energies ε_k are degenerate at densities $g(\boldsymbol{k})$, which are in large numbers, as estimated from a spherical volume of radius $|\boldsymbol{k}|$ in the reciprocal space. Therefore, the partition function can be expressed by

$$Z_k = g(k)\exp\left(-\frac{\varepsilon_k}{k_B T}\right) = g(k)\exp\left(-\frac{\hbar\omega_k}{2k_B T}\right)\sum_{n_k}^{\infty}\exp\left(-\frac{n_k\hbar\omega_k}{k_B T}\right),$$

where the infinite series on the right converges if $\frac{\hbar\omega_k}{k_B T} < 1$. Actually, this condition is fulfilled at any practical temperatures T lower than the melting point, so that Z_k can be expressed as

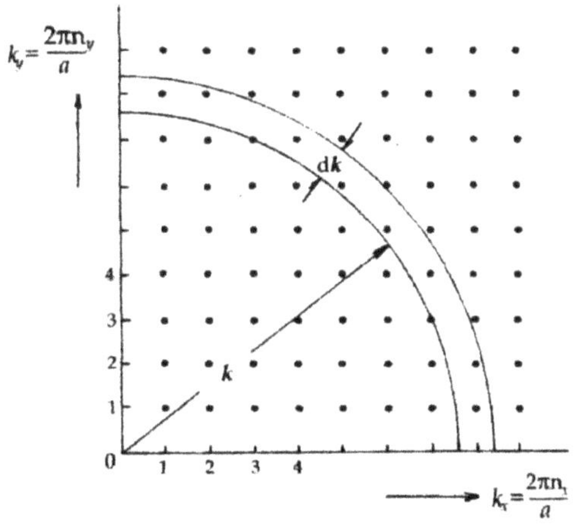

Figure 1.2. Two-dimensional reciprocal lattice, where (k_x, k_y) indicates a lattice point. Two-quarter circles of radii k and $k + dk$ represent surfaces of constant energies ε_k and ε_{k+dk}, respectively, valid at small $|\boldsymbol{k}|$.

$$Z_k = \frac{g(k)\exp\left(-\frac{\hbar\omega_k}{k_B T}\right)}{1 - \exp\left(-\frac{\hbar\omega_k}{k_B T}\right)}.$$

The total partition function is given by $Z = \prod_k Z_k$, so that $\ln Z = \sum_k \ln Z_k$. To calculate Helmholz' free energy $F = k_B T \sum_k \ln Z_k$, it is convenient to obtain $\ln Z_k$, that is

$$\ln Z_k = -\frac{\hbar\omega_k}{2} + k_B T \ln g(k) - k_B T \ln\left\{1 - \exp\left(-\frac{\hbar\omega_k}{k_B T}\right)\right\}.$$

By definition, we have the relation $F = U - TS = U + T\left(\frac{\partial F}{\partial T}\right)_V$, hence $U = -T^2\frac{\partial}{\partial T}\left(\frac{F}{T}\right)$ is the formula to obtain U. Accordingly,

$$U = U_o + \sum_k \frac{\hbar\omega_k}{\exp\frac{\hbar\omega_k}{k_B T} - 1} \quad \text{where } U_o = \frac{1}{2}\sum_k \hbar\omega_k,$$

Therefore, the specific heat C_V can be expressed as

$$C_V = \left(\frac{\partial U}{\partial T}\right)_V = k_B \sum_k \frac{\left(\frac{\hbar\omega_k}{k_B T}\right)^2 \exp\frac{\hbar\omega_k}{k_B T}}{\left(\exp\frac{\hbar\omega_k}{k_B T}\right)^2}. \tag{1.17a}$$

For calculating (1.17a), the summation has to be carried out in the reciprocal space, for which the number of phonon states ε_k need to be specified. In anisotropic crystals, the surface of a constant ε_k is not spherical, as shown in figure 1.3(a), but a closed surface in complex shape. Nevertheless, the summation in (1.17a) can be evaluated by assuming continuous distribution of $\boldsymbol{k}$ states. In this case, we use the relations between volume- and surface-elements denoted by $d^3\boldsymbol{k} = d\boldsymbol{k}.\ d\boldsymbol{S} = dk_\perp dS$, as illustrated in figure 1.3(b). Writing that

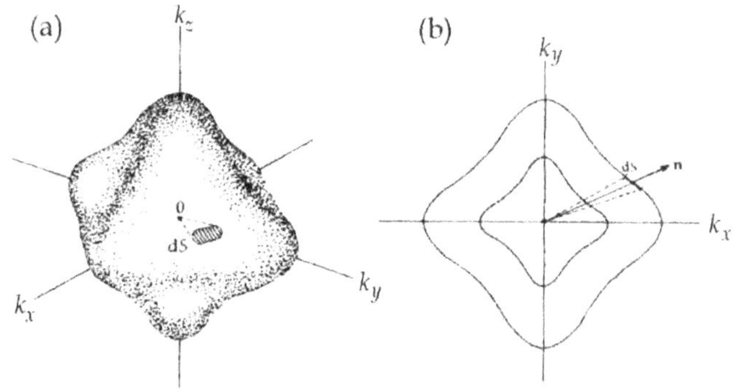

Figure 1.3. (a) A typical constant-energy surface in three-dimensional reciprocal space; dS is a differential surface element. (b) Two-dimensional k_x–k_y-plane.

$$d\omega_k = \frac{d\varepsilon_k}{\hbar} = \frac{1}{\hbar}|\nabla_k \varepsilon_k| dk_\perp = v_g dk_\perp,$$

where v_g is the group velocity of propagation, the summation of (1.17*a*) can be replaced by integration, and

$$C_V = k_B \int_{\omega_k} \frac{\left(\frac{\hbar\omega_k}{k_B T}\right)^2 \exp\frac{\hbar\omega_k}{k_B T}}{\left(\exp\frac{\hbar\omega_k}{k_B T} - 1\right)^2} \mathcal{D}(\omega_k) d\omega_k \quad \text{where } \mathcal{D}(\omega_k) = \left(\frac{L}{2\pi}\right)^3 \oint_S \frac{dS}{v_g}, \tag{1.17b}$$

is the density of phonon states on the surface S.

1.6 Approximate phonon distributions

1.6.1 Einstein's model

At elevated temperatures, thermal properties of a monatomic crystal are considered to be dominated by n phonons of energy $\hbar\omega_0$. Originally proposed by Einstein, we disregard all frequencies other than the dominant mode ω_0. Known as Einstein's model, this assumption is characterized by $\mathcal{D}(\omega_0) = 1$, equation (1.17*b*) and the internal energy U can be expressed as

$$C_V = 3Nk_B \frac{\xi^2 \exp\xi}{(\exp\xi - 1)^2} \quad \text{and} \quad U = 3Nk_B\left(\frac{\xi}{2} + \frac{\xi}{\exp\xi - 1}\right),$$

respectively, where $\xi = \frac{\Theta_E}{T}$ with $\Theta_E = \frac{\hbar\omega_0}{k_B}$ that is called Einstein's temperature. It is noted that in the limit $\xi \to 0$, the above specific heat C_V approaches the Dulong–Petit law $C_V \to 3Nk_B$ when T is increased.

According to the equipartition theorem in statistical mechanics, the internal energy U at high temperatures can be determined by the thermal energy $\frac{1}{2}k_B T$ given to a mass particle vibrating with two degrees of freedom independently, hence the total U of $3N$ particles is equal to $3N \times 2 \times \frac{1}{2}k_B T = 3Nk_B T$. Thus, the Dulong–Petit laws, $C_V = 3Nk_B$ and $U = 3Nk_B T$, are consistent with the Einstein model for $T > \Theta_E$.

1.6.2 Debye's model

At low temperatures, longitudinal vibrations at low frequencies are dominant modes, which are characterized by a non-dispersive relation $\omega = v_g k$. Here the group velocity v_g can be assumed as constant on the nearly spherical surface of ε_k in $\boldsymbol{k}$-space, as shown in figure 1.2. The density function of (1.17*b*) can be expressed as

$$\mathcal{D}(\omega) = \left(\frac{L}{2\pi}\right)^3 \frac{4\pi\omega^2}{v_g^3}. \tag{1.18a}$$

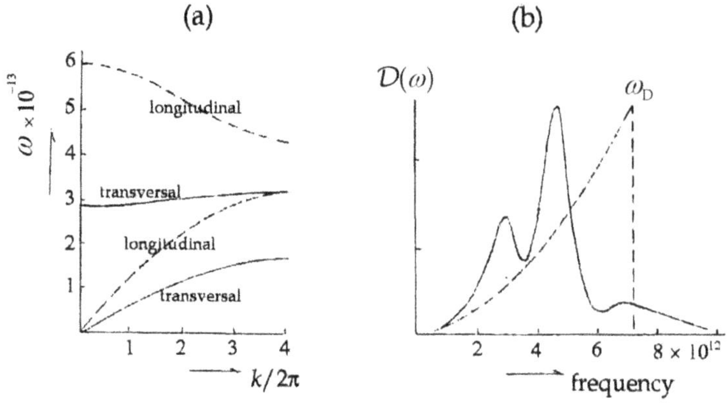

Figure 1.4. (a) An example of typical dispersion curves. Longitudinal and transverse dispersions are indicated by solid and broken curves, respectively. The solid curve shows the observed density functions $\mathcal{D}(\omega)$, which are compared to the broken curve calculated with Debye's model.

as shown by the dotted curve in figure 1.4(b), while comparing with practical frequency distributions in figure 1.4(a) and (b). Determined by the normalization $\int_0^{\omega_D} \mathcal{D}(\omega)d\omega = 3N$, Debye's distribution (1.18*a*) can be replaced by

$$\mathcal{D}(\omega) = \frac{9N}{\omega_D^3}\omega^2. \qquad (1.18b)$$

Using (1.18*b*), U and C_V for Debye's model are given by

$$U = 3Nk_BT\int_0^{\omega_D}\left(\frac{\hbar\omega}{2} + \frac{\hbar\omega}{\exp(\hbar\omega/k_BT) - 1}\right)\frac{3\omega^2}{\omega_D^3}d\omega$$

and

$$C_V = 3Nk_B\int_0^{\omega_D}\frac{\exp(\hbar\omega/k_BT)}{(\exp(\hbar\omega/k_BT) - 1)^2}\left(\frac{\hbar\omega}{k_BT}\right)^2\frac{3\omega^2}{\omega_D^3}d\omega,$$

respectively. Defining Debye's temperature by $\hbar\omega_D/k_B = \Theta_D$, and a variable $\xi = \hbar\omega_D/k_BT$ in analogy to Einstein's model, Debye's U and C_V are expressed as

$$U = \frac{9}{8}Nk_B\Theta_D + 9Nk_BT\left(\frac{T}{\Theta_D}\right)^3\int_0^{\frac{\Theta_D}{T}}\frac{\xi^3}{\exp\xi - 1}d\xi \qquad (1.19a)$$

and

$$C_V = 9Nk_B\left(\frac{T}{\Theta_D}\right)^3\int_0^{\frac{\Theta_D}{T}}\frac{\xi^4\exp\xi}{(\exp\xi - 1)^2}d\xi. \qquad (1.19b)$$

Although only approximate at low temperatures, Debye's model was found to explain experimental results well, particularly at very low temperatures, where we

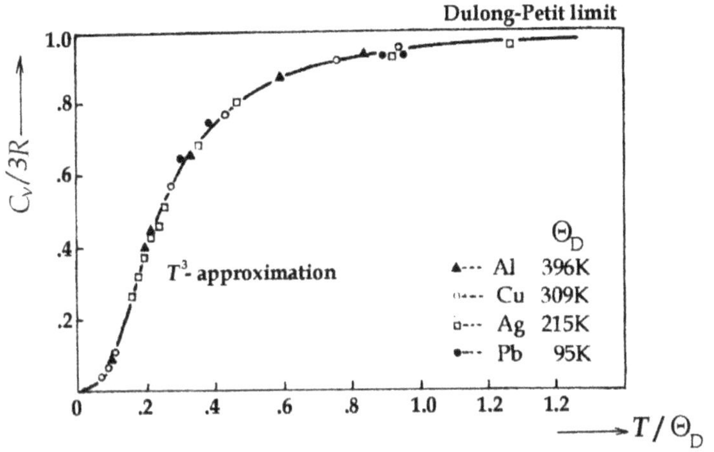

Figure 1.5. Experimental specific heat $C_V/3R$ versus T/Θ_D of representative metals. $R/N k_B$ is the molar gas constant, and values of Θ_D for representative metals are shown in the bottom right corner. The curve exhibits the T^3-law and Dulong–Petit limit in low- and high-temperature regions, respectively.

have the relation $C_V \propto T^3$, known as the T^3-law, as derived from (1.19*b*) in the limit $\Theta_D/T \to \infty$. The integration of (1.19*a*, *b*) can be performed by using the relation

$$\int_0^\infty \frac{\xi^3 d\xi}{\exp\xi - 1} = \int_0^\infty d\xi \left(\xi^3 \sum_{s=1}^\infty \exp(-s\xi) \right) = 6 \sum_{s=1}^\infty \frac{1}{s^4} = \frac{\pi^4}{15},$$

where the last sum over s is obtained from standard reference. Applying this result to (1.19*a*), we obtain

$$U \approx 3\pi^4 N k_B T^4 / 5\Theta_D$$

and

$$C_V \approx \frac{12\pi^4}{5} N k_B \left(\frac{T}{\Theta_D} \right)^3 \propto T^3 \quad \text{for } T < \Theta_D \propto T^3.$$

Figure 1.5 shows a comparison of experimental results with T^3-law and Dulong–Petit law at low- and high-temperature regions, respectively. Table 1.1 shows experimentally determined values of Debye's temperature Θ_D for representative metallic solids.

Table 1.1. Debye's temperatures Θ_D determined by thermal and elastic experiments. Data from [4].

	Fe	Al	Cu	Pb	Ag
Thermal	453	398	315	88	215
Elastic[a]	461	402	332	73	214

[a] Calculated with elastic data at room temperature.

1.7 Phonon correlations

Quantizing the lattice vibration field, we assume a gas of phonons ($\hbar\omega_k$, $\hbar \boldsymbol{k}$) for thermal properties of a stable lattice. A large number of phonons exist in excited states, so that crystals behave as in gaseous states. However, phonons are not exactly independent particles, but correlated quantum-mechanically at low temperatures in particular, owing to their property as *unidentifiable* particles.

Although undetermined dynamically, the number of phonons can be specified in thermodynamics by the surface kept at temperature T, where phonon energies are exchanged with heat from the surroundings. In equilibrium crystals, the number of phonons n_k at a k-state can be either one of 1, 2, ..., $3N$, therefore, the Gibbs function can be specified by the thermodynamic probability function $g(p, T, n_k)$ that is distributed within a large number of phonon states $\varepsilon_k = n_k \hbar\omega_k$.

It is significant to note that finding correlated phonon states among these states is an *exclusive event*, if their correlation distance is short, which is realistic in crystals. Writing probabilities for finding k-state and k'-state as $g_k(p, T, n_k)$ and $g_{k'}(p, T, n_{k'})$, respectively, the probability for finding both together should be determined by the sum

$$g(p, T, n_k + n_{k'}) = g_k(p, T, n_k) + g_{k'}(p, T, n_{k'}) \qquad \text{where } n_k + n_{k'} = \text{constant},$$

indicating that phonons are symmetrically correlated, owing to their boson character. Considering small variations δn_k and $\delta n_{k'}$ in the correlating process, we have the relations

$$\delta n_k = -\delta n_{k'} \qquad \text{and} \qquad (\delta g)_{p,T} = \left(\frac{\partial g_k}{\partial n_k}\right)_{p,T} \delta n_k + \left(\frac{\partial g_{k'}}{\partial n_{k'}}\right)_{p,T} \delta n_{k'},$$

from which we can derive a common quantity between g_k and $g_{k'}$, i.e.

$$\left(\frac{\partial g_k}{\partial n_k}\right)_{p.T} = \left(\frac{\partial g_{k'}}{\partial n_{k'}}\right)_{p,T},$$

which is the *chemical potential.* Denoted by μ, it is a conventional notation to express a change of the Gibbs function due to adiabatic fluctuations among correlated phonons as $\delta G = -\mu\delta n$.

Phonon gas in the general thermodynamic environment can be discussed with the Gibbs function $G(p, T, n)$, which determines the equilibrium by $\delta G \geqslant 0$. Hence, the equilibrium condition can be expressed as

$$\delta G = \delta U - T\delta S + p\delta V - \mu\delta n,$$

where the variations $\delta U = U_o - \varepsilon$, $\delta n = n_o - n$ and $\delta V = 0$ give rise to a change in δG for the entropy change

$$\Delta S = S(U_o, n_o) - S(U_o - \varepsilon, n_o - n) = -\left(\frac{\partial S}{\partial U_o}\right)_{n_o} \varepsilon - \left(\frac{\partial S}{\partial n_o}\right)_{U_o} n.$$

From the equilibrium determined by $\delta G = 0$, we obtain the relations

$$\left(\frac{\partial S}{\partial U_0}\right)_{n_0} = \frac{1}{T} \quad \text{and} \quad \left(\frac{\partial S}{\partial n_0}\right)_{U_0} = -\frac{\mu}{T},$$

hence $\Delta S = \frac{\varepsilon - \mu n}{T}$. Writing statistical probabilities for the states (U_0, n_0) and $(U_0 - \varepsilon,\ n_0 - n)$ as g_0 and g, respectively, we have the relation $g/g_0 = \exp(-\Delta S/k_B)$, thereby obtaining the relation

$$g = g_0 \exp\frac{\mu n - \varepsilon}{k_B T}. \tag{1.20}$$

This is the Gibbs factor in statistical mechanics of an *open system*, while the phonon system is always open to internal quantum-mechanical correlations among phonons. By defining a parameter $\gamma = \exp\frac{\mu}{k_B T}$, the Gibbs factor can be modified as

$$g = g_0 \gamma^n \exp\frac{-\varepsilon}{k_B T},$$

where γ can be interpreted as the access probability for one elemental correlation [5], while the Boltzmann factor determines access to phonon energy $\varepsilon = n\varepsilon_0$, where $\varepsilon_0 = \frac{1}{2}\hbar\omega_0$, so-called zero-point energy.

To determine thermodynamic properties, we calculate the (grand) partition function

$$Z_N = \sum_{n=0}^{N} \gamma^n \exp\left(-\frac{n\varepsilon_0}{k_B T}\right) = \sum_{n=0}^{N} \left\{\gamma \exp\left(-\frac{\varepsilon_0}{k_B T}\right)\right\}^n.$$

The series on the right can converged for $\gamma \exp(-\varepsilon_0/k_B T) < 1$, taking $N \to \infty$, hence

$$Z_\infty = \frac{1}{1 - \exp\left(\frac{-\varepsilon_0}{k_B T}\right)},$$

allowing us to calculate the average number of phonons as

Theorem:

$$\langle n \rangle = \gamma \frac{\partial \ln Z_\infty}{\partial \gamma} = \frac{1}{\frac{1}{\gamma}\exp\left(\frac{-\varepsilon_0}{k_B T}\right) - 1} = \frac{1}{\exp\frac{\varepsilon_0 - \mu}{k_B T} - 1}, \tag{1.21}$$

This is known as the *Bose–Einstein distribution*. Noting that ε_0 and μ are temperature independent and $\varepsilon_0 > \mu$, therefore $\langle n \rangle \approx \exp(-\varepsilon_0/k_B T)$ that is the Boltzmann factor. In contrast, if $\varepsilon_0 = \mu$, we have $\langle n \rangle = \infty$, which, however, is by no means condensation, because n maximum is theoretically undetermined. This singular behavior may appear, if phonon correlations signified by $\varepsilon_o = \mu$ at a critical temperature. Nevertheless, in practical crystals, $\langle n \rangle$ can be estimated by the Boltzmann statistics for $\varepsilon_o > \mu$, as discussed in section 1.6 for the specific heat of metals.

It is interesting to note that the parameter Θ_D for Debye's model signifies the specific temperature for phonon correlations that exist if $T < \Theta_D$. In contrast, the absence of correlations for $T > \Theta_D$ allows us to use the Boltzmann statistics for gaseous phonons. Hence, the total number of *correlated phonons* in a crystal can be roughly estimated from the value of Debye's parameter Θ_D. However, we should realize that the Bose–Einstein condensation is a different issue from phonon correlations.

Exercises

1. On the basis of the foregoing theory of phonons, argue that the lattice structure is invariant under constant temperature, being stable by harmonic forces.
2. Show that the Bose–Einstein condensation occurs if the chemical potential μ is extremely low compared with the lattice vibration energy ε_o. This may be a condition justified by a low value of Debye' parameter Θ_D for metals. Discuss the issue.
3. Discuss why phonons are regarded as constituting a gas thermodynamically. In that case, why are they independent quasi-particles, but correlated at low temperatures?

References

[1] Tinkham M 1964 *Group Theory and Quantum Mechanics* (New York: McGraw-Hill)
[2] Knox R S and Gold A 1964 *Symmetry in the Solid State* (New York: Benjamin)
[3] Kittel C 1963 *Quantum Theory of Solids* (New York: John Wiley)
[4] Kittel C 1086 *Introduction to Solid State Physics* 6th edn (New York: John Wiley)
[5] Kittel C and Kroemer H 1980 *Thermal Physics* (San Francisco, CA: Freeman)

IOP Publishing

Solitons in Crystalline Processes (2nd Edition)
Irreversible thermodynamics of structural phase transitions and superconductivity
Minoru Fujimoto

Chapter 2

Displacive order variables in collective mode and adiabatic Weiss' potentials

Crystals constitute a group of condensed matters composed of a large number of identical ions or molecular groups [1]. The *correlation energy* is essential for the structural stability in equilibrium crystals, where identical constituents are arranged in a lattice of *permutable identical particles*. On the other hand, pertaining to thermodynamic environment, the stability is ascribed to the structural invariance for geometric operations, i.e. rotation, translation and others in the symmetry group.

If the local symmetry is disrupted by *finite intrinsic displacements* at lattice sites, the structure is destabilized, however the *strain energy* is released spontaneously to the surroundings with decreasing temperature to obtain new strain-free structure, as stated by the principle for stability proposed by Born and Huang [2]. In this chapter, we discuss the origin of *displacive variables in collective mode in finite magnitude* that are responsible for a structural change. It is important that such displacements determine the *new lattice symmetry* independent of lattice vibrations at temperatures below T_c, while they are averaged out to zero at all temperatures above T_c.

2.1 One-dimensional ionic chain

The lattice excitation in a stable crystal is primarily sinusoidal in small amplitude, characterized by the wavevector $\boldsymbol{k}$ and frequency ω. Following Born and von Kármán, we consider a one-dimensional infinite chain of two different masses, carrying positive and negative charges, which are alternatively arranged, as shown in figure 2.1. Disregarding *Coulomb interactions*[note1] between charges in long distances, we considered harmonic interactions between neighboring masses, $m_+ > m_-$. In this

[note1] Coulomb's static force is in long range, and ignorable if the neutral crystal is uniformly polarized. The depolarization field may act on ionic charges, but not on ionic masses. Therefore, we consider only mass correlations for lattice dynamics. A similar discussion can be found in chapter 13 for conduction electrons in metals.

doi:10.1088/978-0-7503-2572-1ch2

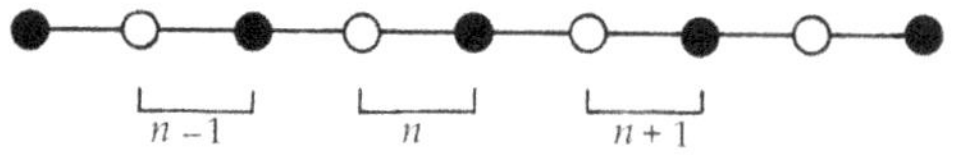

Figure 2.1. A dipolar chain lattice. Plus and minus ions are located in pairs at sites $n-1, n, n+1$ in the chain.

model, we assume only nearest-neighbor interactions expressed by a potential $\phi(r - r_0)$, where r_0 and r are the static lattice constant and dynamic distance, respectively. Taking positions along the x direction, the equations of motion can be written for ionic masses m_+ and m_- as

$$m_+\ddot{u}_n^+ = -\phi'\left(r_0 + u_n^- - u_n^+\right) + \phi'\left(r_0 + u_n^+ - u_{n-1}^-\right) = \phi''\left(u_n^- + u_{n-1}^+ - 2u_n^+\right)$$

and

$$m_-\ddot{u}_n^- = -\phi'\left(r_0 + u_{n+1}^- - u_n^+\right) + \phi'\left(r_0 + u_n^- - u_n^+\right) = \phi''\left(u_{n+1}^+ + u_n^+ - 2u_n^-\right),$$

where the suffixes indicate lattice sites, as shown in the figure. Here ϕ' and ϕ'' are the first and second derivatives with respect to the position x, whereas $\ddot{u}_n^\pm$ are time-derivatives of the second order.

Noting that such an ionic arrangement is stable, and letting $u_n^\pm = u_0^\pm \exp i(kx - \omega t)$ and $|x_{n+1}^0 - x_n^0| = r_0$, we can derive steady relations between u_0^+ and u_-^0, namely

$$\{m_+\omega^2 - 2\phi''(r_0)\}u_0^+ + \phi''(r_0)\{1 + \exp(-ikr_0)\}u_0^- = 0 \tag{2.1}$$

and

$$\phi''(r_0)\{1 + \exp(ikr_0)\}u_0^+ + \{m_-\omega^2 - 2\phi''(r_0)\}u_0^- = 0.$$

Eliminating u_0^+ and u_0^- from (2.1), we have the secular equation

$$\begin{vmatrix} m_+\omega^2 - 2\phi''(r_0) & \phi''(r_0)\{1 + \exp(-ikr_0)\} \\ \phi''(r_0)\{1 + \exp(ikr_0)\} & m_-\omega^2 - 2\phi''(r_0) \end{vmatrix} = 0,$$

which can be solved for ω^2, and

$$\omega^2 = \frac{\phi''(r_0)}{m_+m_-}\left\{(m_+ + m_-) \pm \sqrt{(m_+ + m_-)^2 - 4m_+m_-\sin^2\frac{kr_0}{2}}\right\}. \tag{2.2}$$

From (2.1) and (2.2), the amplitude ratio can be written as

$$\frac{u_0^+}{u_0^-} = \frac{-m_-\{1 + \exp(-ikr_0)\}}{(m_+ - m_-) \pm \sqrt{(m_+ + m_-)^2 - 4m_+m_-\sin^2\frac{kr_0}{2}}}. \tag{2.3}$$

For the + sign in (2.2) and (2.3), we have $\omega^2 \neq 0$ and

$$\omega = \sqrt{\frac{2(m_+ + m_-)\phi''(r_0)}{m_+m_-}} \neq 0 \quad \text{at} \quad k = 0.$$

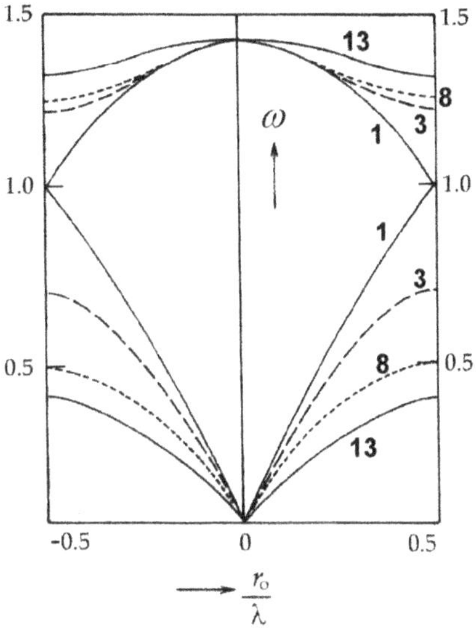

Figure 2.2. Dispersion curves of one-dimensional bipolar lattice calculated in section 3.1 for various values of m_-/m_+. Frequency gaps between optic and acoustic modes are apparent at all values of the mass ratio, except for $m_-/m_+ = 1$. From [3], © 2010. With permission of Springer.

On the other hand, for the − sign we obtain

$$\omega \approx k\sqrt{\frac{\phi''(r)}{2(m_+ + m_-)}} \quad \text{for } k \approx 0.$$

These two ± cases represent different spectra that are called *optic* and *acoustic* modes, respectively, and are linearly independent in harmonic approximation.

It is notable that the optic mode is characterized by the *reduced mass* defined as $\mu = \frac{m_+ m_-}{m_+ + m_-}$, for which there is the relation $m_+ u_o^+ + m_- u_o^- = 0$ at $k = 0$, as verified by (2.3). Therefore, the center of mass coordinates of two adjacent ions are set to zero for the optic mode, representing the relative motion of the ionic pair. In contrast, acoustic mode represents the *total mass* $m_+ + m_-$ located at the center-of-mass coordinate.

Figure 2.2 illustrates curves of ω versus r_0/λ, equivalent to k, as $\lambda = 2\pi/k$, plotted for various values of mass-ratio m_-/m_+, showing all significant dispersive curves deviating from the linear relation $\omega \propto k$. Significant here is the frequency gap between optic and acoustic modes, except for $m_-/m_+ = 1$, which is, however, hardly practical, because such displacements in crystals are unlikely under normal thermodynamic conditions. Nevertheless, for a small reduced mass for $m_-/m_+ < 1$, the gap in dispersion relation is considerable, suggesting that $m_\pm$ are not necessarily heavy ionic masses in practice.

While ionic charges $\pm e$ are disregarded in the foregoing discussion, such a relative charge displacement $u_n^+ - u_n^-$ in ionic crystals should generate an electric dipole moment $p_n = e(u_n^+ - u_n^-)$, which can be studied by applying a time-dependent electric field; hence called an optic mode, such a charged displacement $e(u_n^+ - u_n^-)$ is also called an *order variable* or *displacive order variable*. On the other hand, the acoustic mode describes sound propagation, for which harmonic displacements of $m_+ + m_-$ are responsible.

In the forgoing arguments, we assumed that the interatomic potential $\phi\ (r - r_0)$ is unspecified. It is clear that the displacement $r - r_0$ is not necessarily harmonic. If harmonic, the lattice remains in equilibrium with its surroundings, where all order variables and associated quantities are thermally averaged out. Accordingly, order variables should be characterized as *displacive vectors* in periodic structure, rather than *anharmonic oscillators*.

2.2 Practical examples of displacive order variables

The order parameter should be identified in a given system, constituting a basic subject of experimental studies. If specified by an electric or magnetic moment, the local displacement can be detected with an applied electric or magnetic field, respectively; otherwise not directly identifiable, unless verified by diffuse x-ray diffraction pattern at the transition threshold, where microscopic evidence can be found for such displacements. As discussed in the following examples, order variables are mostly related to partial movements in constituent molecular complexes, which may not precisely be identifiable with respect to the geometrical structure. Nevertheless, we consider that such displacements along a specific symmetry axis occur to lower strain energies in crystals. However, ionic displacements in the lattice are unlikely under practical conditions, because of their heavy masses; on the other hand, *partial displacements* with reduced masses can practically be responsible for symmetry changes in most observed cases.

In the following are shown examples of typical partial displacements, where order variables can be visualized geometrically from the molecular configuration.

Example 1 *Perovskites*

In perovskite crystals [1], the constituent in chemical formula ABO_3 is composed of a negative octahedral complex $(BO_6)^{2-}$ that is surrounded by eight positive ions A^+ at the corner of a cubic cell, as illustrated in figure 2.3. In the bi-pyramidal structure of $(BO_6)^{2-}$, there is an internal degree of freedom, where the central B^{4+} ion can be displaced in the direction 1 to 1′ on the z-axis, as shown in figure 2.3(a), or the group can be rotated around the z-axis connecting 1 and 1′ in figure 2.3(b). Related by inversion between 1 and 1′ with respect to the center of the structure, such linear and angular displacements can be considered for binary transitions in perovskites, $BaTiO_3$ and $SrTiO_3$, respectively. In the former, the center of the complex shifts parallel to the z-axis, generating an electric dipole moment, whereas in the latter the structure is twisted around the z-axis, keeping the center position unchanged. Physically significant in these cases is that the inversion violates the symmetry at minimum energy level, thereby allowing us to define the order variable as $\sigma_z(t)$, as a function of time t. It is notable that such a variable is detectable at a finite amplitude

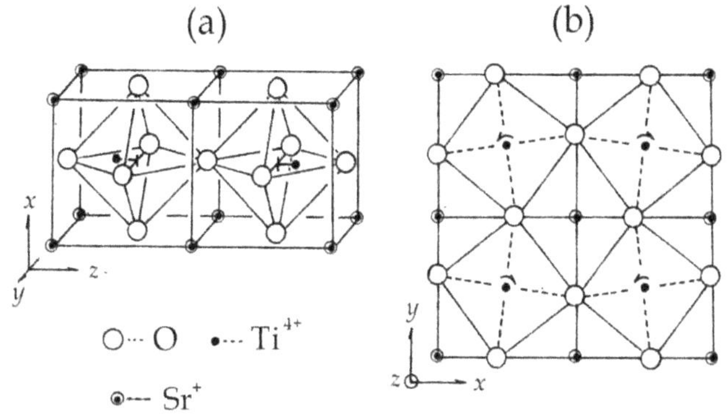

Figure 2.3. Cell structure in perovskite crystals $BaTiO_3$and $SrTiO_3$. In cases (a) and (b), Ti^{4+} ions inside bipyramidal $TiO_6{}^{2+}$ complexes are in displacive and rotational motion, respectively.

by experiments in timescale t_0, if sufficiently shorter than the timescale of inversion t, otherwise undetectable as the time average $\langle\sigma_z\rangle_t$ vanishes for $t > t_0$. (See section 1.6.) Therefore, symmetry axis x and y remain unchanged, while $\langle\sigma_z(t)\rangle_t$ emerges from zero along the z-axis at T_c, making the z-axis unique to characterize the cubic-to-tetragonal transition.

Example 2 *Tris-sarcosine calcium chloride (TSCC)*

Sarcosine is an amino acid $H_2C–NH_2–CH_2–COOH$, crystallizing with $CaCl_2$ in quasi-orthorhombic structure as TSCC, when growing from aqueous solutions at room temperature. Crystals thus obtained are twinned at room temperature in quasi-hexagonal form, consisting of three triangular domains with monoclinic axes approximately in common, if viewed by a polarizing microscope. (See figure 11.6 in chapter 11.) Although slightly monoclinic along the a-axis, a single domain specimen can be easily prepared from a twin sample for studying ferroelectric order that occurs at 120 K under atmospheric pressure. Figure 2.4(a) shows an x-ray result [3], where a quasi-hexagonal molecular arrangement is evident in the b–c plane. Figure 2.4(b) illustrates a $Ca^{2+}–(O_2C^-)_6$ complex in a TSCC crystal, where the Ca^{2+} ion is surrounded by six O^- of sarcosines. As explained in chapter 8, using a paramagnetic probe substituting for Ca^{2+}, the order variable σ_z can be identified from the impurity spectra as associated with a CaO_6 complex in the b-direction.

Example 3 *Potassium di-hydrogen phosphate (KDP)*

Figure 2.5 shows the molecular arrangement in orthorhombic crystals of KDP [4]. Tetrahedral PO_4^{3-} ions are linked to four neighbors via *hydrogen bonds* marked **H** in the figure. Along each hydrogen bond, a proton H^+ oscillates between two oxygen ions, forming an oscillating dipole. In this model, σ_z can be identified as associated with these four protons, which should be related with deformed phosphates as well.

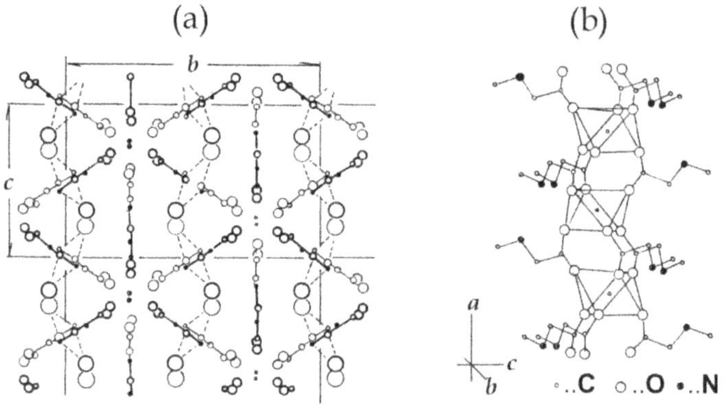

Figure 2.4. Molecular arrangement in a TSCC crystal. (a) A view in the *b–c*-plane. (b) Structure around a Ca^{2+} ion surrounded by six O^- ions of sarcosine molecules. The order variable is associated with the bipyramidal complex composed of six sarcosines. From [3], © 2010. With permission of Springer.

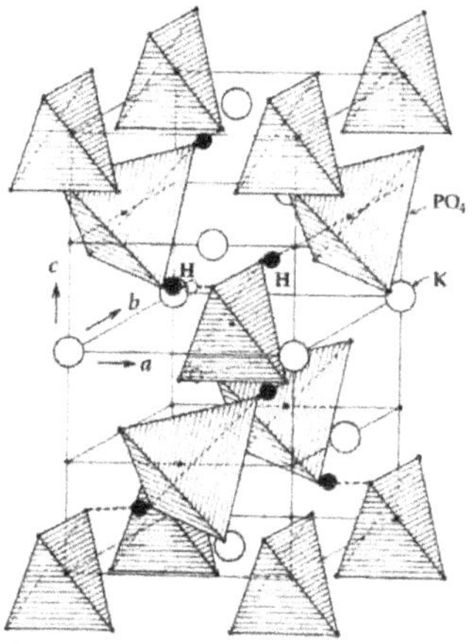

Figure 2.5. Structure of a KDP crystal. The order variable can be associated with a pyramidal PO_4^{3+} with four protons in the vicinity; two close protons and two other distant ones. From [3], © 2010. With permission of Springer.

2.3 The Born–Oppenheimer approximation and adiabatic Weiss' potentials

Considering various types of displacements, the basic model for binary order needs to be established, with the constituent mass M that is separated into a small m and a large $M - m \approx M$ in finite distance at a lattice site. Referring to *Newton's action–reaction principle*, such a separation requires an internal force originating from the lattice. Born and Huang [2] worked out this problem, considering that m and M are the electron and the constituent ion from which m is originated. Such a model can be applied to binary displacements in general, if m is a sufficiently small portion of M, as in the example of a Ti^{4+} in the octahedral TiO_6^{2-} complex. Although imprecise geometrically, such a model is useful to deal with critical phenomena, for which the displacement of m can be discussed in Born–Oppenheimer's approximation with a parameter $\kappa = (m/M)^{1/4}$ related to the mass-ratio m/M.

Simplifying the problem, we denote coordinates of m and M in a one-dimensional lattice. For the particle m in the crystal, the Hamiltonian can be written as

$$\mathcal{H}_o = -\sum_n \frac{\hbar^2}{2m}\frac{\partial^2}{\partial x_n^2} + \Delta U \quad \text{where } \Delta U = \sum_n \Delta U_n,$$

represents the potential energy due to mutual correlations among these particles. We consider that the Hamiltonian $\mathcal{H}_o$ is perturbed by M as

$$\mathcal{H} = \mathcal{H}_o - \sum_n \frac{\hbar^2}{2M}\frac{\partial^2}{\partial X_n^2}.$$

Using the small parameter κ-related to the mass ratio by $\kappa^4 = m/M$, $\mathcal{H}$ can be re-expressed in the form

$$\mathcal{H} = \mathcal{H}_o + \kappa^4 \mathcal{H}_1 \quad \text{where } \mathcal{H}_1 = -\sum_n \frac{\hbar^2}{2m}\frac{\partial^2}{\partial x_n^2}, \tag{2.4}$$

Omitting the suffix n for brevity, the unperturbed and perturbed wavefunctions can be written as $\phi_i(x, X)$ and $\psi_i(x, X)$, respectively, corresponding to eigenvalues ε_i and E_i at a representative site in a crystal. Namely,

$$\mathcal{H}_o \phi_i(x, X) = \varepsilon_i \phi_i(x, X) \tag{2.5a}$$

and

$$\mathcal{H}\psi_i(x, X) = E_i \psi_i(x, X). \tag{2.5b}$$

Corresponding with the displacement x of m, the displacement X from the lattice point X_o is assumed to be expressed as $X - X_o = \kappa u$, where u represent lattice displacement, and we expand (2.5a) with respect to κ for the following *asymptotic approximation*. First, expanding ε_i, ϕ_i and $\mathcal{H}_o$ for a small value of κ as

$$\varepsilon_i(x, X_o + \kappa u) = \varepsilon_i^{(0)} + \kappa\varepsilon_i^{(1)} + \kappa^2\varepsilon_i^{(2)} + \cdots,$$
$$\phi_i(x, X_o + \kappa u) = \phi_i^{(0)} + \kappa\phi_i^{(1)} + \kappa^2\phi_i^{(2)} + \cdots$$

and

$$\mathcal{H}_o\left(x, \frac{\partial}{\partial x}, X_o + \kappa u\right) = \mathcal{H}_o^{(0)} + \kappa\mathcal{H}_o^{(1)} + \kappa\mathcal{H}_o^{(2)} + \cdots,$$

where $\phi_i^{(0)}$, $\phi_i^{(1)}$, $\phi_i^{(2)}$, … are mutually orthogonal. Using these in (2.5a) and setting the coefficients of terms κ, κ^2, κ^3, … zero, we obtain

$$\left(\mathcal{H}_o^{(0)} - \varepsilon_i^{(0)}\right)\phi_i^{(0)} = 0, \tag{2.6a}$$

$$\left(\mathcal{H}_o^{(0)} - \varepsilon_i^{(0)}\right)\phi_i^{(1)} = -\left(\mathcal{H}_o^{(1)} - \varepsilon_i^{(1)}\right)\phi_i^{(0)}, \tag{2.6b}$$

$$\left(\mathcal{H}_o^{(0)} - \varepsilon_i^{(0)}\right)\phi_i^{(2)} = -\left(\mathcal{H}_o^{(1)} - \varepsilon_i^{(1)}\right)\phi_i^{(1)} - \left(\mathcal{H}_o^{(2)} - \varepsilon_i^{(2)}\right)\phi_i^{(0)}, \tag{2.6c}$$

For convenience, re-writing the term $\kappa^4\mathcal{H}_1$ in (2.4) as $\kappa^2\mathcal{H}_1^{(2)}$ where $\mathcal{H}_1^{(2)} = -\kappa^2\sum_u \frac{\hbar^2}{2m}\frac{\partial^2}{\partial u^2}$, (2.4) can be expressed asymptotically

$$\mathcal{H} = \mathcal{H}_o^{(0)} + \kappa\mathcal{H}_o^{(1)} + \kappa^2\left(\mathcal{H}_o^{(2)} + \mathcal{H}_1^{(2)}\right) + \cdots \tag{2.7}$$

$$\cdots\cdots\cdots\cdots\cdots\cdots\cdots\cdots\cdots\cdots\cdots ,$$

For the perturbed equation (2.7), E_i and ψ_i can be expanded in a similar manner to the above perturbed equation (2.7), that is

$$E_i = E_i^{(0)} + \kappa E_i^{(1)} + \kappa^2 E_i^{(2)} + \cdots \quad \text{and} \quad \psi_i = \psi_i^{(0)} + \kappa\psi_i^{(1)} + \kappa^2\psi_i^{(2)} + \cdots.$$

Substituting these expansions into (2.6*b*), we obtain

$$\left(\mathcal{H}_o^{(0)} - \varepsilon_i^{(0)}\right)\psi_i^{(0)} = 0, \tag{2.8a}$$

$$\left(\mathcal{H}_o^{(0)} - \varepsilon_i^{(0)}\right)\psi_i^{(1)} = -\left(\mathcal{H}_o^{(1)} - E_i^{(1)}\right)\psi_i^{(0)}, \tag{2.8b}$$

$$\left(\mathcal{H}_o^{(0)} - \varepsilon_i^{(0)}\right)\psi_i^{(1)} = -\left(\mathcal{H}_o^{(1)} - E_i^{(1)}\right)\psi_i^{(1)} - \left(\mathcal{H}_o^{(2)} + \mathcal{H}_1^{(2)} - E_i^{(2)}\right)\psi_i^{(0)}, \tag{2.8c}$$

$$\cdots\cdots\cdots\cdots\cdots\cdots\cdots\cdots\cdots\cdots\cdots ,$$

where we used the relation $E_i^{(0)} = \varepsilon_i^{(0)}$. Comparing (2.6*a*) and (2.8*a*), we have $\phi_i^{(0)}(x, X_o) = \psi_i^{(0)}(x, X_o)$, so that $u = 0$. Therefore, for $u \neq 0$, we can write

$$\psi_i^{(0)}(x, u) = \chi^{(0)}(u)\phi_i^{(0)}(x), \tag{2.9}$$

where $\chi^{(0)}(u)$ is an arbitrary function of u. Using this, we can solve the inhomogeneous equation (2.8*b*), if the integral $\int\phi_i^{(0)}(\mathcal{H}_o^{(0)} - \varepsilon_i^{(0)})\psi_i^{(0)}\mathrm{d}x = -\chi^{(0)}\int\phi_i^{(0)}(\mathcal{H}_o^{(1)} - E_i^{(1)})\phi_i^{(0)}\mathrm{d}x$ can vanish. However, as

$$\int\phi_i^{(0)}\left(\mathcal{H}_o^{(0)} - \varepsilon_i^{(0)}\right)\phi_i^{(0)}\mathrm{d}x = -\int\phi_i^{(0)}\left(\mathcal{H}_o^{(1)} - \varepsilon_i^{(1)}\right)\phi_i^{(0)}\mathrm{d}x = 0$$

from (2.6*b*), the integral is zero. On the other hand, as an expansion term of $\varepsilon(X_o + \kappa u)$, $\kappa\varepsilon_i^{(1)} = \kappa\left(\frac{\partial\varepsilon_i}{\partial X}\right)_{X=X_o} u_i$, is zero, if $u_i = 0$, therefore $\mathcal{H}_o^{(1)} = 0$. In addition, $E_i^{(1)} = \varepsilon_i^{(1)} = 0$ are required for $\mathcal{H}_o^{(1)} - E_i^{(1)} = 0$. Therefore, in the first-order accuracy of wavefunction $\psi_i = \psi_i^{(0)} + \kappa\psi_i^{(1)}$, we have

$$\psi_i^{(1)}(x, u) = \chi^{(0)}(u)\phi_i^{(1)}(x) + \chi^{(1)}(u)\phi_i^{(0)}(x), \tag{2.10}$$

where $\chi^{(0)}(u)$ and $\chi^{(1)}(u)$ are another set of functions of u.

Substituting (2.9) and (2.10) for $\chi^{(0)}$ and $\chi^{(1)}$ in (2.8*c*), taking $E_i^{(1)} = 0$ into account, we arrive at

$$\left(\mathcal{H}_o^{(0)} - \varepsilon_i^{(0)}\right)\phi_i^{(2)} = -\mathcal{H}_o^{(1)}\left(\chi^{(0)}\phi_i^{(1)} + \chi^{(1)}\phi_i^{(0)}(2)\right) - \left(\mathcal{H}_o^{(2)} + \mathcal{H}_1^{(2)} - E_i^{(2)}\right)\chi^{(0)}\phi_i^{(0)}.$$

Subtracting $\chi^{(1)} \times$ (2.6*a*) and $\chi^{(0)} \times$ (2.8*a*) from this formula, we obtain

$$\left(\mathcal{H}_0^{(0)} - \varepsilon_i^{(0)}\right)\left(\psi_i^{(2)} - \chi^{(2)}\phi_i^{(2)} - \chi^{(1)}\phi_i^{(1)}\right) = -\left(\mathcal{H}_1^{(2)} + \varepsilon_i^{(2)} - E_i^{(2)}\right)\chi^{(0)}\phi_i^{(0)}, \quad (2.11)$$

which is soluble, if $\int\phi_i^{(0)}(\mathcal{H}_1^{(2)} + \varepsilon_i^{(2)} - E_i^{(2)})\chi^{(0)}\phi_i^{(0)}\mathrm{d}x = 0$ is satisfied, which is indeed zero, if $\chi^{(0)}(u)$ is determined by

$$\left(\mathcal{H}_1^{(2)} + \varepsilon_i^{(2)} - E_i^{(2)}\right)\chi^{(0)}(u) = 0,$$

implying that the perturbed motion of M is harmonic and independent of x.

Comparing (2.6*a*) with (2.8*a*), we can write $(\mathcal{H}_0^{(2)} - \varepsilon_i^{(2)})(\chi^{(2)}\phi_i^{(0)}) = 0$, where $\chi^{(2)}$ is an arbitrary function of u, the function $\psi_i^{(2)} - \chi^{(2)}\phi_i^{(2)} - \chi^{(1)}\phi_i^{(1)}$ solved in the above (2.11) should be equivalent to $\chi^{(2)}\phi_i^{(2)}$. Therefore, we can express that

$$\psi_i^{(2)}(x, u) = \chi^{(0)}(u)\phi_i^{(2)}(x) + \chi^{(1)}(u)\phi_i^{(1)}(x) + \chi^{(2)}(u)\phi_i^{(0)}(x) \quad (2.12)$$

At this point, Born and Huang postulated replacing these $\phi(x)$-functions for higher-order approximation by

$$\phi_i^{(1)}(x) \to \phi_i^{(0)}(x) + \kappa\phi_i^{(1)}(x) \quad \text{and} \quad \phi_i^{(2)}(x) \to \phi_i^{(0)}(x) + \kappa\phi_i^{(1)}(x) + \kappa^2\phi_i^{(2)}(x)$$

to indicate *progressing anharmonicity*, thereby expressing (2.12) as

$$\psi_i^{(2)}(x, u) = \{\chi^{(0)}(u) + \kappa\chi^{(1)}(u) + \kappa^2\chi^{(2)}(u)\}\left\{\phi_i^{(0)}(x) + \kappa\phi_i^{(1)}(x) + \kappa^2\phi_i^{(2)}(x)\right\}, \quad (2.13)$$

although additional small terms such as $\kappa^3\chi^{(1)}(u)\phi_i^{(2)}(x)$ and some others may be required. Nevertheless, called the *adiabatic approximation*, (2.12) and (2.13) indicate the essential feature of the Born–Oppenheimer approximation for separating nuclear motion from order variables. It is realized that (2.13) is consistent with *condensates*, in that these functions of ϕ and χ are combined in *coherent motion* in periodic structure, which can be extended to coherence in the whole crystal in later discussions.

Theorem: In general, the perturbed Hamiltonian is given by

$$\mathcal{H}_i = \mathcal{H}_0 + \kappa^2\mathcal{H}_0^{(1)} + \kappa^2\left(\mathcal{H}_1^{(2)} + \Delta U_i\right), \quad (2.14a)$$

where

$$\Delta U_i = E_i^{(2)}(u) + \kappa E_i^{(3)}(u) + \kappa^3 E_i^{(4)}(u) + \cdots. \quad (2.14b)$$

Noting that (2.14*b*) is a function of u only, ΔU_i represents the *adiabatic potential* in the lattice in thermodynamic environment, exerting a driving force $-\nabla_u(\Delta U_i)$ on $\psi_i(x, u)$.

Thus, in *adiabatic Born–Oppenheimer's approximation*, an internal potential ΔU_i in the field theory for the order variables can be interpreted as originating from correlations among displacements u_i in a given crystal, which are *implicit in a relative coordinate system* in thermodynamic environment. The force $-\nabla_u(\Delta U_i)$ can therefore be regarded as the origin of the *Weiss field*, which is the *mesoscopic concept established in adiabatic approximation* for thermodynamics in crystals.

2.4 The Bloch theorem for collective order variables

In section 2.3, a partial displacement at a lattice site was discussed in Born–Oppenheimer approximation, while it is significant that such local displacements below T_c occur with the periodic lattice *synchronously if the crystal is in thermal equilibrium*. Considering that the displacement for $T < T_c$ is expressed as $u_n = u_o \exp i(\phi_n + \Delta\phi_n)$ in the periodic structure, the phase uncertainties $\Delta\phi_n$ are inevitable and *randomly distributed* among lattice sites, signifying distorted structure. According to Born–Huang's principle, which will be formulated in chapter 3, minimal lattice strains should be achieved in equilibrium, as expressed by a *phasing process* $\Delta\phi_n \rightarrow 0$ to establish a new symmetrical structure, where the Bloch theorem is important to deal with such a phasing of $\boldsymbol{\sigma}_n$ for mesoscopic processes in sizable crystals where surfaces can be disregarded.

However, notable at this stage is that *no such wave interferences would occur in infinite crystals.* Although familiar in solid-state theory, the Bloch theorem is *essential for collective modes of order variables*, and so briefly reviewed in section 2.4.2, prior to discussing transitions between crystalline phases.

2.4.1 Reciprocal lattice and renormalized coordinates

In equilibrium crystals, we assume that microscopic events such as displacements $\boldsymbol{u}_n(\boldsymbol{r}, t)$ are identical at almost all sites n in infinite crystals except defects and surfaces, which are considered as invariant of inversion at all times. A periodic structure is postulated in such a crystal of sufficiently large size.

Disregarding the time-dependence, these displacements can be determined by x-ray diffraction within the timescale, referring to the lattice vector

$$\boldsymbol{R} = n_1\boldsymbol{a}_1 + n_2\boldsymbol{a}_2 + n_3\boldsymbol{a}_3, \tag{2.15}$$

which is the basic translation in the direction $\boldsymbol{n} = (n_1, n_2, n_3)$ along unit vectors $\boldsymbol{a}_1$, $\boldsymbol{a}_2$ and $\boldsymbol{a}_3$. Using (2.11), the spatial invariance of $\boldsymbol{u}(\boldsymbol{r}_n, t)$ can be expressed by

$$\boldsymbol{u}(\boldsymbol{r}_n, t) = \boldsymbol{u}(\boldsymbol{r}_n + \boldsymbol{R}, t). \tag{2.16a}$$

Such displacements are considered as caused by excitations in the lattice that can be expressed by Fourier's series

$$\boldsymbol{u}(\boldsymbol{r}_n, t) = \sum_{\boldsymbol{G}} \boldsymbol{u}_G \exp i(\pm\boldsymbol{G}\cdot\boldsymbol{r}_n - \omega_n t). \tag{2.16b}$$

If $\boldsymbol{G}$ is determined by a value as related to the periodic boundary conditions, the Fourier amplitude is given by $\boldsymbol{u}_G = \boldsymbol{u}(\boldsymbol{r}_n, \boldsymbol{t})\exp\{-i(\mp\boldsymbol{G}\cdot\boldsymbol{r}_n - \omega_n t)\}$. Hence we can omit $\sum_G$ from (2.12*b*), and inversion symmetry $\boldsymbol{r}_n \rightarrow -\boldsymbol{r}_n$ can be replaced by wavevector inversion $\boldsymbol{G} \rightarrow -\boldsymbol{G}$.

Further, realizing that the periodic boundary is steady in crystals, we can apply (2.12*a*) to $\boldsymbol{r}_n = 0$ and $t = 0$, and obtain $\boldsymbol{u}(0, 0) = \boldsymbol{u}_G$ and $\exp i\boldsymbol{G} \cdot \boldsymbol{R} = 1$. Accordingly,

$$\boldsymbol{G} \cdot \boldsymbol{R} = 2\pi \times (0 \text{ or integer}). \tag{2.17}$$

indicating that $\boldsymbol{G} \perp \boldsymbol{R}$. Defining next a set of reciprocal vectors as

$$\boldsymbol{a}_1{}^* = \frac{2\pi}{\Omega}(\boldsymbol{a}_2 \times \boldsymbol{a}_3),\ \boldsymbol{a}_2{}^* = \frac{2\pi}{\Omega}(\boldsymbol{a}_3 \times \boldsymbol{a}_1),\ \boldsymbol{a}_3{}^* = \frac{2\pi}{\Omega}(\boldsymbol{a}_1 \times \boldsymbol{a}_2) \text{ and } \Omega = (\boldsymbol{a}_1, \boldsymbol{a}_2, \boldsymbol{a}_3),$$

is the volume of a cell made by three vectors $\boldsymbol{a}_1$, $\boldsymbol{a}_2$ and $\boldsymbol{a}_3$, the vector $\boldsymbol{G}$ can be written from (2.17) as

$$\boldsymbol{G} = h\boldsymbol{a}_1{}^* + k\boldsymbol{a}_2{}^* + l\boldsymbol{a}_3{}^* \quad \text{where } hn_1 + kn_2 + ln_3 = 2\pi \times (0 \text{ or integer}), \tag{2.18}$$

Here (h, k, l) is another set of integers in the reciprocal space, corresponding to (n_1, n_2, n_3) in the crystal space.

An orthorhombic crystal is a simple example, where the elemental volume is given by $\Omega = a_1a_2a_3$, hence the volume of a large crystal is expressed as $V = L_1L_2L_3$, where $L_1 = n_1a_1$, $L_2 = n_2a_2$ and L_3n_3, so that $h = (2\pi/L_1)n_1$, $k = (2\pi/L_2)n_2$ and $l = (2\pi/L_3)n_3$. Realizing that the reciprocal vectors $\boldsymbol{a}_1{}^*$, $\boldsymbol{a}_2{}^*$ and $\boldsymbol{a}_3{}^*$ are defined as orthonormal, $\boldsymbol{G}^2$ is invariant in the reciprocal space, representing the kinetic energy of lattice excitation, which is regarded as driven by the adiabatic potential energy $-\Delta U_G(u_n)$, where u_n is a function of the phase $\phi_n = \boldsymbol{G} \cdot \boldsymbol{r}_n - \omega t$. Restricted to the range $-\pi/2 < \phi_n < \pi/2$ by the Born–von Kármán boundary condition, ϕ_n can nevertheless be replaced by an arbitrary angle $\phi = \boldsymbol{G} \cdot \boldsymbol{r} - \omega t$ in the range $-\pi/2 < \phi < \pi/2$ for the equilibrium states, which we shall call the *renormalized phase* determined by the *renormalized position* r. Expressing the phase as $\phi = G(r - vt)$ at $v = \omega/G$, u_n is invariant for the translation $r \rightarrow r - vt$ within $-\pi/2 < \phi < \pi/2$, which is nothing but Galilean transformation. Because of such invariance, thermodynamic variables can generally be a function of the renormalized variables in finite crystals, with no reference to lattice points.

Renormalized coordinates are significant in crystals of finite size, and particularly important for thermodynamics, in that heat exchanges with surroundings need to be discussed. As mentioned in the introduction, crystal surfaces can be defined by the ends of Brillouin zone, namely $-\pi/2 \leqslant \phi$ and $\phi \leqslant \pi/2$. Just a mathematical convenience, this definition is adequate for thermodynamics, while the surface properties are negligible in sizable crystals.

2.4.2 The Bloch theorem

Theorem: Such displacements as $u(\boldsymbol{r}, t)$, occurring identically at all lattice sites, can be described as collection motion of mass particles m moving with a wavefunction

$\psi(\boldsymbol{r}, t)$ in renormalized space-time in finite crystals. We therefore write, following Kittel's textbooks [5, 6], the wave equation

$$\mathcal{H}\psi(\boldsymbol{r}) = \left\{\frac{p^2}{2m} + U(\boldsymbol{r})\right\}\psi(\boldsymbol{r}) = E\psi(\boldsymbol{r}) \tag{2.19}$$

for a steady state, signified by E. Here, $U(\boldsymbol{r})$ is the potential that satisfies the periodicity of the lattice, i.e. $U(\boldsymbol{r}) = U(\boldsymbol{r} + \boldsymbol{R})$, so that we have $\psi(\boldsymbol{r}) = \psi(\boldsymbol{r} + \boldsymbol{R})$, if the momentum $\boldsymbol{p}$ can be ignored.

The periodicity along the direction of a symmetrical $\boldsymbol{a}$ axis, can be expressed by $\boldsymbol{R} = n\boldsymbol{a}$, where $n = 1, 2, 3, \ldots, N$. N is the number of sites on the chain in length $L = Na$, so that if we write $\psi(\boldsymbol{r} + n\boldsymbol{a}) = c^n\psi(\boldsymbol{r})$, $c^N = 1$ for $n = N$, meaning that $c = \exp i(2\pi N) = \exp i(GaN)$, whereas for a general $n < N$ we have

$$\psi(\boldsymbol{r} + n\boldsymbol{a}) = \{\exp i\boldsymbol{G}\cdot(n\boldsymbol{a})\}\psi(\boldsymbol{r}) = \exp i\boldsymbol{G}\cdot(\boldsymbol{r} + n\boldsymbol{a})\psi(\boldsymbol{r})\exp(-i\boldsymbol{G}\cdot\boldsymbol{r}).$$

Therefore,

$$\psi(\boldsymbol{r} + \boldsymbol{R}) = \exp i(\boldsymbol{r} + \boldsymbol{R})\varphi_G(\boldsymbol{r}) \quad \text{where } \varphi_G(\boldsymbol{r}) = \exp(-i\boldsymbol{G}\cdot\boldsymbol{r})\psi(\boldsymbol{r}),$$

hence the Fourier transformation of $\psi(\boldsymbol{r} + \boldsymbol{R})$ can generally be expressed as

$$\psi(\boldsymbol{r} + \boldsymbol{R}) = \sum_{\boldsymbol{G}}\varphi_G \exp i\boldsymbol{G}\cdot(\boldsymbol{r} + \boldsymbol{R}), \tag{2.20}$$

which is called the *Bloch theorem*.

On the other hand, if $\psi(\boldsymbol{r})$ is propagating at a wavevector $\boldsymbol{k}$, we have to deal with the wavefunction $\psi(\boldsymbol{r}) = \varphi(\boldsymbol{r})\exp i\boldsymbol{k}\cdot\boldsymbol{r}$, so that

$$\psi(\boldsymbol{r} + \boldsymbol{R}) = \sum_{\boldsymbol{k}+\boldsymbol{G}}\varphi_{k+G}\exp i(\boldsymbol{k} + \boldsymbol{G})\cdot(\boldsymbol{r} + \boldsymbol{R}) = \sum_{\boldsymbol{k}}\varphi_k \exp i(\boldsymbol{k} + \boldsymbol{G})\cdot\boldsymbol{r}, \tag{2.21}$$

where

$$\varphi_k = \varphi_{k+G}(\boldsymbol{r})\exp i\boldsymbol{k}\cdot\boldsymbol{R} \tag{2.22}$$

Using quantum-mechanical operator $\boldsymbol{p} = -i\hbar\nabla_r$ on the wavefunction $\psi(\boldsymbol{r})$, the differential equation $\left\{\frac{p^2}{2m} + U(\boldsymbol{r})\right\}\psi(\boldsymbol{r}) = E\psi(\boldsymbol{r})$ can be manipulated by the following calculation

$$\begin{aligned}\boldsymbol{p}\psi(\boldsymbol{r}) = -i\hbar\nabla_r\{\varphi(\boldsymbol{r})\exp i\boldsymbol{k}\cdot\boldsymbol{r}\} &= \boldsymbol{p}\varphi(\boldsymbol{r})\times\exp i\boldsymbol{k}\cdot\boldsymbol{r} + \varphi(\boldsymbol{r})\times(-i\hbar\boldsymbol{k}\exp i\boldsymbol{k}\cdot\boldsymbol{r})\\ &= \exp i\boldsymbol{k}\cdot\boldsymbol{r}(\boldsymbol{p} - i\hbar\boldsymbol{k})\varphi(\boldsymbol{r})\\ &= \exp i\boldsymbol{k}\cdot\boldsymbol{r}(\hbar^2\boldsymbol{k}^2 + \boldsymbol{p}^2 - i\hbar\boldsymbol{k}\cdot\boldsymbol{p})\varphi(\boldsymbol{r})\\ &= \exp i\boldsymbol{k}\cdot\boldsymbol{r}(\boldsymbol{p} - i\hbar\boldsymbol{k})^2\varphi(\boldsymbol{r})\end{aligned},$$

resulting in

$$\left\{\frac{(\boldsymbol{p} - i\hbar\boldsymbol{k})^2}{2m} + U(\boldsymbol{r})\right\}\varphi(\boldsymbol{r}) = E\varphi(\boldsymbol{r}),$$

which is the same as

$$\left\{\frac{\hbar^2}{2m}\left(\nabla_r^2 - 2i\boldsymbol{k}\cdot\nabla_r\right) + U(\boldsymbol{r})\right\}\varphi(\boldsymbol{r}) = \left(E - \frac{\hbar^2 k^2}{2m}\right)\varphi(\boldsymbol{r}).$$

Accordingly, energy surfaces in the $\boldsymbol{k}$-space are given by $\varepsilon_k = E - \frac{\hbar^2 k^2}{2m} = \text{const.}$ that is characterized by

$$\varepsilon_k = \varepsilon_{k+G} \tag{2.23}$$

Writing φ_k in (ii) as $\varphi_k(\boldsymbol{r} + \boldsymbol{R})$, we have

$$\varphi_k(\boldsymbol{r} + \boldsymbol{R}) = (\exp i\boldsymbol{k}\cdot\boldsymbol{R})\,\varphi_k(\boldsymbol{r}), \tag{2.24a}$$

where $\exp i\boldsymbol{k}\cdot\boldsymbol{R} = \boldsymbol{T}$ can be defined as the *translation operator* for (2.19) to be expressed as

$$\varphi_k(\boldsymbol{r} + \boldsymbol{R}) = \boldsymbol{T}\varphi_k(\boldsymbol{r}). \tag{2.24b}$$

In practical crystals, the unit cell needs further to be specified by the *point group* that is composed of rotation, reflection and other elemental operations. In such operations $\boldsymbol{S}$, the structure is retained as invariant for space inversion $\boldsymbol{r} \to -\boldsymbol{r}$, hence we have the relation

$$\boldsymbol{S}\varphi_k(\boldsymbol{r}) = \varphi_k(\boldsymbol{S}^{-1}\boldsymbol{r}). \tag{2.25}$$

It is important to realize that all symmetries are valid as long as the crystal is in thermodynamic environment, otherwise they become inhomogeneous by broken symmetry in non-equilibrium, so that transitions can take place by interactions with phonon.

2.4.3 The Brillouin zone

The dynamics of order variables in collective motion in equilibrium crystals should be conservative characterized by eigenvalues (2.23) determined by periodic conditions. Consequently, we can deal with their propagation effectively for the wave vector to be restricted in the region $|\boldsymbol{k}| \leqslant |\boldsymbol{G}|/2$, constituting a polyhedral space at the origin $\boldsymbol{G} = 0$ in the reciprocal lattice, called first *Brillouin zone*. A stable structure in thermal and adiabatic equilibrium, the excitation energy ε_k can be analyzed as a function of wavevector $\boldsymbol{k}$ in the first zone sufficiently, as the other zones are in repetition.

Figure 2.6(a) shows the first zone in one dimension, centered at $k_x = 0$ and bordered at $k_x = \pm\frac{\pi}{a}$, where a is the lattice constant, signified by reflection symmetry $\Delta k_x \rightleftarrows -\Delta k_x$, and

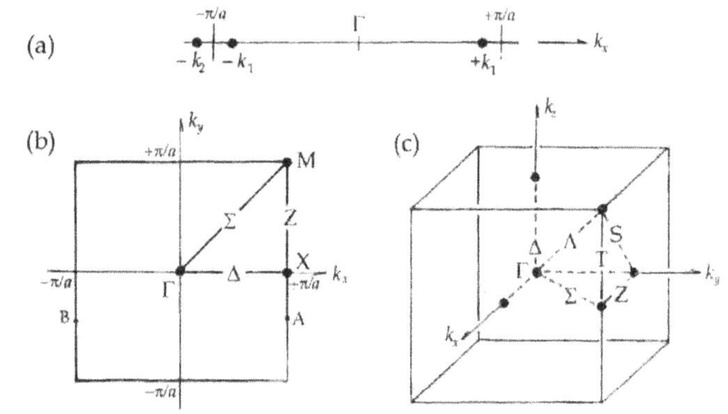

Figure 2.6. (a) Brillouin zone in one dimension: $-\pi/a \leqslant k_x \leqslant \pi/a$, and Γ is the zone center. (b) Two-dimensional Brillouin zone in the k_xk_y-plane. (c) Brillouin zone in a cubic crystal.

$$\psi_{k_x}\left(\frac{\pi}{a}\right) = \psi_{k_x}\left(-\frac{\pi}{a}\right).$$

Inside the zone, we have a standing wave

$$\psi_{k_x}(x) = \cos\frac{\pi x}{a}\,\varphi_{k_x}(x),$$

and at the boundaries

$$\nabla_x \varepsilon_k = 0 \quad \text{for} \quad k_x = \pm\frac{\pi}{a} \tag{2.26}$$

In figure 2.6(b), a square zone in two dimensions is illustrated. Symmetry operations are expressed by $4mm$: a fourfold axis of two mirror planes, m_x and m_y, and two diagonal planes, m_d and $m_{d'}$. Specific points Γ, M, X and specific lines Δ, Z, Σ in the zone are indicated in the figure; the point Γ is at the origin, where all operations transfer to themselves. Point M, X can be transformed to itself, or other corners of the square. Specified by the same $|\boldsymbol{k}|$, all corner points are equivalent. Point X is invariant under operations 2_z, m_x, m_y; a successive reflection m_x and rotation 2_z carries the point $(\frac{\pi}{a}, 0)$ to $(-\frac{\pi}{a}, 0)$. The lines Δ, Σ and Z are invariant under mirror reflection m_x, m_d and m_y, respectively. It is noted that by m_x a point $(\frac{\pi}{a}, -k_y)$ on the otherwise inhomogeneous by broken symmetry in non-equilibrium, so that transitions can take place by interactions with phonon.

The lines Δ, Σ and Z are invariant under mirror reflection m_x, m_d and m_y, respectively. It is noted that by m_x a point $(\frac{\pi}{a}, -k_y)$ on the line Z can be carried to $(-\frac{\pi}{a}, -k_y)$, implying that equation (2.17) can be applied to all points on the zone boundaries. Figure 2.6(c) shows a simple cubic lattice, where specific points and lines, similar to the square lattice, indicate basic features of the Brillouin zone, confirming with the group operations $\frac{4}{m}$, $\bar{3}$ and $\frac{2}{m}$, at four specific points R, M, X, Γ and six specific lines Δ, Σ, T, Σ, Z and Λ. For more details of the group theory, interested readers are referred to standard textbooks [7, 8].

Exercises

1. Discuss the reason why the order variable should be signified by the *reduced mass*.
2. Why is the timescale of observation important in thermodynamics? Discuss the issue.
3. Why do we need to apply the Bloch theorem to order variables. Discuss how we come up with the concept of Weiss field, as the necessary quantity for dynamical order variables in finite crystals. Do you consider that the Weiss field is an acceptable concept in infinite crystals?
4. In thermodynamics, the area outside crystals is regarded as the surroundings. Despite this, we often can consider the phonon system as the heat reservoir. Discuss the issue and limitation of the assumption.

References

[1] Megaw H D 1973 *Crystal Structure: A Working Approach* (Philadelphia, PA: Saunders)
[2] Born M and Huang K 1968 *Dynamical Theory of Crystal Lattices* (London: Oxford University Press)
[3] Fujimoto M 2010 *Thermodynamics of Crystalline States* (Berlin: Springer)
[4] Ziman J M 1958 *Models of Disorder* (London: Cambridge University Press)
[5] Kittel C 1964 *Introduction to Solid State Physics* 6th edn (New York: Wiley)
[6] Kittel C 1963 *Quantum Theory of Solids* (New York: Wiley)
[7] Tinkham M 1964 *Group Theory and Quantum Mechanics* (New York: McGraw-Hill)
[8] Knox R S and Gold A 1964 *Symmetry in the Solid State* (New York: Benjamin)

IOP Publishing

Solitons in Crystalline Processes (2nd Edition)
Irreversible thermodynamics of structural phase transitions and superconductivity
Minoru Fujimoto

Chapter 3

Pseudospin clusters and the Born–Huang principle: coherent order-variables as solitons in crystals

Approaching the critical temperature T_c from higher temperatures, order variables defined by partial displacements of constituents at lattice sites initiate *clustering* in the short range by lowering temperature for the displacements to be *elastically* compatible with *new lattice symmetry*. Recovering thermally from the *strain energy*, the clustered group can propagate through periodic crystals. In phase with the nearest and next-nearest neighbors, the cluster mode of order variables is generally *incommensurate* with the lattice period along the direction of wavevector determined by new lattice symmetry. Such correlated clusters constitute *seed-condensates* in the new symmetry analogous to condensation phenomena, propagating along the direction for minimum strains in crystals.

In the second edition, the important concepts of solitons and their correlations are defined to discuss the properties of order variables in the mesoscopic phase.

3.1 Pseudospins for binary displacements

3.1.1 Binary displacements

Determined by the space group in whole crystals with surfaces, the lattice sites are all identical, where an additional point group configures molecules and ionic complexes in the lattice structure. In stable structure the constituents are primarily in harmonic motion, however, in a destabilized lattice a portion of the constituent can be mobile from one site to another, as specified by the internal degree of freedom of point symmetry, independent of lattice vibrations.

In perovskite crystals, such mobile portions were identified in diffuse x-ray diffraction patterns at the critical temperature T_c. Amongst displacive order variables during phase transitions in perovskites, $BaTiO_3$ and $SrTiO_3$ exhibit typical structural changes, as illustrated in figures 2.3(a) and 2.3(b) in chapter 2. In the former, we consider that the Ti^{4+} ion moves at fast rate for $T > T_c$ between positions 1 and 1′ in the octahedral TiO_6^{2-} along the z-direction at equal probabilities. On the other hand, in the latter, the TiO_6^{2-}-complex rotates at fast rate by small positive and negative angles occuring between 1 and 1′ around the z-axis at equal probabilities for $T > T_c$. In these cases, the motion of an order variable can be expressed by a pseudospin vector $\boldsymbol{\sigma}_n$ parallel to the z axis, which may be considered as a *pseudospin in finite magnitude*.

For $\boldsymbol{\sigma}_n$ and $\boldsymbol{\sigma}_m$ at different sites n and m are primarily independent for $T > T_c$, but become *correlated in slower motion* by an *internal adiabatic potential* emerged at T_c, hence $\boldsymbol{\sigma}_n \neq \boldsymbol{\sigma}_m$ at different sites $m \neq n$ are correlated for $T < T_c$.

3.1.2 Ising's model of a pseudospin at T_c

Inversion of a pseudospin cluster $\boldsymbol{\sigma}_n$ is essentially quantum mechanical *tunnelling* through a potential barrier ΔU_n, which is assumed to occur *along a preferable symmetry axis* below the critical temperature T_c. Considering *random* directions for $T > T_c$, however, inversion axes become parallel to a new symmetric axis of the lattice to attain lower *strain energy* for $T < T_c$, which is tentatively called the z-axis for convenience of the discussion to follow, while it is essentially determined by the *new crystalline phase*. In this section, the tunnelling motion is convenient for statistical arguments of binary inversion in *terrestrial nature in particular*, which is therefore discussed as Ising's model, after Blinc and Zeks [1].

Omitting site index n for brevity, we consider that $\boldsymbol{\sigma}(z)$ is perturbed by $\Delta U(z)$ in the z-direction, which is the inversion axis. Writing the unperturbed equation for the corresponding wave function $\varphi_o(z)$ as

$$\mathcal{H}_o\varphi_o = \varepsilon_o\varphi_o,$$

where the eigenvalue ε_o is considered as positive, representing the inversion energy of $\boldsymbol{\sigma}(z)$. The energy ε_o is doubly degenerate for $\varphi_o(+z)$ and $\varphi_o(-z)$, however, split into two by the perturbation potential $U(z)$ centered at $z = 0$. Therefore, for a sufficiently small $|U(z)|$ at T_c, the perturbed wavefunctions can be expressed in *symmetric* and *antisymmetric* combinations of $\varphi_o(\pm z)$, i.e.

$$\psi_+(z) = \frac{\varphi_o(z) + \varphi_o(-z)}{\sqrt{2}} \quad \text{and} \quad \psi_-(z) = \frac{\varphi_o(z) - \varphi_o(-z)}{\sqrt{2}},$$

which are normalized as

$$\psi_-{}^*\psi_+ + \psi_+{}^*\psi_- = \varphi_o{}^*(+z)\varphi_o(+z) + \varphi_o{}^*(-z)\varphi_o(-z) = 1.$$

Here, $\varphi_o{}^*(+z)\varphi_o(+z) = p_o(+z)$ and $\varphi_o{}^*(-z)\varphi_o(-z) = p_o(-z)$ are probabilities at $+z$ and $-z$, respectively, and hence the normalization is expressed by $p_o(+z)+p_o(-z) = 1$. And the perturbed energy is given by the eigenvalues $\varepsilon_\pm$, as expressed by

$$\varepsilon_{\pm} = \varepsilon_{o} \pm \frac{U(0)}{2}\left(\psi_{+}^{*}\psi_{+} - \psi_{-}^{*}\psi_{-}\right), \tag{3.1}$$

where $\psi_{+}^{*}\psi_{+} = p(+z)$ and $\psi_{-}^{*}\psi_{-} = p(-z)$ can be interpreted as the perturbed probabilities.

Using these tunneling properties, we can define the pseudospin vector by its components

$$\sigma_{z} = \psi_{+}^{*}\psi_{+} - \psi_{-}^{*}\psi_{-} = p(+z) - p(-z),$$
$$\sigma_{x} = \varphi_{o}^{*}(+z)\varphi_{o}(-z) + \varphi_{o}^{*}(-z)\varphi_{o}(+z)$$

and

$$\sigma_{y} = \varphi_{o}^{*}(+z)\varphi_{o}(-z) - \varphi_{o}^{*}(-z)\varphi_{o}(+z).$$

It is noted that we have the following *commutation* relations, making $(\sigma_{x}, \sigma_{y}, \sigma_{z})$ a vector quantity, i.e.

$$[\sigma_{x}, \sigma_{y}] = i\sigma_{z}, \quad [\sigma_{y}, \sigma_{z}] = i\sigma_{x} \text{ and } [\sigma_{z}, \sigma_{x}] = i\sigma_{y}, \tag{3.2}$$

thereby calling a vector $\boldsymbol{\sigma}$ a pseudospin by analogy of the conventional spin ½. For a small potential $|U(z)|$, the motion of $\boldsymbol{\sigma}$ appears as fluctuations like a quantum-mechanical tunneling, known as a tunneling model. However, it should be noted that in the tunneling model the pseudospin is statistically defined for motion between finite displacements, which should be different from harmonic displacements at infinitely small amplitudes. Accordingly, the pseudospin represents distinctively different motion from phonons, as characterized by a finite amplitude. Considering motion statistically, however, it can represent *an order variable as classical vector of finite amplitude, accompanying an adiabatic potential* $U(z)$ that is unidentified in the above but significant in the following.

3.1.3 Pseudospin correlations below T_c

In crystals, pseudospin clusters $\boldsymbol{\sigma}_n$ and $\boldsymbol{\sigma}_m$ at sites n and m should be correlated with each other below T_c, particularly if they are located at near distance in finite amplitude. Although imprecise from diffraction patterns, from the known cell structure in some examples, the correlations are related clearly with distributed electron densities. Accordingly, considering nearest neighbors n and $n+1$, the interaction energy can generally be expressed as

$$\mathcal{H}_{n,n+1} = \frac{1}{2}\sum_{\alpha\beta\gamma\delta}\left\langle \psi_{n,\alpha}^{*}\psi_{n,\beta}\right| V_{\alpha\beta\gamma\delta}\left|\psi_{n+1,\gamma}^{*}\psi_{n+1,\delta}\right\rangle, \tag{3.3}$$

where $V_{\alpha\beta\gamma\delta}$ is the correlation tensor between the densities $\psi_{n,\alpha}^{*}\psi_{n,\beta}$ and $\psi_{n+1,\gamma}^{*}\psi_{n+1,\delta}$, and the indexes α, β and γ, δ are + or − at sites n and $n+1$, respectively. By symmetry, elements of $V_{\alpha\beta\gamma\delta}$ are related as

$$V_{++--} = V_{--++}, V_{+-+-} = V_{-+-+} = V_{-++-}, \text{.....etc.}$$

Manipulating pseudospin components defined in section 3.2, we acquire the following expressions:

$$\psi_{n,+}{}^*\psi_{n,+} = \frac{1}{2}(1 + \sigma_{nz}), \qquad \psi_{n,+}{}^*\psi_{n,-} = \frac{1}{2}(\sigma_{n,x} - \sigma_{n,y}),$$

$$\psi_{n+1,+}{}^*\psi_{n+1,+} = \frac{1}{2}(1 + \sigma_{n+1,z}), \quad \psi_{n+1,+}{}^*\psi_{n+1,-} = \frac{1}{2}(\sigma_{n+1,x} - \sigma_{n+1,y}), \ etc.$$

Using these results, (3.3) can be expressed in simplified form

$$\mathcal{H}_{n,n+1} = -J_{n,n+1}\sigma_{n,z}\sigma_{n+1,z} - K_{n,n+1}\sigma_{n,x}\sigma_{n+1,x},$$

where

$$J_{n,n+1} = 2V_{+-+-} \quad \text{and} \quad K_{n,n+1} = 2V_{+-+-} - V_{++++} - V_{----}.$$

The total Hamiltonian of two correlated pseudospins can then be written as

$$\begin{aligned}\mathcal{H} &= \mathcal{H}_n + \mathcal{H}_{n+1} + \mathcal{H}_{n,n+1}\\ &= \varepsilon_0 - \frac{U(0)}{2}(\sigma_{n,z} + \sigma_{n+1,z}) - J_{n,n+1}\sigma_{n,z}\sigma_{n+1,z} - K_{n,n+1}\sigma_{n,x}\sigma_{n+1,x}.\end{aligned}$$

Assuming $V_{+-+-} = V_{++++} = V_{----}$ for a symmetrical interaction, we have specifically

$$K_{n,n+1} = 0,$$

and the correlation energy is determined as

$$\mathcal{H}_{n,n+1} = -J_{n,n+1}\sigma_n\sigma_{n+1}, \qquad (3.4a)$$

which is in the general form of *pseudospin correlations.*

We consider that such pseudospin vectors $\boldsymbol{\sigma}_n = (\sigma_{nx}, \sigma_{ny}, \sigma_{nz})$ emerge with local potentials $U_n(z)$ in the transition region, then forming correlated *clusters* of $\boldsymbol{\sigma}_n$ at the critical temperature T_c, violating macroscopic lattice symmetry. It is a logical assumption for pseudospins of displacement vectors that correlation energy (3.4*a*) can be characterized in the *scalar product* $\boldsymbol{\sigma}_n \cdot \boldsymbol{\sigma}_{n+1}$ of two vectors $\boldsymbol{\sigma}_n$ and $\boldsymbol{\sigma}_{n-1}$; namely expressing as

Theorem:

$$\mathcal{H}_{m,n} = -J_{m,n}\boldsymbol{\sigma}_m \cdot \boldsymbol{\sigma}_n, \qquad (3.4b)$$

which is the basic expression for dealing with the correlation between pseudospins in vector character.

3.1.4 Boson statistics for modulated pseudospins

Expressing lattice displacements by pseudospin vectors, $\boldsymbol{\sigma}_n$ cannot be identical to a phonon vector. To avoid confusion with lattice vibrations, we should clarify the mathematical argument, considering *bifurcation processes* during transition from random to mesoscopic states, while the bifurcation disappears on lowering temperature. Experimentally, such displacements are known to occur after phase transitions, i.e. for $T \leqslant T_c$, constituting the basic subject of discussions in this book.

In the presence of a positive potential $U(z \sim 0) = \frac{1}{2}z^2$ as in the previous section, we have an equation of motion induced by the modulation energy $\Delta U(z)$, which has either internal or external origins at a specific mesoscopic wave vector $\boldsymbol{q}$, and can be written as

$$\left(-\frac{1}{2}\frac{\mathrm{d}^2}{\mathrm{d}t^2} + \frac{1}{2}z^2 - \Delta U(z)\right)\psi_q(z) = \varepsilon_q \psi_q(z), \tag{3.5}$$

where $\varepsilon_q = \frac{E_q}{\hbar\omega_q}$ with respect to phonons $E_q \propto \hbar\omega_q$ is the eigenvalue at $\boldsymbol{q}$ that is perturbed by a potential $\Delta U(z) = \gamma z$ for a mesoscopic displacement z, for which $\psi_q(z)$ is the wave function and the parameter γ a finite constant for the displacement that occurs in a direction specified by the wavevector $\boldsymbol{q}$. Figure 3.1 illustrates that the potential $\frac{1}{2}z^2$ shifts $U(0)$ at 0 to $U(z)$ at z, as illustrated. Here, ω_q and E_q are unperturbed frequency and eigenvalue, respectively, at a particular wave vector $\boldsymbol{q}$ in the periodic lattice, which should be characterized by *inversion* $\boldsymbol{q} \rightleftarrows -\boldsymbol{q}$ *in finite crystals*.

However, placing it in a periodic lattice, the timescale of varying displacement in (3.5) at $\boldsymbol{q}$ cannot be the same as the phonon vector that keeps the lattice *stable*, so that ε_q is determined in a timescale with respect to the mesoscopic lattice modulation along a specific direction of $\pm\boldsymbol{q}$. Hence, we redefine it as $\varepsilon_q = \frac{E_q}{\lambda_q}$, where λ_q signifies an *elastic energy* due to minimum distortion of a crystal at a temperature for $T < T_c$, which can be specified by a *soliton number in finite lattice* in later discussions.

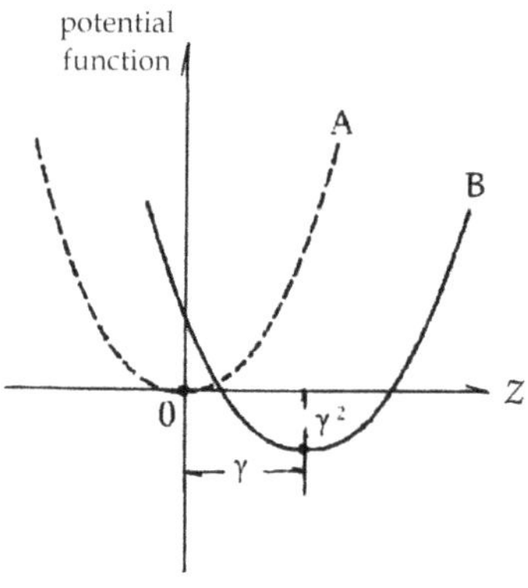

Figure 3.1. The potential energy curve for a finite displacement. A and B are, respectively, the original and displaced harmonic potentials $U(0)$ and $U(z)$, which are shifted by a potential $z^2/2$. The parameter γ is the displacement constant defined in (3.7a).

We define *creation and annihilation operators* b_q and $b_q^\dagger$ by writing

$$z = \frac{1}{\sqrt{2}}\left(b_q + b_q^\dagger\right) \quad \text{and} \quad \frac{\mathrm{d}}{\mathrm{d}z} = \frac{1}{\sqrt{2}}\left(b_q - b_q^\dagger\right),$$

Equation (3.5) can then be written as

$$\left\{b_q^\dagger b_q - \gamma\left(b_q^\dagger + b_q\right)\right\}\psi_q = \left(\varepsilon_q - \frac{1}{2}\right)\psi_q, \tag{3.6}$$

however, we note that these b_q and $b_q^\dagger$ do not constitute commutation relations of boson operators as for harmonic phonons.

Therefore, assuming the commutation relation $\left[b_q, b_q^\dagger\right] = 1$ to be for the parameter $\gamma = 0$ that is mathematically identical to phonons, but a new set of operators $\tilde{b}_q$ and $\tilde{b}_q^\dagger$ can be defined as

$$b_q = \tilde{b}_q + \gamma \quad \text{and} \quad b_q^\dagger = \tilde{b}_q^\dagger + \gamma \tag{3.7a}$$

for *finite displacements* to occur with a nonzero $\gamma \neq 0$, instead of the shifting position $z \neq 0$, thereby the operators $\tilde{b}_q^\dagger$ and $\tilde{b}_q$ are considered as the *Boson commutation relation.*

$$\left[\tilde{b}_q, \tilde{b}_q^\dagger\right] = \tilde{b}_q\tilde{b}_q^\dagger - \tilde{b}_q^\dagger\tilde{b}_q = 1 \tag{3.7b}$$

for *such finite displacements.* Thus, the finite displacement field *in the mesoscopic states can be characterized by the Boson statistics* in momentum space. However, such a nonzero $\gamma \neq 0$ at finite temperature T is considered as distributed in the critical region by *bifurcation*, approaching a constant afterwards.

With a nonzero $\gamma \neq 0$, pseudospins for finite displacements can be characterized by $\tilde{b}_q^\dagger$ and $\tilde{b}_q$ in periodic crystals, signifying they obey Boson statistics, but similar to *infinitesimal* phonon displacements that are signified by conventional $b_q^\dagger$ and b_q.

Haken [2] demonstrated that such modified operators $\tilde{b}_q^\dagger$ and $\tilde{b}_q$ can be applied to transforming wave functions as $\psi\left(b_q, b_q^\dagger\right) \rightarrow \tilde{\psi}\left(\tilde{b}_q, \tilde{b}_q^\dagger\right)$, thereby (3.6) is expressed as

$$\left(\tilde{b}_q^\dagger\tilde{b}_q - \gamma^2\right)\tilde{\psi}_q = \tilde{\varepsilon}_q\tilde{\psi}_q \quad \text{and} \quad \tilde{\varepsilon}_q = \varepsilon_q - \gamma^2 \quad \text{for} \quad \gamma \neq 0 \tag{3.7c}$$

representing finite displacements. This equation can alternatively be written as

Theorem:

$$\left(\tilde{b}_q^\dagger\tilde{b}_q\right)\tilde{\psi}_q = \lambda_q\tilde{\psi}_q \quad \text{and} \quad \lambda_q = \tilde{\varepsilon}_q - \frac{1}{2} \tag{3.7d}$$

where λ_q is the eigenvalue of (3.7*c*) for the equation for finite displacements at $\boldsymbol{q}$; whereas for $\gamma = 0$, in particular, (3.7*c*) can be reduced to

$$\left(b_q^\dagger b_q\right)\psi_q = \left(\varepsilon_q - \frac{1}{2}\right)\psi_q, \tag{3.7e}$$

that is valid for a harmonic oscillator at *infinitesimal amplitude*.

Comparing (3.7*d*) with (3.7*e*), the lattice displacement is clearly distinct from lattice vibrations. Such a finite displacement shares the zero-point energy in common with a harmonic oscillator, because both are signified by $\gamma = 0$ at $T = 0$ K, which is nevertheless obscured experimentally by unavoidable space-time uncertainties in crystals.

The pseudospin $\boldsymbol{\sigma}_q$ for the finite lattice displacement at a specific $\boldsymbol{q}$ is thus characterized by boson operators $\tilde{b}_q^\dagger$ and $\tilde{b}_q$ in one-dimension, which are, however, different from phonons that are characterized by *random distribution* in momentum space. Nevertheless, considering partial displacements of *small masses* in one-dimension, there should be no significant problem other than *semantic definition* of the order variable.

Theorem: Now that operators $\tilde{b}_q^\dagger$ and $\tilde{b}_q$ are operated for inversion $\boldsymbol{q} \rightleftarrows -\boldsymbol{q}$ to use in all equilibrium crystals, where the interaction $\mp J\boldsymbol{\sigma}_q \cdot \boldsymbol{\sigma}_{-q}$ from (3.4*a*) signifies either stable correlated domains or entropy production for separation, respectively, that is fundamental *for the internal energy $U_o(p, T)$ of crystals to be invariant of the symmetry group.*

Probabilities for these shifting events can be expressed as an operator

$$\boldsymbol{P} = \frac{1}{2}(1 + \boldsymbol{\sigma}_q \cdot \boldsymbol{\sigma}_{-q}), \tag{3.7f}$$

which should be commutable with the Hamiltonian of multiple $\boldsymbol{\sigma}_q$, so that $\boldsymbol{P} = 1$ can be assigned to the system where parallel interactions $\boldsymbol{\sigma}_q \| \boldsymbol{\sigma}_{-q}$ are postulated for thermal stability. The lowest eigenstate of a *multi-boson system* is always *symmetric in thermal equilibrium with respect to exchanging quasi-particles*, i.e. $(\boldsymbol{\sigma}_{+q} + \boldsymbol{\sigma}_{-q})/\sqrt{2}$, which is significant for *finite collective displacements in a modulated lattice in finite crystals*; therefore, $\boldsymbol{\sigma}_{\pm q}$ are named distinctively as *solitons.* It is interesting to note that the finite parameter γ in (3.7*a*) can be arbitrary for *symmetric coherent correlations over the mesoscopic area*, which are then referred to as *topological correlations.*

3.2 The Born–Huang principle and pseudospin clusters

In the above tunneling model, we considered a positive potential $U_n(z)$, which should exist in real crystals at T_c. Representing the lattice, $U_n(z)$ is responsible for generating classical displacements $\sigma_n(z)$ along a certain z-axis; thereby a change in lattice symmetry can be attributed to $\Delta U_n(z)$ in a given crystal. In this case, the emerging collective function $\sigma_n(z)$ at T_c should occur with a lattice excitation in the whole crystal as determined by the Bloch theorem.

Unlike independent $\boldsymbol{\sigma}_n$ for $T > T_c$, clustered pseudospins $\boldsymbol{\sigma}_n$ for $T < T_c$ are in *collective motion* in the first Brillouin zone $\boldsymbol{G} = 0$, restricted by new lattice symmetry, but the transition process to coherent motion of clusters is obscured by *bifurcation*, exhibiting *anomalies*. In this case, considering a Fourier series

$$\sigma_n = \sum_q \sigma_q \exp i(\boldsymbol{q} \cdot \boldsymbol{r}_n - \tilde{\omega}\, t_n)$$

for $T < T_c$, the Fourier transform

$$\sigma_q = \sum_n \sigma_n \exp(-i\boldsymbol{q} \cdot \boldsymbol{r}_n + \tilde{\omega} t_n) \tag{3.8}$$

is assumed for collective motion of $\boldsymbol{\sigma}_n$ that is driven by the force $-\nabla\, U(\boldsymbol{r}_n, t_n)$, where the potential $U(\boldsymbol{r}_n, t_n)$ corresponds to an *adiabatic thermodynamic potential*. Experimentally however, $U(\boldsymbol{r}_n, t_n)$ is unidentifiable in practice due to the *bifurcation* region of transitions.

Further, we should consider spatial inversion $\boldsymbol{r}_n \to -\boldsymbol{r}_n$ along the direction of propagation, allowing us to consider the inversion $\boldsymbol{q} \to -\boldsymbol{q}$ for the phase inversion in (3.5), disregarding time inversion for thermodynamic description. Hence, collective motion takes place between $\boldsymbol{\sigma}_{+q}$ and $\boldsymbol{\sigma}_{-q}$, signified by the phase conversion between $\phi_n = \pm\boldsymbol{q} \cdot \boldsymbol{r}_n - \tilde{\omega} t_n$ at lattice sites n.

In periodic crystals, ϕ_n can be replaced by a *renormalized phase* ϕ in the Brillouin zone, as ϕ_n and ϕ these are both in the same ranges $-\pi/2 \leqslant \phi_n$ and $\phi \leqslant \pi/2$. The phase $\phi = \pm\boldsymbol{q} \cdot \boldsymbol{r} - \tilde{\omega} t$ can therefore be considered as a continuous variable, thereby the variable $\boldsymbol{\sigma}_q = \boldsymbol{\sigma}_q\,(\phi)$ in *collective motion* is convenient for thermodynamic arguments. It is important to realize at this point that the collective $\boldsymbol{\sigma}_q(\phi)$ must be subjected to a practical observation in realistic timescale determined by $|\boldsymbol{q}|$.

It is significant to realize that the lattice is locally *strained* by mesoscopic displacements $\boldsymbol{\sigma}_n(\phi)$ at lattice sites n. Writing $\boldsymbol{S}$ for the strain tensor in a crystal, the *strain energy* can be expressed with respect to the displacements as

$$\langle \sigma_n(\phi) | \boldsymbol{S} | \sigma_{n'}(\phi') \rangle = \langle \sigma_q | \boldsymbol{S} | \sigma_{q'} \rangle \exp i(\phi - \phi') \tag{3.9}$$

Born and Huang proposed in their book on lattice dynamics [3] that the strain energy should be kept to a minimum to maintain stable structures, which can be interpreted for the strain matrix (3.9) to be minimized when $\phi = \phi'$ for equilibrium states[note1].

Theorem: It is noted that such $\boldsymbol{\sigma}_n$ and correlated $\boldsymbol{\sigma}_{n'}$ *in phase* are consequent on their *boson properties in favor of symmetric combinations for lower correlation energies*, signified by *symmetric exchange operator* $\boldsymbol{P} = 1$. We therefore logically consider that

[note1] The phasing for minimum strains should quantum-mechanically be equivalent to *the Bose statistics* of pseudospins $\sigma(\phi)$ to obtain coherent clusters in crystals. However, here the classical phasing is an adequate description for Landau's relaxation in [4], which will be attributed to an adiabatic potential emerging at T_c in the soliton theory.

pseudospins should be *symmetrically clustered* in a short range to minimize strains in local structure by phasing to form clustered complexes, as required by boson statistics.

Though not precisely formulated, Born–Huang's principle constitutes an essential guideline for the strain energy in crystals to be removed for the lattice to be *in thermal equilibrium, which is consistent with boson properties of symmetric displacements*. In this book, we shall therefore consider such *phase coherence* as due to *the Born–Huang principle*[note2] for such phasing to acquire new symmetry.

The transition process for $T < T_c$ indicates clearly energy transfer from the order variables to the lattice, for which Landau [3] wrote the equation

$$i\hbar\frac{\partial\boldsymbol{\sigma}_n}{\partial t} = [\mathcal{H}, \boldsymbol{\sigma}_n], \tag{3.10}$$

where $\mathcal{H}$ is a Hamiltonian for clustering pseudospins $\boldsymbol{\sigma}_n$ to be signified by their eigenvalues – ε_n for *thermal stability in the structure of lower symmetry*. To observe the $\boldsymbol{\sigma}_n$ as a classical vector at T_c, it is necessary to specify the timescale of observation t_o to visualize the time-average of $\langle\boldsymbol{\sigma}_n\rangle_t$.

Landau assumed the threshold for the average order $\langle\boldsymbol{\sigma}_n\rangle_t$ to be *coherently increased* by transferring its energy ε_n to the surroundings. Describing with a *phasing process described by an exponential increase to lower symmetry*, he wrote

$$\frac{\partial\langle\boldsymbol{\sigma}_n\rangle_t}{\partial t} = \frac{1}{t_o}\int_0^{t_o}\frac{\partial\boldsymbol{\sigma}_n}{\partial t}\mathrm{d}t = \frac{\langle\boldsymbol{\sigma}_n\rangle_t}{\tau},$$

where τ is an increasing phasing time. Re-expressing (3.10) with this relation, this can be written as

$$\frac{i\hbar\langle\boldsymbol{\sigma}_n\rangle_t}{\tau} = [\langle\mathcal{H}\rangle_t, \langle\boldsymbol{\sigma}_n\rangle_t],$$

where the *commutator* on the right can be replaced by the uncertainty relation

$$\langle\Delta\mathcal{H}_n\rangle_t\langle\Delta\boldsymbol{\sigma}_n\rangle_t \approx \langle\varepsilon_n\rangle_t\langle\Delta\boldsymbol{\sigma}_n\rangle_t.$$

In Landau's approach, at T_c the average energy $\langle\varepsilon_n\rangle_t$ can be determined by the average thermal energy $\mathrm{k_B}T_c$, so that we obtain the relation

$$\frac{\langle\Delta\boldsymbol{\sigma}_n\rangle_t}{\langle\boldsymbol{\sigma}_n\rangle_t} = \frac{\hbar}{\tau\mathrm{k_B}T_c}. \tag{3.11}$$

The average $\langle\boldsymbol{\sigma}_n\rangle_t$ can be detected as a classical vector quantity, if there is a negligible uncertainty $\langle\Delta\boldsymbol{\sigma}_n\rangle_t \approx 0$ as compared with $\langle\boldsymbol{\sigma}_n\rangle_t$; otherwise, we have a case

---

[note2] 'The Born–Huang principle' is somewhat unconventional nomenclature, while Born and Huang proposed this principle on page 226 in their classic book [3]. Nevertheless, the principle has been quoted often in literature, with no reference to the authors. Since we need to refer to it so frequently, we call it their principle.

signified by a large quantum-mechanical uncertainty, i.e. $\langle\Delta\sigma_n\rangle_t > \langle\sigma_n\rangle_t$. Expressing the criterion in these cases with respect to τ and T_c, Landau presented the condition:

$$\tau T_c > \hbar/k_B \quad \text{or} \quad \tau T_c < \hbar/k_B$$

corresponding to classical or quantum-mechanical case that can be specified by the ratio $\hbar/k_B \sim 10^{-11}$ sec K.

In perovskites, phase transitions occur in the temperature range 100 K < T_c < 200 K, and the estimated value of τ is about 0.5×10^{-13} sec, so that τT_c is roughly 500 times larger than $\hbar/k_B$. Hence, $\langle\sigma_n\rangle_t$ is a classical vector. On the other hand in KDP crystals, $T_c \sim 1000$ K and $\tau \sim 10^{-13}$ sec, so that $\tau T_c \sim 10^{-11} \sim \hbar/k_B$; hence $\langle\sigma_n\rangle_t$ is barely quantum mechanical.

However, regardless of the transition character, $\langle\sigma_n\rangle_t$ becomes commutable with $\langle\mathcal{H}_n\rangle_t$ at T_c, thereby signifying *clustered order variables* compatible with inversion symmetry at lattice sites. We shall call such clustered variable $\boldsymbol{\sigma}_n$ *pseudospins* in analogy of a quantum-mechanical spin variable of $s_n = ½$. As will be clarified later, unlike a quantum-mechanical spin, the pseudospin $\boldsymbol{\sigma}_n$ is signified by finite amplitude.

Nonetheless, critical fluctuations $\langle\sigma_n\rangle_t$ and $\langle\Delta\mathcal{H}_n\rangle_t$ are subjected to uncertainties at site n, arising from *mesoscopic* space-time uncertainties of a clustered pseudospin mode determined in the whole crystal. Due to elastic properties of a crystal, the critical value of ε_n is not easily evaluated, but can be assumed to be of the order of $k_B T_c$ in equilibrium, to keep Landau's interpretation valid. It is interesting to note in any case that the transition anomaly is not exactly quantum-mechanical, but related to the time τ as well. Notice that the timescale of collective $\langle\sigma_n\rangle_t$ determined by τ should be longer than the quantum timescale determined by the Planck constant $\hbar$. According to Landau's discussion, the transition anomaly cannot always be attributed to quantum-mechanical uncertainty, but also classical phasing in the lattice. Experimentally, the time scale of observation is a significant factor, as demonstrated by magnetic resonance experiments at different frequencies, which will be discussed in chapter 6.

3.3 Properties of pseudospin clusters

We have postulated that correlated pseudospins in a short range become clustered in phase to minimize strain energy in crystals. Using Born–Huang's principle as the law of nature, we consider *clustered pseudospins in the lattice space* by analogy to *seeds* in vapor condensation phenomena. In thermodynamics, the structural change occurs as signified by symmetry-difference between the two, so that the new phase can start with clusters with the new symmetry.

Accordingly, when the lattice is sufficiently stressed by forming seeds, the strain energy in the whole crystal should be released for the surroundings to be equilibrium via phonon scatterings. A sharp rise in the specific heat $C_p(T)$ with lowering temperature is observed at T_c after a temperature T^*, as shown in figure 3.2, signifying an energy transfer to the lattice [5]. In this context, the cluster is required for initiating a modified structure in new symmetry. On the other hand, the gradual decrease of $C_p(T)$ for $T < T_c$ should be attributed to another phasing mechanism, which will be discussed later.

In this section, we calculate an energy for forming a cluster as a *seed* for transition, and discuss its properties, which allows further studies on propagation in later chapters. We consider first the interaction between $\boldsymbol{\sigma}_n$ at site n and $\boldsymbol{\sigma}_m$ at the neighboring site m, which is given by

$$\mathcal{H}_n = -\sum_m J_{nm}\boldsymbol{\sigma}_n \cdot \boldsymbol{\sigma}_m \tag{3.12}$$

to obtain a lower energy by symmetric arrangement $\boldsymbol{\sigma}_n \| \boldsymbol{\sigma}_m$ for stability, owing to *boson statistics* for pseudospins.

Those pseudospin vectors in collective motion are in *finite amplitudes* and expressed by *classical displacement vectors* $\boldsymbol{\sigma}_n = \sigma_o \boldsymbol{e}_n$ and $\boldsymbol{\sigma}_m = \sigma_o \boldsymbol{e}_m$, where the amplitude is assumed to be constant at temperatures close to T_c, and $\boldsymbol{e}_n$, $\boldsymbol{e}_m$ are unit vectors at sites n and m *that are normalized in finite crystals.*

Writing a specific wavevector $\boldsymbol{q} \to -\boldsymbol{q}$ for inversion of the cluster, we express for symmetric and antisymmetric combinations as

$$\boldsymbol{e}_{n+} = \boldsymbol{e}_{\boldsymbol{q}} \exp i(\boldsymbol{q} \cdot \boldsymbol{r}_n - \tilde{\omega} t_n) + \boldsymbol{e}_{-\boldsymbol{q}} \exp i(-\boldsymbol{q} \cdot \boldsymbol{r}_n - \tilde{\omega} t_n),$$

$$\boldsymbol{e}_{n-} = \boldsymbol{e}_{\boldsymbol{q}} \exp i(\boldsymbol{q} \cdot \boldsymbol{r}_n - \tilde{\omega} t_n) - \boldsymbol{e}_{-\boldsymbol{q}} \exp i(-\boldsymbol{q} \cdot \boldsymbol{r}_n - \tilde{\omega} t_n)$$

at site n, and similar relations for $\boldsymbol{r}_m$ at site m as well. The interaction energy (3.12) at a site n is then given by

$$\mathcal{H}_n = -2\sigma_o^2 \sum_m J_{nm} \boldsymbol{e}_{\boldsymbol{q}} \cdot \boldsymbol{e}_{-\boldsymbol{q}} \exp i\{\boldsymbol{q} \cdot (\boldsymbol{r}_n - \boldsymbol{r}_m) - \tilde{\omega}(t_n - t_m)\}.$$

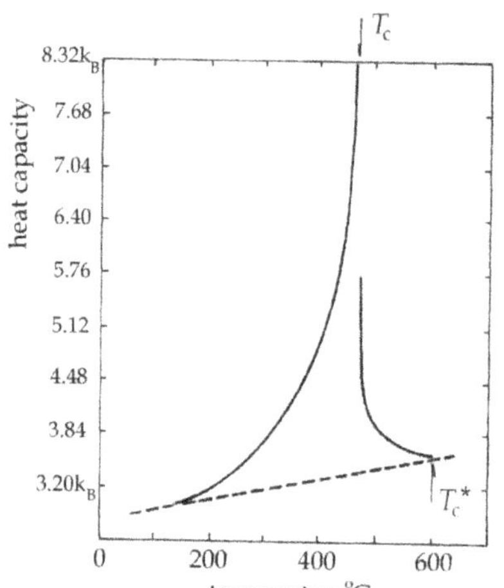

Figure 3.2. A transition anomaly in C_V observed in β-brass, signified by a sharp rise at T_c, narrow rising ranges from the threshold T_c^* and a gradual decay for $T < T_c$. Reproduced from [5], copyright 2010. With permission of Springer.

The observable energy in timescale t_o can therefore be calculated as

$$\langle \mathcal{H}_n \rangle_t = \frac{1}{2t_o} \int_{-t_o}^{+t_o} \mathcal{H}_n \mathrm{d}(t_n - t_m) = \sigma_o^2 \Gamma \boldsymbol{e}_{+\boldsymbol{q}} \cdot \boldsymbol{e}_{-\boldsymbol{q}} \sum_m J_{nm}(\boldsymbol{q}) \exp i\boldsymbol{q} \cdot (\boldsymbol{r}_n - \boldsymbol{r}_m),$$

where

$$J(\boldsymbol{q}) = \sum_m J_{nm} \exp i\boldsymbol{q} \cdot (\boldsymbol{r}_n - \boldsymbol{r}_m), \tag{3.13a}$$

and

$$\Gamma = \frac{1}{2t_o} \int_{-t_o}^{+t_o} \exp\{-i\tilde{\omega}(t_n - t_m)\} \mathrm{d}(t_n - t_m) = \frac{\sin \tilde{\omega} t_o}{\tilde{\omega} t_o} \tag{3.13b}$$

is the time correlation function, which is close to 1, if $\tilde{\omega} t_o < 1$.

To minimize $\langle \mathcal{H}_n \rangle_t$, the value of the function $J(\boldsymbol{q})$ should be restricted by the condition $\nabla_{\boldsymbol{q}} J(\boldsymbol{q}) = 0$, by which the wavevector $\boldsymbol{q}$ can be specified for the *stable cluster in new symmetry*.

At this point, we pay attention to a significant feature of $\boldsymbol{e}_{\pm\boldsymbol{q}}$, as related to the normalization.

First, as $\boldsymbol{\sigma}_{\pm\boldsymbol{q}}$ and $\boldsymbol{e}_{\pm\boldsymbol{q}}$ are *real* vectors, we have the relation

$$\boldsymbol{e}_{-\boldsymbol{q}}{}^* = \boldsymbol{e}_{+\boldsymbol{q}} \quad \text{and} \quad \boldsymbol{e}_{+\boldsymbol{q}}{}^* = \boldsymbol{e}_{-\boldsymbol{q}},$$

hence

$$\boldsymbol{e}_{+\boldsymbol{q}}{}^* \cdot \boldsymbol{e}_{-\boldsymbol{q}} + \boldsymbol{e}_{-\boldsymbol{q}}{}^* \cdot \boldsymbol{e}_{+\boldsymbol{q}} = 2\boldsymbol{e}_{\pm\boldsymbol{q}} \cdot \boldsymbol{e}_{\pm\boldsymbol{q}} = 2.$$

On the other hand,

$$\boldsymbol{e}_n{}^* \cdot \boldsymbol{e}_n = \boldsymbol{e}_{+\boldsymbol{q}} \cdot \boldsymbol{e}_{-\boldsymbol{q}} + \frac{1}{2}\left\{\boldsymbol{e}_{+\boldsymbol{q}}^2 \exp(2i\boldsymbol{q} \cdot \boldsymbol{r}_n) + \boldsymbol{e}_{-\boldsymbol{q}}^2 \exp(-2i\boldsymbol{q} \cdot \boldsymbol{r}_n)\right\}.$$

Therefore, the wavevector $\boldsymbol{q}$ should satisfy either

$$\exp(2i\boldsymbol{q} \cdot \boldsymbol{r}_n) = \exp(-2i\boldsymbol{q} \cdot \boldsymbol{r}_n) \quad \text{or} \quad \exp(4i\boldsymbol{q} \cdot \boldsymbol{r}_n) = 1, \tag{3.14a}$$

otherwise

$$\boldsymbol{e}_{+\boldsymbol{q}}^2 = \boldsymbol{e}_{-\boldsymbol{q}}^2 = 0 \tag{3.14b}$$

The case (3.14*a*) implies that the wavevector is either $\boldsymbol{q} = 0$ at any position or $\boldsymbol{q} = \pm\frac{\boldsymbol{G}}{2}$ at $\boldsymbol{r}_n = \pm \pi/|\boldsymbol{G}|$, leading to a symmetric or antisymmetric arrangement, respectively. On the other hand, the case (3.14*b*) signified by the relation $e_{q_x}^2 + e_{q_y}^2 + e_{q_z}^2 = 0$ indicates that the pseudospin arrangement can be *incommensurate* with the lattice periodicity, as characterized by a complex vector determined by $|\boldsymbol{q}|$. For example, if $e_{q_x}^2 = 1$ and $e_{q_y}^2 + e_{q_z}^2 = -1$, we have $e_{q_x} = \exp i\varphi$ and $e_{q_y} \pm ie_{q_z} = \exp(\pm i\varphi)$, where φ is an arbitrary angle in the range $0 \leqslant \varphi \leqslant 2\pi$, determining *an incommensurately spiral arrangement along the z-direction.* These cases in a variety of

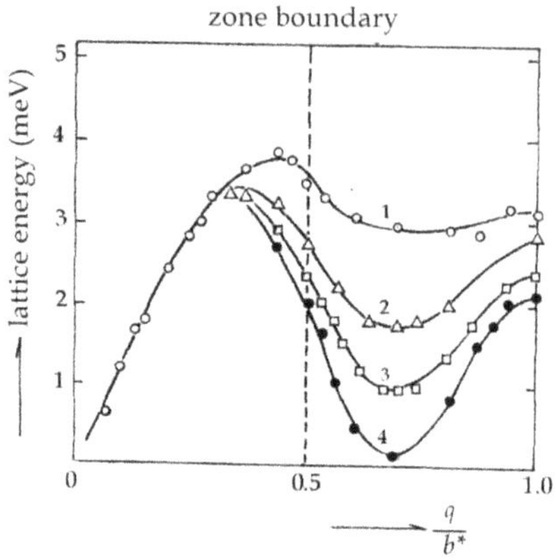

Figure 3.3. Phonon dispersion curves in K_2SeO_4 near the Brillouin zone boundary obtained by neutron inelastic scattering experiments. Curves 1, 2, 3 and 4 were observed at 250, 175, 145 and 130 K, respectively. Reproduced from [6], copyright 2010. With permission of Springer.

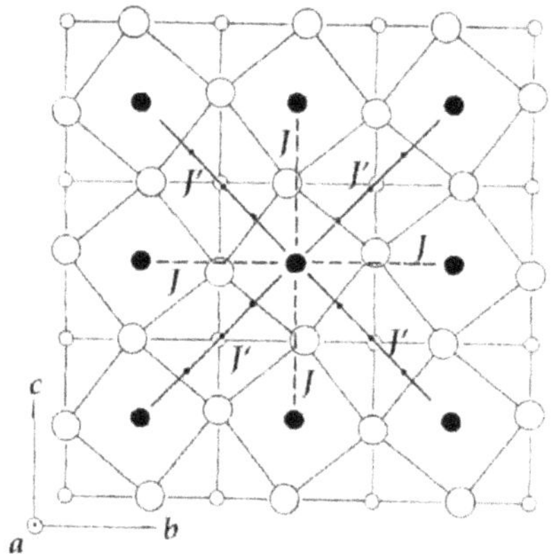

Figure 3.4. A model of the short-range cluster proposed for perovskites. Interaction parameters J and J' are assigned to six nearest neighbors and 12 next-nearest-neighbors, respectively.

types have been found experimentally in practical systems. In the phonon dispersion curve from a K_2SeO_4 crystal, shown in figure 3.3, we see the evidence of *a zone-boundary transition* at $\boldsymbol{q} = \frac{G}{2}$, although not exactly at the zone boundary [6], and there are many reports of incommensurate spin arrangement in the literature of magnetic systems [7, 8].

3.4 Examples of pseudospin clusters

3.4.1 Cubic-to-tetragonal transition in $SrTiO_3$

Figure 3.4 shows the scheme of short-range interactions in a $SrTiO_3$ crystal, where a TiO_6^{2-} at the center is correlated with other TiO_6^{2-} complexes at the nearest- and next-nearest-sites. The strengths of correlation are designated as J and J', respectively. In such a cluster, the correlation function $J(\boldsymbol{q})$ determined by (3.13a) can be written for an orthorhombic crystal of lattice constants a, b and c as

$$J(\boldsymbol{q}) = 2J(\cos q_a a + \cos q_b b + \cos q_c c) + 4J'(\cos q_b b \cos q_c c + \cos q_c c \cos q_a a + \cos q_a a \cos q_b b)^{\cdot}$$

As we deal with cubic-tetragonal transition, it is convenient to assume orthonormal a, b, c for general discussion. To minimize $\langle \mathcal{H}_n \rangle_t$, we maximize $J(\boldsymbol{q})$, hence the equations $\partial J/\partial q_a = 0$, $\partial J/\partial q_b = 0$, and $\partial J/\partial q_c = 0$ are solved to obtain

$$(\sin q_a a)(J + 2J'\cos q_b b + 2J'\cos q_c c) = 0,$$
$$(\sin q_b b)(J + 2J'\cos q_c c + 2J'\cos q_a a) = 0$$

and

$$(\sin q_c c)(J + 2J'\cos q_a a + 2J'\cos q_b b) = 0,$$

respectively.

We can find specific wavevectors $\boldsymbol{q}$ to satisfy these relations, if the values of J and J' are known. Suppose we choose a specific $\boldsymbol{q}_1$ to satisfy

$$\sin(q_{1a}a) = \sin(q_{1b}b) = \sin(q_{1c}c) = 0,$$

we have $\boldsymbol{q}_1 = \left(\frac{\pi l}{a}, \frac{\pi m}{b}, \frac{\pi n}{c}\right)$, where l, m, n are 0 or integers, and $J(\boldsymbol{q}_1) = 6J + 12J'$ due to six nearest- and 12 next-nearest neighbors.

On the other hand, we can take a specific $\boldsymbol{q}_2$ satisfying the relations

$$\cos q_{2b}b + \cos q_{2c}c = -1 + \frac{J}{2J'}, \quad \cos q_{2c}c + \cos q_{2a}a = -1 + \frac{J}{2J'},$$
$$\cos q_{2a}a + \cos q_{2b}b = -1 + \frac{J}{2J'},$$

indicating for an arrangement to be incommensurate along the direction $\boldsymbol{q}_2$, if $\left|-\frac{1}{2} + \frac{J}{4J'}\right| \leqslant 1$.

In contrast, if $\boldsymbol{q}_3$ is determined from

$$\sin q_{3a}a = 0, \quad \cos q_{3c}c + \cos q_{3a}a = -1 + \frac{J}{2J'}, \quad \cos q_{3a}a + \cos q_{3b}b = -1 + \frac{J}{2J'},$$

$\boldsymbol{q}_3 = \left(\frac{2\pi}{a}, q_{3b}, q_{3c}\right)$ is an incommensurate vector in two dimensions in the bc plane, where

$$\cos q_{3b}b = \cos q_{3c}c = -2 + \frac{J}{2J'}, \quad \text{where } \left|-2 + \frac{J}{2J'}\right| \leqslant 1.$$

Experimentally, this seems to be the case for $SrTiO_3$, judging from paramagnetic resonance studies.

3.4.2 Monoclinic crystals of TSCC

As shown in figure 3.5(a), the molecular arrangement in a monoclinic TSCC crystal is hexagonal in the bc plane. A plausible cluster of pseudospins consists of a Ca^{2+} ion

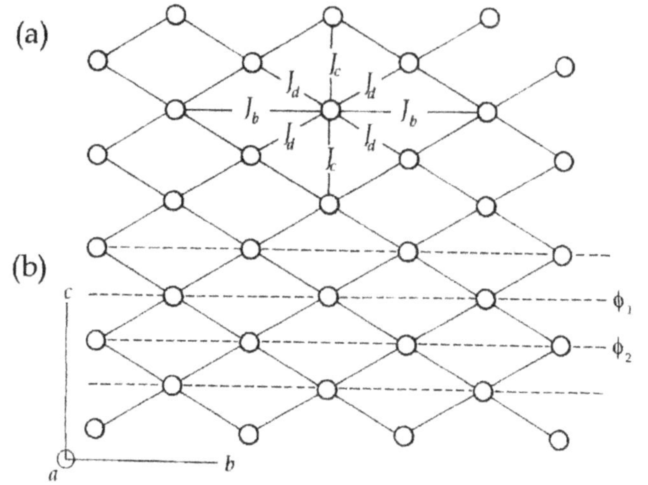

Figure 3.5. Pseudospin lattice in a TSCC crystal. (a) A cluster in the *bc* plane. (b) Parallel adjacent chains of pseudospins along the *b* direction interacting in relation to a phase difference $\phi_1 - \phi_2$, where $0 \leqslant \phi \leqslant 2\pi$.

surrounded in octahedral form by six sarcosine molecules. The figure shows a model of such cluster network in TSCC, where the correlation parameters in the *bc*-plane are indicated by J_b, J_c and J_d for nearest- and next-nearest neighbors, respectively. With these notations, the function $J(\boldsymbol{q})$ can be written as

$$J(\boldsymbol{q}) = 2J_a\cos q_a a + 2J_b\cos q_b b + 2J_c\cos q_c c + 4J_d\cos \frac{q_b b}{2} \cos \frac{q_c c}{2},$$

from which we derive specific wavevectors $\boldsymbol{q}_1, \boldsymbol{q}_2, \boldsymbol{q}_3$ to satisfy $\nabla_q J(\boldsymbol{q}) = 0$: $\boldsymbol{q}_1 = \left(\frac{\pi l}{a}, \frac{\pi m}{b}, \frac{\pi n}{c}\right)$, where l, m, n are 0 or integers, is commensurate with the lattice and $J(\boldsymbol{q}_1) = 2J_a + 2J_b + 2J_c$;

$$\boldsymbol{q}_2 = \left(\frac{\pi l}{a}, q_{2b}, \frac{\pi n}{c}\right), \quad \text{where } \cos \frac{q_{2b} b}{2} = -\frac{J_d}{2J_b};$$

$$\boldsymbol{q}_3 = \left(\frac{\pi l}{a}, \frac{\pi m}{b}, q_{3c}\right), \quad \text{where } \cos \frac{q_{3c} c}{2} = -\frac{J_d}{2J_c}.$$

With $\boldsymbol{q}_2$ and $\boldsymbol{q}_3$, the waves are incommensurate along *b* and *c* axes, if $\left|\frac{J_d}{2J_b}\right| \leqslant 1$ and $\left|\frac{J_d}{2J_c}\right| \leqslant 1$, respectively.

A TSCC crystal below 120 K is known as ferroelectric along the *b* direction, for which $\boldsymbol{q}_2$ can be considered as responsible, hence

$$J(\boldsymbol{q}_2) = 2J_a + 2J_c + 2J_b\left(1 - \frac{J_d^2}{J_b^2}\right).$$

From the above results, we have some useful remarks. If $J_d = 0$, $J(\boldsymbol{q}_1) = J(\boldsymbol{q}_2)$, on the other hand, if $J_d = -2J_b$, we have $J(\boldsymbol{q}_1) > J(\boldsymbol{q}_2)$, thereby the incommensurate $\mathbf{k}_2$ gives a lower energy than the commensurate $\mathbf{k}_1$. Writing $q_{2b}b/2 = \varphi$ as a variable, the above expression can be expressed as

$$J(\boldsymbol{q}_2) = 2J_a + 2J_c + 2J_b\cos 2\varphi + 4J_d\cos\varphi$$

which is a familiar formula in the theory of magnetism for spiral spin arrangement in one dimension, where φ is the spiral angle [8–10]. A similar situation is encountered in charge density waves, where Rice [11] assumed that the interchain interaction is given by $J_d \propto \cos(\phi_1 - \phi_2)$ between adjacent propagating charge-density wave, suggesting significant phase difference $\phi_1 - \phi_2$, as shown in figure 3.5(b).

3.4.3 Remarks on pseudospin coupling constants

In the foregoing examples, we see that those coupling constants denoted by J_i determine the property of the pseudospin cluster with respect to the structure, directing propagation in a given system. Although related to new lattice symmetry, the value of J_i is theoretically undeterminable from the first principle, with no reference to temperature, as obscured by unavoidable uncertainties near T_c. Nevertheless, it is significant that the new symmetry determines the ordering process with these constants J_i in clusters originated from correlation energies among pseudospins in the new phase. Thus, the pseudospin cluster behaves like a *seed* for ordering in a thermodynamic process along the direction determined by J_i for its structure. We may call this direction the *easy direction* by analogy to an *easy axis of ferromagnetic order.*

Statistically significant is that the pseudospin for a displacement behaves as a *boson particle* in the lattice, hence equation (3.12) expresses negative correlation energy between $\boldsymbol{\sigma}_n$ and $\boldsymbol{\sigma}_m$ arranged symmetrically in phase due to their *boson character*. Accordingly, in new lattice symmetry below T_c the lattice creates a stable cluster with the nearest and next-nearest pseudospins combined in boson correlations, assuming that the coupling energy depends on distance. Minimizing correlations with coherent pseudospins for thermodynamic equilibrium is thus consistent with their boson statistics of order variables.

Exercises

1. We consider a seed of a phase transition to be formed as an initial cluster of order variables. What determines formation of clusters?
2. The phasing process of $\boldsymbol{\sigma}(\phi_n)$ for the cluster expressed by (3.8) is considered with respect to the Born–Huang principle. Discuss the issue, as related to the strained lattice.
3. Clustering should occur spontaneously, as $\langle \mathcal{H}_n \rangle_{\text{short-range}}$ is minimized. Prove this for those examples in section 3.4, confirming the negative ordering energy to be thermally transferred to the lattice in the new symmetry.
4. It is important that minimizing of $J(\boldsymbol{q})$ is necessary with respect to *lattice symmetry*, which is the basic requirement for the system to be in equilibrium

with its surroundings. Needless to say, that is a postulate, but discuss why this has to be considered as fundamental. Binary inversion is also a postulate, but its direction needs to be specified along symmetry axis. Discuss the issue, giving a logical reason for such inversion.

References

[1] Blinc R and Zeks B 1974 *Soft Modes in Ferroelectrics and Antiferroelectrics* (Amsterdam: North Holland)

[2] Haken H 1973 *Quantenfeldtheorie des Festkörpers* ch 6 (Stuttgart: B. G. Teubner)

[3] Born M and Huang K 1968 *Dynamical Theory of Crystal Lattices* (London: Oxford University Press)

[4] Landau L D and Lifshitz E M 1958 *Statistical Physics* (London: Pergamon)

[5] Nix F C and Shockley W 1938 *Rev. Mod. Phys.* **10** 1

[6] Iizumi M, Axe J D, Shirane G and Shimaoka K 1977 *Phys. Rev.* **B15** 4392

[7] Müller K, Berlinger W and Waldner F 1968 *Phys. Rev. Lett.* **21** 814

[8] Kittel C 1986 *Introduction to Solid State Physics* 6th edn (New York: Wiley)

[9] Nagamiya T 1963 *Solid State Physics* vol 20 (New York: Academic)

[10] Walker L R 1963 *Magnetism 1* (New York: Academic)

[11] Rice M J 1978 *Charge Density Waves in Solitons and Condensed Matter Physics* ed A R Bishop and T Schneider (Berlin: Springer)

Chapter 4

The mean-field theories and critical phase fluctuations at transition temperatures

For the condensate model of collective pseudospins with lattice displacements, we consider that Landau's expansion of the Gibbs function can be interpreted as an internal *adiabatic potential*, originating from pseudospin correlations in a crystal. Also, an adiabatic potential is required for condensates to be detected in experiments. In stable crystals, the pseudospin wave should be in phase with an internal adiabatic potential, representing the Weiss field defined in adiabatic approximation.

Critical anomalies in second-order phase transitions are discussed for pinned condensates, which, however, can be associated with random phonons as discussed in this chapter, turning to separation of the whole crystal into domain structure. In addition to intrinsic pinning, extrinsic potentials due to lattice imperfections are important in practice, as they need to be identified to distinguish intrinsic anomalies.

4.1 Landau's theory and Curie–Weiss' law

4.1.1 Landau's theory of binary transitions

In Landau's thermodynamic theory [1] of binary order, the Gibbs potential is defined as a function of external pressure p, temperature T and an *order parameter* η that is traditionally assumed to be the mean-field average $\langle\sigma_n\rangle$ in a crystal. As a macroscopic variable, $\eta = \langle\sigma_n\rangle$ is defined in the range $0 \leqslant \eta \leqslant 1$, where $\eta = 0$ and $\eta = 1$ indicate disordered and perfectly ordered states, respectively, and hence η is a thermodynamic variable $\eta = \eta(p, T)$. In practice, the critical temperature T_c is always found to be lower than the mean-field value T_o predicted by Landau's theory, hence $T_c < T_o$, indicating failure of the Landau theory. Nevertheless, the singularity at T_o determines a binary phase transition theoretically.

In his theory, Landau expanded the Gibbs' function in power series of $\eta(p, T)$. At constant pressure and temperature, the Gibbs' function varies as a function of η, and is invariant for inversion $\eta \rightarrow -\eta$; hence the expansion can be written as

doi:10.1088/978-0-7503-2572-1ch4

$$G(\eta) = G(0) + \frac{1}{2}A\eta^2 + \frac{1}{4}B\eta^4 + \frac{1}{6}C\eta^6 + \cdots, \tag{4.1a}$$

where the coefficients A, B, C, ... are just constants or functions of p and T. However, under a constant volume condition, the expression $\Delta V = 0$ allows us to consider that p is a constant. Under constant volume, the expansion (4.1a) has no odd-power terms of η because of inversion symmetry, and hence we have

$$G(\eta) = G(-\eta) \tag{4.1b}$$

In a crystal in equilibrium with the surroundings, the Gibbs' function should take a minimum value for a variation $\Delta\eta$, so that

$$\Delta G = G(\eta) - G(0) = \left(\frac{\partial G}{\partial \eta}\right)_T \Delta\eta \geqslant 0.$$

For a small $\Delta\eta$, (4.1a) can be truncated at the third term, and written as

$$\Delta G = \frac{1}{2}A\eta^2 + \frac{1}{4}B\eta^4 \geqslant 0.$$

Assuming that the minimum is given by $\eta = \eta_0$, we write $A\eta_0 + B\eta_0^3 = 0$ for $\Delta G(\eta_0) = 0$, and therefore η_0 satisfies either

$$\eta_o = 0 \quad \text{or} \quad \eta_o^2 = -\frac{A}{B}.$$

The first solution $\eta_0 = 0$ corresponds to a disordered state, which is the same as the second solution, if $A = 0$. The latter solution determines partial order, if assuming $A \neq 0$. Hence the coefficient A should generally be temperature dependent, for which Landau proposed the following relations:

$$A = A'(T - T_o) \quad \text{for } T \geqslant T_o \tag{4.2a}$$

and

$$A = A'(T_o - T) \quad \text{for } T < T_o \tag{4.2b}$$

Assuming A' and B are both positive in (4.2b), η_0 can be expressed in parabolic form $\eta_0 \propto \sqrt{T_o - T}$, and the corresponding Gibbs' functions can be specified by

$$G(\eta_o) = G(0) \quad \text{and} \quad \eta_o = 0 \quad \text{for } T \geqslant T_o$$

and

$$G(\eta_o) = G(0) - \frac{3A^2}{4B} \quad \text{and} \quad \eta_o = \pm\sqrt{\frac{A'}{B}(T_o - T)} \quad \text{for } T < T_o.$$

Figure 4.1 shows schematically such a change of the Gibbs function with respect to temperature.

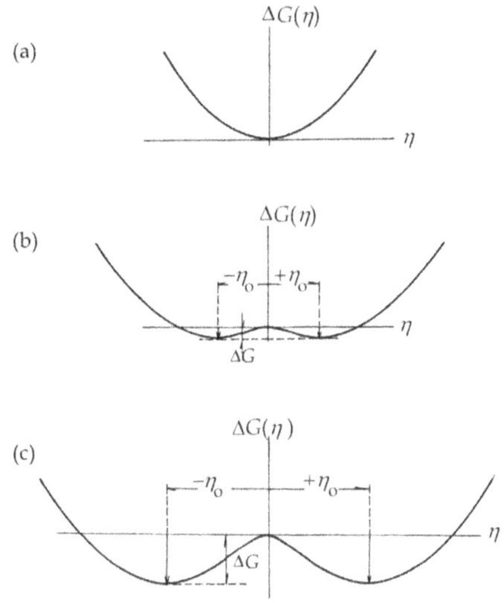

Figure 4.1. Truncated Gibbs' potential. (a) $T = T_c$, and (b) and (c) for $T < T_c$. Binary fluctuations are implied by two-way shifts $\pm\eta_o$ with increasing depths at minima on decreasing temperature.

In the former case, as shown by figure 4.1(a), $\Delta G(\eta_o) = 0$ at all temperatures for $T \geqslant T_o$, whereas in the latter case, the two minima $\pm\eta_o$ in figure 4.1(b) shift symmetrically in parabolic form with increasing depth.

4.1.2 Curie–Weiss law of susceptibilities

Interesting to note is that the concept of the Weiss field is built into Landau's expansion. To minimize Gibbs' function for equilibrium, we used mathematically an arbitrary variation $\delta\eta$, which can, however, be attributed to a variation of the order variable $\delta\sigma_n$ associated with corresponding adiabatic potentials ΔU_n in the condensate model. In this view, Landau's expansion terms can be analyzed by writing

$$\Delta G = \frac{A}{2}\eta^2 + \frac{B}{4}\eta^4 + \cdots = \langle \Delta U_n \rangle = -\eta X_{\text{int}},$$

where

$$X_{\text{int}} = A\eta + B\eta^3 + \cdots = -\left\langle \frac{\partial U(\sigma_n)}{\partial \sigma_n} \right\rangle.$$

Since $\langle U(\sigma_n)\rangle = U(\eta)$, we have $X_{\text{int}} = -\frac{\partial U}{\partial \eta}$, where U and η should be in phase in equilibrium crystals. Accordingly, the quantity X_{int} represents an *internal field* in the mean-field accuracy, acting on η as if applied externally. Such an X_{int} is essentially the same idea as the *Weiss molecular field* in a ferromagnetic crystal (1907). Weiss actually postulated the relation $X_{\text{int}} \propto \eta$, which is consistent with Landau's X_{int} in mean-field accuracy, so generally referred to as the *Weiss field*, arising from internal correlations among order variables.

In the mean-field accuracy, we consider $\Delta G = 0$ for equilibrium at temperatures T above T_0. However, assuming small variations $\delta\eta$ in the close vicinity of T_0, the Curie–Weiss law of susceptibility can be derived as follows.

Writing

$$\Delta G_{T>T_0} = \Delta G(\eta) - \eta X_{\text{int}} \quad \text{and} \quad \left(\frac{\partial \Delta G_>}{\partial \eta}\right)_T = 0,$$

we obtain $A\eta - X_{\text{int}} = 0$, from which the susceptibility formula can be obtained as

$$\chi_{T>T_0} = \frac{\eta}{X_{\text{int}}} = \frac{1}{A} = \frac{1}{A'(T - T_0)}, \qquad (4.3a)$$

using (4.2*a*) for A.

On the other hand, for $T < T_0$

$$\Delta G_{T<T_0} = \frac{1}{2}A\eta^2 + \frac{1}{4}B\eta^4 - \eta X_{\text{int}},$$

therefore from $\left(\frac{\partial \Delta G_<}{\partial \eta}\right)_T = 0$, we obtain $A\eta + B\eta^3 - X_{\text{int}} = 0$. Using the relation $\eta_0^2 = -\frac{A}{B}$ for $X_{\text{int}} = 0$, we can derive the susceptibility formula, for T close to T_0,

$$\chi_{T<T_0} = \frac{\eta_0}{2A} = \frac{1}{2A'(T_0 - T)}, \qquad (4.3b)$$

using Landau's relation (4.2*b*) for the constant A.

Both (4.3*a*) and (4.3*b*) are known as the Curie–Weiss law that can be applied to temperatures close to T_0, showing a singular behavior at T_0. These equations are similar, but noticeably different by factors, $\frac{1}{A}$ and $\frac{1}{2A}$ due to the unharmonic term of B, which are also recognized experimentally in ferroelectric crystals.

In Landau's theory and Curie's law, the critical temperature T_0 is determined in mean-field accuracy, which is, however, found to be very different from the observed critical temperature T_c, indicating failure of the mean-field theory, in addition, the critical anomalies are disregarded in the mean-field approximation. Nevertheless, both of the laws can be regarded as precursory features of the structural phase transitions to be explained by their experimental evidences, together with *soft-mode data*, as will be explained in chapter 7.

4.2 Fluctuations of pseudospin clusters in adiabatic potentials

4.2.1 Initial pinning of pseudospin fluctuations

The Gibbs potential fluctuates with clustered pseudospins under critical conditions near the phase transition, which can be attributed to fluctuating mutual correlations in crystals, so that the volume V cannot be constant in principle under constant p and T. Such fluctuations appear to be sinusoidal in random phase at the threshold temperature T_c* indicated in figure 3.2 in chapter 3, where the emerging potential

$\Delta U_n = \frac{a}{2}\sigma_n^2 + \frac{b}{4}\sigma_n^4$ is dominated by the beginning terms with a small positive constant b; $a > b > 0$; which is considered as consistent to Landau's expansion in section 4.1.

Below T_c*, displacements $\boldsymbol{\sigma}_n$ in short ranges due to the term of b become clustered to form stable seeds in larger amplitude at a critical temperature T_c, where the inversion $\boldsymbol{\sigma}_n \to -\boldsymbol{\sigma}_n$ at finite amplitude enhanced by the thermal energy $k_B T_c$ can release energies to the lattice in the form of heat, exhibiting a sharp rise in the specific heat curve C_p versus T. Although experimentally evident, such energy conversion cannot be elucidated, unless the lattice strains are considered for the minimum to establish stable structure. Corresponding to the process proposed by Born and Huang, Landau's relaxation of $\langle \boldsymbol{\sigma}_n \rangle_t$ explains the heat transfer to the lattice. Phonon inelastic scatterings should actually be the mechanism for energy dissipation in crystals, elucidating gradual decay in the C_p-curve for $T < T_c$. Nevertheless, anomalies in the vicinity of T_c can be understood with regard to space–time or phase uncertainties in the critical region.

Below T_c*, the clustered pseudospin $\boldsymbol{\sigma}_n$ of condensate is expressed primarily as $\boldsymbol{\sigma}_n = \boldsymbol{\sigma}_q \exp i(\boldsymbol{q} \cdot \boldsymbol{r}_n - \tilde{\omega} t_n)$ *along the symmetry axis determined by coordinates* $\boldsymbol{r}_n$, *where* $n = 1, 2, \ldots$, *for the lowest lattice excitation in finite amplitude* $\boldsymbol{\sigma}_q$, *which is associated with the adiabatic potential* $\Delta U_n(\boldsymbol{\sigma}_n)$. In this case, the Fourier transform $\boldsymbol{\sigma}_q = \boldsymbol{\sigma}_n \exp\{-i(\boldsymbol{q} \cdot \boldsymbol{r}_n - \tilde{\omega} t_n)\}$ represents a collective mode of coherent $\boldsymbol{\sigma}_n$ along the direction determined by these coordinates $\boldsymbol{r}_n$. Therefore, to deal with the space–time variation, it is convenient to use the phase function $\phi_n = \boldsymbol{q} \cdot \boldsymbol{r}_n - \tilde{\omega} t_n$, thereby writing $\Delta\phi_n = \boldsymbol{q} \cdot \Delta\boldsymbol{r}_n - \tilde{\omega}\Delta t_n$ for uncertainties $\Delta\boldsymbol{r}_n$ and Δt_n at site n. In the $\boldsymbol{q}$-space, omitting the index n in the reduced Brillouin zone, the phase variable $\phi = \boldsymbol{q} \cdot \boldsymbol{r} - \tilde{\omega} t$ and its uncertainty $\Delta\phi$, in the range $-\pi/2 \leqslant \phi \leqslant \pi/2$, can be used to describe collective motion. It should be noted here that $\tilde{\omega}$ is distinct from ω with the corresponding $\boldsymbol{q}$ that are responsible for solitons.

Although initially sinusoidal, waves are essentially nonlinear with $\boldsymbol{q}$ not simply proportional to ω, characterized by the nonlinear phase ϕ; hence the collective mode can be signified by $\boldsymbol{\sigma}_q = \boldsymbol{\sigma}_o f(\phi)$, where $\boldsymbol{\sigma}_o$ is a finite amplitude. We have therefore *inversion* symmetry $\boldsymbol{q} \rightleftarrows -\boldsymbol{q}$ between $\phi(+\boldsymbol{q})$ and $\phi(-\boldsymbol{q})$ of collective modes.

At this point, further remarks can be made. The transition at T_c starts to occur generally at a slow rate that is observed as relaxation, however, it can be oscillatory, if occurring at a faster rate, as detected in dielectric crystals.

4.2.2 Critical fluctuations

Experimentally, a second-order phase transition can be observed at the critical temperature T_c determined by specific heat anomalies. Theoretically, it is signified by the clustered pseudospins $\boldsymbol{\sigma}_q = \boldsymbol{\sigma}_o f(\phi)$ pinned by the negative potential $-\Delta U_q(\phi) \approx -\frac{a\boldsymbol{\sigma}_q(\phi)^2}{2}$ at T_c. Resulting from a microscopic phasing process of *condensates*, it should be in sufficiently strong magnitude for transition; and both $\boldsymbol{\sigma}_q(\phi)$ and $+\Delta U_q(\phi)$ are functions of the phase ϕ in the coordinate system moving at a speed $v = \frac{\tilde{\omega}}{|\boldsymbol{q}|}$; constituting for $\phi = 0$ to determine the *initial condition for generating*

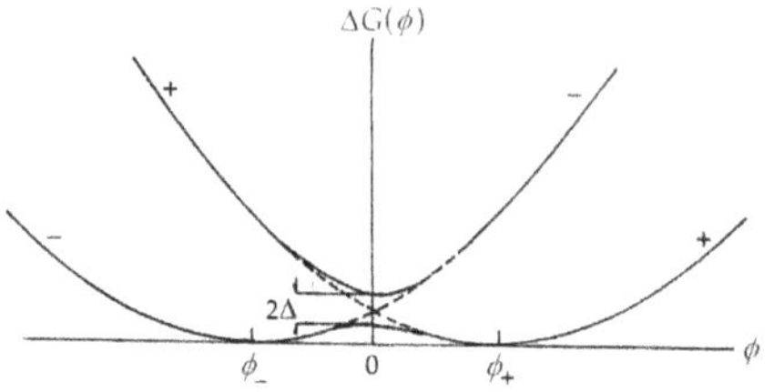

Figure 4.2. Phase fluctuations in Gibbs' potential $\Delta G(\phi)$ near the critical temperature T_c.

nonlinearity. Therefore, σ_k is considered as pinned by $\Delta U_q(0)$ at $\phi = 0$, where the tunneling $\sigma_{\pm q}$ appear to be in fluctuation between $+\Delta\phi$ and $-\Delta\phi$.

Assuming fluctuations between $\Delta U_k(+\Delta\phi)$ and $\Delta U_k(-\Delta\phi)$, as shown in figure 4.2, we consider the adiabatic potential $\Delta U_q(\phi) = \Delta U_q(+\Delta\phi) + \Delta U_q(-\Delta\phi)$ sketched against $\Delta\phi = 0$ in the vicinity of $\phi = 0$. At the crossing point of these, two parabolic potentials $\frac{a}{2}\sigma_q^2(\pm\Delta\phi)$ are noted to be degenerate in this approximation, which are lifted by a symmetric perturbation[note1] expressed as $\Delta U_q(\phi) = Vf(\phi)$, where V is the potential height and $f(\phi)$ a normalized even function in the range $-\pi/2 \leqslant \phi \leqslant \pi/2$. Such a problem of *mesoscopic fluctuations* can be discussed analogously to a *level-crossing* problem in quantum mechanics.

First, we consider a linear combination

$$\sigma = c_+\sigma_+ + c_-\sigma_- \quad \text{and} \quad c_+^2 + c_-^2 = 1,$$

where c_+ and c_- are the normalization factors for $\sigma_\pm = \sigma(\pm\Delta\phi)$. Then, with the perturbation of $Vf(\phi)$, we solve the secular equation

$$\begin{vmatrix} \varepsilon_+ - \varepsilon & \Delta \\ \Delta & \varepsilon_- - \varepsilon \end{vmatrix} = 0,$$

for ε, where $\varepsilon_\pm$ and Δ are calculated as

$$\varepsilon_\pm = \frac{1}{\pi}\int_{-\pi/2}^{\pi/2} \sigma_o{}^* Vf(\phi)\sigma_o \, d\phi = V\sigma_o^2 = \varepsilon_o$$

and

$$\Delta = \frac{1}{\pi}\int_{-\pi/2}^{\pi/2} \sigma_+{}^* Vf(\phi)\sigma_- \, d\phi = \frac{V\sigma_o^2}{\pi}\int_{-\pi/2}^{\pi/2} \exp(-2i\phi)f(\phi)\, d\phi$$
$$= \frac{V\sigma_o^2}{\pi}\int_{-\pi/2}^{\pi/2} f(\phi)\cos 2\phi \, d\phi = \frac{V\sigma_o^2}{2},$$

we obtain

$$\varepsilon = \varepsilon_o \pm \Delta. \tag{4.4a}$$

[note1] According to the Toda theory, such a perturbation can be replaced by an exponential potential, as described in chapter 11, expressing fluctuations in mesoscopic scale, while remaining as a hypothesis at this stage.

Hence, the degeneracy is lifted by the energy gap 2Δ, as indicated in figure 4.2. This is exactly equal to the amount of energy required for inversion of *finite pseudospins* $\boldsymbol{\sigma}_k \to -\boldsymbol{\sigma}_k$ or $\boldsymbol{\sigma}_{+k} \to \boldsymbol{\sigma}_{-k}$ at T_c. According to the above perturbation theory, however eigenfunctions for eigenvalues (4.4*a*) are given by symmetric and antisymmetric combinations of $\boldsymbol{\sigma}_{\pm}$, i.e.

Theorem:

$$\sigma_A = \frac{\boldsymbol{\sigma}_+ + \boldsymbol{\sigma}_-}{\sqrt{2}} \quad \text{and} \quad \sigma_P = \frac{\boldsymbol{\sigma}_+ - \boldsymbol{\sigma}_-}{\sqrt{2}}, \tag{4.4b}$$

which we shall call *amplitude and phase modes*, respectively, representing *phase fluctuations* in field-theoretical approximation.

Regarding the finite energy splitting 2Δ between $\boldsymbol{\sigma}_A$ and $\boldsymbol{\sigma}_P$, we assumed the emergence of a positive potential $V(0)$ at T_c, constituting the *initial condition* for developing nonlinearity on lowering temperature. At this point, it is interesting to note that such a finite energy gap arises theoretically from *Toda*'s correlation *potentials* among pseudospins in crystals, as will be discussed in chapter 11.

4.2.3 Energy transfer to the lattice at T_c

Writing $\sigma_A = \sigma_o \cos\phi$ and $\sigma_P = \sigma_o \sin\phi$, equations (4.4*b*) indicate that σ_A and σ_P are components of a classical vector $\boldsymbol{\sigma} = (\sigma_A, \sigma_P)$ composed of *transverse* and *longitudinal* components with respect to inversion axis. In this expression, the critical fluctuations can classically be described in a complex form

$$\mp i\sigma_A + \sigma_P = \sigma_o \exp i\left(\phi \pm \frac{\pi}{2}\right), \tag{4.5}$$

where the imaginary part $i\sigma_A$ indicates that $\sigma_A = \sigma_o$ at $\phi = 0$, whereas the real part σ_P is determined at $\phi = \mp\pi/2$ and $\sigma_P = 0$ at $\phi = 0$. Nonetheless, such a classical interpretation provides a realistic model for the collective pseudospin motion in crystals; thereby the energy exchange with lattice strains can be calculated logically as *work* to rotate transversal $\sigma_\perp = \sigma_A$ by 180°.

Mathematically the variable $\boldsymbol{\sigma}$ is a classical vector consistent with the nonlinear lattice dynamics of order variables (chapter 11), indicating that two-component waves for $T < T_c$ confirm initial conditions for the two-components; hence the vector model is logical for calculating energy transfer to the lattice.

The clustered pseudospin vector $\boldsymbol{\sigma}$ in the lattice can be reacted by the adiabatic potential via phonon scatterings in the crystal. We write phonon wavefunctions at wavevectors $\boldsymbol{k}$ and $\boldsymbol{k}'$ in a modulated crystal as $\psi_k(\boldsymbol{q}, \tilde{\omega}_q) \propto \exp i(\boldsymbol{q}\cdot\boldsymbol{r} - \tilde{\omega}_q t)$ and $\psi_{k'}(\boldsymbol{q}', \tilde{\omega}_{q'}) \propto \exp i(\boldsymbol{q}'\cdot\boldsymbol{r} - \tilde{\omega}_{q'} t')$, respectively, and find that the matrix elements $\langle\psi_k(\boldsymbol{q}, \tilde{\omega}_q)|\sigma_{A,P}|\psi_{k'}(\boldsymbol{q}', \tilde{\omega}_{q'})\rangle$ are the quantities significant for energy exchange via elastic and inelastic scatterings of phonons.

It is noted that for $\boldsymbol{k} \neq \boldsymbol{k}'$, denoting thermal averages by the brackets $\langle\cdots\rangle$, we have

$$\left\langle \psi_k(\boldsymbol{q}, \tilde{\omega}_q)|\sigma_P|\psi_{k'}(\boldsymbol{q}', \tilde{\omega}_{q'})\right\rangle \neq 0 \qquad \text{for scatterings} \quad \boldsymbol{q} + \boldsymbol{q}' = 0 \tag{4.6}$$

and

$$\left\langle \psi_k(\boldsymbol{q}, \tilde{\omega}_q)|\sigma_A|\psi_{k'}(\boldsymbol{q}', \tilde{\omega}_{q'})\right\rangle \neq 0 \qquad \text{for scatterings} \quad \boldsymbol{q} + \boldsymbol{q}' \neq 0;$$

signifying the energy relations $\Delta\varepsilon_{q,q'} = 0$ and $\Delta\varepsilon_{q,q'} \neq 0$, respectively. Accordingly, the inelastic scatterings are responsible for finite energy transfer $\Delta\varepsilon_{q,q'}$ between σ_A and surrounding phonons, which is the case of $\boldsymbol{q}' = -\boldsymbol{q}$. Denoting wavevectors $\boldsymbol{K}, \boldsymbol{K}'$ for phonons, we should have the relations $\boldsymbol{K} \pm \boldsymbol{q} = \boldsymbol{K}' \mp \boldsymbol{q}$, hence $\boldsymbol{K}' - \boldsymbol{K} = \pm 2\boldsymbol{q} \neq 0$, i.e. inelastic phonon scatterings for $\Delta\varepsilon_{+q} = -\Delta\varepsilon_{-q}$ and for $2\Delta\varepsilon_{\pm q} > 0$ to be applied to the σ_P and σ_A modes, respectively, as illustrated in figure 4.2.

Energy transfer processes can be described by taking such non-zero elements as a perturbation $\boldsymbol{H}' = \lambda\sigma_A$ between unperturbed phonon Hamiltonians $\mathcal{H}_q(\boldsymbol{K})$ and $\mathcal{H}_{-q}(\boldsymbol{K}')$ [3]. Therefore, we have $\boldsymbol{H}_{-q}(\boldsymbol{K}) = \boldsymbol{H}_q(\boldsymbol{K}) + \lambda\sigma_A$ and $\mathcal{H}_{-q}(\boldsymbol{K}') = \mathcal{H}_q(\boldsymbol{K}') + \lambda\sigma_A$; and the time-dependent Schrödinger equation can be written as

$$i\hbar\dot{\psi}_{-q}(\boldsymbol{K}) = \boldsymbol{H}_{-q}(\boldsymbol{K})\psi_{-q}(\boldsymbol{K}).$$

Writing $\psi_{-q}(\boldsymbol{K}) = a_q(t)u_q(\boldsymbol{K})\exp(-i\tilde{\omega}_q t)$, this equation can be solved as

$$i\hbar\dot{a}_{-q}\exp(-i\tilde{\omega}_{-q}t) = a_q\exp(-i\tilde{\omega}_q t)\frac{\lambda}{2t_o}\int_{-t_o}^{+t_o} u_{-q}^*(\boldsymbol{K})\,\sigma_A u_q(\boldsymbol{K})\mathrm{d}(-2t),$$

hence

$$\dot{a}_{q'} = \frac{\lambda}{i\hbar}\sum_q \langle -\boldsymbol{q}|\sigma_A|\boldsymbol{q}\rangle_k a_q \exp i\left(\frac{\Delta\varepsilon_{-q,q}}{\hbar}t\right), \tag{4.7}$$

where

$$\langle -\boldsymbol{q}|\sigma_A|\boldsymbol{q}\rangle_k = \frac{1}{2t_o}\int_{-t_o}^{+t_o} u_{-q}^*(\boldsymbol{K})\sigma_A u_q(\boldsymbol{K})\mathrm{d}(-2t) \tag{4.8}$$

Equation (4.8) is modified for practical observation in timescale t_o, but integrating for $\boldsymbol{H}'$ to apply at arbitrary time t, namely $\mathcal{H}' = 0$ at $t = -\infty$, we have

$$\begin{aligned} a_{-q}(t) &= \frac{\lambda}{i\hbar}a_q(0)\int_{-\infty}^{t}\langle -\boldsymbol{q}|\sigma_A(t')|\boldsymbol{q}\rangle \exp\frac{i\tilde{\omega}_{-q,q}t'}{\hbar}\mathrm{d}t' \\ &= -\langle -\boldsymbol{q}|\lambda\sigma_A|\boldsymbol{q}\rangle\frac{\exp(i\tilde{\omega}_{-q,q}t) - 1}{\hbar\tilde{\omega}_{-q,q}}, \end{aligned}$$

where $a_k(0) = 1$ is assumed for simplicity; $\langle -\boldsymbol{q}|\lambda\sigma_A|\boldsymbol{q}\rangle$ is independent of $\boldsymbol{K}$, hence the suffix is dropped. Therefore,

$$a_{-q}(t)^2 = \frac{4\langle -\boldsymbol{q}|\lambda\sigma_A|\boldsymbol{q}\rangle^2 \sin^2 \frac{\tilde{\omega}_{-q,q}t}{2}}{\hbar^2 \tilde{\omega}_{-q,q}^2}, \tag{4.9}$$

which is analyzed with a familiar curve in optics characterized by a strong peak at the center, as shown in figure 4.3, giving a probability for energy transfer during inversion $\boldsymbol{q} \rightleftarrows -\boldsymbol{q}$, and hence the relation $a_{-q}(t)^2 = a_q(t)^2$ gives a finite probability for the energy transfer from σ_A to the lattice.

An energy exchange between different wavevectors $\boldsymbol{q}$ and $\boldsymbol{q}'$ can take place in adiabatic processes, while the specific inversions $\pm\boldsymbol{q} \rightleftarrows \mp\boldsymbol{q}$ for nonzero elements play significant roles in the process at $T = T_c$ with maximum probability proportional to $\lambda^2\langle \pm\boldsymbol{q}|\sigma_A| \mp \boldsymbol{q}\rangle^2$, which are characterized by $|\Delta \boldsymbol{q}| = 2|\boldsymbol{q}| \neq 0$, corresponding to $\sin^2 \frac{\tilde{\omega}_{q,-q}t}{2} / \tilde{\omega}_{q,-q}^2 = \frac{t^2}{4}$. Indicated in the figure, the singularity originating from such inversion exhibits a sharp loss of energy to the lattice, as observed in C_p-curve at T_c.

Subjecting to random phonon scatterings at T_c, the matrix element of (4.6) can be modified as

$$\begin{aligned}&\langle n_k\hbar\omega_o|\lambda^2\langle \pm\boldsymbol{q}|\sigma_A|\mp\boldsymbol{q}\rangle^2|n_{k'}\hbar\omega_o\rangle \\ &= \lambda^2\{\bar{n}_k k_B T_k \langle \pm\boldsymbol{q}|\sigma_A| \mp \boldsymbol{q}\rangle^2 \bar{n}_{k'} k_B T_{k'}\} \to \lambda^2 n_c^2 (k_B T_c)^2 \langle \pm\boldsymbol{q}|\sigma_A| \mp \boldsymbol{q}\rangle^2,\end{aligned} \tag{4.10}$$

assuming $\lim(T_k, T_{k'}) \to T_c$ that corresponds to $\lim(\bar{n}_k, \bar{n}_{k'}) \to n_c$. These $\bar{n}_k$, $\bar{n}_{k'}$ and n_c are average phonon numbers at T_k, $T_{k'}$ and the critical point T_c, respectively; λ^2 can be determined by the relation

Theorem:

$$k_B T_c = 2\Delta, \tag{4.11}$$

representing the energy gap between those states of σ_A and σ_P, implying for the amount of energy transfer to the lattice to be determined by the values of n_c and T_c. For $T > T_c$ and $T < T_c$, the states of $\varepsilon_A > \varepsilon_P$ and $\varepsilon_P < \varepsilon_A$ are characterised respectively by shifting the symmetric state $\varepsilon_A = (\varepsilon_+ + \varepsilon_-)/2$ to below ε_P.

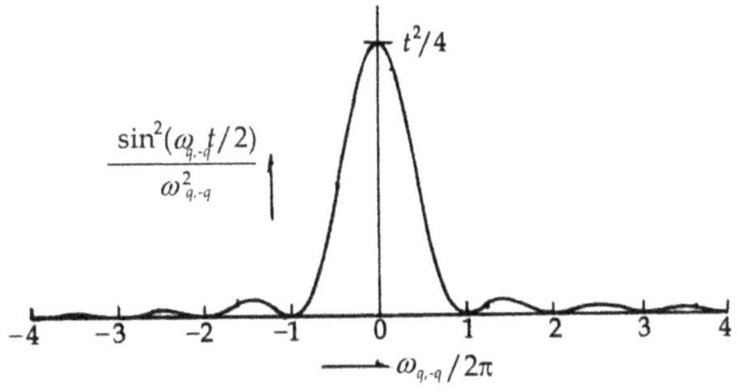

Figure 4.3. A relative intensity spectrum of energy transfer to the lattice in (4.9).

Experimentally, the relation (4.11) is adequately used for a large number $n_c = n(T_c)$ to evaluate *entropy production* to the lattice. It is notable that such a singular behavior of the pseudospin cluster at T_c is attributed to *adiabatic inversion* $\boldsymbol{q} \rightleftarrows -\boldsymbol{q}$, constituting the initial condition for mesoscopic domain formation, as will be discussed in chapter 12.

At a positive energy gap 2Δ in this case, the phase mode $\sigma_P(\phi)$ determines the lower energy state between $\sigma_P(+\Delta\phi)$ and $\sigma_P(-\Delta\phi)$ separated by $2\Delta\phi$. Hence, the antisymmetric σ_P should correspond to the lower state in the lattice, while the symmetric σ_A is *no longer stable at the upper state*, owing to *thermal interaction* with the lattice. In contrast, the antisymmetric σ_P remains stable for $T < T_c$, leading to *domain structure,* while the symmetric σ_A decays in the lattice after transition.

In comparison to the statistical order–disorder theory, where the gap 2Δ is neglected, thereby the binary transition is continuous and of second-order in Ehrenfest's classification, whereas the phase transition is of first-order with finite 2Δ, which is small but significant, as will be discussed in chapter 12. Nevertheless, the presence of gap 2Δ is substantiated by neutron inelastic scatterings and magnetic resonance experiments, as described in part II[note2]. Binary fluctuations of collective pseudospins are characterized by *phase inversion* $+\phi \rightleftarrows -\phi$ in space-time in mesoscopic scale[note3].

4.3 Observing critical phase anomalies

Critical anomalies described by (4.4b) are expressed in terms of a phase variable ϕ, but can be studied in practical experiments by scanning the phase variation with respect to renormalized space $\boldsymbol{r}$ or with respect to time t. In practical experiments, it is important to consider the timescale t_o of measurements that should be sufficiently long as compared with the characteristic time $\frac{2\pi}{\tilde{\omega}}$ of fluctuations, specifying for the condition $\tilde{\omega}t_o > 2\pi$ to resolve fluctuation spectra.

Considering a sampling experiment of timescale t_o, the time averages of order variables can be expressed by

$$\langle\sigma_A\rangle_t = \frac{\sqrt{2}\sigma_o}{2t_o}\int_{-t_o}^{+t_o}\cos(\boldsymbol{q}\cdot\boldsymbol{r} - \tilde{\omega}t)\,dt = 2\sqrt{2}\sigma_o\frac{\sin\tilde{\omega}t_o}{\tilde{\omega}t_o}\cos(\boldsymbol{q}\cdot\boldsymbol{r}) \qquad (4.12a)$$

and

$$\langle\sigma_P\rangle_t = \frac{2\sqrt{2}\sigma_o}{2t_o}\int_{-t_o}^{+t_o}\sin(\boldsymbol{q}\cdot\boldsymbol{r} - \tilde{\omega}t)\,dt = 2\sqrt{2}\sigma_o\frac{\sin\tilde{\omega}t_o}{\tilde{\omega}t_o}\sin(\boldsymbol{q}\cdot\boldsymbol{r}), \qquad (4.12b)$$

[note2] The gap 2Δ is attributed to the soliton potential according to the soliton theory discussed in later chapters.

[note3] Observed transition anomalies are often discussed, attributed to Heisenberg's quantum-mechanical uncertainties due to visual observations, which is, however, incorrect in principle, because the space–time of mesoscopic scale should be involved in displacement waves. Therefore, uncertainties in this case should also be referred to with an energy of the order of Δ per sec, besides Planck's constant in quantum theory, as the transition is classified as bifurcation. In fact, this view of structural phase transitions will be found to be consistent with superconducting transitions in thermodynamic principles, as explained in chapters 14 and 15.

where the amplitudes are reduced by the factor $\Gamma = \frac{\sin \tilde{\omega} t_0}{\tilde{\omega} t_0}$ from the maximum $2\sqrt{2}\sigma_0$; but $\Gamma \to 1$ if $\omega t_0 \to 1$. Accordingly, such time averages can be studied for the spatial phase $\phi_s = \boldsymbol{q} \cdot \boldsymbol{r}$ if the condition $\tilde{\omega} t_0 > 1$ is met. The sinusoidal phase in the critical region can usually be monitored by conventional spatial scanning methods.

Expressed as complex quantities, $\langle \sigma_A \rangle_t$ and $\langle \sigma_P \rangle_t$ constitute a complex vector $i\sigma_A + \sigma_P$, where σ_A and σ_P represent longitudinal and transverse components, whose spatial phase is distributed in the range $-\pi/2 \leqslant \phi_s \leqslant \pi/2$, and the amplitude σ_0 is temperature dependent. Define a linear variable ξ by writing $\xi = \cos \phi_s$ and $\sqrt{1 - \xi^2} = \sin \phi_s$, the components σ_A and σ_P can be expressed as

$$\sigma_A = \int_{-1}^{+1} f_A \frac{\sigma_0}{\xi} d\xi \quad \text{and} \quad \sigma_P = \int_{-1}^{+1} f_P \frac{\sigma_0}{\sqrt{1 - \xi^2}} d\xi, \tag{4.13}$$

where f_A and f_P are functions of ϕ_s, thereby these integrands determine the densities of distribution, as illustrated in figure 4.4.

The presence of two fluctuations described by (4.7) is well substantiated in a number of binary systems, as will be discussed in detail in chapters 5 and 6. In scattering and magnetic resonance experiments, both σ_A and σ_P modes are clearly distinctive, where the former gives evidence of energy transfer to the lattice, while the latter determines the domain structure.

Measurements of $\boldsymbol{\sigma}_A$ and $\boldsymbol{\sigma}_P$ can analyze sampling experiments, monitoring responses from a selected portion of a crystal. On the other hand, the specific heat is a thermodynamic quantity of a crystal as a whole. Figure 4.5 summarises schematically a comparison of observed C_p-curve with the mean-field analysis, showing a clear discrepancy, indicating failure of the mean-field idea, as marked 1, 2 and 3 in the figure. The decay in C_p for $T < T_c$ depends on the nonlinear properties of $\boldsymbol{\sigma}(\phi)$, which cannot be analyzed at this stage, leaving it to later discussions in chapter 9.

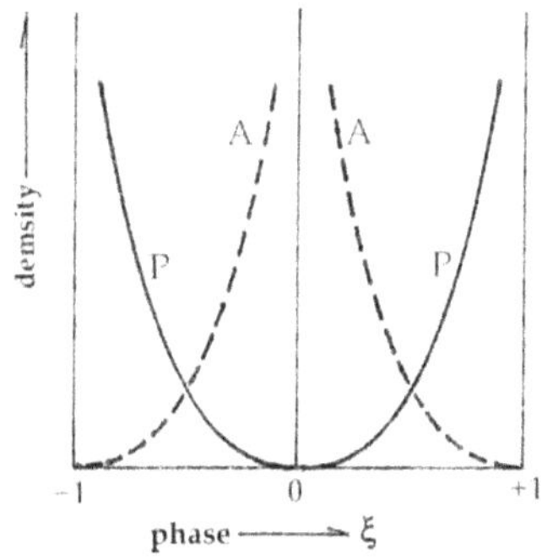

Figure 4.4. Intensity distribution of A- and P-modes of fluctuations. Reproduced from [4], copyright 2010. With permission of Springer.

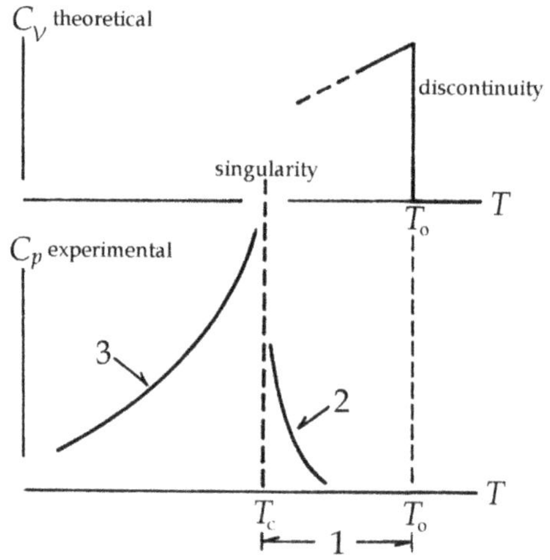

Figure 4.5. A comparison of transition anomalies of theoretical C_V and experimental C_p. Significant differences are indicated by the areas 1, 2 and 3. Reproduced from [4], copyright 2010. With permission of Springer.

4.4 Intrinsic and extrinsic pinning

In the foregoing, the Landau theory was interpreted in terms of the substantiated concept of pinning by an adiabatic potential. In practical crystals, however, we cannot ignore surfaces and imperfections that should be represented by pinning potentials. While purified to nearly defect-free samples, practical crystals are by no means perfect for testing theories. Nevertheless, the order variables and potentials are observable from carefully prepared samples, provided that imperfections are identified. Extrinsic pinning is also usable when an electric or magnetic field is applied externally. Experimentally, extrinsic pinning can be distinguished from internal origins in idealized samples, because the former is independent from the lattice periodicity, whereas the latter is characterized as Galilean invariant.

4.4.1 Point defects

A practical crystal normally contains some unavoidable imperfections and impurities, occupying lattice sites, disrupting translational symmetry, which are called *point defects*. Order variables can be pinned at these sites, and detected as standing waves. Although pinning is necessary for observation, the potential should be distinguished from internal origin in practice. Denoting defect sites by $\boldsymbol{r}_i$, the defect potential is expressed as negative $-V(\boldsymbol{r}-\boldsymbol{r}_i)$ to indicate that it is localized in the vicinity, where a pseudospin $\boldsymbol{\sigma}_i = \boldsymbol{\sigma}_o \exp(\pm i\phi_i)$ is pinned symmetrically with respect to $\boldsymbol{r}_i$. As perturbed by the negative potential of a point defect, the symmetric $\boldsymbol{\sigma}_A = \frac{\sigma_+ + \sigma_-}{\sqrt{2}}$ makes $-V(\boldsymbol{r}-\boldsymbol{r}_i)$ for stable pinning, which is written as

$$V(\phi) = -V_o \cos\phi, \qquad \text{where} \qquad -\pi/2 \leqslant \phi \leqslant \pi/2, \tag{4.14}$$

Here, $\phi = 0$ gives the minimum, and V_o is a redefined magnitude.

Surfaces may be regarded as many point defects, in which case the pinning potential should be characterized as positive, and signified by antisymmetric $\boldsymbol{\sigma}_P$, if surfaces offer active pinning sites. However, a stable potential may be defined as

$$V_S(\phi) = -V_o \sin\left(\phi - \frac{\pi}{2}\right), \tag{4.15}$$

which is theoretically logical, but not certain experimentally.

4.4.2 Electric field pinning

We usually apply a uniform electric field $\boldsymbol{E}$, hence a crystal is acted by $V = -\sum_n \boldsymbol{\sigma}_n \cdot \boldsymbol{E}$ adiabatic potential. Accordingly, as the pinning potential, $V(\phi)$ is antisymmetric with respect to the phase ϕ. Therefore, the pinned $\boldsymbol{\sigma}(\phi)$ cannot be stable energetically, so that the phase ϕ shifts to $\phi - \frac{\pi}{2}$, to be stabilized with $\boldsymbol{E}$ in thermodynamic environment. In fact, such a mechanism is substantiated by a magnetic resonance experiment [2], hence the pinning potential is given by

$$V_E\left(\phi - \frac{\pi}{2}\right) = -\sigma E \sin\left(\phi - \frac{\pi}{2}\right) = -\sigma E \cos\phi. \tag{4.16}$$

The result confirm that the Born–Huang principle governs the observed process.

4.4.3 Surface pinning

Surfaces of a crystal can be regarded as if a group of point defects to interact with order variables, where the field is antisymmetric like those by an electric field, but only the inside needs to be considered. Mathematically, pseudospin waves reflect in the inside space of a crystal, setting a standing wave. Therefore, writing it as $(\boldsymbol{\sigma}_{-k} - \boldsymbol{\sigma}_{+k})_{\text{surface}}/\sqrt{2}$, the pining potentials can be expressed as

$$V_{\text{surface}}(\text{P}) = -\sigma_o \text{P} \sin\phi_\text{P} \quad \text{and} \quad V_{\text{surface}}(\text{A}) = -\sigma_o \text{A} \cos\phi_\text{A} \tag{4.17}$$

for stability at $\phi_\text{P} = -\pi$ and $\phi_\text{A} = 0$, respectively.

Practical surface conditions can theoretically be ignored when using the Bloch theorem for idealized dynamics. On the other hand, in practice they play a significant role in finite crystals to exchange heat energies with surroundings. Dividing crystals into integral number of Brillouin zones, surfaces can nevertheless be represented by both potentials $V_{\text{surface}}(\text{P})$ and $V_{\text{surface}}(\text{A})$, which are considered as *necessary addenda* for practical crystals to be discussed for thermodynamic equilibrium.

Exercises

1. Discuss the origin of the adiabatic potential $V(\phi)$. Why does it have to be a function of ϕ?

2. In section 4.2.2, we consider that the adiabatic potential $V(\phi)$ is symmetrically distributed as proportional to $\pm f(\phi)\,\mathrm{d}\phi$, regarding space–time uncertainties of the phase variable ϕ. Discuss the origin of the distribution.
3. Experimentally, how can extrinsic pinnings be distinguished from the intrinsically pinned fluctuations?
4. Discuss how some energy can be transmitted from order variables to the lattice. Is the lattice considered as the surroundings?

References

[1] Landau L D and Lifshitz E M 1958 *Statistical Physics* (London: Pergamon)

[2] Fujimoto M, Jerzak S and Windsch W 1986 *Phys. Rev.* **B34** 1668

[3] Schiff L I 1955 *Quantum Mechanics* Ch 8 (New York: McGraw-Hill)

[4] Fujimoto M 2010 *Thermodynamics of Crystalline States* (Berlin: Springer)

Part II

Experimental studies on critical anomalies and soft modes

The collective order variable is evident in experimental studies of many types, providing evidence for collective $\boldsymbol{\sigma}(\phi)$, where the soft mode represents their response susceptibilities to internal and applied fields.

IOP Publishing

Solitons in Crystalline Processes (2nd Edition)
Irreversible thermodynamics of structural phase transitions and superconductivity
Minoru Fujimoto

Chapter 5

Scattering experiments on critical anomalies

The properties of the condensate in crystals can be studied from the renormalized pseudospin mode pinned by intrinsic potentials in crystals. Condensates are primarily longitudinal, propagating in the direction of the wavevector, signified by two-components, A and P, as a function of temperature. Using suitable probes, the transition anomaly can be sampled in detail by various experiments; e.g. photon probes are sensitive to temporal variations, and neutrons can be utilized as probes to study lattice displacements. In this chapter, such experimental techniques as x-ray diffraction, light and neutron scatterings are summarized with results relevant to critical anomalies.

5.1 X-ray diffraction

The crystal structure is determined usually by x-ray diffraction studies. Scattered by a crystal, a collimated x-ray beam exhibits a diffraction pattern on a photographic plate placed perpendicular to the beam. The pattern from elastic x-ray *photon* scatterings is due to distributed electron densities in a crystal, leaving nuclear masses intact. In the critical region, the structure is modulated along symmetry axes, exhibiting broadened diffraction spots in anomalous shape that are called *diffuse diffraction*.

In early crystallographic studies, Bragg established the concept of *crystal planes*, similar to optical reflection, signifying the periodic structure in normal crystals. A collimated beam of x-ray reflects from a large number of parallel crystal planes, showing interference pattern on a photographic plate placed nearby. Figure 5.1 illustrates a practical arrangement that is commonly used for diffraction analysis.

Considering that a collimated beam of a monochromatic x-ray of a wavevector $\boldsymbol{K}_o$ and frequency ω_o is incident on a crystal plane characterized by the normal vector $\boldsymbol{G}_o$, scattered wavevector $\boldsymbol{K}$ and frequency ω are determined by the Bragg law

$$\boldsymbol{K}_o - \boldsymbol{K} = \pm\boldsymbol{G}, \tag{5.1a}$$

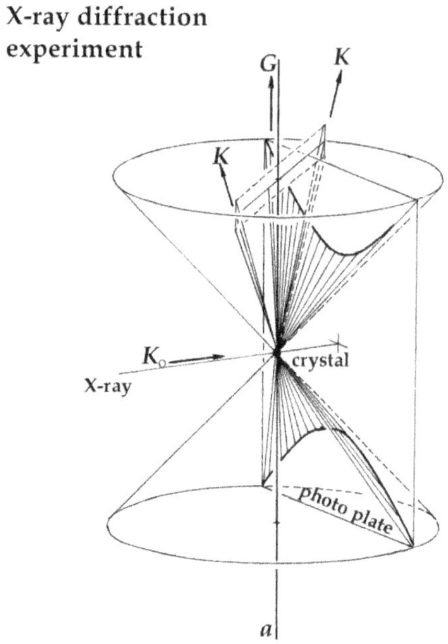

Figure 5.1. A schematic setup for x-ray diffraction experiments. The incident beam is indicated by $\boldsymbol{K}_o$, and the diffracted beam is all distributed on a conical surface as shown. The modulation axis is indicated by $\boldsymbol{G}\|\boldsymbol{a}$.

where

$$|\boldsymbol{K}_o| = |\boldsymbol{K}| \quad \text{and} \quad \hbar\omega_o = \hbar\omega, \tag{5.1b}$$

expressing the conservation of momentum and energy in elastic scatterings. Here $\boldsymbol{G} = n\boldsymbol{a}$ is the lattice translation vector in normal direction to parallel crystal planes, where n is an integer and $\boldsymbol{a}$ is the spacing. We notice from (5.1b), only the magnitude $|\boldsymbol{G}|$ is significant for elastic scatterings, and the diffraction takes place in all directions determined by (5.1a). All wavevectors K of diffracted rays are on conical surfaces of $\boldsymbol{a}$, as shown in figure 5.1, for which the Laue law was formulated, providing a useful method for constructing diffraction patterns. In the figure, illustrated also is a lattice modulation along the a axis, modifying the pattern.

Assuming that a differential charge element $\rho(\boldsymbol{r}_o)\mathrm{d}^3\boldsymbol{r}$ scatters an incident photon field $E_o \exp i(\boldsymbol{K}\cdot\boldsymbol{r}_o - \omega t_o)$ at the point $\boldsymbol{r}_o$ at time t_o, the scattered field is spherical and expressed by

$$A(\boldsymbol{r}, t) \propto \int_{V(r_o)} \mathrm{d}^3\boldsymbol{r}\,\rho(\boldsymbol{r}_o)E_o \exp i(\boldsymbol{K}_o\cdot\boldsymbol{r}_o - \omega t_o)\frac{\exp i\{\boldsymbol{K}\cdot(\boldsymbol{r}-\boldsymbol{r}_o) - \omega(t-t_o)\}}{|\boldsymbol{r}-\boldsymbol{r}_o|}.$$

Accordingly, the scattered amplitude at a large distance $r \gg r_o$ is given approximately as proportional to $\frac{1}{r}\exp i(\boldsymbol{K}\cdot\boldsymbol{r} - \omega t)$. Assuming such an amplitude at a large distance A_o for brevity, we can write

$$\frac{A_o}{E_o} \propto \int_{V(r_o)} \mathrm{d}^3\boldsymbol{r}\,\rho(\boldsymbol{r}_o)\exp i\{(\boldsymbol{K}_o - \boldsymbol{K})\cdot\boldsymbol{r}_o\} \tag{5.2}$$

Elastic scatterings from a rigid crystal can be specified by (5.1*a*), hence $\exp i(\pm \boldsymbol{G} \cdot \boldsymbol{r}_o) = 1$ in (5.2) and A_o/E_o is determined by the maximum density at $\boldsymbol{r}_o$.

In practice, however, a collimated x-ray beam strikes a finite area on the crystal plane, where the reflected intensity originates from all scatterers in the impact area. Letting positions of scatterers be $r_{o,m}$, where $m = 1, 2, \ldots$, equation (5.2) can be modified as

$$\frac{A_o}{E_o} = \sum_m f_m(G)\exp(-iG \cdot r_{om}) \qquad \text{where} \qquad f_m(G) = \int_{V(r_o)} \rho(r_{om})\mathrm{d}^3\boldsymbol{r} \tag{5.3}$$

is called the *atomic form factor*. However, electronic densities in crystals are overlapped between neighbors, so that usually (5.3) gives overestimated results. In this context, a redefined factor for independent contributions is often desirable.

Replacing $\boldsymbol{r}_{om}$ by a continuous variable $\boldsymbol{s} - \boldsymbol{r}_{om}$, we can write

$$\int_{V(s)} \rho(\boldsymbol{s})\exp(-i\boldsymbol{G} \cdot \boldsymbol{s})\mathrm{d}^3\boldsymbol{s} = f_m(\boldsymbol{G})\exp(-i\boldsymbol{G} \cdot \boldsymbol{r}_{om}),$$

where

$$f_m(\boldsymbol{G}) = \int_{V(s)} \rho(\boldsymbol{s} - \boldsymbol{r}_{om})\exp\{-i(\boldsymbol{s} - \boldsymbol{r}_{om})\}\mathrm{d}^3(\boldsymbol{s} - \boldsymbol{r}_{om}) \tag{5.4}$$

Using (5.4), the *structural form factor* can be defined from (5.3) as

$$S(\boldsymbol{G}) = \frac{A_o}{E_o} = \sum_m f_m(\boldsymbol{G})\exp\frac{2\pi i(hx_m + ky_m + lz_m)}{\Omega},$$

thereby the field $E_G(\boldsymbol{r})$ of a scattered beam at a distant point $\boldsymbol{r}$ is expressed as

$$\frac{E_G(\boldsymbol{r})}{E_o} = \frac{S(\boldsymbol{r})}{r}\exp i(\boldsymbol{K} \cdot \boldsymbol{r} - \omega t),$$

and the scattered/incident intensity ratio is given by

$$\frac{I(\boldsymbol{G})}{I_o} = \frac{1}{r^2}S^*(\boldsymbol{G})S(\boldsymbol{G}). \tag{5.5}$$

5.2 Diffuse diffraction from a modulated lattice

We consider a crystal that is modulated in the *a* direction. The periodicity is modified by the wavevectors $\boldsymbol{G} \pm \boldsymbol{q}$ of displacements, and the corresponding energies are given by $\mp\Delta\varepsilon$, so that the scatterings are basically *inelastic*. In this case, the conservation laws are expressed as

$$\boldsymbol{K}_o - \boldsymbol{K} = \boldsymbol{G} \quad \text{and} \quad \varepsilon(\boldsymbol{K}_o + \boldsymbol{q}) - \varepsilon(\boldsymbol{K} - \boldsymbol{q}) = \Delta\varepsilon \approx 0. \tag{5.6}$$

Equation (5.2) can then be revised as

$$\frac{E_G(\boldsymbol{r})}{E_o} \propto \frac{\exp i(\boldsymbol{K}\cdot\boldsymbol{r} - \Delta\omega t)}{r} \times \sum_m \{f(\boldsymbol{G}+\boldsymbol{q})\exp i(\boldsymbol{q}\cdot\boldsymbol{r}_m) + f(\boldsymbol{G}-\boldsymbol{q})\exp i(-\boldsymbol{q}\cdot\boldsymbol{r}_m)\},$$

where

$$f(\boldsymbol{G}\pm\boldsymbol{q}) = \int_{V(\boldsymbol{r}_o)} \mathrm{d}^3\boldsymbol{r}\rho(\boldsymbol{r}_o)\exp i(\boldsymbol{G}\pm\boldsymbol{q}).\ \boldsymbol{r}_o.$$

Assuming $|\boldsymbol{q}| < |\boldsymbol{G}|$, hence $f(\boldsymbol{G}\pm\boldsymbol{q}) \approx f(\boldsymbol{G})$, the phase fluctuation factor in the integral, $\exp i(\boldsymbol{G}\pm\boldsymbol{q})\cdot\boldsymbol{r}_o$, can be separated into amplitude- and phase-modes, as discussed in chapter 4, so the corresponding intensity ratio is expressed as

$$\frac{I(\boldsymbol{G}\pm\boldsymbol{q})}{I_o} = \frac{I(\boldsymbol{G})}{I_o} + \frac{2\,|f(\boldsymbol{G})|^2}{r^2}\sum_{m.n}[\cos\{\boldsymbol{q}\cdot(\boldsymbol{r}_m - \boldsymbol{r}_n)\} + \sin\{\boldsymbol{q}\cdot(\boldsymbol{r}_m - \boldsymbol{r}_n)\}],$$

where the first term on the right arises from elastic scatterings, and the second term is due to inelastic scatterings. Observable intensities are related as

$$\frac{\langle I(\boldsymbol{G}\pm\boldsymbol{q})\rangle - \langle I(\boldsymbol{G})\rangle}{I_o} = \frac{2\,|f(\boldsymbol{G})|^2}{r^2}\frac{1}{S}\int_S \{\langle\cos\phi\rangle + \langle\sin\phi\rangle\}\mathrm{d}S,$$

where $\phi = \boldsymbol{q}\cdot(\boldsymbol{r}_m - \boldsymbol{r}_n)$ represents the fluctuating phase.

Assuming a rectangular area $S = L_xL_y$, for which we write $\phi = \boldsymbol{q}\cdot\boldsymbol{r}$ for brevity, the symmetric part can be calculated as

$$\frac{1}{S}\int_S \langle\cos\phi\rangle\mathrm{d}S = \left(\frac{1}{L_x}\int_0^{L_x}\langle\cos\phi\rangle\mathrm{d}x\right)\left(\frac{1}{L_y}\int_0^{L_y}\mathrm{d}y\right)$$
$$= \frac{1}{qL_x}\int_{\phi_1}^{\phi_2}\cos\phi\ \mathrm{d}\phi = \frac{2}{\phi_2 - \phi_1}\sin\frac{\phi_2 - \phi_1}{2}\cos\frac{\phi_2 + \phi_1}{2}.$$

where $\phi_2 - \phi_1 = qL_x$ and $\phi_2 + \phi_1 = 0$. Similarly calculating antisymmetric part, we obtain the result for the total anomaly, that is

$$\frac{\langle\Delta I(\boldsymbol{G})\rangle}{I_o} = \frac{\sin qL_x}{qL_x}(\langle\cos qx\rangle + \langle\sin qx\rangle) \tag{5.7}$$

for diffuse intensities of x-ray diffraction composed of A- and P-modes of fluctuation. However, lacking sufficient resolution in diffraction pattern, such a broadened spots and lines are difficult to analyze experimentally.

5.3 Neutron inelastic scatterings

Although difficult in x-ray scatterings, the modulated structure can be resolved in neutron inelastic scatterings, because neutron scatterers are heavy nuclei or magnetic spins. If $\boldsymbol{K}_o = \boldsymbol{G}$ in (5.1*a*), obviously we have $\boldsymbol{K} = 0$, meaning no elastic scatterings.

If the target is modulated however, $\boldsymbol{G} \pm \boldsymbol{q} = \Delta\boldsymbol{K}$, where $\Delta\boldsymbol{K}$ represents *inelastic scatterings* that occur in all directions. Therefore, inelastic scatterings can be detected in a direction perpendicular to $\boldsymbol{G}$ efficiently, avoiding possible contributions from elastic scatterings.

The zone-boundary transition at $\pm\boldsymbol{G}/2$ in $SrTiO_3$ or magnetic MnF_2 crystals is of particular interest, as thermal neutrons are available from modern nuclear reactors in this range of wavevectors. Figure 5.2 illustrates an experimental setup, where the modulation q can be detected for inelastic scattering in a direction for $\Delta\boldsymbol{K}\perp\boldsymbol{K}_o$, with minimal elastic background. Sketched in figure 5.3 is a *triple-axes spectrometer*, consisting of monochrometer, goniometer and analyzer that are all rotatable individually around their axes in parallel. The monochrometer and analyzer are diffraction devices using crystals of known lattice constants, thereby a selected neutron wavelength can be measured by adjusting angles, θ and θ', as indicated in the figure. A sample crystal is mounted on the goniometer, which is rotated to scan around a symmetry axis.

The conservation laws apply to the impact between neutrons and scatterers, that is

$$\boldsymbol{K}_o - \boldsymbol{K} = \boldsymbol{G}_z \pm \boldsymbol{q} \quad \text{and} \quad \varepsilon(\boldsymbol{K}_o) - \varepsilon(\boldsymbol{K}) = \pm\Delta\varepsilon(\boldsymbol{q}), \tag{5.8}$$

where $\boldsymbol{G}_z$ is a reciprocal vector at the critical point near $\pm\frac{G}{2}$, so written as shifted near the zone boundary, and $\Delta\varepsilon(\boldsymbol{k})$ is the transferred energy by impact, which can be expressed as $\hbar\omega$ for convenience.

Analogous to x-ray photons, the neutron beam can be represented by a wave-function $\psi \approx \exp i\phi$ similar to a field $E \approx \exp i\phi$. Therefore, assuming $|\boldsymbol{r} - \boldsymbol{r}_{om}| \approx r$ and $t_{om} \approx t_o$ for the target group, scattered neutrons by the laws (5.8) can be expressed as

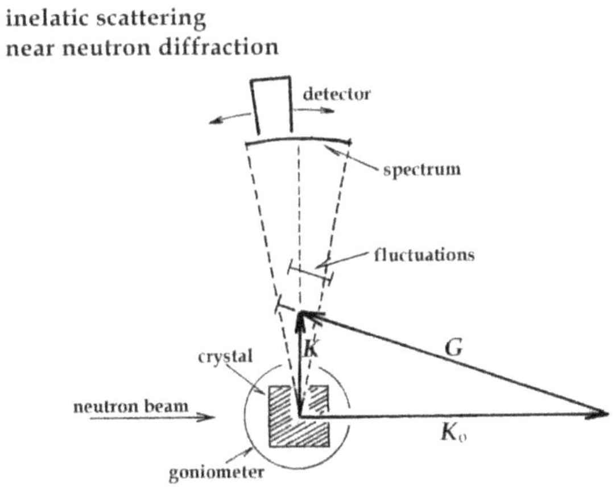

Figure 5.2. An experimental arrangement for neutron inelastic scattering. For an incident beam $\boldsymbol{K}_o$, no inelastic scattering takes place in the perpendicular direction, where only inelastic scattering $\Delta\boldsymbol{K}$ can be detected for $\boldsymbol{K}\perp\boldsymbol{K}_o$. In such an apparatus, $\Delta\boldsymbol{K}$ can be scanned to observe anomalies.

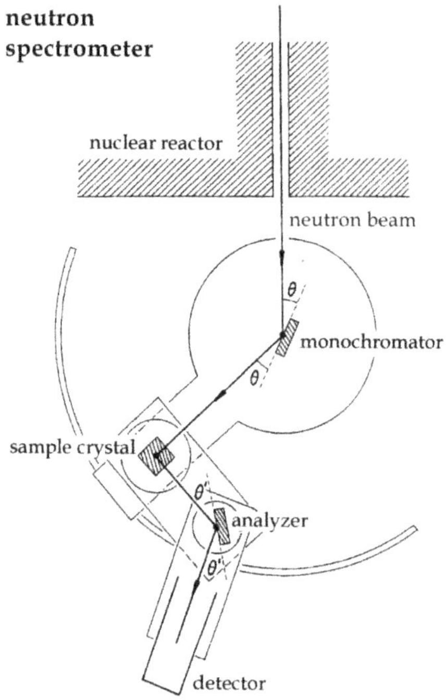

Figure 5.3. Triple-axis spectrometer for neutron scattering experiments. The monochromator, goniometer and analyzer are all rotatable around their axes in parallel.

$$\psi(\boldsymbol{r}, t) \approx \int_{V(\boldsymbol{r}_0)} \mathrm{d}^3\boldsymbol{r}_0\, \rho(\boldsymbol{r}_0) \frac{\exp i\{\mp \boldsymbol{q}\cdot(\boldsymbol{r}-\boldsymbol{r}_0) - \Delta\omega(t-t_0)\}}{|\boldsymbol{r}-\boldsymbol{r}_0|}$$
$$\times \exp i\{(\boldsymbol{G}_z \pm \boldsymbol{q})\cdot \boldsymbol{r}_0 \mp (\Delta\omega)t_0\},$$

and the structure factor can be defined as

$$f(\boldsymbol{G}_z \pm \boldsymbol{q}) = \int_{V(\boldsymbol{r}_0)} \mathrm{d}^3\boldsymbol{r}_0\, \rho(\boldsymbol{r}_0) \exp i(\boldsymbol{G}_z \pm \boldsymbol{q})\cdot \boldsymbol{r}_0.$$

the net wavefunction can be expressed as

$$\Psi(\boldsymbol{G}_z \pm \boldsymbol{q}) \propto \frac{\exp i(\Delta\boldsymbol{K}\cdot\boldsymbol{r})}{r} \sum_m \{f(\boldsymbol{G}_z + \boldsymbol{q})\exp i\phi_m + f(\boldsymbol{G}_z - \boldsymbol{q})\exp(-i\phi_m)\},$$

where $\phi_m = \boldsymbol{q}\cdot\boldsymbol{r}_{0m}$. Writing $f_m(\boldsymbol{G}_z \pm \boldsymbol{q}) = f(\boldsymbol{G}_z \pm \boldsymbol{q})\exp(\pm i\phi_{0m})$, the scattered intensity can then be given by

$$I(\boldsymbol{G}_z \pm \boldsymbol{q}) = |\Psi(\boldsymbol{G}_z \pm \boldsymbol{q})|^2 \propto \left\langle \sum_m |f_m(\boldsymbol{G}_z \pm \boldsymbol{q})|^2 \right\rangle$$
$$+ \left\langle \sum_{m \neq n} \{f_m{}^*(\boldsymbol{G}_z + \boldsymbol{q}) f_n(\boldsymbol{G}_z + \boldsymbol{q}) + f_m{}^*(\boldsymbol{G}_z - \boldsymbol{q}) f_n(\boldsymbol{G}_z - \boldsymbol{q})\} \right\rangle,$$

where the first- and second-terms on the right correspond to elastic and inelastic scatterings, respectively.

Expressing that $\boldsymbol{r}_{om} - \boldsymbol{r}_{on} = \boldsymbol{r}$ as before, the latter can be written as

$$\langle \Delta I(\boldsymbol{G}_z \pm \boldsymbol{q}) \rangle \propto \langle \exp i\boldsymbol{G}_z \cdot \boldsymbol{r} \{ |f(\boldsymbol{G}_z + \boldsymbol{q})|^2 \exp i\phi + |f(\boldsymbol{G}_z - \boldsymbol{q})|^2 \exp(-i\phi) \} \rangle, \quad (5.9)$$

where $\phi = \boldsymbol{q} \cdot \boldsymbol{r}$ is the modulated space phase. At this point, considering $|\boldsymbol{q}| < |\boldsymbol{G}_z|$ for a weak modulation, we assume

$$|f(\boldsymbol{G}_z \pm \boldsymbol{q})|^2 = |f(\boldsymbol{G}_z)|^2 \pm 2i\boldsymbol{q} \cdot \nabla_q |f(\boldsymbol{G}_z)|^2,$$

which modifies (6.9) as

$$\langle \Delta I(\boldsymbol{G}_z \pm \boldsymbol{q}) \rangle = \mathrm{A}\langle \cos\phi \rangle + \mathrm{P}\langle \sin\phi \rangle,$$

where

$$\mathrm{A} \propto \int_{V(r)} |f(\boldsymbol{G}_z)|^2 \cos(\boldsymbol{G}_z \cdot \boldsymbol{r}) \mathrm{d}^3\boldsymbol{r} \quad \text{and} \quad \mathrm{P} \propto \int_{V(r)} 2\boldsymbol{q} \cdot \nabla_q |f(\boldsymbol{G}_z)|^2 \cos(\boldsymbol{G}_z \cdot \boldsymbol{r}) \mathrm{d}^3\boldsymbol{r}.$$

These fluctuations are observable, when pinned by the zone boundary potential at $\pm\boldsymbol{G}_z$ that are equal to $\pm\boldsymbol{G}/2$.

Figure 5.4 shows scanned spectra of neutron inelastic scatterings observed from magnetic MnF_2 crystals at the critical Neél temperature T_N, composed of amplitude- and phase-modes in various ratios, depending on the direction $\boldsymbol{q}$ of modulation [1].

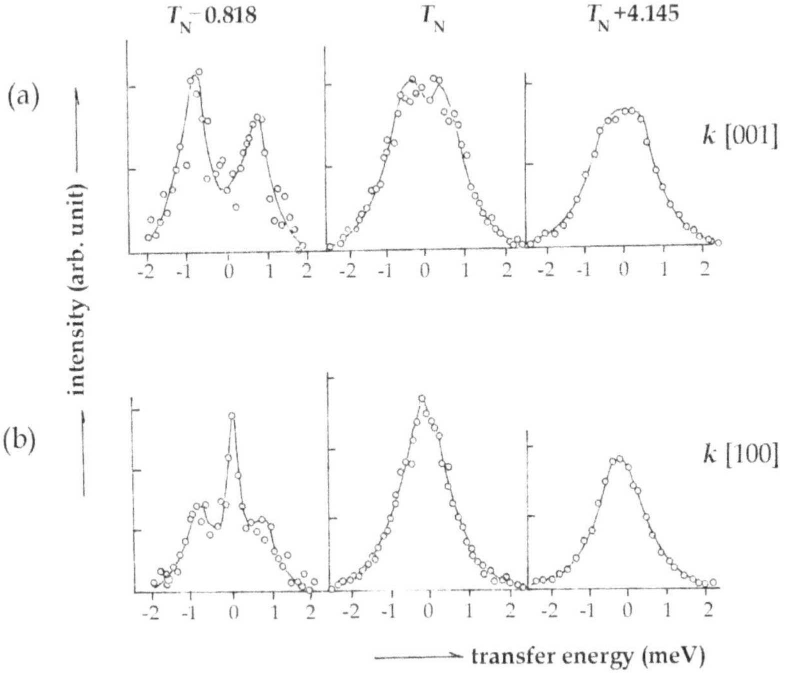

Figure 5.4. Inelastically scattered neutrons from magnetic MnF_2 crystals detected at Neél temperature T_N, where A- and P-peaks are recognized in different proportions depending on the scattering geometry with respect to Crystal axes. Reproduced from [1] with permission from American Physical Society.

Anomalous distributions in different scattering geometries are clear evidence of coexisting A and P modes predicted in the above theory.

5.4 Light scattering experiments

Using intense coherent light waves from a *laser* oscillator, we can detect feeble inelastic scatterings from a crystal in enhanced intensity, permitting detailed studies on scattered photon spectra. Incident light induces dielectric fluctuations arising from random impacts with constituents, producing primarily the *Rayleigh radiation* that is attributed to elastic collisions. Inelastic scatterings are detectable with applied high-intensity radiation, yielding information about structural changes in dielectric crystals. Laser spectroscopy is a useful technique for investigating structural modulation. However, performing in a fixed scattering geometry, experiments are limited to studying temporal variations of spectra, unlike scanning experiments with a triple-axis spectrometer for neutron inelastic scatterings. Nevertheless, the light scattering techniques have provided valuable information on transition anomalies.

5.4.1 Brillouin scatterings

When sound waves in one dimension are standing in liquid, an incident laser light beam in perpendicular direction generates a diffraction pattern characterized by modulated liquid densities. In dielectric crystals, similar diffractions can be observed, for which induced dielectric fluctuations are responsible. Known as the *Brillouin scatterings*, such a modulated diffraction is caused by the photo-elastic property of a crystal, which can be studied as inelastic impact between laser photons and phonons.

Figure 5.5(a) illustrates an experimental layout for Brillouin scatterings, showing that an intense laser light of wavevector $\boldsymbol{K}_0$ polarized along the

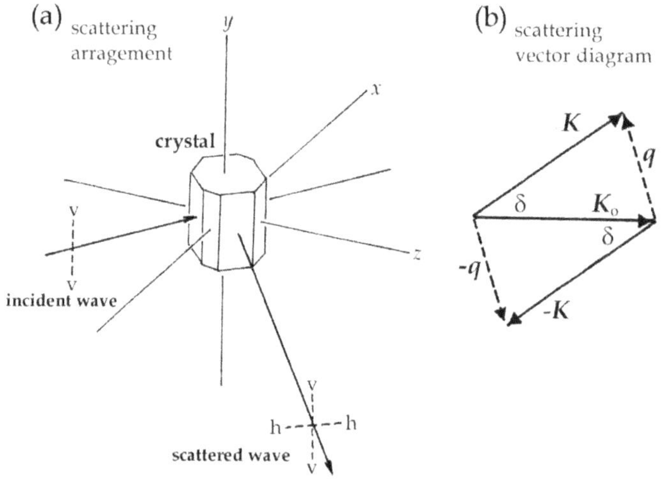

Figure 5.5. (a) A light-scattering arrangement. Incident light polarized in the vertical direction v···v is scattered, after which it is polarized in the vertical and horizontal directions v···v as well as h···h, respectively. (b) A general light-scattering geometry, where $\pm\boldsymbol{q}$ are wavevectors of the displaced lattice.

y direction incident on the (101) plane of the sample crystal, and that the scattered light of wavevector K is detected in the perpendicular direction [$\bar{1}01$]. In this case, the *lattice displacement wave* of wavevector $\pm\boldsymbol{q}$ should participate in the scattering process, as shown in figure 5.5(b), illustrating the scattering geometry for $\boldsymbol{K}_o$, $\boldsymbol{K}$ and $\pm\boldsymbol{q}$. Due to photo-elastic properties related with crystal symmetry, the scattered displacement waves are not only longitudinal but also transversal, and polarized in two directions, indicated as v–v and h–h in the figure. We therefore write $\Delta p_v = \alpha_v E$ and $\Delta p_h = \alpha_h E$, where α_v, α_h are polarizabilities, and E is the electric field of light.

Expressing the electric field of incident light at a scatterer located at $(\boldsymbol{r}_o, t_o)$ as $\boldsymbol{E}_o \exp i(\boldsymbol{K}_o \cdot \boldsymbol{r}_o - \omega t_o)$, the induced polarization is given by $\Delta\boldsymbol{p}(\boldsymbol{r}_o, t_o) = (\alpha)\boldsymbol{E}_o \exp i(\boldsymbol{K}_o \cdot \boldsymbol{r}_o - \omega t_o)$, where (α) is the polarizability tensor. Analogous to section 5.1, the scattered electric field at the space-time $(\boldsymbol{r}, t)$ can be written as

$$\boldsymbol{E}(\boldsymbol{r}, t) \approx \frac{\exp i(\boldsymbol{K} \cdot \boldsymbol{r} - \omega t)}{r} \times \iint \Delta\boldsymbol{p}(\boldsymbol{r}_o, t_o) \cdot \boldsymbol{E}_o \exp i\{(\boldsymbol{K}_o - \boldsymbol{K}) \cdot \boldsymbol{r}_o - (\omega_o - \omega)t_o\} d^3\boldsymbol{r}_o dt_o. \tag{5.10}$$

If $\omega = \omega_o$, (5.10) is maximum at all directions of $\boldsymbol{K}$, where $K = K_o$, of elastic collisions, known as *Rayleigh's scatterings*. Denoting the scattering angle by $\boldsymbol{K}_o \cdot \boldsymbol{r}_o = \frac{2\pi r_o}{\lambda} \cos\varphi$, where λ is the wavelength, Rayleigh's intensity $I_R(K)$ can be expressed for $\lambda \gg r_o$ as

$$\frac{I_R(K)}{I(K_o)} \propto \frac{2\pi^2 v^2}{r^2 \lambda^4}(1 + \cos^2\varphi),$$

where v is the speed of light in air, and the two terms on the right correspond to contributions of the component fields $\boldsymbol{E}_v$ and $\boldsymbol{E}_h$.

For inelastic Brillouin scatterings, the conservation laws are

$$\boldsymbol{K}_o - \boldsymbol{K} = \pm\boldsymbol{q} \quad \text{and} \quad \omega_o - \omega = \mp\Delta\omega,$$

where $\pm\boldsymbol{q}$ and $\mp\Delta\omega$ are wavevectors and frequency shifts of lattice-displacement waves $\boldsymbol{\sigma}$. While such inelastic scatterings with phonons are always observed in crystals, which are however distinguishable from classical displacements as phonon scatterings are isotropic.

For anisotropic scatterings, the angle δ between K and $\boldsymbol{K}_o$ is called the scattering angle, with which the magnitude of Q is given by

$$|\boldsymbol{q}| = 2|\boldsymbol{K}_o|\sin\frac{\delta}{2},$$

and the intensity can be calculated as

$$I_B(\boldsymbol{K}, \pm\boldsymbol{q}) \propto \frac{1}{r^2}\iint(\Delta p_i \Delta p_j)\exp i\{\pm\boldsymbol{q}\cdot(\boldsymbol{r}_o' - \boldsymbol{r}_o) \mp \Delta\omega(t_o' - t_o)\}d^3(\boldsymbol{r}_o' - \boldsymbol{r}_o)d(t_o' - t_o). \tag{5.11}$$

This is similar in form to the intensity formula derived for neutron inelastic scatterings. Accordingly, if the wavevector q and frequency $\Delta\omega$ represent a specific displacement of interest, considering $\langle\Delta p_i \Delta p_j\rangle$ like the structure factor in the neutron case, we can similarly define the phases of fluctuations $\phi = \boldsymbol{q}\cdot\boldsymbol{r} - \Delta\omega t$ and $-\phi = -\boldsymbol{q}\cdot\boldsymbol{r} + \Delta\omega t$, where $\boldsymbol{r} = \boldsymbol{r}_o' - \boldsymbol{r}_o$ and $t = t_o' - t_o$ for the present case. Then, we have the intensity formula for Brillouin scatterings identical to neutron scattering:

$$\langle I_B(\pm\boldsymbol{q})\rangle_t = A\langle\cos\phi\rangle_t + P\langle\sin\phi\rangle_t,$$

where the factors A and P can be specified as before in neutron scatterings.

Referring to figure 5.5(a), we consider that the displacement and driving electric field are given by u_v and $E_v\exp(-i\Delta\omega t)$, and that dielectric property is represented by a complex dipole moment $\Delta p_v = (\alpha' \pm i\alpha'')E_v = eu_v$. Assuming an effective mass m, we write the equation of motion:

$$m\left(\ddot{u}_v + \gamma\dot{u}_v + \omega_o^2 u_v\right) = eE_v\exp(-it\Delta\omega),$$

whose steady-state solution is

$$u_v^o = \frac{eE_v/m}{\omega_o^2 - \Delta\omega^2 \mp i\gamma\Delta\omega}.$$

Hence

$$\Delta p_v = \frac{e^2E_v/m}{\omega_o^2 - \Delta\omega^2 \mp i\gamma\Delta\omega} = (\alpha' \mp i\alpha'')E_v \quad\text{and}\quad \alpha' + i\alpha'' = \frac{e^2/m}{\omega_o^2 - \Delta\omega^2 \mp i\gamma\Delta\omega}.$$

Equation (5.11) can therefore be expressed as

$$\langle I_B(\pm\boldsymbol{G})\rangle_t \propto \frac{E_v^2}{r^2}\{(\alpha'\alpha_v)\langle\cos\phi\rangle_t + (\alpha''\alpha_v)\langle\sin\phi\rangle_t\},\ \text{for } 0 \leqslant \phi \leqslant 2\pi, \tag{5.12}$$

where these terms represent A and P modes, respectively.

In the Brillouin spectra, however the symmetric A-mode is overlapped with Rayleigh line, but the P-mode can be identified, if they are resolved in separation larger than the width of the Rayleigh line.

Experimentally, to keep incident photons away from the detector, $\delta = 90°$ is normally selected, but Rayleigh's scatterings at $\boldsymbol{q} = 0$ cannot be avoided in practice, to obscure emerging Brillouin spectra. Nevertheless, we consider that $|\boldsymbol{K}_o| = \frac{2\pi n}{\lambda}$, where n is the optical index of refraction, and the Brillouin frequency ν_B is measured by frequency shifts $\Delta\nu_B = \nu_B - \nu_R$ from the Rayleigh line ν_R. Therefore, the Brillouin shifts $\pm\Delta\nu_B$ are determined by

$$\Delta\nu_B = \pm\frac{2\pi v}{\lambda}\sin\frac{\delta}{2},$$

where v is the speed of light in the crystalline medium. Assuming $n \approx 1.5$ for the dielectric property, $v = 2 \times 10^3$ m s^{-1}, and $\lambda = 514.5$ nm for example, the Brillouin shift can be estimated from the above formula as $\Delta\nu_B \simeq 6 \times 10^9$ Hz $= 0.2$ cm^{-1}.

A Fabry–Perot interferometer, sketched in figure 5.6, is commonly used for Brillouin measurements, where the device consists of parallel semi-transparent planes with a narrow gap d that enhance interference after multiple reflections. Referring to the figure, we can write the interference condition for an incident angle ψ as

$$2nd\cos\psi = m\lambda,$$

where m is an integer, and n the index of refraction of the air adjustable by applying pressure for scanning spectra. Usually the normal incidence is specified by $\psi = 0°$, in which case we have $\frac{1}{\lambda} = \frac{m}{2d}$. Assuming that the wavelength $\lambda - \Delta\lambda$ gives positive interference at $m + 1$, we therefore have $\frac{1}{\lambda - \Delta\lambda} = \frac{m+1}{2d}$, hence the resolvable range in spectra can be determined by

$$\frac{1}{\lambda} - \frac{1}{\lambda - \Delta\lambda} = \frac{1}{2d},$$

which is called the free spectral range of the spectrometer. Corresponding to $\pm\boldsymbol{q}$, a pair of Brillouin lines appear under normal conditions of dielectric crystals, as illustrated by the diagram in figure 5.8(a).

Figure 5.7(a) shows typical Brillouin spectra, where the frequency shifts $\pm\Delta\nu_B$ may lead to lattice instability with decreasing temperature, as illustrated by figure 5.7(b). Physically, high dielectric excitations can modify the lattice structure, as a natural consequence, depending on photo-elastic properties of a crystal. On the other hand, enhanced excitations should increase displacement being dispersive and

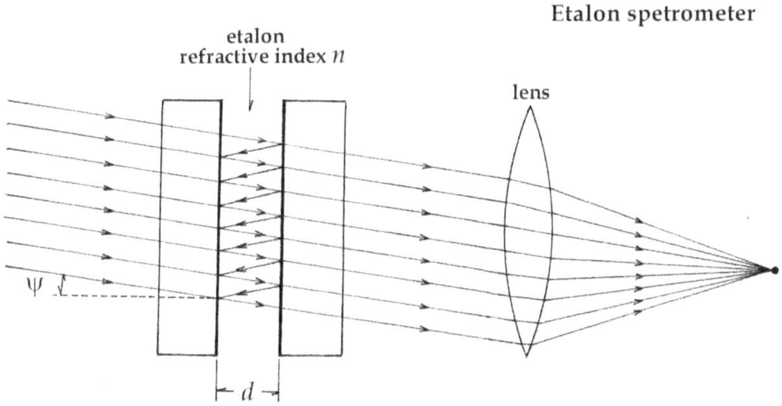

Figure 5.6. A Fabry–Perot interferometer.

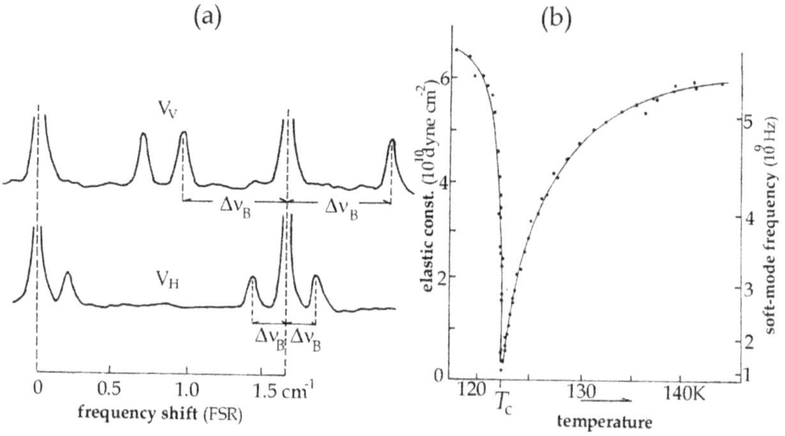

Figure 5.7. (a) Brillouin spectra from KDP crystals at room temperature observed in the v···v and h···h directions. (b) Soft-mode frequencies versus temperature. Reproduced from [2].

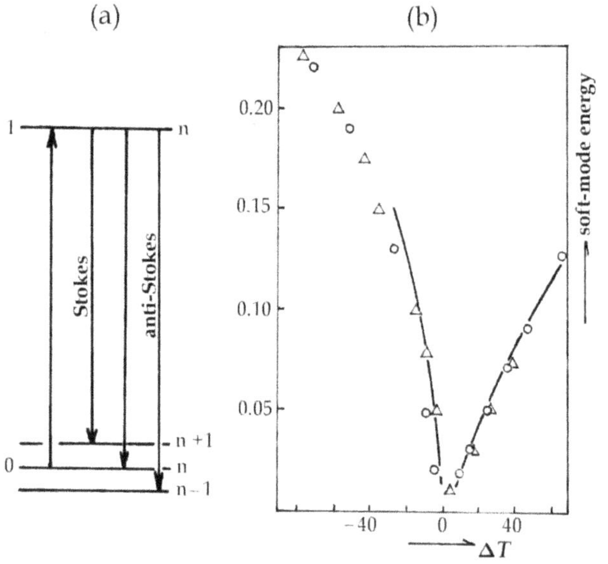

Figure 5.8. (a) Raman transitions with Stokes and anti-Stokes lines. (b) Soft modes observed from ferroelastic LaP_5O_{14} and $La_{0.5}Nd_{0.5}P_5O_{14}$. Reproduced from [3].

dissipative in a thermodynamic environment, as related to the photo-elastic tensor as indicated by (5.12), with respective to the structural symmetry.

Figure 5.8(b) shows a typical result of Brillouin studies from a ferroelastic crystal LaP_5O_{14} and $La_{0.5}Nd_{0.5}P_5O_{14}$ [2, 3]. Frequency shifts $\Delta\nu_B$ depended on scattering geometry, exhibiting a parabolic temperature-dependence, which is clear evidence of a *soft mode*, while the amplitude mode was undetected, because of an intense central Rayleigh line at the center. Figure 5.9 shows another soft-mode frequency from the Brillouin studies of ferroelastic $LiNH_4C_4H_4O_6H_2O$ (LAT) crystals [4].

Experimentally, Brillouin scatterings occur not only by lattice displacements but also by regular phonons of harmonic displacements. The latter is found in all crystallographic directions and temperature-independent, but the lattice displacement takes place only in specific directions and is temperature-dependent, which should be identified prior to successful measurements.

5.4.2 Raman spectroscopy of soft modes

Light scatterings can also be observed from ions in crystals, independent of dielectric response from the lattice. This is of particular interest if ions are associated with lattice displacements. In this case, we can expect a frequency shift in scattered light, similar to the Brillouin scatterings. Such scatterings were discovered by Raman (1928), offering information of lattice displacements distinguished from the dielectric response. In this section, following [5], Placzek's theory of Raman scatterings is discussed with representative examples.

Considering a constituent ion characterized by energy levels ε_0 and ε_1, hence, we have $\mathcal{H}\varphi_o = \varepsilon_0\varphi_o$ and $\mathcal{H}\varphi_1 = \varepsilon_1\varphi_1$, where $\varepsilon_0 < \varepsilon_1$ for convenience. For such an ion, wavefunctions in a crystal can be written as $\varphi_0\chi$ and $\varphi_1\chi$, where χ represents the lattice in the adiabatic approximation, and linearly polarized light can be decomposed into two circularly polarized components in opposite directions, i.e.

$$\boldsymbol{E}\cos\Omega t = \frac{1}{2}\boldsymbol{E}_+\exp i\Omega t + \frac{1}{2}\boldsymbol{E}_-\exp(-i\Omega t).$$

The Schrödinger equation for the perturbed ion can be written as

$$\left\{\mathcal{H} - \frac{1}{2}\boldsymbol{p}\cdot\boldsymbol{E}_+\exp i\Omega t - \frac{1}{2}\boldsymbol{p}\cdot\boldsymbol{E}_-\exp(-i\Omega t)\right\}\Psi = i\hbar\frac{\partial\Psi}{\partial t},$$

where $\boldsymbol{p}$ is a dipole moment induced in the ion. The perturbed wavefunction in the ground state can be expressed by

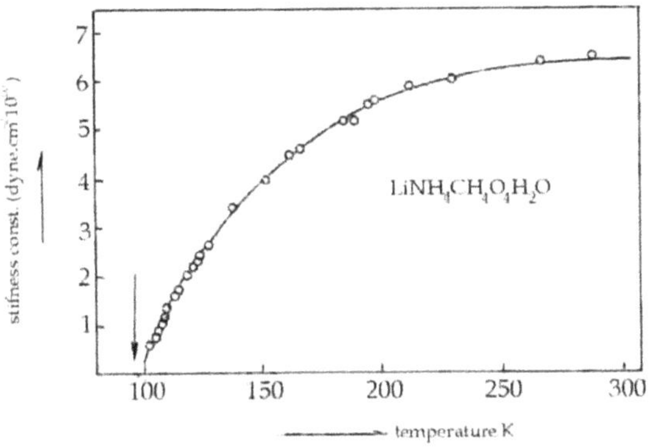

Figure 5.9. The soft-mode character of Brillouin scatterings observed from ferroelastic LAT crystals. Reproduced from [4] with permission from American Physical Society.

$$\Psi_0 = \{\psi_0 + \psi_{0+}\exp i\Omega t + \psi_{0-}\exp(-i\Omega t)\}\exp\left(-it\frac{\varepsilon_0 + n\hbar\omega}{\hbar}\right),$$

accordingly in the first order approximation we have

$$\{\mathcal{H} - (\varepsilon_0 + n\hbar\omega \pm \hbar\Omega)\}\psi_{0\pm} = \boldsymbol{p} \cdot \boldsymbol{E}_{\pm}\psi_0, \tag{5.13a}$$

connecting $\psi_{0\pm}$ to ψ_0. Therefore, we can write

$$\psi_0 = \varphi_0\chi_0 \quad \text{and} \quad \psi_{0\pm} = \varphi_0(c_+\chi_+ + c_-\chi_-) + c_0\psi_0, \quad \text{where } c_\pm = \frac{\langle\pm|\boldsymbol{p}|0\rangle}{\hbar(\pm\omega \pm \Omega)},$$

A dipole component in the ground state can therefore be expressed as

$$p_i(t) = \langle 0|p_i|0\rangle + \sum_j\{\alpha_{ij}(\Omega)E_{+j}\exp i\Omega t + \alpha_{ij}(-\Omega)E_{-j}\exp(-i\Omega t)\} \tag{5.13b}$$

where

$$\begin{aligned}\alpha_{ij}(\Omega) = &\frac{\langle 0|p_i|+\rangle\langle +|p_j|0\rangle}{\hbar(\omega + \Omega)} \\ &+ \frac{\langle 0|p_j|+\rangle\langle +|p_i|0\rangle}{\hbar(\omega - \Omega)} + \frac{\langle 0|p_i|-\rangle\langle -|p_j|0\rangle}{\hbar(-\omega + \Omega)} + \frac{\langle 0|p_j|-\rangle\langle -|p_i|0\rangle}{\hbar(-\omega - \Omega)}\end{aligned} \tag{5.13c}$$

constitute a polarizability tensor with respect to x–y axes in the plane of $\boldsymbol{E}_\pm$. Note that the relations $\alpha_{ij} = \alpha_{ji}^*$ and $\alpha_{ij} = \alpha_{ij}^*$ can be verified, the tenor (α_{ij}) is real and *hermitic*.

Similar expressions can be written for the excited state ε_1, that is

$$\Psi_1 = \left\{\frac{\langle +|\boldsymbol{p}|0\rangle \cdot \boldsymbol{E}_+}{\hbar(\omega + \Omega)}\exp i\Omega t + \frac{\langle -|\boldsymbol{p}|0\rangle \cdot \boldsymbol{E}_-}{\hbar(-\omega - \Omega)}\exp(-i\Omega t)\right\}\exp\left(-it\frac{\varepsilon_1 + n\hbar\omega}{\hbar}\right),$$

from which the induced probability for an emission process can be calculated as

$$\begin{aligned}\langle\Psi_o^*|\boldsymbol{p}|\psi_1\rangle &= \exp\frac{i\Delta\varepsilon_{01}t}{\hbar} \times \int\left\{c_0^*\langle\psi_0^*|\boldsymbol{p}|\psi_1\rangle + c_+^*\langle\psi_+^*|\boldsymbol{p}|\psi_1\rangle + c_-^*\langle\psi_-^*|\boldsymbol{p}|\psi_1\rangle\right\}\mathrm{d}v \\ &= c_0^*\langle\psi_0^*|\boldsymbol{p}|\psi_1\rangle\exp\frac{i\Delta\varepsilon_{01}t}{\hbar} \\ &\quad + \frac{\langle 0|\boldsymbol{p}^*|\pm\rangle\langle\pm|\boldsymbol{p}|0\rangle}{\hbar(\pm\omega + \Omega)}\exp it\left(\Omega \pm \frac{\Delta\varepsilon_{o1}}{\hbar}\right) \\ &\quad + \frac{\langle 0|\boldsymbol{p}^*|\pm\rangle\langle\pm|\boldsymbol{p}|0\rangle}{\hbar(\pm\omega - \Omega)}\exp it\left(-\Omega \pm \frac{\Delta\varepsilon_{o1}}{\hbar}\right).\end{aligned} \tag{5.13d}$$

The first term on the right is independent of the phonon energy $\hbar\omega$, whereas the second and third terms correspond to Raman emission processes at $\Omega = \Delta\omega_{01} \pm \omega$, signifying transitions $\Delta n = \pm 1$. Therefore, illustrated in figure 5.8(a), two satellite lines appear in Raman spectra, corresponding to emission and absorption of photons, which are traditionally called Stokes and anti-Stokes lines, respectively.

Placzek's theory deals with the Raman scatterings from an atom that occupies a lattice site, but is considered to be represented by phonons so that electro-elastic properties are neglected. Therefore, we need to consider that a polar displacement $\boldsymbol{\sigma}_n$ is defined as

$$e\boldsymbol{\sigma}_{\pm} = e(\beta)\boldsymbol{E}\exp(\pm i\,\Omega t) \quad \text{or} \quad \boldsymbol{\sigma}_{\pm} = (\beta)\boldsymbol{E}\exp(\pm i\Omega t),$$

where (β) is an electro-elastic tensor that is equivalent to the polarization tensor $(\alpha)/e$, where (α) is the polarizability tensor in isotropic media defined by (iii).

In particular, if we consider $\boldsymbol{\sigma}_{\mathrm{A}} = \frac{1}{\sqrt{2}}(\boldsymbol{\sigma}_{+} + \boldsymbol{\sigma}_{-})$ and $\boldsymbol{\sigma}_{\mathrm{P}} = \frac{1}{\sqrt{2}}(\boldsymbol{\sigma}_{+} - \boldsymbol{\sigma}_{-})$, the electro-elastic tensor should be written as

$$(\beta) = \begin{pmatrix} \beta_{\mathrm{AA}} & \beta_{\mathrm{AP}} \\ \beta_{\mathrm{PA}} & \beta_{\mathrm{PP}} \end{pmatrix} \quad \text{where} \quad \beta_{\mathrm{AA}} = \frac{\langle 0|\sigma_{\mathrm{A}}|\pm\rangle^2}{2\Delta \mp \hbar\Omega},$$

$$\beta_{\mathrm{PP}} = \frac{\langle 0|\sigma_{\mathrm{P}}|\pm\rangle^2}{-2\Delta \pm \hbar\Omega} \quad \text{and} \quad \beta_{\mathrm{AP}} = \frac{\langle 0|\sigma_{\mathrm{A}}|\pm\rangle\langle\pm|\sigma_{\mathrm{P}}|0\rangle}{\pm 2\Delta \mp \hbar\Omega},$$

depending on the energy separation 2Δ at T_c.

Hence, corresponding to (iii), we have the relation

$$(\beta) = \frac{\partial(\beta)}{\partial\sigma_{\mathrm{A}}}\sigma_{\mathrm{A}} + \frac{\partial(\beta)}{\partial\sigma_{\mathrm{P}}}\sigma_{\mathrm{P}} + \cdots,$$

where $\partial(\beta)/\partial\sigma$ is a significant factor to determine Raman line intensities observed when $\boldsymbol{\sigma}\|\boldsymbol{E}$, which is therefore called the Raman tensor. Particularly, both Stokes and anti-Stokes lines are maximum when $\boldsymbol{\sigma}_{\mathrm{P}}\|\boldsymbol{E}$, from which the soft mode behavior can be analyzed with respect to the susceptibility $\chi = \sigma_{\mathrm{P}}/E$.

The Raman tensor is generally determined with regard to point symmetry, for which Loudon [6] tabulated for all point groups. However, it is important to determine appropriate *Raman-active symmetry*. Center symmetry in a ferroelectric crystal is not Raman-active, while piezoelectric symmetry is Raman-active, because ferroelectric crystals are piezoelectric [7]. For practical study, the Raman activity should therefore be examined prior to experiments.

References

[1] Schulhof M P, Heller P, Nathans R and Linz A 1970 *Phys. Rev.* B **1** 2403

[2] Brody E M and Cummins H Z 1968 *Phys. Rev. Lett.* **31** 1263
Brody E M and Cummins H Z 1974 *Phys. Rev.* **9** 179

[3] Toledano J C, Errandonea E and Jaguin J P 1976 *Solid State Comm.* **20** 905

[4] Burns G and Scott B A 1973 *Phys. Rev.* B **7** 3088

[5] Born M and Huang K 1968 *Dynamical Theory of Crystal Lattices* p 204 (London: Oxford University Press)
[6] Loudon R 1964 *Adv. Phys.* **13** 423
Loudon R 1963 *Proc. Roy. Soc.* A **275** 218
Loudon R 1963 *Roy. Phys. Soc.* **82** 393
[7] Nakamura T 2000 *Ferroelectrics and Structural Phase Transitions* (in Japanese) (Tokyo: Shokabo) ch 4

Chapter 6

Magnetic resonance studies on critical anomalies

In this chapter, we review magnetic resonance studies on critical anomalies in structural transitions. Limited to selected crystals, many successful results were reported in the literature, where the *spin-Hamiltonian* is allowed to confirm the predicted behavior of pseudospins. To deal with collective pseudospin dynamics, Bloch's theory of *nuclear induction* is fundamental to magnetic resonance, and its extended use for paramagnetic probes can be used to analyze observed anomalies. Critical anomalies are due to a modulated lattice structure by order variables in collective motion, exhibiting nonlinear features in representative examples. Observed resonance spectra were always characterized by symmetric and antisymmetric fluctuations, providing the visual image of critical fluctuations that is consistent with scattering results.

6.1 Magnetic resonance

The magnetic resonance technique can be applied to either nuclear magnetic moments or impurity magnetic ions to monitor the local field in a crystal undergoing a structural transition. In the nuclear resonance, naturally abundant magnetic isotopes can be used as spin probes, whereas chemical magnetic impurities with unpaired electrons are embedded in crystals at low density as paramagnetic probes. For the latter, we need to determine if the amount of doping is structurally negligible for the modulated structure. If judged as negligible, however, the method provides useful information about structural changes. On the other hand, nuclear probes can be used to study possible structural changes, although usable isotopes are limited in their natural abundance.

6.1.1 Nuclear magnetic resonance and relaxation

Although familiar in molecular beam experiments, the magnetic resonance in condensed matter is established with Bloch's theory for nuclear induction in condensed matter [1], which is also applicable to paramagnetic resonance technique.

doi:10.1088/978-0-7503-2572-1ch6

Nuclear magnetic moments are basically independent from the non-magnetic lattice, however, their time-dependent feature can be studied, subjecting them to energy transfer to the lattice in thermodynamic environment, as discussed by the Bloch theory.

Microscopically, *nuclear magnetic moments* μ_n in an applied magnetic field $\boldsymbol{B}_0 \| z$ are in Larmor's precession around the z-axis at a constant frequency $\omega_L = \gamma B_0$, where γ is the gyromagnetic ratio, as shown in figure 6.1(a). In precession, the z-component μ_z is constant, while the perpendicular component $\mu_\perp$ is rotating at the frequency ω_L. In magnetic resonance, we apply an oscillating magnetic field in small amplitude $B_1 \propto \exp i\omega t$ as $\boldsymbol{B}_1 \perp \boldsymbol{B}_0$, hence exerting a torque $\boldsymbol{\mu}_n \times \boldsymbol{B}_1$ on μ_n. If $\omega = \omega_L$, the angle of precession increases by the torque, called magnetic resonance, in which case, only a part of B_1 proportional to $\exp i\omega t$ is essential.

Considering a large number of microscopic $\boldsymbol{\mu}_n$ in a crystal, Bloch assumed that the corresponding macroscopic magnetization is given by $\boldsymbol{M} = \sum_n \boldsymbol{\mu}_n$ all in-phase, which is an acceptable assumption in a high magnetic field, as we impose the same boundary condition on all $\boldsymbol{\mu}$ in a crystal. Bloch wrote the equation of motion for a macroscopic magnetization, as

$$\frac{dM_z}{dt} = -\frac{M_z - M_0}{T_1} \quad \text{and} \quad \frac{dM_\perp}{dt} = -\frac{M_\perp}{T_2} \tag{6.1}$$

The first equation signifies a relaxation process toward thermal equilibrium with the lattice, for which T_1 is called the *spin-lattice relaxation time*. As M_z is the macroscopic quantity, this equation describes energy transfer to the lattice, T_1 should correspond to thermal relaxation in timescale. On the other hand, the second equation in (6.1) can be considered for synchronization of microscopic $\mu_\perp$ until all become in-phase,

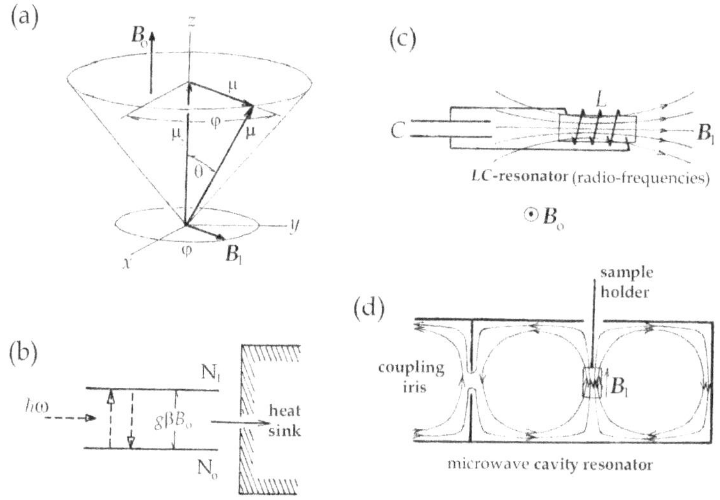

Figure 6.1. (a) Larmor's precession in static and oscillating magnetic fields, $\boldsymbol{B}_0$ and $\boldsymbol{B}_\perp$, respectively. (b) Magnetic resonance absorption in a thermal experiment. (c) An LC resonator for experiments at radio-frequencies. (d) A microwave resonator for paramagnetic resonance.

for which the characteristic phasing time is specified by T_2, and called *spin–spin relaxation time*. Bloch assumed that mutual spin–spin interactions are responsible for the phasing. Writing the magnitude of spin–spin interaction as ΔB, Bloch's assumption can be stated as

$$\frac{1}{T_2} \approx \gamma \Delta B \ll \gamma B_1, \tag{6.2}$$

which is known as the condition for a *slow passage*.

In the presence of two magnetic fields applied as $\boldsymbol{B}_o \perp \boldsymbol{B}_1$, Bloch's equations (6.1) are modified as

$$\frac{\mathrm{d}M_\pm}{\mathrm{d}t} \pm \gamma B_o M_\pm + \frac{M_\pm}{T_2} = -i\gamma B_1 \exp(\mp i\omega t)$$

and

$$\frac{\mathrm{d}M_z}{\mathrm{d}t} + \frac{M_z - M_o}{T_1} = \frac{i\gamma B_o}{2}\{M_+\exp(-i\omega t) + M_-\exp(i\omega t)\}, \tag{6.3}$$

where $M_\pm$ are transverse components of M synchronized with the rotating field $B_1 \exp(\mp i\omega t)$.

A steady solution under a slow passage condition (6.2) is given by

$$M_z = \text{const.} \quad \text{and} \quad M_\pm = \frac{\gamma B_1 M_z \exp(\mp i\omega t)}{\mp\omega + \gamma B_o + iT_2^{-1}}.$$

Substituting these into (7.3), we obtain

$$\frac{M_z}{M_o} = \frac{1 + (\omega - \omega_L)^2 T_2^2}{1 + (\omega - \omega_L)^2 + \gamma^2 T_1 T_2} \quad \text{and} \quad \frac{M_\pm}{M_o} = \frac{\{(\omega - \omega_L)T_2 + i\}\gamma B_1 \exp(\mp i\omega t)}{1 + (\omega - \omega_L)^2 T_2^2 + \gamma^2 B_1^2 T_1 T_2}.$$

If $(\gamma B_1 T_1 T_2)^2 \ll 1$, from the first relation we have $M_z \approx M_o$, for which the angle θ can be calculated from

$$\tan\theta = \frac{M_\pm}{M_o} \approx \frac{\gamma B_1 T_2}{1 + (\omega - \omega_L)^2 T_2^2}.$$

Defining the high-frequency susceptibility by $M_\pm = \chi(\omega) B_1 \exp(\mp i\omega t)$ and $M_o = \chi_o B_o$, the real and imaginary parts of $\chi(\omega)$ can be written as

$$\frac{\chi'(\omega)}{\chi_o} = \frac{\omega_L(\omega - \omega_L)}{(\omega - \omega_L)^2 + \delta\omega^2 + \gamma^2 B_1^2 T_1 \delta\omega} \quad \text{and}$$
$$\frac{\chi''(\omega)}{\chi_o} = \frac{\omega_L \delta\omega}{(\omega - \omega_L)^2 + \delta\omega^2 + \gamma^2 B_1^2 T_1 \delta\omega}, \tag{6.4}$$

where $\delta\omega = 1/T_2$ for convenience, indicating the linewidth. In magnetic resonance experiments, either $\chi'(\omega_L)$ or $\chi''(\omega_L)$ is measured with spectrometers, exhibiting dispersion and absorption signals at the resonance frequency ω_L.

Equations (6.4) are the basic formula for observing magnetic resonance, which can be detected by scanning the frequency ω of the high-frequency field $\boldsymbol{B}_1$. However, in practice, using Larmor's relation $\omega_L = \gamma B_o$, it is more convenient to scan the applied field B_o at a fixed frequency ω_L, instead of constant B_o. As described in later sections, most experiments were performed for $\chi(B_o)$ by field scan.

6.1.2 Paramagnetic resonance with impurity probes

Unlike nuclear magnetic moments located virtually at all unit cells in a sample crystal, for paramagnetic resonance experiments, we normally use magnetic impurity ions accommodated at the same specific locations of unit cells by doping at low density, if sufficiently within detectable intensity of a practical spectrometer.

A paramagnetic moment of impurity ions is associated with the angular momentum J of electrons in the incompletely filled shell in a magnetic ion, i.e. $\mu = \gamma J = \gamma(\boldsymbol{L} + \boldsymbol{S})$, where $\boldsymbol{L}$ and S are total orbital and spin angular momenta, respectively. Its magnetic energy in an applied field $\boldsymbol{B}_o$ is given by $\varepsilon_m = -\gamma\hbar J_m B_o$, where $J_m = m, m-1, \ldots, -m$ are magnetic quantum numbers, for which the selection rules $\Delta m = \pm 1$ apply. Accordingly, transition between these magnetic levels can take place by magnetic resonance, if

$$\hbar\omega_L = \hbar\gamma B_o = \varepsilon_m - \varepsilon_{m-1}.$$

Figure 6.1(b) shows such a quantum view of magnetic resonance. For an ion represented by unpaired electrons, Larmor's frequency is usually expressed as $\omega_L = g\beta B_o$, where $\beta = \gamma/2\hbar$ and g are called *Bohr's magneton* and *Landé's factor*, respectively. For a single electron in particular, we have $g = 2$.

Probabilities for the transitions induced by magnetic resonance are given by

$$w_{m,m-1} = \frac{\pi B_1^2}{2\hbar^2}\langle m|\mu_\perp|m-1\rangle^2 f(\omega), \quad \text{where} \int f(\omega)d\omega = 1,$$

and $f(\omega)$ corresponds to $1/T_2$ of Bloch's theory. It is noted that at radio- or microwave frequencies in practical experiments, induced transitions are predominant, while the probability for spontaneous transitions is negligible. Further, in a thermodynamic environment at temperature T, the transitions between ε_m and ε_{m-1} are enhanced by populations N_m and N_{m-1}, which are determined as proportional to the Boltzmann factors, $\exp(-\varepsilon_m/k_BT)$ and $\exp(-\varepsilon_{m-1}/k_BT)$, respectively. Hence, macroscopic energy loss W by magnetic resonance can be calculated by

$$\begin{aligned} W &= w_{m,m-1}(\hbar\omega_L)(N_m - N_{m-1}) = w_{m,m-1}(\hbar\omega_L)N_m\{1 - \exp(-\hbar\omega_L/k_BT)\} \\ &\approx \frac{N_m\pi\omega_L^2 B_1^2}{k_BT}\langle m|\mu_\perp|m-1\rangle^2 f(\omega_L) = \frac{1}{2}\omega_L\chi''(\omega_L)B_1^2. \end{aligned}$$

Therefore

$$\chi''(\omega_L) = \frac{N_m\pi\omega_L}{k_BT}\langle m|\mu_\perp|m-1\rangle^2 f(\omega_L), \tag{6.5}$$

which indicates that the intensity of magnetic resonance can be significantly large at low temperatures.

The magnetic resonance signified by a complex susceptibility can be measured from a sample crystal placed in an *inductor* $L = \chi(\omega)L_o$, where L_o is the inductance of the empty coil. The current through L_o produces an oscillating field B_1 inside the space. Using a conventional laboratory magnet of $B_o \sim 10^4$ Gauss, the frequency of about $1 \sim 10$ MHz is convenient for nuclear resonance, whereas microwave frequencies of 9–35 GHz are suitable for paramagnetic resonance. Accordingly, an LC resonator or a cavity resonator can be used for these measurements, respectively, as illustrated in figure 6.1(c) and (d), where the sample crystal is placed at a location of B_1 in maximum strength. Such a loaded LC resonator and a microwave cavity are placed in bridge circuits for impedance measurements. The impedance of a test resonator can be expressed as

$$Z = R + i\omega L + \frac{1}{i\omega C} = R + \omega\chi'' L_o + i\left(\omega\chi' L_o - \frac{1}{\omega C}\right),$$

whose real and imaginary parts can be selected for measuring χ'' and χ' separately by adjusting the bridge. Figure 6.2 shows a schematic circuit diagram of commonly used magnetic resonance spectrometer, consisting of a standard microwave bridge and a laboratory magnet.

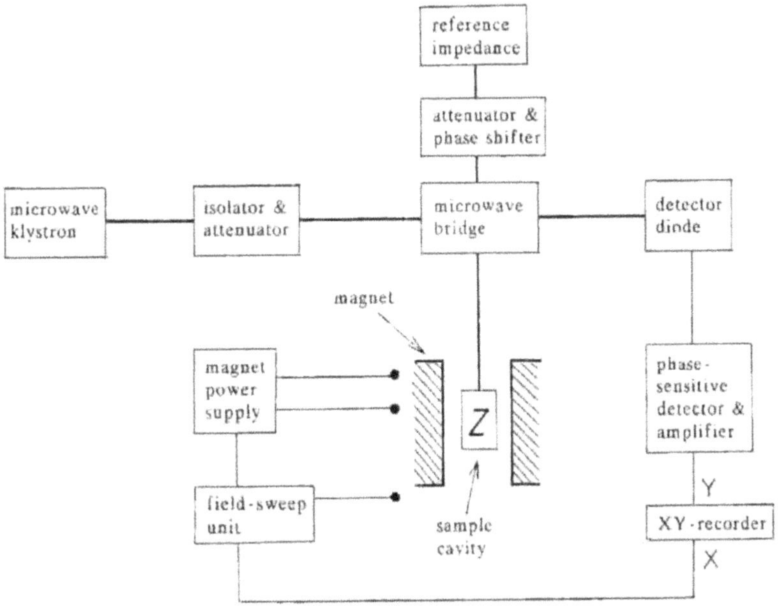

Figure 6.2. A microwave bridge spectrometer for paramagnetic resonance.

6.1.3 The spin-Hamiltonian and a crystal field

Although magnetic impurity in low density is primarily independent of the non-magnetic host lattice, an impurity ion modifies local symmetry of the lattice site, where an additional potential should be considered to perturb the state of the impurity ion. Such a potential called the *crystal field* should be regarded as a reaction of the lattice against accommodating an impurity ion.

Therefore, the impurity state in a crystal cannot be the same as in free state, but perturbed by the crystal field. Here, we postulate a crystal field in orthorhombic symmetry for simplicity that is expressed by

$$V(x, y, z) = Ax^2 + By^2 + Cz^2, \tag{6.6}$$

where the coordinates x, y and z are consistent with the lattice symmetry. The coefficients A, B and C are arbitrary, however, they are restricted by the fact that the potential must satisfy the *Laplace equation* $\nabla^2 V = 0$, so that we have the relation $A + B + C = 0$.

Most paramagnetic ions utilized for doping are from transition elements in iron and rare-earth groups, whose electronic states are characterized by total orbital and total spin angular momenta $\boldsymbol{L}$ and S, called Russell–Saunders' coupling. In this case, L is quantized by the crystal field, while S is primarily free from it as it is magnetic. Nevertheless, we should consider the magnetic interaction between L and S in the next accuracy. It is helpful to consider a classical image where L is quantized along a crystal field axis, around which S is in precession, such as in the Larmor case. The former is called orbital quenching, while the latter is described by the spin–orbital coupling $\lambda \boldsymbol{L} \cdot \boldsymbol{S}$, where λ is a spin–orbit constant. Experimentally, the crystal field energy is much larger than the spin–orbit coupling, supporting the above view.

With respect to the orbital momentum L quenched along x, y, z axes, the spin–orbit coupling energy can be expressed in the first-order approximation as

$$E_{LS}^{(1)} = \lambda(L_x S_x + L_y S_y + L_z S_z) = \lambda \boldsymbol{L} \cdot \boldsymbol{S}, \tag{6.7}$$

assuming λ is a scalar quantity. Then the second-order calculations give rise to

$$E_{LS}^{(2)} = \lambda^2 \sum_{ij} S_i S_j \frac{1}{\Delta\varepsilon_{ij}} \left(\int_V \psi_o{}^* L_i \psi_o \mathrm{d}V \right) \left(\int_V \psi_o{}^* L_j \psi_o \mathrm{d}V \right),$$

which is determined by off-diagonal elements of L_i, L_j and energy gap $\Delta\varepsilon_{ij}$. We express this result in convenient tensor form

$$E_{LS}^{(2)} = \sum_{i,j} S_i D_{ij} S_j = \langle S | \boldsymbol{D} | S \rangle, \tag{6.8}$$

where $\boldsymbol{D}$ is called the *fine structure tensor*, for which we can verify the relation trace

$$D = \sum_i D_{ii} = 0 \quad \text{or} \quad D_{xx} + D_{yy} + D_{zz} = 0. \tag{6.9}$$

Physically, the fine structure tensor signifies that the charge cloud of an impurity ion is deformed in ellipsoidal shape by the orthorhombic crystal field, which is otherwise spherical, as illustrated in figure 6.3(a). Such a deformed charge can be interpreted as an *electric quadrupole moment*, similar to nuclear quadrupole moment. Nevertheless, the D tensor is traditionally known as the fine-structure tensor.

In magnetic resonance spectroscopy, the magnetic moment associated with the angular momentum J should have a Zeeman energy in a given magnetic field $\boldsymbol{B}_0$, which is expressed by the Hamiltonian

$$\mathcal{H}_Z = -\beta(\boldsymbol{L} + 2\boldsymbol{S}) \cdot \boldsymbol{B}_0,$$

where β is Bohr's magneton, and the factor 2 arises from the nature of electron spin. However, in the first-order approximation, the orbital L is quenched by the crystal field, i.e. $\langle \boldsymbol{L} \rangle = 0$, so that $\langle \mathcal{H}_Z \rangle^{(1)} = -2\beta \boldsymbol{S} \cdot \boldsymbol{B}_0$, where the magnetic moment is $\boldsymbol{\mu} = 2\beta \boldsymbol{S}$, which is called a spin only case, and the *Landé factor* $g = 2$. Considering the spin–orbit coupling as the perturbation, the Zeeman energy for $B_0 \| z$ is modified as

$$E_z = -g_z \beta S_z B_0, \quad \text{where } g_z = 2(1 + 2\lambda/\Delta\varepsilon),$$

in the first-order approximation, whereas for $B_0 \| x$, and y we obtain

$$E_{x,y} = -g_{x,y} \beta S_{x,y} B_0, \quad \text{where } g_{x,y} = 2(1 - \lambda/\Delta\varepsilon),$$

where the crystal field is assumed as uniaxial along the z-axis, for simplicity. In general, however, such an anisotropic g-factor should be a tensor quantity, therefore the Zeeman energy should be written as

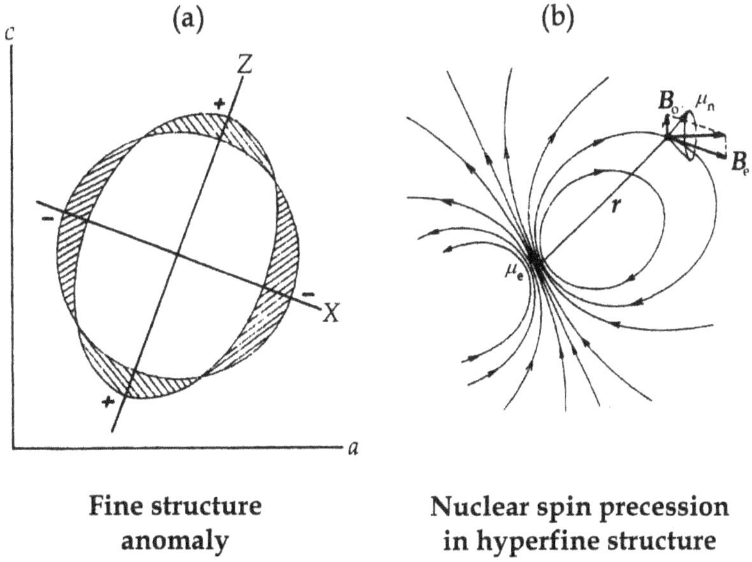

Figure 6.3. (a) Symmetric distortion of a quadratic hyperfine tensor, illustrated in two dimensions. (b) The magnetic hyperfine field $\boldsymbol{B}_e$ at a nuclear moment μ_n that is in precession around the field $\boldsymbol{B}_0 + \boldsymbol{B}_e$. Reproduced from [2], copyright 2010. With permission of Springer.

$$\mathcal{H}_Z = -\beta\langle \boldsymbol{S}|\boldsymbol{g}|\boldsymbol{B}_o\rangle,$$

where $\boldsymbol{g}$ is a tensor with respect to the crystal field. On the other hand, for experimental convenience, we write $\boldsymbol{S} = S\boldsymbol{n}$ to specify the unit vector for spin direction, expressing $\mathcal{H}_Z$ as

$$\mathcal{H}_Z = -g_n\beta\boldsymbol{S}\cdot\boldsymbol{B}_o, \quad \text{where } g_n^2 = \langle \boldsymbol{n}|\boldsymbol{g}^2|\boldsymbol{n}\rangle, \tag{6.10}$$

Accordingly, the Hamiltonian for magnetic resonance experiments can be given by

$$\mathcal{H} = -g_n\beta\boldsymbol{S}\cdot\boldsymbol{B}_o + \langle \boldsymbol{S}|\boldsymbol{D}|\boldsymbol{S}\rangle, \tag{6.11}$$

which is known as the *spin-Hamiltonian.*

6.1.4 Hyperfine interactions

The interaction between the electronic magnetic moment μ_e and nuclear magnetic moment μ_n located within the orbital can be determined from paramagnetic resonance spectra, yielding significant structural information in the vicinity. As illustrated in figure 6.3(b), such an interaction, known as the hyperfine interaction, can be expressed not only by classical dipole–dipole interaction, but is also contributed to quantum-mechanically by the electronic density at the nucleus, i.e. the contact interaction. The hyperfine interaction can then be described by a quadratic tensor $\langle\mu_e|\boldsymbol{A}|\,\mu_n\rangle$, however, trace $\boldsymbol{A} \neq 0$, because of the contact interaction. Writing these magnetic moments $\langle\mu_e| = \beta\langle\boldsymbol{S}|\boldsymbol{g}$ and $|\mu_n\rangle = \gamma_n|\boldsymbol{I}\rangle$, the hyperfine energy can be expressed as

$$\mathcal{H}_{\mathrm{HF}} = \beta\gamma_n\langle\boldsymbol{S}|\boldsymbol{g}\boldsymbol{A}|\boldsymbol{I}\rangle.$$

Since the spin S is quantized along $\boldsymbol{B}_o$ at quantum number M, defining $\beta\langle\boldsymbol{S}|\boldsymbol{g}\boldsymbol{A} = \langle\boldsymbol{B}_e|$ as the magnetic field of electron at the nucleus, we have

$$E_{\mathrm{HF}}^{(1)} = -\gamma_n\boldsymbol{B}_e\cdot\boldsymbol{I}, \quad \text{where } B_e = \beta M\sqrt{\langle\boldsymbol{n}|\boldsymbol{g}\boldsymbol{A}\boldsymbol{A}^\dagger\boldsymbol{g}^\dagger|\boldsymbol{n}\rangle}\,.$$

This indicates that the nuclear moment $\gamma_n\boldsymbol{I}$ is in precession around the field $\boldsymbol{B}_e$, if larger than B_0. In this case, letting the nuclear magnetic quantum number be m, the above hyperfine energy can be re-expressed as

$$E_{\mathrm{HF}}^{(1)} = -g_n\beta K_n Mm, \quad \text{where } g_n^2K_n^2 = \langle\boldsymbol{n}|\boldsymbol{g}\ \boldsymbol{A}\ \boldsymbol{A}^\dagger\boldsymbol{g}^\dagger|\boldsymbol{n}\rangle \tag{6.12}$$

Here, the constant K_n is in energy units, but is noted to be more convenient in field units. Expressing in field units, we can write $E_{\mathrm{HF}}^{(1)} = -K_nMm$ and $K_n^2 = \langle\boldsymbol{n}|\boldsymbol{K}\boldsymbol{K}^\dagger|\boldsymbol{n}\rangle$, replacing $\boldsymbol{g}\ \boldsymbol{A}$ by $\boldsymbol{g}\ \boldsymbol{A}/g_n = \boldsymbol{K}$.

Further notable is that the symmetries of the $\boldsymbol{K}^2$ and $\boldsymbol{g}^2$ tensors are not necessarily consistent with each other.

6.2 Magnetic resonance in modulated crystals

The magnetic resonance spectra of transition ions are generally complicated, due to the fact that the applied field is not strong enough to eliminate forbidden transitions.

Nevertheless, anomalies can be detected from g-factors, fine- and hyperfine structures in selected directions of $\boldsymbol{B}_o$, although complete analysis is not always possible.

We assume that a probe spin S can be modulated in a mesoscopic state by collective motion of order variables $\sigma(\phi)$ in the following way.

Theorem:

$$\boldsymbol{S}' = \boldsymbol{\alpha}\cdot\boldsymbol{S}, \qquad \text{where } \boldsymbol{\alpha} = \mathbf{1} + \sigma\mathbf{e} \tag{6.13}$$

Here, **1** and **e** are tensors of identity and local strains in a crystal. The latter deforms symmetry, restricted by trace $\mathbf{e} = 0$. The spin Hamiltonian can then be modified as

$$\mathcal{H}' = -\beta\langle \boldsymbol{S}'|\boldsymbol{g}|\boldsymbol{B}_o\rangle + \langle \boldsymbol{S}'|\boldsymbol{D}|\boldsymbol{S}'\rangle + \langle \boldsymbol{S}'|\boldsymbol{K}|\boldsymbol{I}\rangle.$$

Applying (6.13), this can be transformed to

$$\mathcal{H}' = \mathcal{H} + \mathcal{H}_1, \qquad \text{where } \mathcal{H} = -\beta\langle \boldsymbol{S}|\boldsymbol{g}|\boldsymbol{B}_o\rangle + \langle \boldsymbol{S}|\boldsymbol{D}|\boldsymbol{S}\rangle + \langle \boldsymbol{S}|\boldsymbol{K}|\boldsymbol{I}\rangle$$

and

$$\mathcal{H}_1 = -\sigma\beta\langle \boldsymbol{S}|\mathbf{e}^\dagger\boldsymbol{g}|\boldsymbol{B}_o\rangle + \sigma\langle \boldsymbol{S}|\mathbf{e}^\dagger\boldsymbol{D} + \boldsymbol{D}\mathbf{e}|\boldsymbol{S}\rangle + \sigma^2\langle \boldsymbol{S}|\mathbf{e}^\dagger\boldsymbol{D}\mathbf{e}|\boldsymbol{S}\rangle + \sigma\langle \boldsymbol{S}|\boldsymbol{K}|\boldsymbol{I}\rangle, \tag{6.14}$$

which is essential for anomalies in paramagnetic resonance spectra from modulated phases.

In practice, we analyze anomalies in terms of quantum numbers M and m. The Zeeman and hyperfine anomalies expressed by the first and last terms in (6.14) are determined by $-g_n\beta MB_o$ and K_nMm, respectively, where the spin is linearly modulated by σ at a binary transition, exhibiting symmetrical and antisymmetrical fluctuations. On the other hand, the fine structure term of D is linear and quadratic, characterized by σ and σ^2, respectively.

For the hyperfine term in (6.14), we have

$$\begin{aligned} K'^2_n &= \langle \boldsymbol{n}|\boldsymbol{\alpha}^\dagger\boldsymbol{K}^\dagger\boldsymbol{K}\,\boldsymbol{\alpha}|\boldsymbol{n}\rangle \\ &= \langle \boldsymbol{n}|\boldsymbol{K}^2|\boldsymbol{n}\rangle + \sigma\langle \boldsymbol{n}|\mathbf{e}^\dagger\boldsymbol{K}^2 + \boldsymbol{K}^2\mathbf{e}|\boldsymbol{n}\rangle + \sigma^2\langle \boldsymbol{n}|\mathbf{e}^\dagger\boldsymbol{K}\mathbf{e}|\boldsymbol{n}\rangle. \end{aligned}$$

Therefore, for a binary splitting characterized by $+\boldsymbol{\sigma}$ and $-\boldsymbol{\sigma}$, we have

$$K_n'(+)^2 - K_n'(-)^2 = 2\sigma\langle \boldsymbol{n}|\mathbf{e}^\dagger\boldsymbol{K}^2 + \boldsymbol{K}^2\mathbf{e}|\boldsymbol{n}\rangle$$

and the hyperfine anomaly

$$\Delta K_n' = K_n'(+) - K_n'(-) = \frac{2\sigma}{\overline{K}_n'}\langle \boldsymbol{n}|\mathbf{e}^\dagger\boldsymbol{K}^2 + \boldsymbol{K}^2\mathbf{e}|\boldsymbol{n}\rangle, \tag{6.15}$$

where $\overline{K}_n' = \frac{1}{2}(K_n'(+) + K_n'(-))$. $\Delta K_n' \propto \sigma$, which is analogous to Δg_n.

In contrast, from (6.15) the fine structure anomaly can be written as

$$\mathcal{H}_{1F} = \sigma\langle S_n|\mathbf{e}^\dagger\boldsymbol{D} + \boldsymbol{D}\mathbf{e}|S_n\rangle + \sigma^2\langle S_n|\mathbf{e}^\dagger\boldsymbol{D}\mathbf{e}|S_n\rangle,$$

hence, letting these coefficients be a_n and b_n, the modulation energy is

$$\Delta E_{\mathrm{F}}^{(1)} = (a_n\sigma + b_n\sigma^2)M^2;$$

the magnetic resonance conditions are

$$\hbar\omega = g_n\beta B_{\mathrm{o}} + (D_n \pm \Delta D_n)(2M + 1), \quad \text{where } \Delta D_n = a_n\sigma + b_n\sigma^2 \qquad (6.16a)$$

Characterized by $\Delta D_n \rightarrow -\Delta D_n$, a binary splitting of a resonance line is often recognized, signifying $\sigma \rightarrow -\sigma$, where the symmetric and antisymmetric modes appear at the critical temperature. In this case, these fine structure anomalies are simply proportional to amplitude mode σ_{A} and phase mode σ_{P}, respectively, hence

$$\Delta D_n(\mathrm{A}) = 2a_n\sigma_{\mathrm{A}} \quad \text{and} \quad \Delta D_n(\mathrm{P}) = 2a_n\sigma_{\mathrm{P}} \qquad (6.16b)$$

If otherwise, fine structure anomalies are described by (6.16*b*) that are contributed by the σ^2 term as well, showing a complex shape. Figure 6.4 illustrates these cases, where the fluctuations are determined by a sinusoidal phase ϕ; $\sigma_{\mathrm{A}} \propto \cos\phi$, $\sigma_{\mathrm{P}} \propto \sin\phi$ but generally proportional to $a\cos\phi + b\cos^2\phi$. The illustration made on first derivatives in the figure was just a matter of technical convenience.

Experimentally, it is important to realize that such an anomaly as (6.16*b*) proportional to σ is significant for domain structure in crystals below T_c to be identified by inversion of $\sigma_{\mathrm{P}} \rightarrow -\sigma_{\mathrm{P}}$ consistent with lattice symmetry, as shown by the examples in section 6.3. In addition, σ_{A} is signified by energy conversion to the lattice, responsible for its thermal decay, as discussed in section 4.2.3 of chapter 4.

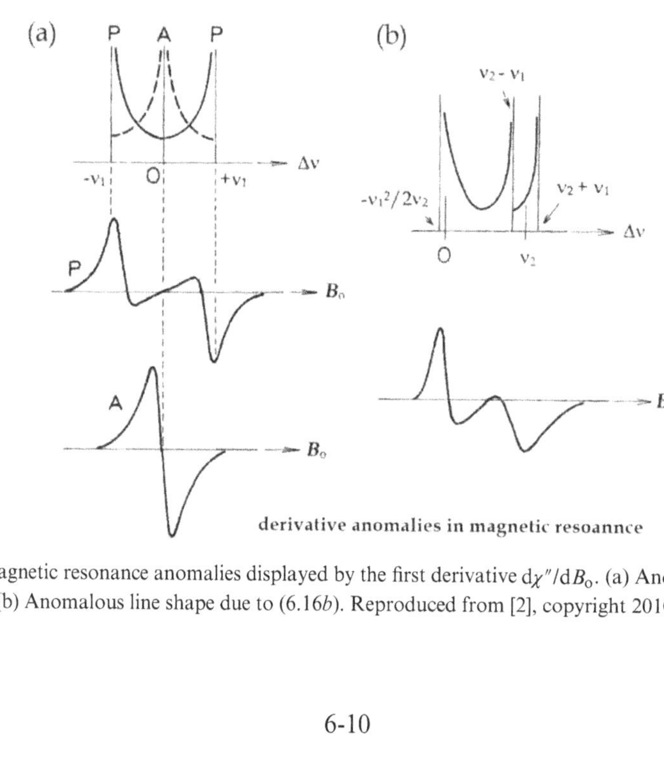

Figure 6.4. Magnetic resonance anomalies displayed by the first derivative $\mathrm{d}\chi''/\mathrm{d}B_{\mathrm{o}}$. (a) Anomalous line shape due to (6.15). (b) Anomalous line shape due to (6.16*b*). Reproduced from [2], copyright 2010. With permission of Springer.

6.3 Examples of transition anomalies

6.3.1 Mn^{2+} spectra in TSCC

The ferroelectric phase transition in TSCC crystals at $T_c = 130$ K was thoroughly investigated with paramagnetic impurities, Mn^{2+}, Cr^{3+}, Fe^{3+}, VO^{2+}, etc. Trivalent impurities showed complex spectra associated with charge defects, however, there are no such problems with divalent ions. Mn^{2+} ions are particularly useful in TSCC, because simple spectral analyses can be performed. Figure 6.5(a) shows a Mn^{2+} spectrum with B_o applied in the *bc* plane, where two lattice sites in the unit cell give identical spectra at the critical temperature. The *g*-tensor and the hyperfine tensor of a ^{55}Mn nucleus ($I = 5/2$) were found to be isotropic, while in all directions the spectra were dominated by the fine structure, as shown in figure 6.5(b) [2]. Figure 6.6(a) shows a solid curve of observed D in the *bc* plane for $T > T_c$, which is split into two broken curves for $T < T_c$, exhibiting anomalous lines in *ferroelectric domains*. Figure 6.6(b) shows changing line shape of resonance lines of 9.2 and 35 GHz microwaves for $\Delta M = 1$ and $\Delta m = 0$, in comparison to decreasing temperature [3]. It is noted that the lineshape and splitting are distinctively different

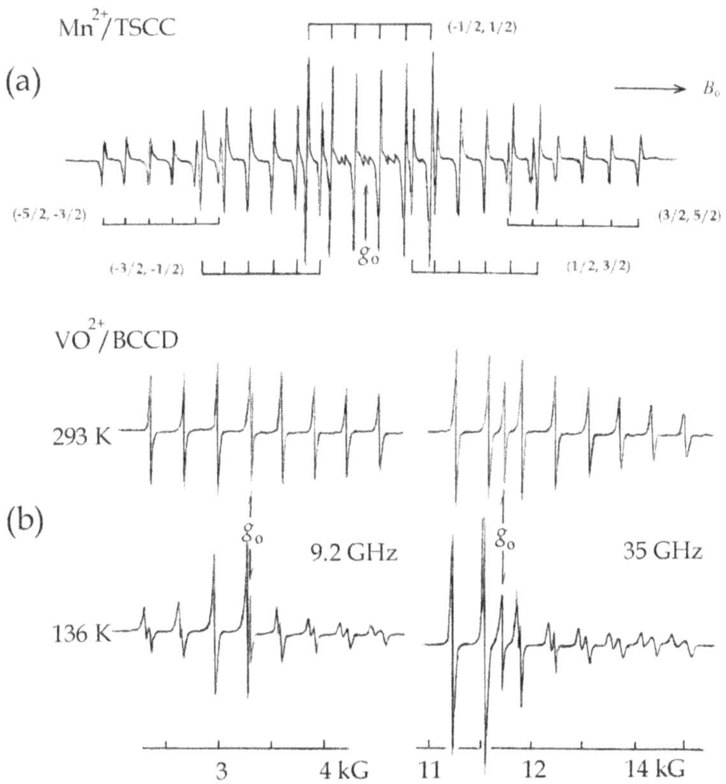

Figure 6.5. (a) A typical Mn^{2+} spectrum from a TSCC crystal. Reproduced with permission from [3]. (b) Typical VO^{2+} spectra from BCCD crystals, observed by microwaves of 9.2 and 35 GHz. Reproduced with permission from [4].

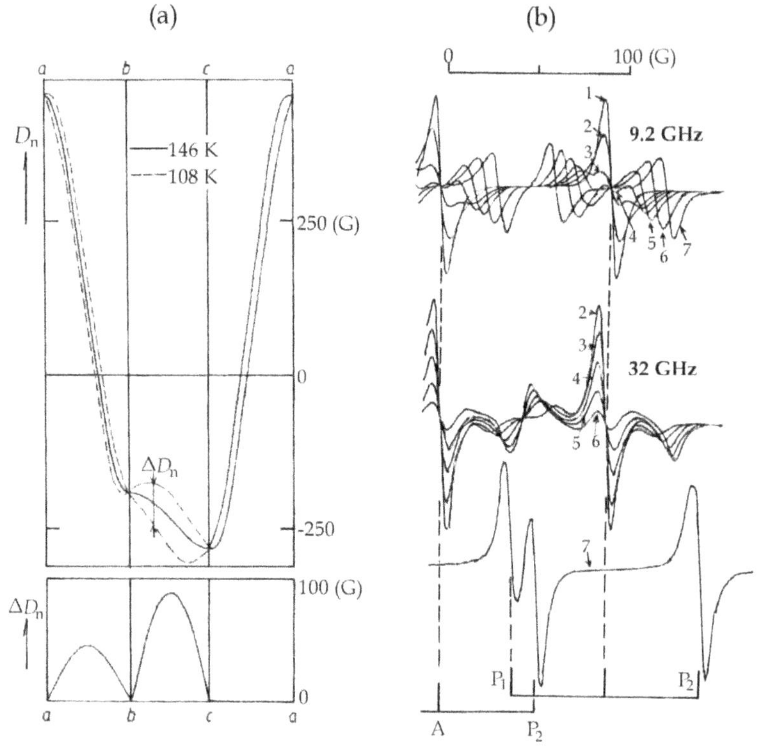

Figure 6.6. (a) An angular dependence of the fine-structure constant D_n of Mn^{2+} spectra in TSCC. (b) Critical anomalies in ^{55}Mn-hyperfine lines. Here, the splitting to domain lines exhibits anomalous shapes, becoming normal splitting with decreasing temperature, as indicated by 1, 2, … and 7. Reproduced with permission from [3].

when observed in different timescales, and the pair of domain lines P are stable, while the decaying central A-line is temperature-dependent, clearly indicating energy transfer to the lattice.

6.3.2 Mn^{2+} spectra in BCCD

Crystals of *betaine calcium chloride dihydrate* (BCCD) exhibit sequential phase transitions, as illustrated in figure 6.7(c), showing commensurate and incommensurate phases, as indicated by C and I, where soft modes were observed at thresholds of some transitions. Mn^{2+} impurities substituted for Ca^{2+} ions showed spectra to distinguish these phases, which, however, were too complex for complete analysis. Nevertheless, aided by a large fine structure, Mn^{2+} ions in a large crystal field exhibit both allowed and forbidden lines in BCCD. However, spectra along symmetry axes showed only allowed transitions; furthermore, corresponding to maximum fine structure, transition anomalies were clearly resolved on hyperfine lines 5/2, with fluctuations of $a\sigma + b\sigma^2$ type, as shown in figure 6.7(b).

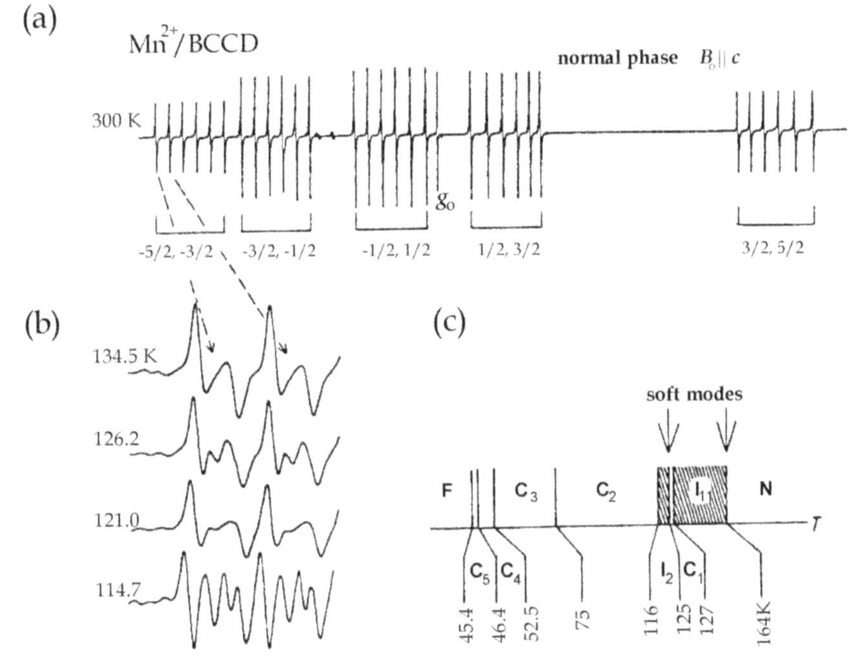

Figure 6.7. (a) A Mn^{2+} spectrum is the normal phase of BCCD for $B_0 \| c$. Reproduced with permission from [3]. (b) Anomalous splitting of the lowest hyperfine lines in Mn^{2+} spectra in BCCD, observed with decreasing temperature. (c) Sequential phase transitions in BCCD crystals. N, I, C and F indicate the normal, incommensurate, commensurate and ferroelectric phases, respectively. Reproduced with permssion from [4].

6.3.3 VO^{2+} spectra in BCCD

VO^{2+} ions substituted for Ca^{2+} in a BCCD crystal show simple spectra without fine structure, dominated by the hyperfine structure of ^{51}V nucleus ($I = 7/2$). Eight hyperfine lines observed with microwaves at 9.2 and 35 GHz exhibited different shapes and separations, which were analyzed with (6.16*a*) and (6.16*b*). These g_n and K_n tensors are both anisotropic, as shown in figure 6.8(a) [4], and anomalies ΔK_n can be explained by (6.15). Figure 6.8(b) shows observed temperature dependence of a hyperfine line.

6.3.4 Comments on the temperature-dependence of critical spectra

It is noted from above examples that the critical splittings showed broader lineshapes than predicted by *sinusoidal* distributions, particularly prominent at temperatures close to T_c, as shown in the VO^{2+} spectra of figure 6.8(b). Such distributions of A and P modes should be attributed to the *cnoidal potential*, referring to the soliton theory, which, however, change back to the sinusoidal P with decreasing temperature, consistent with the above analysis, indicating that the order variable is given as the *Bargmann's solution*, $\sigma = \sigma_0 \exp i\phi$, where the amplitude σ_0 and phase ϕ are finite and cnoidal, respectively (see chapter 10). Thus, the effect of

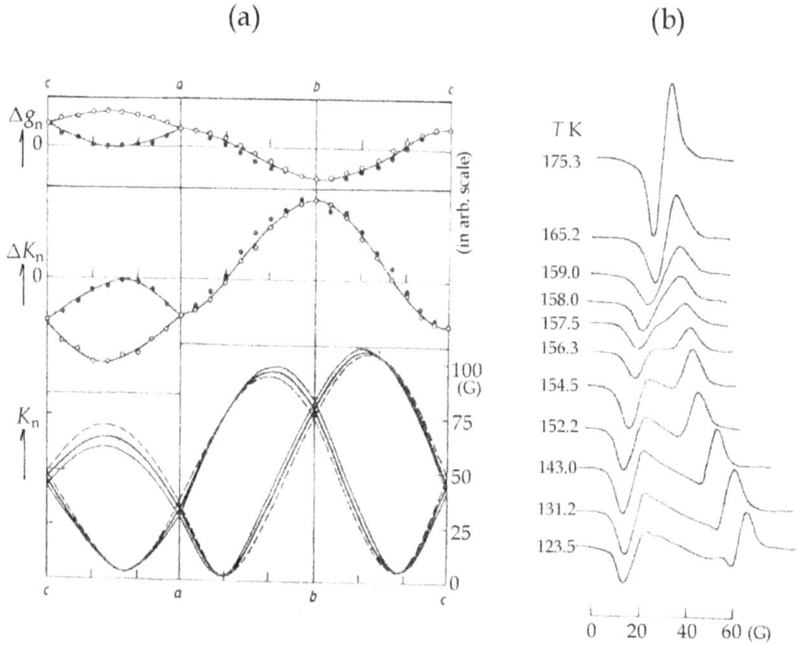

Figure 6.8. (a) Angular dependences of Δg_n, ΔK_n and K_n in VO^{2+} spectra from BCCD crystals at 130 K. Ref. [5]. (b) Variable line-shapes of a ^{51}V hyperfine line with decreasing temperature.

nonlinearity, while reversible at temperatures close to T_c, is evident in all magnetic resonance spectra, which is noticeably consistent with scattering results.

Further, as inferred from Landau's analysis in section 3.2, the critical anomalies are observed in different spectra, as shown in examples in section 6.3; those recorded at 9.2 and 32 GHz microwaves showed significantly different line-shape, depending on timescales of experiments.

References

[1] Bloch F 1946 *Phys. Rev.* **70** 460
[2] Fujimoto M 2010 *Thermodynamics of Crystalline States* (Berlin: Springer)
[3] Fujimoto M, Jerzak S and Windsch W 1986 *Phys. Rev.* B **34** 1668
[4] Fujimoto M and Kotake Y 1989 *J. Chem. Phys.* **90** 532
[5] Fujimoto M and Kotake Y 1989 *Chem. Phys.* **91** 6671

Chapter 7

Soft modes of lattice displacements

Exhibiting structural changes, *soft modes* observed in dielectric experiments are characterized by temperature-dependent frequencies and thermal relaxation. In dielectric crystals, soft modes were detected by applying time-dependent electric field at the Brillouin-zone center, whereas in cell-doubling perovskite crystals soft modes were detected with time-dependent neutron inelastic scatterings at zone boundaries. Measured as a function of temperature under constant-pressure conditions, observed frequencies are naturally temperature-dependent under constant volume conditions. Showing an isothermal profile of collective displacements, soft modes in dielectric crystals represent the response from the adiabatic electric potential, which can be analyzed with respect to the Born–Huang principle. Reviewing observed soft modes from a variety of systems, we discuss in this chapter symmetry change, which is a significant feature of order variables, representing critical anomalies during structural phase transitions in crystals.

In the second edition, we discuss the fact that observed soft-modes represent dynamic lattice displacements, as the significant evidence.

7.1 The Lyddane–Sachs–Teller relation in dielectric crystals

In ionic crystals, the spontaneous polarization $\boldsymbol{P}(\boldsymbol{r}, t)$ is originated from charge displacements in the lattice, which is related to the lattice excitation. Detected with an applied electric field $\boldsymbol{E} = \boldsymbol{E}_\mathrm{o} \exp(-i\omega t)$, where $\boldsymbol{E}_\mathrm{o}$ and ω are the amplitude and frequency, respectively, the polarization $\boldsymbol{P}(\boldsymbol{r}, t)$ occurs not only by an ionic polarization $\alpha\boldsymbol{E}$ where α is the *ionic polarizability*, but is also associated with an ionic mass displacement $\boldsymbol{u}(\boldsymbol{r}, t)$, representing the *piezoelectric* property of dielectric crystals. It should be realized that such intrinsic quantities as $\boldsymbol{P}(\boldsymbol{r}, t)$ and $\mathbf{u}(\boldsymbol{r}, t)$ represent collective motion of microscopic variables in *finite amplitudes*, which are expressed at low frequencies by continuous functions of space-time in crystals.

doi:10.1088/978-0-7503-2572-1ch7

Following Elliott and Gibson [1], we write the relations

$$\boldsymbol{P} = a\boldsymbol{u} + b\boldsymbol{E}, \tag{7.1}$$

which was proposed originally by Born and Huang [2], and

$$\frac{\partial^2 \boldsymbol{u}}{\partial t^2} = a'\boldsymbol{u} + b'\boldsymbol{E} \tag{7.2}$$

describing the lattice displacement $\boldsymbol{u}$ due to a sufficiently weak $\boldsymbol{E}$, where a, b, a' and b' are constant coefficients. It is significant that virtually all dielectric crystals are *electro-elastic*, because the polarization due to charge displacements naturally strains the structure; in contrast, the absence of such an elastic effect makes gaseous and liquid dielectrics different from solids. Therefore, non-zero coefficients a and a' are independent parameters that are significant for crystalline dielectrics.

The vector $\boldsymbol{u}$ for a mass displacement can generally be either *irrotational* or *rotational*, signified by curl $\boldsymbol{u} = 0$ or curl $\boldsymbol{u} \neq 0$, respectively. The former applies to the present case of *laminar flow*, namely characterized by div $\boldsymbol{u} \neq 0$. Furthermore, when $\boldsymbol{E}$ is proportional to $\exp(-i\omega t)$, both $\boldsymbol{P}$ and $\boldsymbol{u}$ should be proportional to $\exp i(\boldsymbol{q}\cdot\boldsymbol{r} - \omega t)$ in crystals, where $\boldsymbol{q}$ is the wavevector determined by $\omega = v|\boldsymbol{q}|$, where v is the speed of propagation. Therefore, expressing $\boldsymbol{P} = \boldsymbol{P}_o \exp(-i\omega t)$ and $\boldsymbol{u} = \boldsymbol{u}_o \exp(-i\omega t)$, the amplitudes $\boldsymbol{P}_o$ and $\boldsymbol{u}_o$ are such spatial functions as proportional to $\exp i\boldsymbol{q}\cdot\boldsymbol{r}$. And (7.1) and (7.2) can be written as

$$\boldsymbol{P}_o = a\boldsymbol{u}_o + b\boldsymbol{E}_o \quad \text{and} \quad -\omega^2\boldsymbol{u}_o = a'\boldsymbol{u}_o + b'\boldsymbol{E}_o$$

Eliminating $\boldsymbol{u}_o$ from these, we obtain

$$\boldsymbol{P}_o = \left(b + \frac{ab'}{-a - \omega^2}\right)\boldsymbol{E}_o.$$

To derive the dielectric response function $\varepsilon(\omega)$, we combine this with the standard formula

$$\boldsymbol{D}_o = \varepsilon(\omega)\boldsymbol{E}_o = \varepsilon_o\boldsymbol{E}_o + \boldsymbol{P},$$

from which we obtain

$$\varepsilon(\omega) - \varepsilon_o = b + \frac{ab'}{-a' - \omega^2}.$$

Writing $\varepsilon(0)$ and $\varepsilon(\infty)$ for the dielectric function at $\omega = 0$ and $\omega = \infty$, respectively, we have relations $a' = -\omega_o^2$, $b = \varepsilon(\infty) - \varepsilon_o$ and $ab'/\omega_o^2 = \varepsilon(0) - \varepsilon(\infty)$, so that

$$\varepsilon(\omega) = \varepsilon(\infty) + \frac{\varepsilon(0) - \varepsilon(\infty)}{1 - (\omega/\omega_o)^2}. \tag{7.3}$$

Next, we consider that the field $\boldsymbol{E}$ is composed of longitudinal and transverse components, i.e. $\boldsymbol{E} = \boldsymbol{E}_L + \boldsymbol{E}_T$, hence $\boldsymbol{u} = \boldsymbol{u}_L + \boldsymbol{u}_T$. For components of $\boldsymbol{u}$, we have spatial relations

$$i\boldsymbol{q} \times \boldsymbol{u}_{oL} = 0 \quad \text{or} \quad \boldsymbol{q} \| \boldsymbol{u}_{oL} \text{ for the longitudinal component}$$

and

$$i\boldsymbol{q} \cdot \boldsymbol{u}_{oT} = 0 \quad \text{or} \quad \boldsymbol{q} \perp \boldsymbol{u}_{oT} \text{ for the transverse component.}$$

While keeping the relations $\boldsymbol{E}_L \| \boldsymbol{u}_{oL}$ and $\boldsymbol{E}_T \| \boldsymbol{u}_{oT}$, the formula div $\boldsymbol{D} = 0$, expressing the absence of charge, is applied to insulating crystals, where we have a relation div $\boldsymbol{P} = -\varepsilon_o$div $\boldsymbol{E}$. For a transverse wave, we have then div $\boldsymbol{E}_T = 0$ and hence div $\boldsymbol{u}_T = 0$, whereas for longitudinal components we have the relations

$$\text{div}\, \boldsymbol{u}_L = -\frac{\varepsilon_o + b}{a} \text{div}\, \boldsymbol{E}_L \neq 0 \quad \text{and} \quad -\omega_L^2 \boldsymbol{u}_L = a' \boldsymbol{u}_L + b' \boldsymbol{E}_L.$$

Therefore, equations of motion for the two components are written as

$$\frac{\partial^2 \boldsymbol{u}_L}{\partial t^2} = \left(a' + \frac{ab'}{\varepsilon_o + b} \right) \boldsymbol{u}_L = -\frac{\varepsilon(0)}{\varepsilon(\infty)} \omega_o^2 \boldsymbol{u}_L$$

and

$$\frac{\partial^2 \boldsymbol{u}_T}{\partial t^2} = a' \boldsymbol{u}_T = -\omega_o^2 \boldsymbol{u}_T,$$

for which we have

$$\omega_{oL} = \omega_o \sqrt{\frac{\varepsilon(0)}{\varepsilon(\infty)}} \quad \text{and} \quad \omega_{oT} = \omega_o,$$

therefore the following applies.

Theorem:

$$\frac{\omega_{oL}^2}{\omega_{oT}^2} = \frac{\varepsilon(0)}{\varepsilon(\infty)}, \tag{7.4}$$

which is known as the Lyddane–Sachs–Teller (LST) relation.

Figure 7.1 shows the dielectric response function $\varepsilon(\omega)$ plotted against ω/ω_{oT}, which is signified by the empty gap between $\varepsilon(0)$ and $\varepsilon(\infty)$, the singularity at $\omega = \omega_{oT}$ and $\varepsilon(\omega_{oL}) = 0$. Further, we can confirm that the LST relation (7.4) can be compatible with the Landau theory. Denoting the transition temperature by T_o in mean-field accuracy, we write the Curie–Weiss law for the susceptibility as $\varepsilon_<(0) \approx \varepsilon_o \chi_<(0)$ as $\chi_<(0) = C/(T_o - T)$ for $T < T_o$, and obtain $\omega_{oT}^2/\omega_{oL}^2 \propto (T_o - T)$, which should give rise to the *soft-mode frequency* $\tilde{\omega}_o^2 \propto (T_o - T)$ of the *lattice displacement* $\boldsymbol{u}$ in the mean-field approximation.

It is important to realize that the above derivation of LST formula was derived for a homogeneous dielectrics, but in crystals both $\boldsymbol{P}_o$ and $\boldsymbol{u}_o$ proportional to exp $i\boldsymbol{q}.\, \boldsymbol{r}$ are independent of $\boldsymbol{r}$, if $\boldsymbol{q} = (0, \pm \boldsymbol{G}/2)$ are multiplied by $\boldsymbol{r} =$ (any position, $\pm\pi/|\boldsymbol{G}|$),

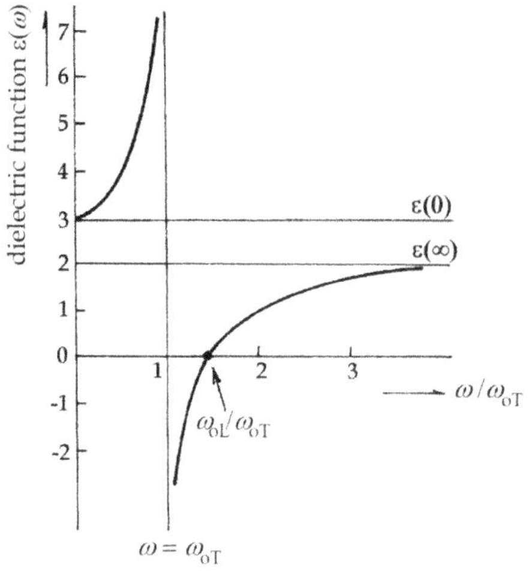

Figure 7.1. Dispersion curve for a dielectric function $\varepsilon(\omega)$. There is a forbidden frequency gap between $\varepsilon(0)$ and $\varepsilon(\infty)$. Reprodued from [3], copyright 2010. With permission from Springer.

respectively. Accordingly, the soft mode cannot be specified by the frequency only, but signified by $\boldsymbol{q} = 0$ or $\boldsymbol{q} = \pm\boldsymbol{G}/2$ as well, representing *Brillouin zone center or boundaries*, respectively. Dielectric experiments of soft modes can therefore be performed at zone center $\boldsymbol{G} = 0$, because we need to apply an external field $\boldsymbol{E}$; on the other hand neutron *inelastic scatterings* [2] can be studied from scatterings $\Delta\boldsymbol{K} = \pm\boldsymbol{G}/2$ at the zone boundaries.

At this point, we notice that the origin $\boldsymbol{u} = 0$ cannot be considered in equations (7.1) and (7.2) quantum mechanically. However, if we consider $\boldsymbol{E}_o = 0$ in (7.2), we have $\partial^2\boldsymbol{u}/\partial t^2 = a'\boldsymbol{u} = -\omega_0^2\boldsymbol{u}$ that is mathematically identical to a harmonic oscillator where the spring force is given by $-\omega_0^2\boldsymbol{u}$. Not necessarily representing lattice vibrations, in this case the displacement $\boldsymbol{u}$ signifies the properties of fluctuations in the lattice indistinguishable from phonons.

7.2 Soft modes in perovskite oxides

Soft modes in dielectric crystals in isothermal processes are considered for the response function of collective order variables $\boldsymbol{\sigma}$ from the adiabatic potential originating from an applied field and or internal field of lattice correlations. Therefore, soft modes should be originated from the critical fluctuations of $\boldsymbol{\sigma}_A$ and $\boldsymbol{\sigma}_P$ at the transition, representing *initial* conditions for nonlinear variations of the condensate composed of $\boldsymbol{\sigma}$ and $\boldsymbol{u}$. Generated by a weak adiabatic potential, soft modes represent developing lattice displacements in propagation, being observed at *zone-center* in dielectric crystals and at *zone-boundaries* in perovskite crystals. In the latter experiments, the intensity of neutron inelastic scatterings in angular setting can be studied as a function of temperature.

In general, soft modes of σ_A and σ_P are both characterized by temperature-dependent frequencies as determined approximately by (7.14) and damping constant associated with spatial inversion $\boldsymbol{q} \rightleftarrows -\boldsymbol{q}$ of displacements. Signified by these two components, mesoscopic order variable $\boldsymbol{\sigma}(\phi)$ exhibits a vector quantity determined by the phase ϕ; indicating *dipolar* properties in dielectric crystals.

For crystals of $PbTiO_3$ [4], softening transverse optic, transversal acoustic and longitudinal acoustic modes were reported as observed by the neutron inelastic scattering experiments at paraelectric temperature 510°C that is slightly higher than the critical point, as illustrated in figure 7.2(a). Here a transverse mode shows a well-defined spectrum that is modified by the interaction with acoustic mode.

On the other hand, a typical temperature-dependent frequency was also detected by Raman scatterings [5] in the ferroelectric phase, as shown in figure 7.3(b). In $PbTiO_3$, it is particularly significant that the ferroelectric phase is *Raman-active* because of piezoelectric properties, where elastic displacements are essential for soft modes to be detected as in figure 7.2(c).

Figure 7.3 summarizes the result of magnetic resonance samplings on $SrTiO_3$ and $LaTiO_3$ crystals [6], showing a *soft shear mode* is responsible as identified from Fe^{3+} *paramagnetic resonance spectra*. Figure 7.3(a) is a model of the lattice structure, where the central Ti^{4+} ion is replaced by a paramagnetic Fe^{3+} ion, accompanying oxygen vacancies in either at A or at B positions as indicated in the figure. In the

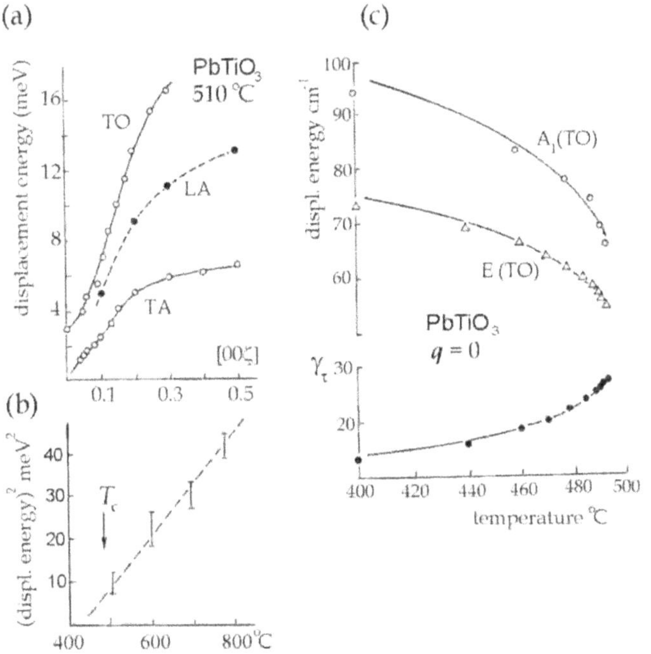

Figure 7.2. Soft mode spectra from ferroelectric $PbTiO_3$ crystals. (a) Neutron inelastic scattering in the paraelectric phase. (b) Soft-mode spectra in the ferroelectric (piezoelectric) phase. (c) Soft-mode spectra and temperature dependence of γ_τ in the ferroelectric phase. Reproduced with permission from [4].

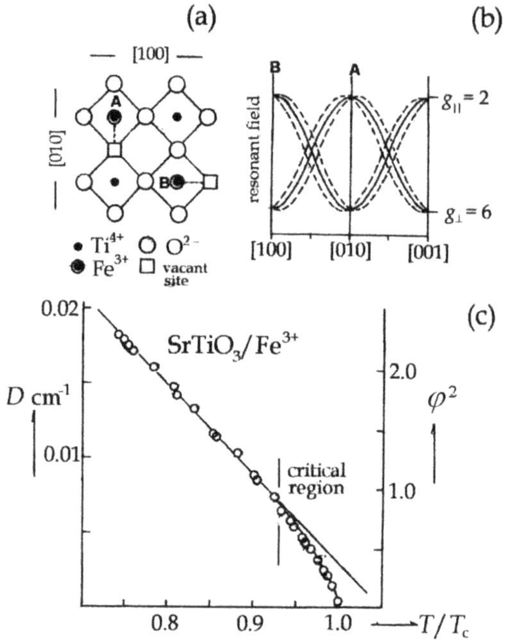

Figure 7.3. Soft-mode spectra of the zone-boundary transitions of $SrTiO_3$ crystals. (a) Structural model for zone-boundary fluctuations. (b) Variations in a shear angle φ measured with the fine structure parameter D of Fe^{3+} spectra. (c) Soft-mode spectra determined by φ^2 versus temperature. Reproduced with permission from [6].

magnetic resonance spectra of Fe^{3+} impurity ions, the fine-structure parameter D is a varying function of angle of rotation of an applied magnetic field, as plotted in figure 7.3(b). Here, the solid curves for $T > T_c$ were found to split into two broken curves for $T < T_c$, similar to Mn^{2+} hyperfine lines in figure 6.6(b) of chapter 6. The angular dependence of the spectrum was analyzed by $D \propto \varphi^2$ in magnetic resonance spectra [7], where φ is the angle of rotation around the axis [001]. Shown in figure 7.3(c), a temperature dependence of the angular displacement in a typical form $\varphi^2 \approx T_c - T$ in reasonably good accuracy. In this case, we can logically assume that the angle φ represents rotational shear strains in the lattice, as observed in the paramagnetic resonance spectrum of Fe^{3+} probes.

7.3 Lattice response to collective pseudopins

7.3.1 Energy dissipation of soft modes

In the condensate model, the oscillatory motion described by the wave equation for displacements $\boldsymbol{u}$ should be driven by an adiabatic potential ΔU, hence the dielectric function $\varepsilon(\omega)$ is referred to as the response function. The function ΔU can generally be expressed for a condensate by a power series expanded with respect to components of displacement vectors $(\boldsymbol{u}_L, \boldsymbol{u}_T)$ as in a general form

$$\Delta U = V^{(1)}(u_\mathrm{L} + u_\mathrm{T}) + V^{(2)}\left(u_\mathrm{L}^2 + u_\mathrm{T}^2\right) + V^{(3)}\left(u_\mathrm{L}u_\mathrm{T}^2 + u_\mathrm{T}u_\mathrm{L}^2\right) + V^{(4)}\left(u_\mathrm{L}^4 + u_\mathrm{T}^4\right) + \cdots, \tag{7.5}$$

where $V^{(1)}$, $V^{(2)}$, $\cdots$ are arbitrary coefficients, and the expansion is written for a case characterized by local symmetry. In the thermodynamic environment however, it should be reduced to a form compatible with the symmetry of equilibrium structure, signified by the absence of odd-power terms. Accordingly, we have

$$\Delta U_\mathrm{eq} = \Delta U_\mathrm{L} + \Delta U_\mathrm{T},$$

where

$$\Delta U_\mathrm{L} = V^{(2)}u_\mathrm{L}^2 + V^{(4)}u_\mathrm{L}^4 + \cdots \quad \text{and} \quad \Delta U_\mathrm{T} = V^{(2)}u_\mathrm{T}^2 + V^{(4)}u_\mathrm{T}^4 + \cdots. \tag{7.6}$$

By virtue of Born–Huang's principle, all the odd-power terms in (7.5) are assumed to be responsible for energy transfer to the surroundings, since they are signified by off-diagonal elements with respect to discrete phonon states. Expressing a displacement component by $u_{L,T}(\boldsymbol{q}, \Delta\omega)$, inelastic scatterings of phonons can take place between $\langle \boldsymbol{K}', \Omega'|$ and $|\boldsymbol{K}, \Omega\rangle$ via off diagonal matrix elements with respect to a finite $\boldsymbol{q}$, e.g. $\langle \boldsymbol{K}', \Omega'|\Delta U_{L,T}|\boldsymbol{K}, \Omega\rangle = V^{(1)}\langle \boldsymbol{K}'|u_{L,T}(\boldsymbol{q})|\boldsymbol{K}\rangle \exp i\Delta\omega \cdot t$, assuming conservation law $\boldsymbol{K}' - \boldsymbol{K} = \pm\boldsymbol{q}$ and $\Omega' - \Omega = \mp\Delta\omega$. For such scattering processes depending on $\boldsymbol{q}$ determined with lattice symmetry, the probability can be written as

$$P = 2\left(V_{L,T}^{(1)}\right)^2 \langle \boldsymbol{K}'|\boldsymbol{u}_{L,T}(\boldsymbol{q})|\boldsymbol{K}\rangle^2 \frac{1 - \cos(t\Delta\omega)}{(t\Delta\omega)^2}, \tag{7.7}$$

which should be time-averaged for a soft mode signified by a relaxation time τ as expressed by $\langle P\rangle_\tau$, determining the damping constant in this transfer process by $\gamma_\tau \propto 1/\langle P_\tau\rangle$ in mesoscopic timescale τ.

Such an off-diagonal element originating from an *odd-power potential* can be attributed to two different temperatures T' and T, using statistical equipartition theorem. Therefore, the probability $\langle P_\tau\rangle$ and damping constant γ_τ determined by (7.7) can be assigned to a transition characterized by a temperature change $\Delta T = T' - T$.

Assuming $\Delta U_\mathrm{eq} \approx \frac{1}{2}m\omega_\mathrm{o}^2 u_\mathrm{T}^2$ for a small driving potential for transverse displacement wave $u_\mathrm{T}(x, t)$, we can write a forced oscillator equation

$$m\left(\frac{\mathrm{d}^2 u_\mathrm{T}}{\mathrm{d}t^2} + \gamma_\tau \frac{\mathrm{d}u_\mathrm{T}}{\mathrm{d}t} + \omega_\mathrm{o}^2 u_\mathrm{T}\right) = F_\mathrm{T}\exp(-i\tilde{\omega}t), \tag{7.8}$$

where $F_\mathrm{T}\exp(-i\tilde{\omega}t)$ is either a *time-dependent internal force*, or *externally applied*. Writing the resonance frequency $\tilde{\omega} = \omega_\mathrm{o}$ for the transverse displacement u_T, the parameter γ_τ is the damping constant to describe the process of energy dissipation. Equation (7.8) deals with the lattice response in terms of ω_o and γ_τ, providing spectral data for equations (7.1) and (7.2).

Figure 7.4 shows typical parabolic temperature dependence of ω_o^2 versus T_c [8], however, showing a slight discrepancy from the mean-field theory at temperatures

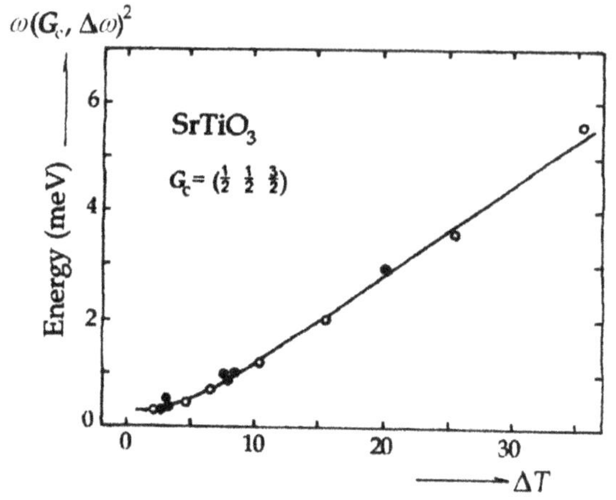

Figure 7.4. A plot of ω_0^2 as a function of $T_c - T$ obtained from neutron inelastic scattering experiments on $SrTiO_3$. Reproduced with permission from [8].

close to the critical T_c. The susceptibility is measured for a weak field F_T, however yielding useful data in practice, where deviations from the parabolic curve can be attributed significantly to nonlinearity of soft modes.

7.3.2 Susceptibility analysis of soft modes

Against an effective driving force F_T, (7.8) has a steady solution for a sufficiently small F_T. Therefore, studies of susceptibilities with a small F_T provide useful information for the *lattice displacement.* Experimentally, the electric field $E_0 \exp(-i\tilde{\omega}t)$ with a weak amplitude E_0 is applied externally to study dielectric soft modes. Although rather unclear in neutron experiments, the presence of a weak internal field F_T is evident in results of neutron scatterings[note1], which are analyzed conveniently with susceptibility formula.

Assuming $u_T = u_{oT} \exp(-i\tilde{\omega}t)$ for a transverse displacement, (7.8) can be written as

$$m\left(-\tilde{\omega}^2 - i\tilde{\omega}\,\gamma_\tau + \omega_0^2\right)u_{oT} = F_T,$$

from which we can define the complex susceptibility

$$\chi(\tilde{\omega}) = \frac{mu_{oT}}{F_T} = \frac{1}{\omega_0^2 - \tilde{\omega}^2 - i\tilde{\omega}\gamma_\tau} \qquad (7.9a)$$

Writing this as $\chi(\tilde{\omega}) = \chi'(\tilde{\omega}) - i\chi''(\tilde{\omega})$, the real and imaginary parts of (7.9*a*) are expressed as

$$\chi'(\tilde{\omega}) = \frac{\omega_0^2 - \tilde{\omega}^2}{\left(\omega_0^2 - \tilde{\omega}^2\right)^2 + \gamma_\tau^2\tilde{\omega}^2} \quad \text{and} \quad \chi''(\tilde{\omega}) = \frac{\tilde{\omega}\gamma_\tau}{\left(\omega_0^2 - \tilde{\omega}^2\right)^2 + \tilde{\omega}^2\gamma_\tau^2}, \qquad (7.9b)$$

[note1] The internal field F_T represents the *soliton potential energy* (chapter 12), verified theoretically as consistent with those arguments in section 7.4.

respectively. Figure 7.5 shows these functions plotted against $\omega_0 - \tilde{\omega} = \Delta\tilde{\omega}$, where $\chi'(\tilde{\omega})$ is dispersive and $\chi''(\tilde{\omega})$ is a dumbbell-shaped symmetric curve, for which the relation $\gamma < |\omega_0^2 - \tilde{\omega}^2|/\tilde{\omega}^2 \approx 2\Delta\tilde{\omega}$ is assumed. If assumed otherwise, $\chi''(\tilde{\omega}) \sim 1/\tilde{\omega}\,\gamma_\tau$ for $\gamma_\tau > 2\Delta\tilde{\omega}$, which shows no visible peak, and is called *overdamped* as against the *underdamped* case in the former. In contrast to the dispersion $\chi'(\tilde{\omega})$, the imaginary part $\chi''(\tilde{\omega})$ represents energy dissipation.

Experimentally, the characteristic frequency ω_0 of a soft mode decreases toward zero with decreasing temperature, hence equation (7.8) becomes dominated by the damping term, and is written as

$$m\left(\gamma_\tau \frac{du_T}{dt} + \omega_0^2 u_T\right) = F_T \exp(-i\tilde{\omega}t). \tag{7.10a}$$

Writing $\omega_0^2/\gamma_\tau = 1/\tau'$ and $F_T/m\gamma_\tau = F'$ for convenience, (7.10*a*) can be expressed as

$$\frac{du_T}{dt} + \frac{u_T}{\tau'} = F' \exp(-i\tilde{\omega}t), \tag{7.10b}$$

where τ' is the thermal *relaxation time*. Characterized by τ', the susceptibility formula, known as *Debye's relaxation*, can be written from (7.10*b*), that is

$$\chi_D(\tilde{\omega}) = \frac{1}{-i\tilde{\omega} + \tau'^{-1}} = \frac{\tau'}{1 + \tilde{\omega}^2\tau'^2} - \frac{i\tilde{\omega}\tau'}{1 + \tilde{\omega}^2\tau'^2} \tag{7.11}$$

Although derived as a response function from the lattice, Debye's susceptibility (7.11) is not necessarily for internal adiabatic potentials, but is often considered for

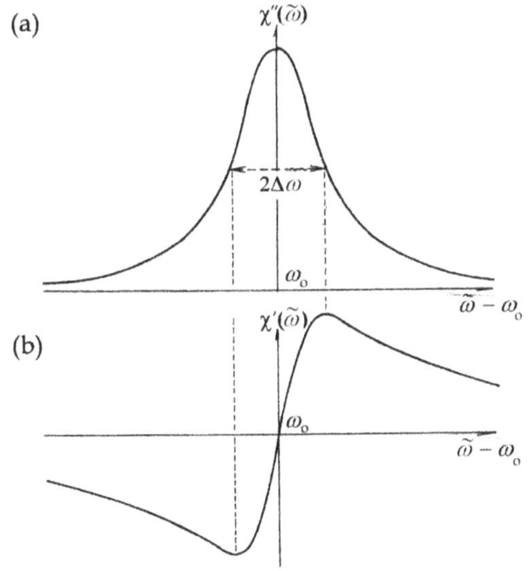

Figure 7.5. A complex susceptibility $\chi(\omega) = \chi'(\omega) - i\chi''(\omega)$. (a) The imaginary part $\chi''(\omega)$, (b) the real part $\chi'(\omega)$.

unavoidable imperfections as well. Figure 7.6 shows examples of dielectric functions observed in ferroelectric $BaTiO_3$ crystals by Petzelt and his coworkers [9], showing the presence of soft modes in the paraelectric phase above T_c.

7.3.3 Central peaks

In soft mode spectra observed by neutron inelastic scatterings at zone boundaries, a sharp absorption peak was found at $\omega \approx 0$. Figure 7.7 [2] shows examples of soft mode spectra observed by neutron scatterings, exhibiting a sharp line at near zero frequency. Known as a *central peak*, such an absorption can be analyzed with Debye's formula (7.11), but its unusually narrow width has been the subject of much speculation. Nevertheless, these central peaks can likely be attributed to extrinsic lattice imperfections in crystals.

Assuming that a soft mode $u_T(x, t)$ that coexists with unknown mode $v(x, t)$ near zero frequencies, we can write equations of motion

$$\frac{d^2u_T}{dt^2} + \gamma\frac{du_T}{dt} + c\gamma'\frac{dv}{dt} + \omega_o^2u_T = \frac{F}{m}\exp(-i\tilde{\omega}t)$$

and

$$\frac{dv}{dt} + \frac{v}{\tau} = F'\exp(-i\tilde{\omega}t),$$

to obtain a steady solution, where c is the coupling constant between u_T and v. We can thus derive the susceptibility for the soft mode u_T coupled with *Debye's relaxator* v as

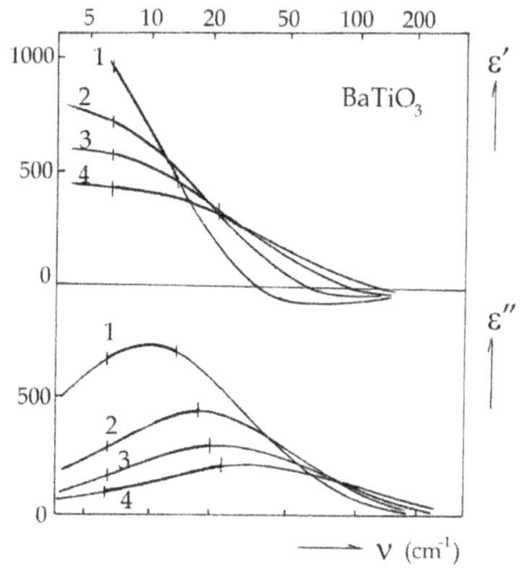

Figure 7.6. Dielectric spectra of $\varepsilon'(\omega)$ and $\varepsilon''(\omega)$ from $BaTiO_3$ measured with the backward-wave technique. Curves 1, 2, 3 and 4 were obtained at temperatures 474, 535, 585 and 667 K. Reproduced with permission from [9].

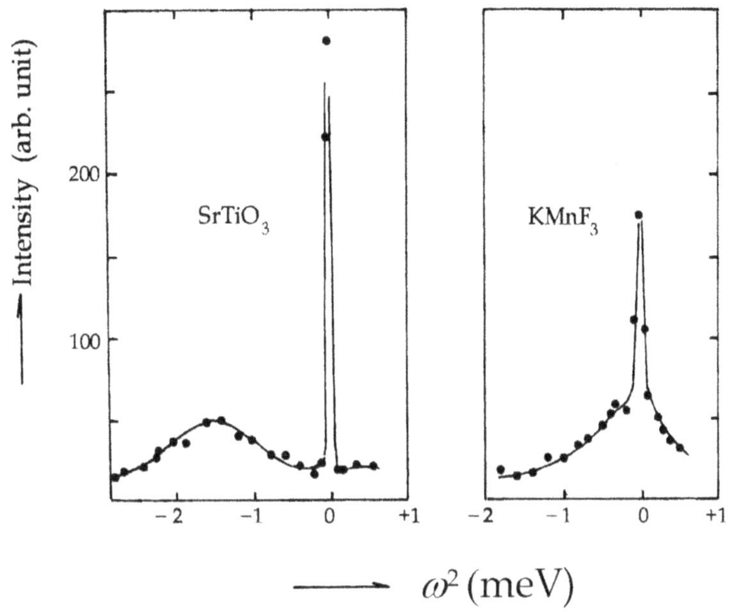

Figure 7.7. Soft-mode spectra from $SrTiO_3$ and $KMnF_3$, including sharp central peaks at $\omega = 0$. Perproduced from [2] with permission from American Physical Society.

$$\chi(\omega) = \frac{mu_{\text{oT}}}{F} = \left(\omega_0^2 - \tilde{\omega}^2 - i\tilde{\omega}\gamma - ic\gamma' F' \frac{\tilde{\omega}\tau}{1 - i\tilde{\omega}\tau}\right)^{-1},$$

which is nevertheless identical to (7.9*a*), if the coupling c is neglected.

However, these two modes are practically independent of each other, in which case the susceptibility function is expressed by the sum of resonant and relaxation contributions, namely

$$\chi(\tilde{\omega}) = \frac{A}{\omega_0^2 - \tilde{\omega}^2 - i\tilde{\omega}\gamma} + \frac{B\tau}{1 - i\tilde{\omega}\tau},$$

where A and B are constants for mixing modes; the second term is assigned to a central peak, arising from a point imperfection.

7.4 Temperature dependence of soft mode frequencies

Temperature-dependent frequencies of a soft mode are typically expressed approximately as $\omega_0^2 \propto T_c - T$, as shown in figure 7.2, which is adequate except for the critical region. Cowley discussed theoretically that it arises from the response function to an adiabatic potential in the lattice. Following [10], such a temperature dependence is described in this section as related to phonon scatterings in the thermodynamic environment.

First, we confirm that any volume change $\beta = \Delta V / V$ of a crystal should accompany a temperature change ΔT. We have the relation

$$\beta = \frac{\Delta V}{V} = -\kappa \Delta p = -\frac{\kappa\gamma}{V} U_{\text{vib}} = -\frac{\kappa\gamma \Delta T}{V},$$

where κ and γ are the *compressibility* and *Grüneisen's constant*, respectively, which are normally assumed to be temperature independent. Therefore, we have $\beta \propto \Delta T$ in any case of $\Delta V \neq 0$, and β is expected to appear in ordering processes in general, where ω_0 is temperature dependent. However, for finite temperature change, these constants vary nonlinearly as related to anisotropic ΔU, so that the temperature-dependence of β is not as simple as being proportional to ΔT.

For dielectric studies, the direction of transverse displacement u_T is parallel to an applied field $\boldsymbol{E}$. If $\boldsymbol{E} = 0$, the direction of u_T is arbitrary in equilibrium crystals. Expressing such unspecified u_T in complex form $u_{Ty} \pm iu_{Tz}$ with respect to the longitudinal x-axis, Cowley specified the adiabatic potential approximately as given by

$$\Delta U_C = V^{(3)} u_L^2 \left(u_{Ty} \pm iu_{Tz}\right) + V^{(4)} u_L^2 \left(u_{Ty}^2 + u_{Tz}^2\right) \tag{7.12}$$

that is contributed by fourth-order terms compatible with the mean-field theory.

In this case, the damping constant is also defined as complex, so that we can write $\gamma = \Gamma - i\Phi$, and for the transverse displacement, the steady solution can be specified by

$$u_T(\boldsymbol{q}, \Delta\omega) = \frac{F_\pm/m}{-\omega(\boldsymbol{q})^2 + \omega_0^2 - i\omega(\boldsymbol{q})(\Gamma - i\Phi)}.$$

Therefore, the characteristic frequency is shifted from ω_0 to ω $(\boldsymbol{k}, \Delta\omega)$ as given by

$$\omega(\boldsymbol{q}, \Delta\omega)^2 = \omega_0^2 + \omega(\boldsymbol{q})\Phi(\boldsymbol{q}, \Delta\omega), \tag{7.13}$$

where Φ can be calculated with ΔU_C in (7.12). Here, a possible contribution from the volume change $\Phi_0 = (\partial\omega(\boldsymbol{q})/\partial V)\Delta V$ is neglected, assuming $\Delta V \approx 0$. Therefore, the contribution from ΔU_C can be expressed as

$$\Phi(\boldsymbol{q}, \Delta\omega) = \Phi_1(\boldsymbol{q}) + \Phi_2(\boldsymbol{q}, \Delta\omega),$$

where

$$\Phi_1(\boldsymbol{q}) = \frac{\hbar}{N\omega(\boldsymbol{q})} \frac{(2n+1)V^{(4)} u_L^2}{2\omega'} \left\langle -\boldsymbol{K}, \boldsymbol{K} \middle| u_T^2 \middle| \boldsymbol{K}', -\boldsymbol{K}' \right\rangle,$$

where N is the total number of phonons, and n the number of scattered phonons $(\boldsymbol{K}', \omega')$. Here, using the equipartition theorem, we can set $\hbar\omega'(n + 1/2) = k_B T$, and obtain $\Phi_1(\boldsymbol{q}) \propto T$ that gives a temperature dependence to the frequency shift[note2].

[note2] Referring to soliton numbers, transitions $n \to n \pm 1$ involve *adiabatic* energy fluctuations $2\omega'$, the soft mode should be characterized as oscillatory. It is, however, practically correct that the soft mode is observed under isothermal conditions, while these adiabatic transitions are also involved in inelastic phonon scatterings. In any case, the results under internal $F_\pm$ clearly suggest the nonlinearity of the process.

Further calculations yield the result

$$\Gamma(\boldsymbol{q}, \Delta\omega) = \Phi_2(\boldsymbol{q}, \Delta\omega) = \frac{\pi\hbar}{32N\omega(\boldsymbol{q})}\sum_{1,2}\frac{V^{(4)}u_L^2\langle\boldsymbol{q}|u_T|\boldsymbol{K}_1, \boldsymbol{K}_2\rangle^2}{\omega_1\omega_2}$$

$$\times[(n_1 + n_2 + 1)\{-\delta(\omega + \omega_1 + \omega_2) + \delta(\omega - \omega_1 - \omega_2)\}$$
$$-(n_1 - n_2)\{-\delta(\omega - \omega_1 + \omega_2) + \delta(\omega + \omega_1 - \omega_2)\}].$$

Defining two temperatures by $\hbar\omega(\boldsymbol{k})(n_1 + n_2 + 1) = k_B T_1$ and $\hbar\omega(\boldsymbol{k})(n_1 - n_2) = k_B T_2$, the damping factor $\Gamma(\boldsymbol{k}, \Delta\omega)$ indicates that the phonon energy $\hbar\omega(2n_2 + 1)$ is transferred to the surroundings, which is equal to $k_B(T_1 - T_2)$.

It was controversial if the soft mode frequency converged to zero, however, experimentally that is a matter of timescale of observation. No answer can be given, because the transition region is obscured by *uncertainties* arising from bifurcation. Nevertheless, $\omega(\boldsymbol{q})$ at $\boldsymbol{q} = 0$ is signified by the value of $\Delta\omega$, allowing us to write $\omega(0) = \omega_o \pm \Delta\omega$, and hence

$$\tilde{\omega}^2 \approx \omega(0, \Delta\omega)^2 = \omega_o^2 + \Phi_1(0)(\omega_o \pm \Delta\omega).$$

Setting this as equal to zero, we obtain $\omega_o = \pm\Delta\omega/2$ by assuming $\Delta\omega = \pm\Phi_1/4$. So writing $\Phi_1(0)(\omega_o \pm \Delta\omega) = A'T_o$, we can obtain $\omega(0, \Delta\omega)^2 = A'(T_o - T)$, which is Landau's parabolic equation. Hence T_o represents a critical temperature in mean-field approximation.

In contrast to zone-center transitions signified by $\boldsymbol{q} = 0$, for the zone-boundary case we consider $\boldsymbol{q} = \pm\boldsymbol{G}/2$ in the above argument, and equation (7.13) can be expanded for $\boldsymbol{q}$ as

$$\omega(\boldsymbol{q}, \Delta\omega)^2 = \omega(0, \Delta\omega)^2 + \kappa\boldsymbol{q}^2 + \cdots,$$

where κ is a constant. Therefore, using Landau's postulate for $\omega(0, \Delta\omega)^2$, this expansion can be written for a small value of $|\boldsymbol{q}|$ as

$$\tilde{\omega}^2 \approx \omega(\boldsymbol{q}, \Delta\omega)^2 = A'(T - T_c) + \kappa\boldsymbol{q}^2 \quad \text{for} \quad T > T_c$$

and (7.14)

$$\tilde{\omega}^2 \approx \omega(\boldsymbol{q}, \Delta\omega)^2 = A'(T_c - T) + \kappa\boldsymbol{q}^2 \quad \text{for} \quad T < T_c.$$

Equations (7.14) were employed for the analysis of soft modes in neutron inelastic scatterings, showing the results in figure 7.2.

Soft-mode frequencies in binary systems showed mostly a parabolic temperature-dependence, as described approximately by $(T_c - T)^{\frac{1}{2}}$, however, expressed as $(T_c - T)^\beta$ in order to cover deviated parabolic curves numerically, where β is called a *critical exponent*. Although related to the dimensionality of crystals in the statistical scaling theory [11], the problem of critical exponents can be solved by the soliton theory in principle; those interested in the theory of critical exponents are referred to [11], the problem is not discussed in this book. Nevertheless, the critical exponent has so far not been seriously discussed for traditional long-range order with respect to the soliton number.

7.5 Cochran's model of a ferroelectric transition

Temperature-dependent lattice modes were observed in early spectroscopic studies, but it was after Cochran's theory (1960) [12] that such modes became known as soft modes. On the basis of a simple ionic crystal, Cochran showed that the softening can be explained consequent on counteracting short- and long-range interactions in crystals.

In a crystal, if a particle of mass m and charge e is displaced by u, a *hole* of mass m' charged by $-e$ is left behind, creating dipole moment $p = eu$. Dynamically, such a dipole moment in motion is characterized by the reduced mass $\mu = mm'/(m + m')$. Considering an electron and an ion of masses m and M, we have the reduced mass $\mu = mM/(m + M) \approx m$, because of $m \ll M$ in practice. Hence, the dynamics of dipole moment p is signified by a lighter mass of $\mu \approx m$.

Such dipole moments $p_i = eu_i$ located at all lattice sites i are responsible for the macroscopic polarization P parallel to a symmetry axis. If p_i are all aligned in this direction by their mutual correlations, the whole crystal is polarized, where there is the internal field $-P/\varepsilon_0$ acting on p_i. In contrast, if the correlation is dominated by classical interactions of rotatable dipoles, the acting field is Lorentz' field $-P/3\varepsilon_0$ in a crystal in ellipsoidal shape. To be consistent with the condensate model, such internal fields should be equivalent to adiabatic potentials, and Cochran considered both of these polarization fields in his theory of longitudinal and transversal motion.

For transverse motion, we write the dipole moment as $p_i = e(u_{Ti} - u_{Ti})$, which is driven by the field $E + P/3\varepsilon_0$, because there is no polarization component in this direction. In the following, first we consider p_i to occupy a volume V, and $P = p_i/V$. Then, writing the equations of motion of charges e and $-e$ separately as

$$m\frac{d^2u_T^+}{dt^2} = -C\left(u_T^+ - u_T^-\right) + e\left(E + \frac{P}{3\varepsilon_0}\right) \quad \text{and}$$

$$m'\frac{d^2u_T^-}{dt^2} = -C\left(u_T^- - u_T^+\right) - e\left(E + \frac{P}{3\varepsilon_0}\right),$$

we derive the equation for P, that is

$$\mu\frac{d^2P}{dt^2} = \frac{e^2}{V}\left(E + \frac{P}{3\varepsilon_0}\right) - CP. \tag{7.15}$$

For a given field $E = E_0 \exp i\tilde{\omega}_T t$, we write $P = P_0 \exp i\tilde{\omega}_T t$ to obtain a steady solution for the susceptibility function, i.e.

$$\chi(\tilde{\omega}_T) = \frac{P_0}{\varepsilon_0 E_0} = \frac{e^2/\varepsilon_0 V}{C - (e^2/\varepsilon_0 V) - \mu\tilde{\omega}_T^2},$$

showing a singularity at $\mu\tilde{\omega}_T^2 = C - (e^2/\varepsilon_0 V)$, indicating that $\tilde{\omega}_T = 0$ if $C = e^2/\varepsilon_0 V$, signifying the presence of a soft mode.

On the other hand, for a longitudinal polarization $P = \frac{e}{V}(u_L^+ - u_L^-)$, the equation of motion can be written as

$$\mu\frac{d^2P}{dt^2} = \frac{e^2}{V}\left(-\frac{P}{\varepsilon_o} - \frac{P}{3\varepsilon_o}\right) - CP, \qquad \text{for } E = 0. \tag{7.16}$$

The critical frequency, if any, can be determined by $\mu\tilde{\omega}_L^2 = C + (2e^2/3\varepsilon_o V)$, which never becomes zero, besides, the susceptibility is not detectable with $E = 0$.

7.6 Symmetry change at T_c

Symmetry change is a visible signature of structural phase transition, which needs to be confirmed as verified for the theory to be correct. Among crystals in many types, soft modes observed in TSCC and perovskites provided supporting results for symmetry changes at T_c.

Experimentally, soft modes in TSCC, for example, are identified as in B_{2u} and in A_1 symmetry for $T > T_c$ and $T < T_c$, respectively, indicating that their propagation directions are perpendicular to each other. Therefore, we can assume that their diving agents are associated with adiabatic potentials

$$\Delta U_{B_{2u}} = \frac{A}{2}u_x^2 + \frac{B}{4}u_x^4 \quad \text{and} \quad \Delta U_{A_1} = \frac{C}{2}u'^2_y + \frac{D}{4}u'^4_y,$$

where the coordinate axes are taken with the x-axis as the longitudinal direction, the y-axis for the transverse direction and A, B, C and D are constants. Corresponding to inversion $\boldsymbol{K} \to -\boldsymbol{K}$ of wavevectors, we write the coordinate $\boldsymbol{u}'$ in the transitional region as determined by

$$u' = c_x u_x' + c_y u_y', \quad \text{where} \quad c_x^2 + c_y^2 = 1$$

where the transition process can be specified by $c_x \to 0$ and $c_y \to 1$ at T_c. The transitional potential at T_c is expressed as

$$\begin{aligned}\Delta U_{\text{trans}} &= \frac{A}{2}u'^2 + \frac{B}{4}u_x^2 u'^2 \\ &= c_x^2\left(\frac{A}{2}u'^2_x + \frac{B}{4}u'^4_x\right) + c_y^2\left(\frac{A}{2}u'^2_y + \frac{B}{4}u'^2_x u'^2_y\right) \quad \text{and} \quad u'^2_x = -\frac{A}{B},\end{aligned}$$

where the limit of transition gives rise to

$$\lim \Delta U_{\text{trans}} = -\frac{A}{4}u'^2_y = \Delta U_{A1},$$

provided that $C = A/2$ and $D = 0$. Figure 7.8 shows soft-mode dispersions observed in TSCC and $SrTiO_3$[13], where different temperature-dependences between $T > T_c$ and $T < T_c$ are recognized, as related to the coefficients A and $A/2$, (in place of $C = A/2$, $D \neq 0$).

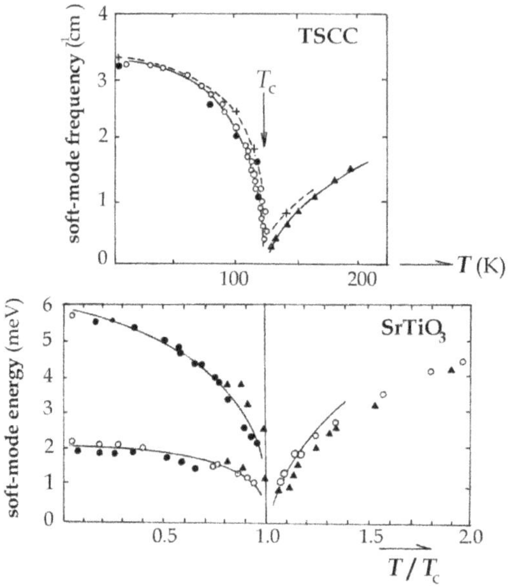

Figure 7.8. Soft-mode frequencies versus temperature: (a) TSCC; (b) $SrTiO_3$. Reproduced from [3], copyright 2010. With permission from Springer.

Exercises

According to section 7.3, a soft mode originates from anharmonic adiabatic potentials $V^{(4)}$, which is however temperature-dependent. This indicates that the $V^{(4)}$ potential cannot be a continuous perturbation, where the discontinuity should lead to energy transfer to the lattice. Related to singularities of pseudospin waves, discuss the issue in a qualitative manner.

Soft modes represent the response from the transverse component of vector waves, whereas scattering and magnetic resonance studies deal with the longitudinal component. Hence, these different methods should complement results on the pseudospin waves. Discuss the issue with respect to the two-component properties of nonlinear waves.

References

[1] Elliott R J and Gibson A F 1975 *An Introduction to Solid State Physics and its Applications* (London: McMillan)

[2] Shapiro S M, Axe J D, Shirane G and Riste T 1972 *Phys. Rev.* **B6** 4332

[3] Fujimoto M 2010 *Thermodynamics of Crystalline States* (Berlin: Springer)

[4] Shirane G, Axe J D, Harada J and Remeika J P 1970 *Phys. Rev.* **B2** 155

[5] Burns G and Scott B A 1978 *Phys. Rev.* **B7** 3088

[6] Kirkpatrick E S, Muller K A and Rubins R S 1964 *Phys. Rev.* **135** A86

[7] Unoki H and Sakudo T 1967 *J. Phys. Soc. Jpn.* **23** 546

[8] Currat R, Müller K A, Berlinger W and Desnoyer F 1978 *Phys. Rev.* **B17** 2938

[9] Petzelt J, Kozlov G V and Volkov V V 1987 *Ferroelectrics* **73** 101

[10] Cowley A and Phys P 1968 **31** 123

[11] Stanley H E 1971 *Introduction to Phase Transitions and Critical Phenomena* (New York: Oxford University Press)
[12] Cochran W 1973 *The Dynamics of Atoms in Crystals* (London: Edward Arnold)
[13] Scott J F 1983 *Raman Spectroscopy of Structural Phase Transitions in Light Scatterings near Phase Transitions* ed H Z Cummins and A P Levanyak (Amsterdam: North-Holland)

Part III

Soliton theory of lattice dynamics

Nonlinear lattice dynamics for $\sigma(\phi)$ are discussed for crystalline processes under thermodynamic conditions, where a new structure occurs at a critical point followed by mesoscopic disorder with entropy production. The mathematical details are discussed to deal with domain structure in general. Toda's soliton lattice shows that the soliton energy composes lattice sites to characterize the Weiss field in phase with corresponding $\sigma(\phi)$.

IOP Publishing

Solitons in Crystalline Processes (2nd Edition)
Irreversible thermodynamics of structural phase transitions and superconductivity
Minoru Fujimoto

Chapter 8

Nonlinear dynamics in finite crystals: displacive waves, complex adiabatic potentials and pseudopotentials

In experimental studies for $T < T_c$, *nonlinearity* is evident from observed results of the line-shape in A- and P-modes from temperature-dependent frequencies of soft-modes. Deviating from sinusoidal variations in stable lattice, nonlinear waves for $T < T_c$ need to be analyzed with respect to the lattice structure for a concerned dynamical system.

Theoretically, in series expansion of Landau's Gibbs function $G(\eta)$ with respect to the order parameter η, the terms $\frac{1}{2}A\eta^2$, $\frac{1}{4}B\eta^4$, etc, can be translated into the adiabatic potentials $U(\boldsymbol{\sigma}_k)$ for the collective pseudospin mode $\boldsymbol{\sigma}_k$ in a mesoscopic lattice, which, however, may not necessarily be compatible mathematically with thermodynamic conditions. Accordingly, we write the wave vector as $\boldsymbol{k}$, instead of $\boldsymbol{q}$ in the previous discussion for displacements in equilibrium crystals.

Using a *truncated Landau potential* in one-dimension, we can obtain nonlinear *longitudinal* propagation of the collective vector wave $\boldsymbol{\sigma}_k(\phi)$ along the direction of $\boldsymbol{k}$ in the Brillouin zone, accompanying the *transversal* component as well. Nonetheless, verifying that the longitudinal component is expressed by elliptic functions of the renormalized phase ϕ, we consider that an expanded Gibbs' function describes the condensate composed of $\boldsymbol{\sigma}(\phi)$ in the nonequilibrium lattice. Actually, the transversal component $\sigma_{k\perp}(\phi)$ is responsible for energy transfer to the lattice, as we confirm here that $\boldsymbol{\sigma}_k(\phi)$ exhibits a *vector wave* of two components driven by the *adiabatic lattice potential* $U(\phi)$ that is primarily *incommensurate* with original lattice periodicity.

In addition, the cyclic *pseudopotential* in crystallography is considered in this chapter as an intrinsic adiabatic lattice potential, where the transverse wave $\boldsymbol{\sigma}_{k\perp}(\phi)$ exhibits phase variations $\Delta\phi$ that can be attributed to *transversal correlations* among neighboring condensates in nonequilibrium crystals, which will be discussed later with the *sine-Gordon equation*. Such nonlinear waves in crystal space are reviewed in this chapter, as sketched from existing data as related with a *distorted lattice in the thermodynamic environment*, constituting preliminaries for the soliton theory of crystalline processes.

8.1 Internal pinning of collective pseudospins

Pseudospins $\boldsymbol{\sigma}_n$ at a site n and their Fourier transform $\boldsymbol{\sigma}_k$ are vector quantities in a periodic lattice that can be regarded to represent a *vector field*, where the propagation is assumed to be continuous. In such continuous media, $\boldsymbol{\sigma}_k$ exhibits infinitesimal amplitude at small $|\boldsymbol{k}|$.

On the other hand, for a collective order variable mode $\boldsymbol{\sigma}_k$ in propagation through modulated crystals, such a truncated adiabatic potential as $\Delta V(\boldsymbol{\sigma}_k) = \frac{1}{2}a\boldsymbol{\sigma}_k^2 + \frac{1}{4}b\boldsymbol{\sigma}_k^4$ is essential for the nonlinearity to be *pinned* by the lattice. Setting the transverse component $\boldsymbol{\sigma}_{k\perp}$ aside, in this section we discuss the longitudinal component $\boldsymbol{\sigma}_{k\parallel}$ in propagation in the direction of $\boldsymbol{k}$. Omitting the suffix $\parallel$, we write the one-dimensional wave equation for σ_k as

$$m\left(\frac{\partial^2}{\partial t^2} - v_o^2\frac{\partial^2}{\partial x^2}\right)\sigma_k(x, t) = -\frac{\partial \Delta V_k}{\partial x} = -\sigma_{ko}k\left(a\sigma_k + b\sigma_k^3\right). \tag{8.1a}$$

Here, $\sigma_k(x, t) = \sigma(kx - \omega t)$ represents the longitudinal component determined by the phase $\phi = kx - \omega t$, where $v_o = \frac{\omega}{k}$ is the speed of propagation. The amplitude σ_{ko} is primarily infinitesimal at a small k, but finite in general.

Krumshansl and Schrieffer [1] have shown that (8.1a) can be solved in a simplified form by re-expressing as

$$\frac{\mathrm{d}^2 Y}{\mathrm{d}\phi^2} + Y - Y^3 = 0, \tag{8.1b}$$

where the following *rescaled variables* are defined as

$$Y = \frac{\sigma_k}{\sigma_{ko}} \quad \text{and} \quad \phi = k(x - vt), \tag{8.1c}$$

where

$$\sigma_{ko} = \sqrt{\frac{|a|}{b}},\; k^2 = \frac{|a|}{m\left(v_o^2 - v^2\right)} = \frac{k_o^2}{1 - \frac{v^2}{v_o^2}} \quad \text{and} \quad k_o^2 = \frac{|a|}{mv_o^2}.$$

Therefore, the relation between k and frequency ω can be written as

$$\omega^2 = v_o^2\left(k^2 - k_o^2\right), \tag{8.2}$$

indicating that the nonlinear propagation is *dispersive*. In (8.2), we notice that $\omega = 0$ for $k = k_o$, but at a frequency ω the speed of nonlinear propagation is determined as

$v < v_0$ for $k > k_0$. Considering the former for $T > T_c$, the latter can be assigned to temperatures for $T < T_c$, where the dispersion relation (8.2) is attributed to the potential proportional to σ_k^3 in (8.1*a*).

For a small σ_k, the higher-order term Y^3 in (8.1*b*) can be ignored, in which case (8.1*b*) is a linear equation with a sinusoidal solution $Y = Y_0 \sin(\phi + \phi_0)$, where Y_0 is finite but small, for which $\phi_0 = 0$ may be chosen as the reference phase. The equation (8.1*b*) can then be solved, as in the following.

Integrating (8.1*b*) once, we obtain

$$2\left(\frac{\mathrm{d}Y}{\mathrm{d}\phi}\right)^2 = (\lambda^2 - Y^2)(\mu^2 - Y^2), \tag{8.3}$$

where

$$\lambda^2 = 1 - \sqrt{1 - \alpha^2} \quad \text{and} \quad \mu^2 = 1 + \sqrt{1 - \alpha^2}.$$

Here $\alpha = \left(\frac{\mathrm{d}Y}{\mathrm{d}\phi}\right)_{\phi=0}$ is a constant of integration determined at $\phi = 0$, providing an *initial* condition. Writing $\xi = \frac{Y}{\lambda}$ for convenience, (8.3) can be re-expressed in integral form as

$$\frac{\phi_1}{\sqrt{2}\kappa} = \int_0^{\xi_1} \frac{\mathrm{d}\xi}{\sqrt{(1 - \xi^2)(1 - \kappa^2\xi^2)}}, \tag{8.4a}$$

where ξ_1 is the upper limit of the variable ξ, and ϕ_1 the corresponding phase. The ratio $\kappa = \frac{\lambda}{\mu}$ is called the *modulus* of the *elliptic integral of the first kind* (8.4*a*). It is noted that the constants λ and μ can be written in terms of the modulus κ, i.e.

$$\lambda = \frac{\sqrt{2}\kappa}{\sqrt{1 + \kappa^2}} \quad \text{and} \quad \mu = \frac{\sqrt{2}}{\sqrt{1 + \kappa^2}}$$

The inverse function of the integral (9.4) can then be written as

$$\xi_1 = \mathrm{sn}\frac{\phi_1}{\sqrt{2}\kappa}, \tag{8.4b}$$

which is known as *Jacobi's elliptic sn-function*, which is an important formula for *nonlinear displacement ξ_1 to be expressed by a practical phase variable ϕ_1 for geometrical lattice periodicity.*

In previous notations, expressing the nonlinear longitudinal mode by σ_1, we have

$$\sigma_1 = \lambda\sigma_0\, \mathrm{sn}\frac{\phi_1}{\sqrt{2}\kappa}, \tag{8.5}$$

which is an elliptic wave modified from a sinusoidal wave by those parameters λ, μ, κ and α. It is significant that the amplitude $\lambda\sigma_o$ and phase ϕ_1 are both determined by the integral in (8.4*a*), which is characteristic for non-linear waves.

Using an angular variable Θ defined effectively by $\xi = \sin\Theta$, (8.4*a*) is expressed as

$$\frac{\phi_1}{\sqrt{2}\kappa} = \int_0^{\Theta_1} \frac{d\Theta}{\sqrt{1-\kappa^2\sin^2\Theta}}, \tag{8.6a}$$

where Θ_1 represents an *effective sinusoidal phase* corresponding to ξ_1, which is convenient for comparing with the *lattice periodicity*. This ξ_1 is a useful variable to express propagation in comparison of a sinusoidal function, i.e.

$$\mathrm{sn}\frac{\phi_1}{\sqrt{2\kappa}} = \sin\Theta_1. \tag{8.6b}$$

In this way, σ_1 can be regarded as the longitudinal component of a classical vector, i.e. $\sigma_1 = \lambda\sigma_o \sin\Theta_1$ with effective amplitude $\lambda\sigma_o$. Jacobi originally called the angle Θ_1 as the *amplitude function* defined as $\Theta_1 = \mathrm{am}\left(\int_0^{\Theta_1}\frac{d\Theta}{\sqrt{1-\kappa^2\sin^2\Theta}}\right)$; nevertheless, we use this definition signifying *effective sinusoidal phase angle* in crystals. Figure 8.1 shows the relation between the elliptic integral $u(\kappa) = \int_0^{\Theta_1}\frac{d\Theta}{\sqrt{1-\kappa^2\sin^2\Theta}}$ and Jacobi's $\Theta_1 = \mathrm{am}\, u$, plotted for $\kappa = 0.5$ and 0.9 in particular.

Except for $\kappa = 1$, the sn-function (8.6*b*) for $0 < \kappa < 1$ is periodic, whose periodicity is determined repeatedly between $\Theta_1 = 0$ and $\Theta_1 = 2\pi$. The integral

$$K(\kappa) = \int_0^{\frac{\pi}{2}} \frac{d\Theta}{\sqrt{1-\kappa^2\sin^2\Theta}}, \tag{8.7}$$

is called the *complete elliptic integral*. We consider that long-range correlations are included implicitly in (8.5) for $0 < \kappa < 1$. The range of $0 < \Theta < \pi$ is defined as the *period* of sn-function that is $4K(\kappa)$.

For $\kappa = 1$ or $\lambda = \mu = 1$, (8.6*a*) can specifically be integrated as

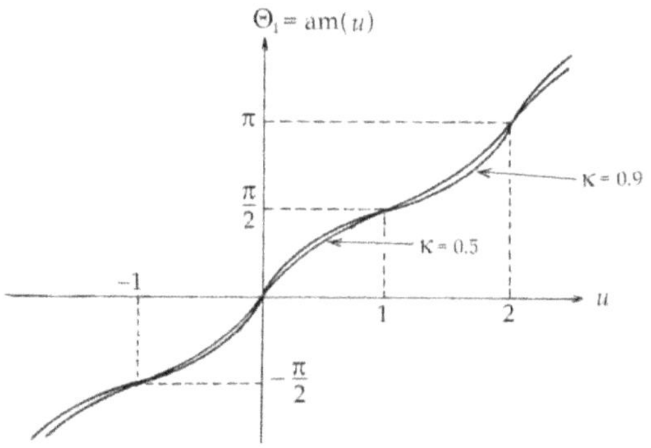

Figure 8.1. Effective phase $\Theta_1(u)$ (Jacobi's amplitude function) versus u.

$$\sigma_l(\phi_1, 1) = \sigma_o \tanh \frac{\phi_1}{\sqrt{2}}, \tag{8.8a}$$

indicating that $\sigma_l \rightarrow \sigma_o$ in the limit of $\phi_1 \rightarrow \infty$, which means the same as equation (0.13) in Bragg–Williams' order–disorder theory discussed in the Introduction. Figure 8.2 illustrates curves of (8.5) and (8.8*a*) for representative values of the modulus κ.

Such $\sigma_l(\phi_1, 1)$ as expressed by (8.8*a*) corresponds to a *completely ordered state* determined in new symmetry of an *infinite* lattice, for which $\sigma_l = \sigma_o$ at all angles, except for $\phi_1 = \infty$ as indicated by $\kappa = 1$. Accordingly, the parameter κ implies the degree of *long-range order* in single domain structure; nonetheless, $\kappa = 1$ can specify equilibrium states in the thermodynamic environment. Furthermore, indicated by Landau's postulate for the truncated potential $\Delta V(\boldsymbol{\sigma}_k)$, there can be a singularity in the dynamical property of the vector $\boldsymbol{\sigma}_k$ at the critical point consistent with specific heat anomalies. The initial value $\alpha = \sigma_{lk}/\sigma_o$ at $\phi_1 = 0$ is undetermined either quantum-mechanically or statistically, because of *bifurcation* in classical description.

Nevertheless, because of the infinite boundary for $\phi_1 \rightarrow \infty$, solution (8.8*a*) for $\kappa = 1$ is *not an acceptable solution for a real crystal in finite size*, which should therefore be characterized by realistic Brillouin boundaries in finite crystals. For that purpose, ellipsoidal variation of (8.5) should also be considered with *transversal components*, as will be discussed next in sections 8.2 and 8.3.

8.2 Transverse components and the cnoidal potential

In the above, the amplitude σ_0 is left undetermined, which is characterized by a finite amplitude that should be determined by thermal interactions with the surroundings at a given temperature. Leaving the temperature-dependence to later discussions, the periodic solution (8.5) for $0 < \kappa < 1$, the expression $\sigma_l = \lambda\sigma_o \sin \Theta_l$ represents geometrically the longitudinal component of an amplitude $\lambda\sigma_0$. Therefore, we can consider the transversal component $\sigma_\perp$ that is written as

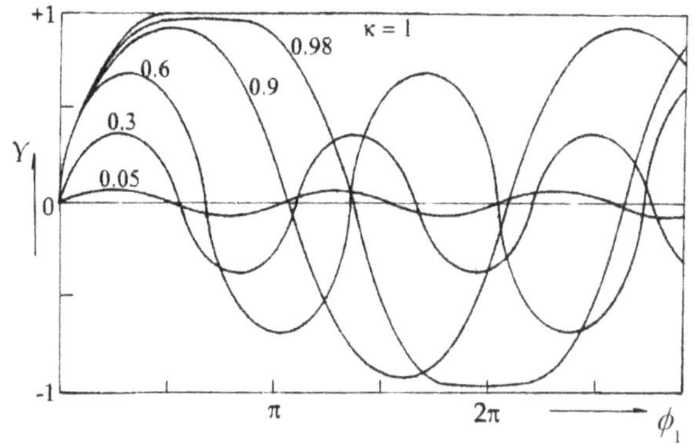

Figure 8.2. Elliptic functions $\mathrm{sn}(\phi/\sqrt{2\kappa})$ versus ϕ, plotted for different moduli κ.

$$\sigma_\perp = \lambda\sigma_o \cos\Theta_l = \lambda\sigma_o \,\mathrm{cn}\frac{\phi_l}{\sqrt{2}\kappa} \qquad \text{for } 0 < \kappa < 1, \tag{8.9a}$$

where we have the relation $\sigma_l^2 + \sigma_\perp^2 = \lambda^2\sigma_o^2$, indicating that (8.9) represents a transversal wave. Therefore, corresponding to (8.8*a*), we have

$$\sigma_\perp(\phi_l, 1) = \pm\sigma_o \,\mathrm{sech}\frac{\phi_l}{\sqrt{2}} \qquad \text{for } \kappa = 1 \tag{8.9b}$$

It is noted that $\sigma_\perp(0) \perp \sigma_l(0)$ with a finite magnitude at $\phi_l = 0$, which is consistent to figure 4.2 for critical fluctuations discussed in chapter 4.

At this point, the potential $\Delta V(\boldsymbol{\sigma})$ should be considered as characterized physically for reflection of elliptic waves. Owing to the transverse component of longitudinally correlated pseudospins, we can consider an adiabatic potential $\Delta V(\sigma_\perp) = \frac{1}{2}a\sigma_\perp^2$, assuming that there are no *cross correlations* with adjacent chains of order variables in a crystal. On the other hand, rotating σ_l across a perpendicular plane from $\phi_l = 0$ to $\phi_l = n\pi$ (*n* is integers), a work by a force $F = -\frac{\partial\Delta V}{\partial x} = -k\frac{\partial\Delta V}{\partial\phi_l}$ is required, which can be calculated as $W = -\int_{-\delta}^{+\delta}\sigma_l\frac{\partial\Delta V}{\partial\phi_l}\mathrm{d}\phi_l$ over the region $-\delta < \phi_l < +\delta$, where 2δ is the effective width of the integrand at $\phi_l = 0$. In the vicinity of $\phi_l = 0$, we consider $\sigma_l(\phi_l) = \sigma_\perp\left(\phi_l + \frac{\pi}{2}\right) \approx \lambda\sigma_o$ and $\frac{\partial\Delta V}{\partial\phi_l} \propto \frac{\mathrm{d}}{\mathrm{d}\phi_l}\mathrm{cn}^2\frac{\phi_l}{\sqrt{2}\kappa}$, so that

$$\frac{\mathrm{d}W}{\mathrm{d}\phi_l} = -2\lambda^2\sigma_o^2\left(\mathrm{cn}\frac{\phi_l}{\sqrt{2}\kappa}\right)\frac{\mathrm{d}}{\mathrm{d}\phi_l}\left(\mathrm{cn}\frac{\phi_l}{\sqrt{2}\kappa}\right).$$

Here, the differentiation can be performed with another elliptic function defined by $\mathrm{d}n^2u = 1 - \kappa^2\mathrm{sn}^2\,u$. For the dn-function for a variable *u*, we have general formula

$$(\mathrm{sn}\,u)' = \mathrm{cn}\,u\,\mathrm{dn}\,u, \quad (\mathrm{cn}\,u)' = -\mathrm{sn}\,u\,\mathrm{dn}\,u \quad \text{and} \quad (\mathrm{dn}\,u)' = -\kappa^2\mathrm{sn}\,u\,\mathrm{cn}\,u.$$

Letting $u = \frac{\phi_l}{\sqrt{2}\kappa}$, with these formulae the above dW can be written as

$$\begin{aligned}\frac{\mathrm{d}W}{\mathrm{d}\phi_l} &= \frac{1}{\sqrt{2}\kappa}\frac{\mathrm{d}W}{\mathrm{d}u} = \frac{2\lambda^2\sigma_o^2}{\sqrt{2}\kappa}\mathrm{cn}\,u\,\mathrm{sn}\,u\,\mathrm{dn}\,u = -\frac{2\lambda^2\sigma_o^2}{\sqrt{2}\kappa^3}(\mathrm{dn}\,u)(\mathrm{dn}\,u)'\\ &= -\frac{\lambda^2\sigma_o^2}{\kappa^2}\frac{\mathrm{d}}{\mathrm{d}\phi_l}\mathrm{dn}^2\frac{\phi_l}{\sqrt{2}\kappa}.\end{aligned}$$

Therefore, we have a formula for the energy $\Delta V(\phi_l, \kappa)\mathrm{d}\phi_l$ to change the direction of $\boldsymbol{\sigma}(\phi_l)$ by $\mathrm{d}\phi_l$, where

$$\Delta V(\phi_1, \kappa) = \mu^2\sigma_o^2 \mathrm{dn}^2\frac{\phi_1}{\sqrt{2}\kappa} \qquad \text{for } 0 < \kappa < 1. \tag{8.10a}$$

Here the factor proportional to $\mathrm{dn}^2\, u$ exhibits the same type of wave form as $\mathrm{sn}^2\, u$ and $\mathrm{cn}^2\, u$, so that (8.10*a*) is generally called a *cnoidal potential* characterized by $0 < \kappa < 1$.

On the other hand, for $\kappa = 1$ we have $\mathrm{dn}\, u = \mathrm{sech}\, u$, hence

$$\Delta V(\phi_1, 1) = \sigma_o^2 \mathrm{sech}^2\frac{\phi_1}{\sqrt{2}} \qquad \text{for } \kappa = 1, \tag{8.10b}$$

representing an adiabatic potential energy for rotating σ_l by 180° to $-\sigma_l$, exhibiting a singular behavior at ∞ that is identical to that at the zone boundary. Nevertheless, the discrepancy in those limits needs to be eliminated for finite crystals, while (8.10*b*) remains valid for inversion $\sigma_l \rightleftarrows -\sigma_l$.

8.3 Finite crystals and the domain structure

It is clear that *real crystals are finite objects*, where there are always *domain boundary planes* and *surfaces in finite distance*[note1], where inversion $\sigma_\perp \rightleftarrows -\sigma_\perp$ takes place in perpendicular directions to longitudinal inversion of σ_l (8.10*b*). *Therefore, the Bragg–Williams' formula* (8.8*a*) *should be revised to meet with the boundary conditions for* $\sigma_\perp$ *in finite crystals.*

While $\sigma_l \perp \sigma_\perp$, the corresponding phase $\phi_l/\sqrt{2}$ is a variable to change from longitudinal σ_l to transversal $\pm\sigma_\perp$, whose boundaries can be specified by $\phi_l/\sqrt{2} \leqslant \pi/2$ on one side and $-\pi/2 \leqslant -\phi_l/\sqrt{2}$ on the other, signifying the *domain wall* composed of two interfacing parallel planes with phase inversion $\phi_1 \rightleftarrows -\phi_1$. We should revise the dynamical phase change in crystals into many smaller angular sections of $-\pi/2 \leqslant \phi_l/\sqrt{2} \leqslant \pi/2$ in repetition, to accommodate domains and surfaces in real crystals in sufficiently large size. In this case, equation (8.8*a*) should be interpreted as modified domain boundaries to be expressed by another phase $\phi_\perp$ at $\phi_\perp/\sqrt{2} = \pm\pi/2$ as characterized by

$$\sigma_\perp(\phi_\perp, 1) = \pm\sigma_o \tanh\frac{\phi_\perp}{\sqrt{2}} \qquad \text{for } \kappa = 1, \tag{8.10c}$$

where $\Delta V(\phi_\perp, 1)$ is proportionally related to $\sigma_\perp^2$, as related to hypothetical work for 180° rotation of $\sigma_\perp$ centered at $\phi_\perp/\sqrt{2} = \pm\pi/2$.[note1]

The above thermodynamic interpretation of propagating phase ϕ_l should be applied to *real crystals with domain boundaries and surfaces*, whereby we need to discuss *entropy production* at the singularity of $\Delta V(\phi_1, 1)$ in a representative section that indicates the lattice strain energy specified by the phase ϕ_1. In this context, re-writing with the conventional phase ϕ_l in place of $\phi_l/\sqrt{2}$, equation (8.10*b*) is expressed as

[note1] See chapter 12 for a more rigorous mathematical description.

$$\Delta V(\phi_1, 1) = \frac{\sigma_0^2}{2}\text{sech}^2\phi_1 \quad \text{at } \phi_1 = \pm\pi/2, \tag{8.10d}$$

representing the *adiabatic potential energy* associated with inversion $\sigma_\perp \rightleftarrows -\sigma_\perp$ in thermodynamic description.

The above argument for domains and surfaces is mathematical, however, taken as the significant fact of finite crystals, *as evidenced experimentally by surface pinning phenomena.* In other words, the domain structure is regarded as a natural consequence of finite crystals, where *the phase ϕ plays the fundamental role to specify generalized coordinates.* All of the following discussions will therefore be made for finite crystals to be discussed with thermodynamic principles.

Figure 8.3(a)–(c), the curves of σ_1, σ_l and $\sigma_\perp$ near $\phi_1 = 0$ are plotted, respectively, against ϕ_1 for $0 < \kappa < 1$ to illustrate their behaviors. Figure 8.3(d) shows the curve of $\sigma_\perp$ for $\kappa = 1$ near domain boundaries and surfaces defined conveniently at phases $\phi_1 \cong \pm\pi/2^{\text{note1}}$ hypothetically.

The variable $\boldsymbol{\sigma}_q(\phi_1)$ is discussed as a classical vector of longitudinal and transverse components, whose amplitude and phase are specified by the modulus $\kappa = 1$. The particular case of $\kappa = 1$ signifies equilibrium condition, where condensates are composed of $\sigma_\perp(1)$ and $\Delta V_\perp(1)$ in phase along a *symmetry direction*, thereby transmitting the latter for *shear strains* to the surroundings to acquire new stability.

In contrast, the non-steady dynamics for $0 \leqslant \kappa < 1$ can basically be unstable to variation. Under certain external conditions, lattice symmetry can be spontaneously changed to equilibrium, where the process $\kappa \to 1$ is *irreversible*, as the Born–Huang principle implies for strain energy (8.10*b*) to be transferred to the surroundings under thermodynamic conditions.

Owing to the presence of domain structure and surfaces, mathematical boundaries $0 \leqslant \phi_1 < \infty$ for elliptic functions need to be replaced by multiple regions $-\pi/2 \leqslant \phi_1 \leqslant \pi/2$ of mesoscopic domains; which is a reasonable assumption for practical

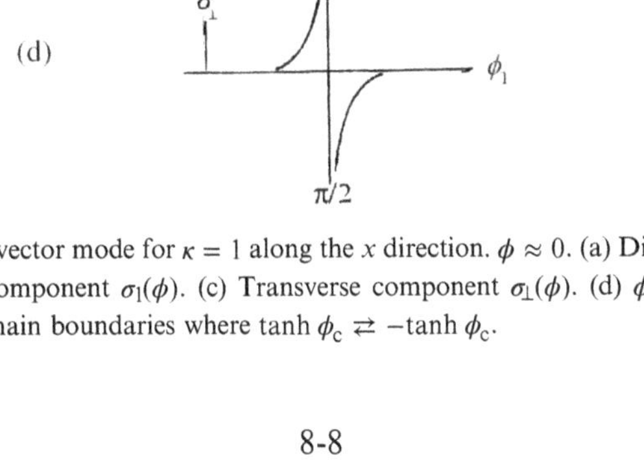

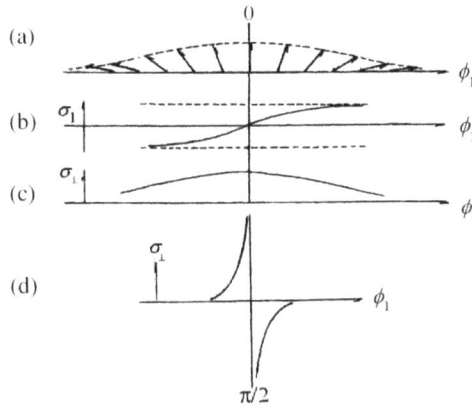

Figure 8.3. A pseudospin vector mode for $\kappa = 1$ along the x direction. $\phi \approx 0$. (a) Distributed vectors in the xz plane. (b) Longitudinal component $\sigma_l(\phi)$. (c) Transverse component $\sigma_\perp(\phi)$. (d) $\phi_c \simeq \pm\pi/2$ and transversal component $\sigma_\perp(\phi_c)$ for domain boundaries where $\tanh\phi_c \rightleftarrows -\tanh\phi_c$.

crystals in sufficiently large size. It is particularly important to express the surfaces, because the heat exchange with surroundings takes place obviously through surface boundaries; theoretical infinite crystals should therefore be abandoned for thermodynamics, where the domain structure is consequent on the finite extent.

Summarizing the above argument on the transversal component, the potential energy $\Delta V_\perp(\phi_1, 1)$ corresponds to classical work for inversion $\sigma_\perp \rightleftarrows -\sigma_\perp$, hence the adiabatic potential should be determined mathematically as a complex function, corresponding to complex $\sigma_1 \pm i\sigma_\perp$, as will be formally discussed in section 8.5. Physically, it is notable that such an energy $\Delta V_1(\phi_\perp, 1)$ signifies *the domain-wall energy and the Weiss field*, indicating essential features of irreversible thermodynamics. In addition, as will be discussed in chapter 18, the soliton theory needs to be extended to *nonthermal processes*, providing a significant approach to different cases than thermodynamic phenomena.

8.4 Lifshitz' incommensurability in mesoscopic phases

The elliptic function for $0 < \kappa < 1$ is periodic. Its period is not always the same with stable lattice periodicity in equilibrium. Generally, such a change in pseudospin mode represents *fluctuations* that are *incommensurate* with the stable period. The collective pseudospins are pinned in phase with the adiabatic potential in dynamic crystals, while the process of $\kappa \rightarrow 1$ accompanies *phase fluctuations* due to space-time uncertainties, dominating the critical region of phase transitions. Dealing with such mesoscopic fluctuations in elastic bifurcation, Lifshitz [2] derived the *incommensurability condition*, assuming for thermodynamic quantities to be dominated by amplitude and phase fluctuations, providing significant criteria to evaluate mesoscopic bifurcation.

For practical analysis, mesoscopic pseudospin modes expressed by elliptic and hyperbolic functions can be simplified by considering classical vector components

$$\sigma_1 = \sigma_0 \sin \Theta_1 \quad \text{and} \quad \sigma_\perp = \sigma_0 \cos \Theta_1$$

of a classical vector $\boldsymbol{\sigma} = (\sigma_1, \sigma_\perp)$, where σ_0 and $\Theta_1(x, t)$ are the effective amplitude and phase. For condensates, these variables are observed as temperature dependent, because the corresponding lattice displacements $(u_1, u_\perp)$ can scatter phonons inelastically. It is noted that the phase variable $\Theta_1(x, t)$ is periodic in the lattice with the period $4K(\kappa)$, which is not necessarily in phase with the lattice. In any case, the observed pseudospins can be described with temperature-dependent amplitude and phase. The Gibbs potential $G(\mathrm{p}, T; \boldsymbol{\sigma})$ can therefore be specified by such a mesoscopic variable $\boldsymbol{\sigma}$, or by σ_0 and Θ_1, taking the minimum value for thermodynamic equilibrium.

Lifshitz assumed such correlation energies as determined statistically by the binary interactions $J_{ij}\boldsymbol{\sigma}_i \cdot \boldsymbol{\sigma}_j$ to derive the thermodynamic condition for incommensurability. Such time-dependent quantities in thermodynamic experiments are usually averaged out in a long timescale t_0 of observation. For sufficiently short t_0, in contrast, we observed them in broadened functions of x, varying between

x_i and x_j on the longitudinal axis, as signified by $\delta x = x_i - x_j$ and $x = \frac{1}{2}(x_i + x_j)$. The observed correlation function can therefore be written as proportional to

$$\left\langle\sigma^*(x_i)\sigma(x_j) + \sigma(x_i)\sigma^*(x_j)\right\rangle_t = 2\langle\sigma^*(x)\sigma(x)\rangle_t + \left\langle\sigma^*(x)\frac{\partial\sigma(x)}{\partial x} - \sigma(x)\frac{\partial\sigma^*(x)}{\partial x}\right\rangle_t \delta x$$
$$+ \left\langle\sigma^*(x)\frac{\partial\sigma(x)}{\partial x} + \sigma(x)\frac{\partial\sigma^*(x)}{\partial x}\right\rangle_t (\delta x)^2$$
$$+ \left\langle\frac{\partial\sigma^*(x)}{\partial x}\frac{\partial^2\sigma(x)}{\partial x^2} - \frac{\partial\sigma(x)}{\partial x}\frac{\partial^2\sigma^*(x)}{\partial x^2}\right\rangle_t (\delta x)^3 + \cdots.$$

The Gibbs potential for such fluctuations is therefore contributed to significantly by such correlation terms as related to δx, i.e.

$$G_L = \frac{iD}{2}\int_0^L \left\langle\sigma^*\frac{\partial\sigma}{\partial x} - \sigma\frac{\partial\sigma^*}{\partial x}\right\rangle_t \frac{dx}{L}$$
$$+ \frac{iD}{2}\int_0^L \left\langle\frac{\partial\sigma^*}{\partial x}\frac{\partial^2\sigma}{\partial x^2} - \frac{\partial\sigma}{\partial x}\frac{\partial^2\sigma^*}{\partial x^2}\right\rangle_t (\delta x)^2\frac{dx}{L} + \cdots \tag{8.11}$$

where $\frac{iD}{2}$ is defined as a constant factor proportional to δx. We can then write

$$G(\sigma) = G(0) + \int_0^L \left\langle\frac{mv_o^2}{2}\left|\frac{\partial\sigma}{\partial x}\right|^2 + \frac{a}{2}|\sigma|^2 + \frac{b}{4}|\sigma|^4\right\rangle_t \frac{dx}{L} + G_L, \tag{8.12}$$

where the upper limit L represents a *sampling length* of the mesoscopic σ as related to practical experiments.

Writing $\sigma = \sigma_o \exp i\phi_1$, $G(\sigma)$ is a function of σ_o and ϕ_1, we have (8.12) re-expressed as

$$G(\sigma_o, \phi_1) = G(0) + \int_0^L \left\langle\frac{a\sigma_o^2}{2} + \frac{b\sigma_o^4}{4} + \frac{mv_o^2}{2}\left(\frac{d\sigma_o}{dx}\right)^2 + \frac{mv_o^2\sigma_o^2}{2}\left(\frac{d\phi_1}{dx}\right)^2\right.$$
$$\left. + D\sigma_o^2\frac{d\phi_1}{dx}\left(1 + \delta x^2\frac{d^2\phi_1}{dx^2}\right)\right\rangle_t \frac{dx}{L},$$

which is minimized for equilibrium by setting $\frac{\partial G}{\partial\sigma_o} = 0$ and $\frac{\partial G}{\partial\phi_1} = 0$ simultaneously. Carrying out these partial differentiations of $G(\sigma_o, \phi_1)$, we obtain the equations

$$a\sigma_o + b\sigma_o^3 + mv_o^2\frac{d^2\sigma_o}{dx^2} + mv_o^2\left(\frac{d\sigma_o}{dx}\right)^2 + 2D\sigma_o\frac{d\phi_1}{dx}\left(1 + \delta x^2\frac{d^2\phi_1}{dx^2}\right) = 0$$

and

$$\left(mv_0^2\frac{d\phi_1}{dx} + D\sigma_0^2\right)\frac{d}{d\phi_1}\left\{\frac{d\phi_1}{dx}\left(1 + \delta x^2\frac{d^2\phi_1}{dx^2}\right)\right\} = 0.$$

We see immediately that the second equation is fulfilled if

$$\frac{d\phi_1}{dx} = -\frac{D}{mv_0^2}. \tag{8.13}$$

Therefore, setting $-\frac{D}{mv_0^2} = q$, (8.13) represents the wavenumber of fluctuations defined independently from the lattice, because both D and mv_0^2 are unrelated constants with the lattice, giving an incommensurate wavevector. As σ_0 is assumed as constant, the first relation can be solved for σ_0^2, i.e.

$$\sigma_0^2 = -\frac{a - \frac{D}{mv_0^2}}{b}.$$

The wave $\sigma_0 \exp i\phi$, observed with a finite amplitude σ_0 in practice, is therefore incommensurate with respect to the lattice period, as generally $D \neq 0$, which is known as the Lifshitz theorem. Incommensurability originates from a non-vanishing displacement $\langle \delta x \rangle_t$ of lattice points, which is responsible for non-zero factor $D \neq 0$ to be an obvious consequence.

8.5 Klein–Gordon equation for the Weiss potential

In the field theory applied to crystals, the order variable can be represented by the *density function*

$$\sigma(\phi_1) = \psi^*(\phi_1)\psi(\phi_1), \tag{8.14a}$$

of the wave function $\psi(\phi_1)$ of the field. Writing the responsible potential for driving $\sigma(\phi_1)$ as $V(\phi_1) = -K^2(\phi_1)\sigma(\phi_1)$, as *inferred from* the analogy with *Weiss' assumption of ferromagnets*, the field equation can be expressed by

$$\frac{\partial^2\sigma}{\partial\tau^2} - v^2\frac{\partial^2\sigma}{\partial x^2} = -K^2(x, \tau)\sigma, \tag{8.14b}$$

where $\sigma(x, \tau)$ is regarded as a *complex variable* to facilitate two components. Equation (8.14*b*) is known as the *Klein–Gordon equation* familiar in the field theory, and v is the speed of propagation. Now that the variable σ is *complex*, (8.14*b*) can be factorized as

$$\frac{\partial\psi}{\partial\tau} - v\frac{\partial\psi}{\partial x} = -iK(x, \tau)\psi^* \quad \text{and} \quad \frac{\partial\psi^*}{\partial\tau} + v\frac{\partial\psi^*}{\partial x} = -iK(x, \tau)\psi,$$

which are considered for *development equations* of ψ and its complex conjugate ψ^*. Writing further $\psi = \psi' + i\psi''$ and $\psi^* = \psi' - i\psi''$, the development equations for ψ' and ψ'' can be expressed as

$$\frac{\partial \psi'}{\partial \tau} - v\frac{\partial \psi'}{\partial x} = K(x, \tau)\psi'' \quad \text{and} \quad \frac{\partial \psi''}{\partial \tau} + v\frac{\partial \psi''}{\partial x} = K(x, \tau)\psi', \tag{8.15}$$

indicating a classical impact with a real potential[note2].

Using Fourier's transformation $\psi' \pm i\psi'' = \{\Psi'(x, k) \pm i\Psi''(x, k)\}\exp(\mp ikx)$ and $K(x, \tau) = K(x)\exp(\mp ikx)$ in crystal space, (8.15) can be re-expressed for Fourier transforms (Ψ', Ψ'');

$$\frac{d\Psi'}{dx} + ik\Psi' = u(x)\Psi'', \frac{d\Psi''}{dx} - ik\Psi'' = -u(x)\Psi' \quad \text{and} \quad u(x) = -\frac{K(x)}{v}.$$

Further, defining the complex amplitude by $\psi_1 = \Psi' \pm i\Psi''$ and $i\psi_2 = \Psi' \mp i\Psi''$, we have

$$\frac{d\psi_1}{dx} \mp iu(x)\psi_1 = \mp k\psi_2 \quad \text{and} \quad \frac{d\psi_2}{dx} \pm iu(x)\psi_2 = \pm k\psi_1. \tag{8.16}$$

Eliminating ψ_2 from (8.16), we have

$$\frac{d^2\psi_1}{dx^2} + \left(k^2 \pm u^2 - i\frac{du}{dx}\right)\psi_1 = 0; \tag{8.17a}$$

eliminating ψ_1 instead, we obtain

$$\frac{d^2\psi_2}{dx^2} + \left(k^2 \pm u^2 + i\frac{du}{dx}\right)\psi_2 = 0. \tag{8.17b}$$

From the above definitions, we can write $\psi_1 = \Psi' \pm i\Psi''$ and $i\psi_2 = \Psi' \mp i\Psi''$, and the wave equation for densities $\sigma_l = \psi_1^2$ and $\sigma_\perp = (i\psi_2)^2$ can be derived as follows.

Using the relation $\psi_1^2 + \psi_2^2 = 0$ that can be obtained from (8.17*a*) and (8.17*b*) with (8.16) by manipulating to find wave equations for σ_l and $\sigma_\perp$, we arrive at

$$\frac{d^2\sigma_l}{dx^2} + (k^2 \pm u^2)\sigma_l = 0 \quad \text{and} \quad \frac{d^2\sigma_\perp}{dx^2} \mp \left(\frac{du}{dx}\right)\sigma_\perp = 0, \tag{8.17c}$$

indicating that *longitudinal* σ_l propagates, whereas *transversal* $\sigma_\perp$ does not along the x-direction.

Summarizing the above argument, the complex displacive waves $\psi_1 + i\psi_2$ originate from an order-variable coupling $-u(\sigma_l + i\sigma_\perp)$ in finite crystals, exhibiting a complex potential energy $-u^2 \mp i\dfrac{du}{dx}$ to determine the amplitude and phase of the nonlinear order variable $\sigma_l(x) + i\sigma_\perp(x)$. Referring to (8.10*a*) and (8.10*b*), the potentials u^2 and $\dfrac{du}{dx}$ in (8.17*c*) can be converted to $u^2 \approx dn^2(\phi_1/\kappa)$ for $0 \leqslant \kappa < 1$

[note2] Equation (8.15) is obtained by a factorized Klein–Gordon equation, which is by no means unique mathematically. However, physically equation (8.15) is the *unique choice* for the impact potential $K(x, \tau)$ to be *real*, implying that the Klein–Gordon equation represents conservative dynamics that is required for a necessary choice in thermodynamics.

and $-\pi/2 \leqslant \phi_1 \leqslant +\pi/2$, changing in the limit of $\kappa \to 1$ to $u^2 \approx \mathrm{sech}^2\,\phi_1$ in the vicinity of $\phi_1 = \pm\pi/2$. As remarked earlier, these potentials have mobility along the direction of $\sigma_\perp$.

In this case, as $\sigma_1 = 0$ at $\phi_1 = \pm\pi/2$, and inversion $\phi_1 \rightleftarrows -\phi_1$ is equivalent to shear strains in opposite rotations $u' \rightleftarrows -u'$, whose soliton energy u'^2 should appear in the first equation in (8.17*c*) as

$$\frac{\mathrm{d}^2\sigma_\perp}{\mathrm{d}\phi_1^2} + (k^2 \pm u'^2)\sigma_\perp = 0, \qquad (8.17d)$$

where the potential u'^2 represents a *pseudopotential* at $\phi_1 = \pm\pi/2$, as will be discussed in the next section.

However, it should be emphasized at this point that we define a classical vector $\sigma_1 + i\sigma_\perp$ based on the physical choice of factor $K(x, t)$ in (8.15). Accordingly, the complex potential $-u^2 \pm i(\mathrm{d}u/\mathrm{d}x)$ is consequent on the conservative dynamics that represents the thermodynamic environment. On the basis of the Klein–Gordon equation to generate displacive waves, the nonlinear waves are clearly described by *two components of complex potential*, as postulated for critical phase fluctuation in chapter 4, supporting $\sigma_\perp^2 \neq 0$ *for phonon inelastic scatterings.*

8.6 Pseudopotentials in mesoscopic phases

The space group can be modified with translational symmetry, signifying a periodic rotation of constituents, along a symmetry axis; for instance, rotations over m-times of the lattice constant [3]. In the presence of a *screw symmetry axis* of m-fold rotation of σ_p by θ_p, as shown in figure 8.4, we can consider a *transversal* potential energy V_m^L, where transversally correlated pseudospin wave $\sigma(\phi)$ can be pinned in crystal space, if the spatial phase of ϕ matches lattice spacing in a *transversal direction.*

Assuming that a lattice with such *pseudosymmetry* is signified by m-times rotations of perpendicular displacements σ_p at lattice points p

$$\sigma_\mathrm{p} = \sigma_\mathrm{o} \exp\left(\pm i\theta_\mathrm{p}\right), \quad \text{where } \theta_\mathrm{p} = \frac{2\pi}{m}\mathrm{p} \quad \text{and} \quad \mathrm{p} = 0, 1, 2, \ldots, m-1, \qquad (8.18)$$

along the m-fold screw axis, we can consider the potential

$$V_m^\mathrm{L}(\theta_\mathrm{o}, \theta_1, \ldots, \theta_{m-1}) \propto \sum_\mathrm{p} \sigma_\mathrm{p} = \sigma_\mathrm{o} \sum_\mathrm{p} \{\exp i\theta_\mathrm{p} + \exp(-i\theta_\mathrm{p})\} = 2\sigma_\mathrm{o} \sum_\mathrm{p} \cos\theta_\mathrm{p}.$$

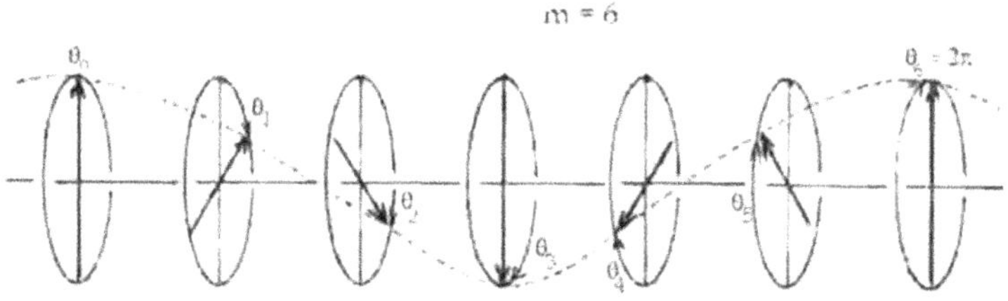

Figure 8.4. Phase variations of a collective pseudospin mode pinned by a pseudopotential $V_m(\phi)$.

Here, these angles θ_p can be re-expressed by the lattice coordinates $x_p = pa_o$ combined with (8.14); $\theta_p = \frac{2\pi}{m}\frac{x_p}{a_o} = G_m x_p$ where $G_m = \frac{2\pi}{ma_o}$, and therefore we have $V_m^L \propto \sum_p \cos(G_m x_p)$.

Thermodynamically, a pseudospin mode $\sigma = \sigma_o \exp i\phi_1$ is considered to be *pinned in phase* by the *adiabatic potential* $V_m^L(\phi_1)$, as *associated with domain structure* where we should have a *phase matching* $G_m x_p = m\phi_1$, so that

$$\cos(m\phi_1) = \frac{1}{2\sigma_o^m}(\sigma^{im\phi_1} + \sigma^{-im\phi_1}).$$

Therefore, the adiabatic potential $V_m^L(\phi_1)$ is expressed as

$$V_m^L(\phi_1) = \frac{2\rho}{m}(\sigma^{im\phi_1} + \sigma^{-im\phi_1}) = \frac{2\rho\sigma_o^m}{m}\cos(m\phi_1), \tag{8.19}$$

where ρ is the proportionality constant. It is noted that $\sigma(\phi_1)$ and $V_m^L(\phi_1)$ move together with the phase ϕ_1, as meant by internal pinning by Weiss' potential $V_m^L(\phi_1)$. At this point, we realize that such a potential $V_m^L(\phi_1)$ as in different symmetry from the space group can be attributed to an additional degree of freedom, so that it is called a *peudopotential,* representing *adiabatic domain boundaries*, arising from mutual correlations among component sites p = m, …, $-m$ along a *transversal* direction, implying the presence of *transversal correlations*.

In a nonequilibrium environment, the collective order variable $\boldsymbol{\sigma}$ is considered to be pinned by the adiabatic potential $V_m(\phi_1)$, fluctuating around the equilibrium position, for which the Gibbs function can be written equivalent to (8.17*d*) as

$$G(\boldsymbol{\sigma}) = \int_0^L \left\{\frac{a\sigma^*\sigma}{2} + \frac{b(\sigma^*\sigma)^2}{4} + \frac{mv_o^2}{2}\frac{\partial\sigma^*}{\partial x}\frac{\partial\sigma}{\partial x} + V_m^L(\phi_1)\right\}\frac{dx}{L}.$$

Replacing the integrand by its time average $\langle\ldots\rangle_t$, we have

$$G(\sigma_o, \phi_1) = \int_0^L \left\langle \frac{a\sigma_o^2}{2} + \frac{b\sigma_o^4}{4} + \frac{mv_o^2}{2}\left(\frac{\partial\sigma_o}{\partial x}\right)^2 + \frac{mv_o^2\sigma_o^2}{2}\left(\frac{\partial\phi_1}{\partial x}\right)^2 + \frac{2\rho\sigma_o^m}{m}\cos(m\phi_1)\right\rangle_t \frac{dx}{L}.$$

Setting $\frac{\partial G(\sigma_o, \phi_1)}{\partial\sigma_o} = 0$ and $\frac{\partial G(\sigma_o, \phi_1)}{\partial\phi_1} = 0$ to determine minimum of $G(\sigma_o, \phi_1)$, we obtain

$$a\sigma_o + b\sigma_o^3 + 2\rho\sigma_o^{m-1}\cos(m\phi_1) + mv_o^2\left\{\sigma_o\left(\frac{d\phi_1}{dx}\right)^2 + \frac{d^2\sigma_o}{dx^2}\right\} = 0 \tag{8.20}$$

and

$$mv_o^2\sigma_o^2\frac{d^2\phi_1}{dx^2} - 2\rho\sigma_o^m\sin(m\phi_1) = 0. \tag{8.21}$$

Integrating the second equation, (8.21) can be modified as $\frac{1}{2}mv_o^2\sigma_o^2\left(\frac{d\phi_1}{dx}\right)^2 + V_m^L(\phi_1) = \text{const.}$, expressing a conservation law that is analogous to a simple pendulum with a finite amplitude. Using abbreviations $\psi = m\phi_1$ and $\xi = \frac{2m\rho\sigma_o^{m-2}}{mv_o^2}$, (8.21) can be simplified as

$$\frac{d^2\psi}{dx^2} - \xi\sin\psi = 0. \tag{8.22a}$$

This is known as the *sine-Gordon equation*. For the conservation law, (8.22*a*) can be integrated as

$$\frac{1}{2}\left(\frac{d\psi}{dx}\right)^2 - \xi\cos\psi = E, \tag{8.22b}$$

where the constant E represents the energy of an equivalent pendulum. On the other hand, the equation (i) determines the amplitude σ_o, which is not constant as in a simple pendulum, depending on the phase ψ. Such a phase-related amplitude is typical for a non-linear oscillation.

Equation (8.22*b*) can be written in an integrated form

$$x - x_o = \int_0^{\psi_1} \frac{d\psi}{\sqrt{2(E + \xi\cos\psi)}},$$

where the upper and lower limit ψ_1 and 0 correspond to coordinates x and x_o. This integral can be expressed by the elliptic integral of the first kind, if the modulus κ is defined by $\kappa^2 = \frac{2\xi}{E + \xi}$, that is, if $\kappa^2 < 1$

$$x - x_o = \frac{\kappa}{\sqrt{\xi}}\int_0^{\Theta_1} \frac{d\Theta}{\sqrt{1 - \kappa^2\sin^2\Theta}},$$

where $\Theta = \frac{\psi}{2}$. Therefore,

$$\sin\Theta_1 = \text{sn}\frac{\sqrt{\xi}(x - x_o)}{\kappa} \qquad \text{for } 0 < \kappa < 1,$$

and

$$\sin\Theta_1 = \tanh\sqrt{\xi}(x - x_o) \qquad \text{for } \kappa = 1.$$

To discuss the phase matching condition, it is convenient to define $x - x_o = \Lambda(\kappa)$ corresponding to the phase difference between $\Theta_1 = 0$ and $\Theta_1 = \pi$, i.e.

$$\Lambda(\kappa) = \frac{\kappa}{\sqrt{\xi}}\int_0^{\pi}\frac{d\Theta}{\sqrt{1-\kappa^2\sin^2\Theta}} = \frac{2\kappa K(\kappa)}{\sqrt{\xi}}, \tag{8.23}$$

where $K(\kappa)$ is the complete elliptic integral of the first kind. Phase matching with the pseudopotential $V_m^{\mathrm{L}} = V_0\sum_{\mathrm{p}}\cos(G_m x_{\mathrm{p}})$ may be stated by $G_m x_{\mathrm{p}} = \Theta_{\mathrm{p}} = \frac{2\pi \mathrm{p}}{m}$, assuming that the pseudospin wave reflects perfectly on the potential barrier V_0.

The magnitude V_0 behaves not only as a simple *resistive* reflector but also as *inductive*, resulting in a phase shift. At the transition temperature, quantum-mechanical uncertainties at lattice sites can be considered as responsible for such a shift, and the pseudopotential can be expressed as

$$V_m^{\mathrm{L}} = -V_0\sum_{\mathrm{p}}\cos\left\{\Theta_{\mathrm{p}}(1-\delta_{\mathrm{p}})\right\}, \tag{8.24}$$

where δ_{p} is called the *incommensuration parameter*, exhibiting transition anomalies due to uncertain phase fluctuations at *p*.

In fact, both potentials $V_m(\phi_1)$ and V_m^{L} in (8.19) and (8.24), respectively, are intrinsic adiabatic potentials often *unrelated with lattice symmetry*, so representing $\pm i\frac{d\sigma_\perp}{d\phi_1}$ for perpendicular component $\sigma_\perp(\phi_1)$ in particular, supported by sine-Gordon equation for ϕ_1. Therefore, *the domain boundaries and surfaces can be signified by inversions*

$$+\phi_1 \rightleftarrows -\phi_1 \quad \text{and} \quad +\Delta\,\phi_1 \rightleftarrows -\Delta\,\phi_1 \tag{8.25}$$

Experimentally notable is that in the presence of a cyclic pseudopotential associated with cyclic domains, nonlinear order variables expressed generally by $\sigma_{k\perp} = \sigma_0 f\{\psi_k(t)\}$ are *phase-modulated* by $V_m^{\mathrm{L}}(\phi_1)$, as described by the sine-Gordon equation (8.20) for the *nonlinear phase* $\psi_k(t)$. The phase modulation will be discussed with the *phase soliton theory* dealing with *transversal correlations at domain boundaries*, as discussed in chapter 12.

Figure 8.5 illustrates the behavior of Θ_{p}, as described by $\sin\frac{2\pi \mathrm{p}}{m} = \mathrm{sn}\frac{\sqrt{\xi}(x-x_0)}{\kappa}$, against $x - x_0$, showing characteristic lengths $\Lambda(\kappa)$ that are determined by (8.21). In the above, we assumed that $\kappa^2 < 1$, which means $E < -\xi$ by definition. On the other hand, if $E > -\xi$ we have a case for $\kappa^2 > 1$. If ζ can be ignored, the sine-Gordon equation can be reduced to an equation for free motion; no pinning occurs if $V_m(\phi_1)$ is disregarded. In this context, the change from $\kappa > 1$ to $\kappa < 1$ with lowering temperature signifies a transition to a mesoscopic phase in commensurate structure. Needless to say, such phase transitions should be initiated with respect to symmetry of the pseudopotential, which is logical but not necessarily confirmed yet by experimental studies.

Normally, the pseudosymmetry is found by crystallographic analysis, but often does not constitute a subgroup of the crystal symmetry. Therefore, it is significant to identify the physical origin of the responsible $\sigma_\perp(\phi_1)$.

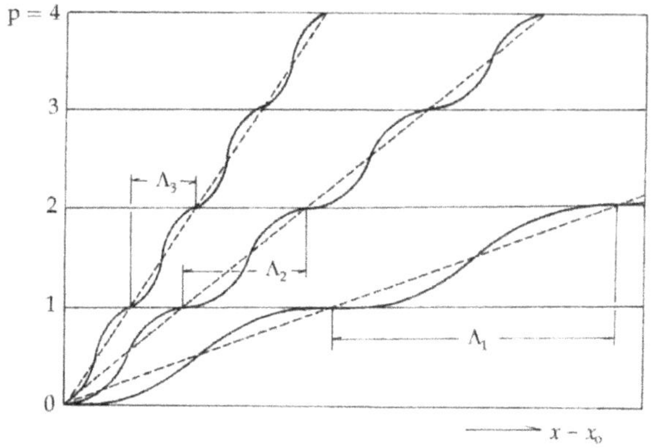

Figure 8.5. (a) A phase function $\phi(x)$ in sine-Gordon equation for a soliton potential $V_3(x)$. (b) Simulated $V_3(x)$ versus renormalized coordinate x. (c) A computed C_3-soliton potential with a small phase difference $\Delta\phi$. Reproduced from [6], copyright 2010. With permission of Springer.

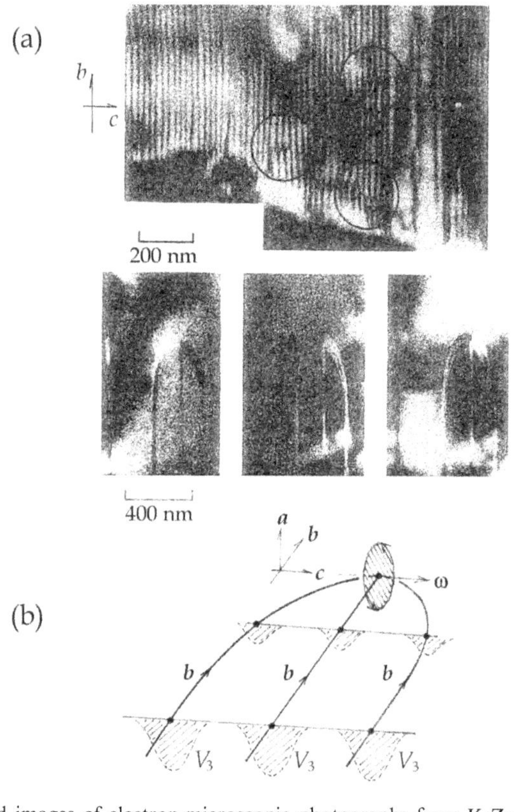

Figure 8.6. (a) Dark-field images of electron-microscopic photographs from K_2ZnCl_4 crystals [5], showing discommensuration lines parallel to C_3-screw axes terminated at vortexes. (b) Illustration of a vortex at the end of discommensuration lines, where v indicates the propagation direction. Reproduced from [6], copyright 2010. With permission of Springer.

Although one-dimensional correlations are a valid assumption for anisotropic crystals, experimentally such a model should be evaluated on sample crystals of high quality characterized by a small defect density. For such phase-locking phase transitions in K_2ZnCl_4 and Rb_2ZnCl_4 crystals, Pan and Unruh [4] reported laminar patterns of such *discommensuration lines* perpendicular to the x direction after transitions recorded by transmission electron microscopy (TEM), which is shown in figure 8.6. Although there were additional 'splitting' and 'vortex-like' patterns, the dominant laminar structure in the dark fields is undeniable evidence for pseudo-potential, as discussed in the above.

Pseudosymmetry, such as screw-axes, is identified in crystals among other crystalline systems. β-$ThBr_4$ crystallized in $I4_1/a\left(C_{4h}^6\right)$ symmetry is a typical case, which is stable above 700 K but *metastable* at lower temperatures, where distorted symmetry of Th-complexes surrounded by eight Br-ions is incommensurately arranged along a C_6 axis with the space group at 4.2 K. The modulated structure was confirmed by numerical analysis of paramagnetic resonance spectra of impurity probes.

Exercises

1. A complex wave is analogous to plane electromagnetic waves in free space and voltage–current transmission through Lecher's wires. In this case, reflection and transmission of two components depend on the target impedance as specified by the coupling. Review the Klein–Gordon equation with respect to the analogy, discuss the nature of the coupling u in section 8.4, which may not necessarily be the same as specified as a real quantity.
2. Show that two maxima of the potential ΔV given by (8.10*b*) and σ of (8.9*b*) are signified by a phase difference $\Delta\phi_1$. Is it physically significant?
3. In section 8.5, the pseudopotentials known in crystallography are discussed for nonlinear waves in crystals, showing that their phase is modulated. Accordingly, the kinks may lead to domain separation. Discuss energy transfer qualitatively, referring to section 8.2.
4. Derive (8.17*c*) from (8.17*a*) and (8.18) for given relations $\psi_1 = \Psi' + i\Psi''$ and $i\psi_2 = \Psi' - i\Psi''$.
5. Discuss the reason why $V_m^{\mathrm{L}}(\phi)$ in (8.19) can be regarded as an adiabatic potential.

References

[1] Krumshansl J A and Schrieffer J R 1975 *Phys. Rev.* **B11** 3535
[2] Blinc R and Levanyak A P 1986 *Incommensurate Phases in Dielectrics* (Amsterdam: North-Holland)
[3] Megaw H D 1973 *Crystal Structures: A Working Approach* (Philadelphia, PA: Saunders)
[4] Pan X and Unruh H-G 1990 *J. Phys. Cond. Matter* **2** 323
[5] Zwanenburg G 1990 *Doctoral Thesis* University of Nijmegen
[6] Fujimoto M 2010 *Thermodynamics of Crystalline States* (Berlin: Springer)

Chapter 9

Opposite Weiss fields for nonlinear order variables and entropy production: the Korteweg–deVries equation for transitions between conservative states

Originating from correlation energies in crystals, the soliton potential represents *responding waves of order variables from the lattice* in Born–Oppenheimer's approximation. It is significant to note that such waves are observable in practical *relative coordinate systems*. Analogous to the *Weiss field* defined originally in ferromagnetic crystals in the mean-field approximation, the adiabatic potential drives nonlinear pseudospin waves in a thermodynamic environment. Expressed by *pulse-shaped* hyperbolic functions determined by the *Korteweg–deVries equation*, the nonlinear waves are characterized by finite *amplitudes and phases*, where traditional *long-range order* can be embedded in mesoscopic soliton potential energies in the thermodynamic environment. In contrast, subjecting to phonon scatterings in the counteracting lattice, an energy dissipation can take place at their singularities in isothermal processes, leading to *entropy production*.

It is essential that such adiabatic potential represents internal energies for the lattice to react against nonlinear pseudospins. The mathematical theory of the Weiss field, as basically consistent with *Newton's action-reaction principle*, is discussed in this chapter following Lamb [1], demonstrating that discrete transitions of phases and entropy productions can take place under *equilibrium conditions in crystals*, constituting a significant criterion in thermodynamics.

9.1 Dispersive equations in asymptotic approximation

In a practical expression $\boldsymbol{\sigma}_k = \boldsymbol{\sigma}_o f(\phi)$ for a cluster mode at a small $|\boldsymbol{k}|$, the phase ϕ is independent of amplitude $\boldsymbol{\sigma}_o$, if sufficiently small. However, for a finite $\boldsymbol{\sigma}_o$, the propagation of the longitudinal component $\boldsymbol{\sigma}_{k\|}$ can be *dispersive* and *dissipative* in

doi:10.1088/978-0-7503-2572-1ch9

anisotropic crystals, depending on the speed of propagation v that is a function of $|\boldsymbol{k}|$. In addition, with an internal potential, the transversal component $\boldsymbol{\sigma}_{k\perp}$ cannot be ignored. In this section, nonlinear waves in an *isotropic* fluid are discussed as a representative example, from which general features of nonlinear propagation can be inferred.

Such *mesoscopic* variables $\boldsymbol{\sigma}_k$ are represented by a classical vector field in the crystal space, whose propagation at long wavelengths is described as in continuous media. In a continuous medium, we can consider the pseudospin density $\rho = |\boldsymbol{\sigma}_k|^2$ along the direction of propagation x and its conjugated current density $j = \rho v$, representing a continuous flow of order at a constant speed v under constant pressure p. On the other hand, if there is a pressure gradient $-\frac{\partial p}{\partial x}$ along the x-direction, the speed v cannot be constant, resulting in speed-dependent amplitude and phase. For a condensate in a crystal, the gradient $-\frac{\partial p}{\partial x}$ corresponds to a reacting force $-\frac{\partial \Delta U}{\partial x}$ arising from the adiabatic potential ΔU, hence we can write the equation

$$\frac{\partial v}{\partial t} + v\frac{\partial v}{\partial x} = -\frac{1}{\rho_0}\frac{\partial p}{\partial x}. \tag{9.1}$$

For a variable speed v and constant $\rho_0 = \boldsymbol{\sigma}_0^2$ if the non-linear term $v\frac{\partial v}{\partial x}$ can be ignored, equation (i) is a linear equation. In practice, the density ρ is not confined to the x-axis, spreading transversally as well, hence we can write $\rho_0 = \rho_x + \rho_z$, assuming ρ_0 as a function of x and z. Disregarding the y direction for simplicity, the *equations of continuity* can be written as

$$\frac{\partial \rho_x}{\partial t} + \rho_0\frac{\partial v}{\partial x} = 0 \tag{9.2}$$

The density ρ_x represents the propagation along the x-axis, assuming that ρ_z does not propagate in the z-direction. In this case, ρ_z is considered to remain in the vicinity of the x-axis, satisfying the equation

$$\frac{\partial^2 \rho_z}{\partial t^2} + \alpha(\rho_z - \rho_0) = p, \tag{9.3}$$

where α is a restoring-force constant.

Equation (9.1) is non-linear because of the term $v\frac{\partial v}{\partial x}$, whereas (9.2) and (9.3) are linear equations. However, (9.1) can be linearized, if considering $v\frac{\partial v}{\partial x}$ as a perturbation. In this case, the components ρ_x and ρ_z are both functions of x and z, for which we have the relation $|\rho_x| = |\rho_z| = \rho_0$, where ρ_0 is constant, as discussed in section 8.4.

Introducing a set of *rescaled variables* $x' = \sqrt{\alpha}\,x$, $t' = \sqrt{\alpha}\,t$ and $\rho' = \frac{\rho - \rho_0}{\rho_0}$, equations (9.1)–(9.3) can be linearized as

$$\frac{\partial v}{\partial t'} + \frac{\partial p'}{\partial x'} = 0, \quad \frac{\partial \rho'}{\partial t'} - \frac{\partial v}{\partial x'} = 0 \text{ and } \frac{\partial^2 \rho'}{\partial t^2} + \rho' = p', \tag{9.4}$$

where $p' = \frac{p}{\rho_o}$. Assuming that these variables, ρ', v and p' are all proportional to $\exp i(kx' - \omega t')$ in the first approximation for $\omega = vk$, we can obtain from (9.4) an equation for their amplitudes. From this, we can derive the dispersion relation

$$\omega^2 = \frac{k^2}{1 + k^2},$$

which can be approximated as $\omega \approx k - \frac{1}{2}k^3$ for a small k, indicating that the speed $v = \omega/k$ is no longer constant in the perturbed state. Writing $k(x' - vt') = \xi$ and $\frac{1}{2}k^3t' = \tau$ in the expression $\exp i(kx' - \omega t') = \exp i\{k(x' - vt') + \frac{1}{2}k^3t'\}$, we perform a coordinate transformation from (x', t') to (ξ, τ) in the perturbed state by differential relations

$$\frac{\partial}{\partial x'} = k\frac{\partial}{\partial \xi} \quad \text{and} \quad \frac{\partial}{\partial t'} = -k\frac{\partial}{\partial \xi} + \frac{1}{2}k^3\frac{\partial}{\partial \tau}. \tag{9.5}$$

Expressing (9.1) and (9.2) in terms of x', t' and p', we obtain

$$\frac{\partial v}{\partial t'} + v\frac{\partial v}{\partial x'} + \frac{\partial p'}{\partial x'} = 0 \quad \text{and} \quad \frac{\partial \rho'}{\partial t'} + \frac{\partial v}{\partial x'} + \frac{\partial(\rho' v)}{\partial x'} = 0.$$

Combining the second relation with (9.4), these equations can be transformed by (9.5) to express with coordinates ξ and τ. That is

$$-\frac{\partial v}{\partial \xi} + \frac{1}{2}k^2\frac{\partial v}{\partial \tau} + v\frac{\partial v}{\partial \xi} + \frac{\partial p'}{\partial \xi} = 0,$$

$$-\frac{\partial \rho'}{\partial \xi} + \frac{1}{2}k^2\frac{\partial \rho'}{\partial \tau} + \frac{\partial v}{\partial \xi} + \frac{\partial(\rho' v)}{\partial \xi} = 0 \tag{9.6}$$

and

$$p' = \rho' + k^2\frac{\partial^2 \rho'}{\partial \xi^2} - k^4\frac{\partial^2 \rho'}{\partial \xi \partial \tau} + \frac{1}{4}k^6\frac{\partial^2 \rho'}{\partial \tau^2}.$$

Here, quantities ρ', p' and $v - v_0$ emerging at T_c can be expressed in power series with respect to small k^2 in an *asymptotic approximation*

$$\rho' = k^2\rho_1' + k^4\rho_2' + \cdots,$$

$$p' = k^2p_1' + k^4p_2' + \cdots$$

and

$$v - v_0 = k^2v_1 + k^4v_2 + \cdots.$$

Substituting these ρ', p' and v' into (9.6), terms of k^2, k^4, ... in the perturbed state can be compared separately. From the terms proportional to k^2, we obtain

$$-\frac{\partial \rho_1'}{\partial \xi}+\frac{\partial v_1}{\partial \xi}=0,\quad -\frac{\partial v_1}{\partial \xi}+\frac{\partial p_1'}{\partial \xi}=0 \quad \text{and } p_1'=\rho_1'.$$

Combining the first and second relations, the third one can be expressed as

$$p_1'=\rho_1'=v_1+\varphi(\tau), \tag{9.7}$$

where $\varphi(\tau)$ is an arbitrary function of τ.

Next, comparing coefficients of the k^4-terms, we obtain

$$-\frac{\partial \rho_2'}{\partial \xi}+\frac{\partial \rho_1'}{\partial \tau}+\frac{\partial v_2}{\partial \xi}+\frac{\partial(\rho_1' v_1)}{\partial \xi}=0,$$

$$-\frac{\partial v_2}{\partial \xi}+\frac{\partial v_1}{\partial \tau}+v_1\frac{\partial v_1}{\partial \xi}+\frac{\partial p_2'}{\partial \xi}=0$$

and

$$p_2'=\rho_2'+\frac{\partial^2 \rho_1'}{\partial \xi^2}.$$

Eliminating ρ_2', v_2 and p_2' from these equations, we arrive at the relations for ρ_1', v_1 and p_1', i.e.

$$\frac{\partial v_1}{\partial \tau}+6v_1\frac{\partial v_1}{\partial \xi}+\frac{\partial^3 v_1}{\partial \xi^3}+\varphi\frac{\partial v_1}{\partial \xi}+\frac{\partial \varphi}{\partial \tau}=0$$

and

$$\frac{\partial \rho_1'}{\partial \tau}+6\rho_1'\frac{\partial \rho_1'}{\partial \xi}+\frac{\partial^3 \rho_1'}{\partial \xi^3}-\varphi\frac{\partial \rho_1'}{\partial \xi}-\frac{\partial \varphi}{\partial \tau}=0.$$

Note that these equations for v_1 and ρ_1' are identical, if the function $\varphi(\tau)$ is chosen to satisfy the relation

$$\varphi\frac{\partial(\rho_1', v_1)}{\partial \xi}+\frac{\partial \varphi}{\partial \tau}=0.$$

Due to the relation (9.7), we notice that p_1' also satisfies this equation for v_1 or ρ_1'. We can therefore write an equation in general form

$$\frac{\partial \sigma}{\partial \tau}+6\sigma\frac{\partial \sigma}{\partial \xi}+\frac{\partial^3 \sigma}{\partial \xi^3}=0, \tag{9.8a}$$

where σ represents one of these variables v_1', ρ_1' and $p_1{}'$ that are all canonical functions of ξ and τ.

Equation (9.8a) is known as the *Korteweg–deVries equation*, where ξ is a *phase of propagation* defined above that is invariant of *Galilean transformations*, and τ is *a basic temporal parameter to evolve non-linearity* that can be independent of ξ. Equation (9.8a) signifies that all these quantities represented by the variable σ

propagate together in phase, as if $\sigma(\xi, \tau)$ is *pinned* by a potential $-\frac{\partial V(\xi)}{\partial \xi} = -\frac{\partial^3 \sigma}{\partial \xi^3}$ in the medium, thereby the propagation can be described by the kinetic equation

$$\left(\frac{d\sigma}{d\xi}\right)_{\xi_0} = \frac{\partial \sigma}{\partial \tau} \pm v \frac{\partial \sigma}{\partial \xi_0} = -\frac{\partial V(\xi_0)}{\partial \xi_0}, \tag{9.8b}$$

with respect to the phase ξ_0 as in the kinetic theory, where $v = 6\sigma$ and $\frac{1}{2}k^3 t' = \tau$. More precisely, equation (9.1*b*) should be written for the transverse component $\sigma_\perp$ of complex σ.

In the nonlinear theory, the relation $V(\xi) \propto \sigma(\xi)$ is derived from a soliton potential energy in the accuracy of k^4 in asymptotic expansion. In this approximation, *collective pseudospins* $\sigma(\xi)$ are driven *in phase* with the adiabatic field $V(\xi)$, increasing non-linearity with decreasing temperature that is characteristically a *soft mode*. In this sense, *the former is pinned by the latter in the moving coordinate system.* The potential $-V(\xi)$ is then equivalent to the *traditional Weiss potential* against $\sigma(\xi)$, so that it can be called the *Weiss field*. Nevertheless, the origin of Weiss field should physically be attributed to the medium to propagate $\sigma(\xi)$, for which Weiss assumed originally a proportionality relation between $-V(\xi)$ and $\sigma(\xi)$ in a *ferromagnet*, being essentially identical to the same asymptotic relation

Theorem:

$$-V(\xi) = k^2 \sigma(\xi). \tag{9.8c}$$

In the above example, one-directional classical waves are discussed in the analogy of a fluid, where the pressure gradient $-\frac{\partial p}{\partial x}$ can be interpreted as related to $\frac{\partial p'}{\partial \xi} = k^2 \frac{\partial^3 \rho_1'}{\partial \xi^3}$ in the accuracy of k^4.

Theorem: In hydrodynamics, the pressure p represents *viscous correlations* in liquid, for which $+\frac{\partial p}{\partial x}$ and $+V(\xi)$ signify distributed strains in fluid, according to *Newton's action–reaction principle.*

Also, notice that the energy gap Δ in section 4.2.2 in chapter 4 should arise from such an internal lattice potential as $+V(\xi)$ in crystals. Here, such nonlinear waves and their *driving potentials* are in-phase, propagating at a speed 6σ to satisfy the Korteweg–deVries equation (9.1*a*), where the third-order derivative term can be related to an *intrinsic adiabatic potential* to be responsible for nonlinearity. Notably, these waves and their driving potentials should be determined by their phase in asymptotic approximation that is *invariant* of $x' \to x - vt$, where the time t is *not necessarily the same as* τ. In this case, the Korteweg–deVries equation (9.1*a*) for nonlinear development plays an identical role to the Klein–Gordon equation (8.14*b*), as discussed in chapter 8.

Implied by the forgoing example is that *any variable associated with nonlinear development may commonly be signified by the Korteweg–deVries equation in adiabatic approximation.* Physically consistent with the classical concept of *condensates* for a phase transition, however, such an interpretation as it stands can only be a mathematical theorem in thermodynamics of crystalline states, unless properly coordinated with *physical reality of finite crystals.*

9.2 The Korteweg–deVries equation

We continue to discuss the order variable $\sigma(\xi)$ where $\xi = k(x' - t')$ for the Korteweg–deVries equation, taking only a longitudinal component as the first approximation. We can dismiss the time t' by writing the function $\sigma(x', t')$ as $\sigma(x)$, which is legitimate in idealized crystals, because the translation $x' - v_0 t' \to x$ is a *Galilean transformation.* This is an equivalent assumption for an idealized crystal in equilibrium with surroundings, as signified by *Born–von Kármán's boundaries*, for the space-time origin in crystals to be located at an arbitrary lattice site in Brillouin zone.

With respect to such a stationary reference for x, writing the longitudinal component as $\sigma(x)$ for simplicity, to represent a collective pseudospin, we have a differential equation

$$\boldsymbol{D}^2\sigma(x) = \lambda_0\sigma(x) \tag{9.9}$$

for basic propagation in general, where $\boldsymbol{D} = \frac{\partial}{\partial x}$ is a *differential operator* and the eigenvalue λ_0 represents a kinetic energy of propagation.

For a progressing order in equilibrium crystals however, an *adiabatic potential energy* $V(x)$ should hinder moving $\boldsymbol{\sigma}(x)$, so that (9.9) is modified as

$$\boldsymbol{L}\sigma = (\boldsymbol{D}^2 + V)\sigma = \lambda\sigma, \tag{9.10}$$

where V represents the role of the Weiss field as proposed for ferromagnetic systems.

We are interested in a wave $\sigma(x)$ for *the eigenvalue* λ to remain constant in a thermodynamic environment, which is required for $\boldsymbol{\sigma}(x)$ to be compatible with thermodynamics. Therefore, for developing order in the progressing parameter τ, we must consider the condition: $\frac{\mathrm{d}\lambda}{\mathrm{d}\tau} = 0$ or $\lambda = \lambda_0$, while initially V is zero at $\tau = 0$. The parameter τ is the real time, but can be the same time t for the space-time coordinates (x, t) in crystals, representing, however, either a temperature change ΔT or a pressure change Δp in thermodynamics. The above τ *is considered as a real variable of developing nonlinearity.*

The non-linearity can evolve with progressing order, as described by the equation

$$\frac{\partial}{\partial\tau}\sigma(x, \tau) = \boldsymbol{B}\sigma(x, \tau), \tag{9.11}$$

where $\boldsymbol{B}$ is called the *developing operator* for spatial order as a function of $\boldsymbol{D}$, and τ the parameter for developing nonlinearity. We therefore call (9.11) the *developing equation.*

If $\boldsymbol{B}$ is determined in a simple form $\boldsymbol{B}_1 = v\boldsymbol{D}$ with a constant c, we obtain $\sigma_\tau = v\sigma_x$ from (9.4), signifying that σ is a function of $x - v\tau$ that is a phase of translation. Also noted is that $\boldsymbol{B}_2 \propto \boldsymbol{D}^2$ cannot be responsible for a non-linear progression, since λ remains constant, so that $\boldsymbol{U}(x, \tau) = \text{const}$. In contrast, if $\boldsymbol{B}$ consists of a third-order derivative $\boldsymbol{D}^3$, the potential energy $\boldsymbol{U}(x, \tau)$ can vary significantly while keeping (9.3) at constant λ, as discussed in the following.

Considering the presence of $V(x, \tau)$, the wave equation can be written as

$$\boldsymbol{L}\sigma(x, \tau) = \left\{\boldsymbol{D}^2 + V(x, \tau)\right\}\sigma(x, \tau) = \lambda\sigma(x, \tau) = \lambda_o\sigma(x, \tau). \tag{9.12}$$

For convenience, shorthanded differentiations are used in the following arguments; e.g. $\frac{\partial V}{\partial x} = V_x$, $\frac{\partial^2 V}{\partial x^2} = V_{xx}$, etc. Differentiating (9.12) with respect to τ, we have

$$\frac{\partial}{\partial\tau}(\boldsymbol{L}\sigma) = \boldsymbol{L}_\tau\sigma + \boldsymbol{L}\sigma_\tau = -V_\tau\sigma + \boldsymbol{LB}\sigma,$$

and

$$\frac{\partial}{\partial\tau}(\lambda\sigma) = \lambda_\tau\sigma + \lambda\sigma_\tau = \lambda_\tau\sigma + \lambda(\boldsymbol{B}\sigma) = \lambda_\tau\sigma + \boldsymbol{BL}\sigma.$$

Therefore,

$$(-V_\tau + [\boldsymbol{L}, \ \boldsymbol{B}])\sigma = \lambda_\tau\sigma,$$

where $[\boldsymbol{L}, \ \boldsymbol{B}] = \boldsymbol{LB} - \boldsymbol{BL}$ is a *commutator*. If the condition $\lambda_\tau = 0$ is fulfilled, this equation can be solved for V, i.e.

$$(-V_\tau + [\boldsymbol{L}, \ \boldsymbol{B}])\sigma(x, \tau) = 0, \tag{9.13}$$

applying particularly to equilibrium crystals.

Assuming $\boldsymbol{B}_3 = a\boldsymbol{D}^3 + v\boldsymbol{D} + b$ with coefficients a, b and v that are functions of x and τ, we have

$$[\boldsymbol{L}, \ \boldsymbol{B}_3]\sigma = (2v_x + 3aV_x)\boldsymbol{D}^2\sigma + (v_{xx} + 2b_x + 3aV_{xx})\boldsymbol{D}\sigma + (b_{xx} + aV_{xxx} + vV_x)\sigma.$$

For this, the coefficients of terms $\boldsymbol{D}^2\sigma$ and $\boldsymbol{D}\sigma$ should vanish, hence we set the relations

$$2v_x + 3aV_x = 0 \quad \text{and} \quad v_{xx} + 2b_x + 3aV_{xx} = 0.$$

Integrating these, the relations can be obtained for the coefficients v and b to be determined. Hence, we then obtain

$$v = -\frac{3}{2}aV_x + C \quad \text{and} \quad b = -\frac{3}{4}aV_x + B$$

where C and B are integration constants.

Consequently, we have $[\boldsymbol{L},\ \boldsymbol{B}_3]\sigma = \left\{\frac{a}{4}(V_{xxx} - 6VV_x) + CV_x\right\}\sigma$, and (9.6) can be expressed as

$$\frac{a}{4}(V_{xxx} - 6VV_x)\sigma + (CV_x - V_\tau)\sigma = 0.$$

Transforming from $(x,\ \tau)$ to $(x - v\tau,\ \tau)$ for *a specific case of* $t = \tau$, and setting $C = 0$ and $a = -4$, this equation can be expressed as

$$V_t - 6VV_x + V_{xxx} = 0, \tag{9.14a}$$

which is the *Korteweg–deVries' equation for the potential* $V = V(x - vt)$ *in Galilean space in thermodynamic equilibrium.* It is noted that the corresponding pseudospin variable $\boldsymbol{\sigma}(x - vt)$ is in-phase with $V = V(x - vt)$, sharing a common feature of propagation at a given eigenvalue. In this case, the *developing equation* is

$$\sigma_t = (-4\boldsymbol{D}^3 + 6V\boldsymbol{D} + 3V_x + B)\sigma, \tag{9.14b}$$

where $\sigma = \sigma(x - vt)$ and v represents the speed of development in a thermodynamic environment. In this case, the time t can be considered for the rate of energy transfer to the heat reservoir, for which *we can consider* $t \propto \Delta T$ *for thermal processes between equilibrium states*, while remaining as a Galilean time for motion in crystals to take place among all identical lattice sites.

It is significant that the Weiss field defined in the mean-field approximation can be determined by solutions of the Korteweg–deVries equation (9.14*a*), providing the necessary condition for nonlinearity to be determined for thermodynamic states.

9.3 Thermodynamic solutions of the Korteweg–deVries equation

Equation (9.7*a*) is analytically soluble for the potential $V = V(x - vt)$. Since such a potential is stationary in the reference frame moving at a constant phase, we require the stationary condition $\frac{\partial V}{\partial(x - vt)} = 0$, and hence $V_t = vV_x$. Therefore, (9.7) can be written as

$$vV_x - 6VV_x + V_{xxx} = 0, \text{ i.e. } vV_x - 3\frac{\mathrm{d}V^2}{\mathrm{d}x} + \frac{\mathrm{d}^3V}{\mathrm{d}x^3} = 0.$$

Integrating this equation,

$$V_{xx} = 3V^2 - vV + a,$$

where a is a constant of integration. Multiplying by V_x, we integrate this equation once more, and obtain

$$V_x^2 = 3V^3 - vV^2 + 2aV + b,$$

where b is another constant. Letting $V = -2f$, for convenience, this can be re-expressed as

$$f_x^2 = -4f^3 + vf + af + b$$

The right side is an algebraic expression of third-order with respect to f, hence the above can be factorized as

$$f_x^2 = -4(f - \alpha_1)(f - \alpha_2)(f - \alpha_3) = 4\varphi(f)$$

Here, α_1, α_2 and α_3 are three roots of the third-order algebraic equation $\varphi(f) = 0$. Illustrated in figure 9.1 are curves of the equation $\varphi(f)$ plotted against f, which can cross the horizontal axis at three points α_1, α_2 and α_3, if these are all real (case A); otherwise crossing at two real roots (case B) or only at one point (not shown). Since our interest is only in the real f_x, we consider the region $\alpha_2 < f < \alpha_3$ of a curve A as significant for our purpose, which is characterized by $-f_x^2 > 0$.

We define the variable g by writing $f - \alpha_3 = -g$ for the above algebraic equation, which can then be expressed as

$$g_x^2 = 2g(\alpha_3 - \alpha_1 - g)(\alpha_3 - \alpha_2 - g).$$

Further, introducing another variable ξ by $g = (\alpha_3 - \alpha_2)\xi^2$, this equation can be written for ξ as

$$\xi_x^2 = \frac{1}{2}(\alpha_3 - \alpha_1)(1 - \xi^2)(1 - \kappa^2\xi^2), \quad \text{where } \kappa^2 = \frac{\alpha_3 - \alpha_2}{\alpha_3 - \alpha_1}.$$

Expressing the phase by $\phi = \sqrt{\alpha_3 - \alpha_1}\,x$, the equation can be further modified as

$$2\xi_\phi^2 = (1 - \xi^2)(1 - \kappa^2\xi^2). \tag{9.15a}$$

Integrating this expression, we obtain

$$\frac{\phi_1}{\sqrt{2}} = \int_0^{\xi_1} \frac{d\xi}{\sqrt{(1 - \xi^2)(1 - \kappa^2\xi^2)}}, \tag{9.15b}$$

where the phase ϕ_1 is determined by the upper limit ξ_1 of this integral. The inverse function of ϕ_1, known as Jacobi's *sn-function*, can be written from (9.15*b*), i.e.

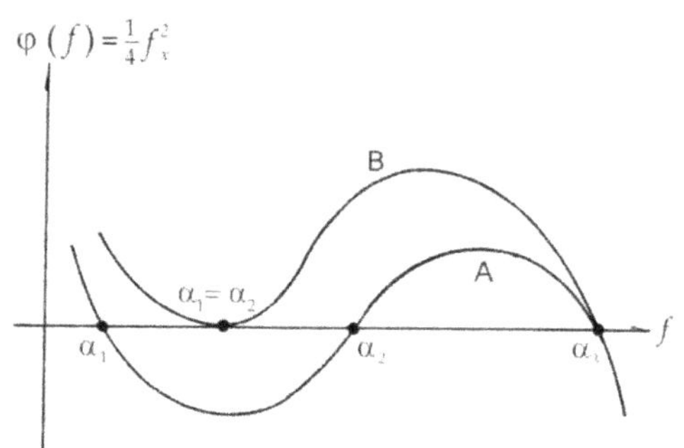

Figure 9.1. For solving the Korteweg–deVries equation. Oscillatory and solitary solutions are represented by curves A and B, respectively.

$$\xi_1 = \mathrm{sn}\left(\frac{\phi_1}{\sqrt{2}};\ \kappa\right). \tag{9.15c}$$

For the *negative potential energy* $-V(\phi_1) = 4f_x^2(\phi_1)$, the stationary solution of (9.15*a*) can then be expressed as

$$-V(\phi_1) = 2\alpha_3 - 2(\alpha_3 - \alpha_2)\mathrm{sn}^2(\sqrt{\alpha_3 - \alpha_1}\ \phi_1;\ \kappa) \text{ for } 0 < \kappa < 1, \tag{9.16a}$$

where the sn^2-function is periodic with the period

$$2K(\kappa) = 2\int_0^1 \frac{\mathrm{d}\xi}{\sqrt{(1-\xi^2)(1-\kappa^2\xi^2)}}.$$

On the other hand, the potential $V(\phi_1)$ in (9.16*a*) has a periodic interval of $\frac{2K(\kappa)}{\sqrt{\alpha_3-\alpha_2}}$, corresponding a finite amplitude $\alpha_3 - \alpha_2$. If $\alpha_2 \to \alpha_1$ for a large ϕ_1, we have curve B for $V(f)$ to correspond to $\varphi(f)$ in figure 9.1, for which $\kappa \to 1$ and $K(1) = \infty$. In this case the lattice potential energy is *real* and expressed as

Theorem:

$$-V(\phi_1) = 2\alpha_3 + 2(\alpha_3 - \alpha_2)\,\mathrm{sech}^2(\sqrt{\alpha_3 - \alpha_1}\ \phi_1) \quad \text{for } \kappa = 1, \tag{9.16b}$$

showing a pulse-shaped potential with height $\alpha_3 - \alpha_2$, propagating with effective phase $\sqrt{\alpha_3 - \alpha_1}\ \phi_1$.

Specifically, if $\alpha_3 - \alpha_2$ is infinitesimal, the modulus κ is almost zero; (9.16*b*) exhibits a sinusoidal variation

$$-V(\phi_1) = 2\alpha_3 - 2(\alpha_3 - \alpha_2)\ \sin^2(\sqrt{\alpha_3 - \alpha_1}\ \phi_1). \tag{9.16c}$$

It is noted that the coefficients in (9.16*a*) and (9.16*b*) are related with the modulus κ. By definition, we have $\alpha_3 - \alpha_2 = \kappa^2(\alpha_3 - \alpha_1)$, and hence the amplitude $\alpha_3 - \alpha_2$ is proportional to κ^2. Adjusting values of α_1, α_2 and α_3, the *directional potential* $V(\phi)$ can be made to be *compatible with the lattice symmetry in thermodynamic equilibrium*, hence representing an *adiabatic potential in crystals.*

Thus, solutions of the Korteweg–deVries equation are expressed by elliptic functions of the propagating phase, which is either oscillatory or solitary in pulse shape, depending on the modulus κ; the *negative solitary potential of* (9.16*b*) *is known as the Eckart potential*, existing in a thermodynamic environment. The pulse-shaped potential behaves like an independent particle, moving freely at a speed v in space-time in a crystal. It is noted that *only the specific case of* $\kappa = 1$ *is compatible with lattice symmetry*, and its *positive counterpart of the condensate* can be unstable for the lattice structure to change to strain-free in new equilibrium, indicating that the potential (9.16*b*) constitutes *thermal boundaries.* On the other hand, elliptic solutions for $0 < \kappa < 1$ should represent strained states of crystals. It is important to realize that only *the steady solutions* (9.16*b*) *and* (9.16*c*) *for* $\kappa = 1$ *represent equilibrium* *lattices signified by symmetry*, while those for $0 \leqslant \kappa < 1$ describe unsteady lattices in *modulated states*.

Jacobi's elliptic functions can be expanded into power series as in Landau's expansion, which can be truncated at a quartic term σ^4. It is therefore logical to consider such expansions in the Landau theory, allowing us to use a potential energy $-V = \frac{a}{2}\sigma^2 + \frac{b}{4}\sigma^4$ truncated from the elliptic expression for small σ in the adiabatic approximation. Arising from the developing operator $\boldsymbol{B}_3$, such Landau's expansion implies dispersive propagation, as discussed in section 8.1 in chapter 8. Consequently, both elliptic and truncated potentials can be used for practical evaluation of mesoscopic variables.

In the foregoing, the Korteweg–deVries equation was considered mathematically for crystals, to deal with nonlinear development, however, the physical nature of lattice symmetry has not been taken into consideration; namely, solution (9.16*b*) is physically *insufficient* as it is, in order to meet boundary conditions for the order variable vector. For that matter, in chapter 8 we already considered the complex vector $\sigma = \sigma_1 + i\sigma_\perp$, corresponding complex potential $\pm u^2 \mp i(\mathrm{d}u/\mathrm{d}x)$, for propagation in periodic domain structure. Particularly, we considered $\sigma = i\sigma_\perp$ on the boundary surface perpendicular to the direction of propagation, so that the potential (9.16*b*) should be expressed as[note1]

$$-iV(\phi_\perp)_c \propto \text{const.} - i\left(\sigma_\perp^2\right)_c, \quad \text{where } \left(\sigma_\perp^2\right)_c \propto \operatorname{sech}^2 \phi_\perp \tag{9.16d}$$

which is represented by the second term on the right, i.e. Eckart's potential.

9.4 Isothermal transitions in the Eckart potential

It is significant that adiabatic potentials are invariant with *Galilean transformation* of order variables density for the Korteweg–deVries equation, as signified by the phase $\phi_\perp = x - vt$. Under the coordinate transformation $x - vt \to x'$, the real potential $-V(x')$ of (9.16*b*) is always observable in-phase with the order variable $\sigma_1(x')^2$ in the transformed system.

Note, however, that we have to set up the boundary conditions for nonlinear waves described by the Korteweg–deVries equation to consider two components. As discussed in section 8.3 for finite crystals, we recognized that the transversal $\sigma_\perp$ is essential for entropy production. Accordingly, disregarding τ for spatial transformation, the counterpart potential $+V(x, \kappa)$ becomes proportional to $\sigma_\perp^2 \approx \operatorname{sech}^2 x$ in the limit $\kappa \to 1$, which is a direction perpendicular to shear strains of a mesoscopic domain boundary to make it *strain-free*.

The *negative* potential energy given by (9.16*b*), known as the *Eckart potential* in nonlinear dynamics, is characterized by a *solitary peak* in crystals that can be written in a form

---

[note1] For a rigorous proof, see chapter 12.

$$V(x) = -V_0 \operatorname{sech}^2\left(\frac{x}{d}\right) \quad \text{for} \quad \kappa = 1, \tag{9.17}$$

excluding the insignificant constant. Here, d is introduced to indicate the width $2d$ of the symmetric peak around $x = 0$, as shown in figure 9.2(a). It is noted that (9.17) originates from (9.16*b*), disregarding the constant term, whose transversal direction is associated with $\sigma_\perp\left(\frac{x}{d}\right)$, representing a *transversal potential.*

In contrast, the potential (9.16*a*) is oscillatory and periodic in space, as expressed by sn^2 for $0 < \kappa < 1$, such a potential is practically the same as cn^2-function, because of the relation $\mathrm{sn}^2 + \mathrm{cn}^2 = 1$. In (8.10*a*) and (8.10*b*) of chapter 8, we showed that the transverse component $\boldsymbol{\sigma}_\perp(x, \kappa)$ is in motion with an adiabatic cnoidal potential

$$V(x, \kappa) = -V_0 \mathrm{dn}^2 u(x, \kappa) \quad \text{for} \quad 0 \leqslant \kappa < 1 \tag{9.18}$$

where V_0 is a proportionality factor. Because of the relation $\mathrm{dn}^2 u = 1 - \kappa^2 \mathrm{sn}^2 u$, (9.18) gives an identical potential

$$V(x, \kappa) = -V_0 \kappa^2 \mathrm{sn}^2 u(x, \kappa) \quad \text{for} \quad 0 \leqslant \kappa < 1, \tag{9.19}$$

disregarding the additive constant. In any case, these squared elliptic functions represent a periodic potential, called a *cnoidal* curve, as sketched in figure 9.2(b), which is periodic with variable amplitude, constituting a *complex potential.* For a physical application, disregarding trivial additive constants, (9.18) and (9.19) express the adiabatic potential.

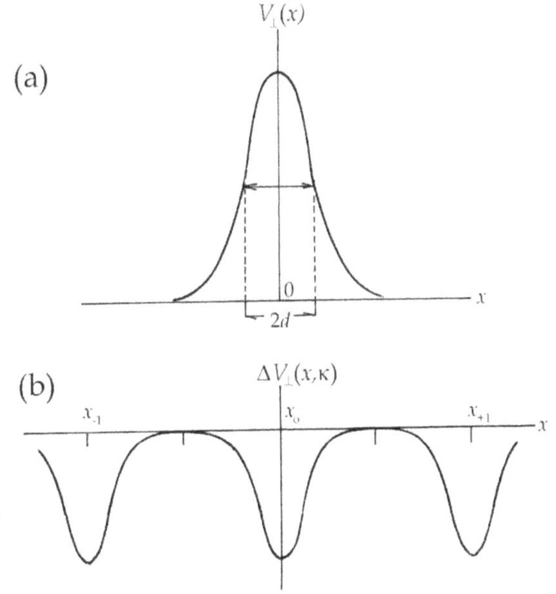

Figure 9.2. (a) An Eckart's potential of a half-width $2d$. (b) A cnoidal potential $\mathrm{cn}^2(\phi/\sqrt{2}\kappa)$ with periodic sech^2–peaks. Reproduced from [2], copyright 2010. With permission of Springer.

At this point, we pay attention to a theorem for an sn^2-function to be replaced by *transversal solitary* sech^2*-peaks* in the cnoidal expansion. Although proven mathematically by the advanced theory of elliptic functions, the rigorous proof is a little too tedious so we leave the details to chapter 12. Namely, we refer to the mathematical formula

Theorem:

$$2\kappa^2\mathrm{sn}^2x = -2a\sum_{\ell=-\infty}^{\ell=+\infty}\mathrm{sech}^2(\sqrt{a}\,x - c\ell) + \text{const.}, \tag{9.20}$$

where

$$a = \frac{\pi^2}{4K'(\kappa)^2},\quad c = \frac{\pi K(\kappa)}{K'(\kappa)} \quad \text{and} \quad K'(\kappa) = K\left(\sqrt{1-\kappa^2}\right).$$

Here, the index l specifies the peak positions at $x_l = c\ell/\sqrt{a}$ for $-\infty < \ell < +\infty$ along the x-axis; $K(\kappa)$ is the complete elliptic integral and $K'(\kappa)$ is defined in the above. However, (9.20) can also be derived from Fourier expansion of (9.19), as will be shown in chapter 11.

Owing to the formula (9.20), the cnoidal potential can be replaced *mathematically* by a periodic array of Eckart potentials located at x_ℓ in crystal space, which should be expressed within *Born–von Kármán boundaries* $-L/2 \leqslant x_\ell \leqslant L/2$, representing a collective expression of internal potentials, *thereby violating lattice symmetry only by uniaxial distortion.* Each sech_ℓ^2-term specified by the index ℓ is therefore *an adiabatic internal potential in new equilibrium at a given temperature* T_ℓ.

Physically it is significant that such a transition between these different sech_ℓ^2-potentials for different ℓ and ℓ' can be interpreted by the Born–Huang principle for $\sigma_l(x, \tau)$ to be responsible for a *thermodynamic change.* In this interpretation, we can say that *pseudospins are ordered in small opposite minidomains* in period $4K(1)$ successively in the mesoscopic state. Such a solution for $\kappa = 1$ in (9.20) can therefore represent a *steady* state in the thermodynamic environment, where the *sech*$_\ell^2$*–potential at* ℓ *is stable potential* V_ℓ *with respect to the lattice symmetry at* T_ℓ.

The soliton lattice is periodic, but generally arranged as *incommensurate* with *translational symmetry.* Nevertheless, the first Brillouin zone to deal $\sigma_\ell(x)$ with Eckart potentials (9.17) is sufficient for *pinning the related dynamics at* T_ℓ, for which surfaces located at both ends of terminal zones are in contact thermally with the surroundings. Unlike in a traditional infinite lattice, a finite soliton lattice is characterized by an *integral number of zones* and two surface elements, essential for heat exchange with surroundings.

However, we should make a remark particularly on liquid crystals, where the liquid lattice is not rigid as in normal crystals, from which the entropy production cannot be the same so that the time-temperature conversion for $+V_\ell$ should not be a valid mechanism, as explained in chapter 18.

9.5 Condensate pinning by the Eckart potentials

In this section, setting the transversal component aside, we discuss the eigenvalue problem of the longitudinal $\sigma_\ell(x, \kappa)$ in the steady Eckerd potential field $V_\ell(x, \kappa)$. Denoting the eigenvalue as λ, the perturbed wave equation is written as

$$\frac{d^2\sigma_\ell(x)}{dx^2} + \left(\lambda + V_o \operatorname{sech}^2 \frac{x}{d}\right)\sigma_\ell(x) = 0.$$

For stable pinning at T_ℓ, the eigenvalue should be negative, so that we write $\lambda = -\mu^2$. Replacing $\frac{x}{d}$ by x, and setting $V_o d^2 = v_o$ and $\mu^2 d^2 = \beta^2$, this equation can be expressed at T_l as

$$\frac{d^2\sigma_\ell(x)}{dx^2} + (-\beta^2 + v_o \operatorname{sech}^2 x)\sigma_\ell(x) = 0. \tag{9.21}$$

This is a differential equation familiar in mathematical physics, whose solution is expressed by *hypergeometric series*. Following Morse and Feshbach [3], we transform (9.21) to the *hypergeometric equation* in the standard form.

By defining the relation $\sigma_\ell(x) = Af(x)\operatorname{sech}^\beta x$, the differential equation for the function $f(x)$ can be obtained as

$$\frac{d^2f}{dx^2} - 2\beta(\tanh x)\frac{df}{dx} + (v_o - \beta^2 - \beta)(\operatorname{sech}^2 x)f(x) = 0.$$

Further, this equation can be modified by another variable $\zeta = \frac{1}{2}(1 - \tanh x)$ as

$$\zeta(1-\zeta)\frac{d^2f}{d\zeta^2} + (1+\beta)(1-2\zeta)\frac{df}{d\zeta} + (v_o - \beta^2 - \beta)f(x) = 0.$$

We define such parameters a, b and c that satisfy the relations

$$a + b = 2c - 1,$$

where

$$c = 1 + \beta \quad \text{and} \quad ab = -v_o + \beta^2 - \beta;$$

that is,

$$a,\ b = \frac{1}{2} + \beta \pm \sqrt{v_o + \frac{1}{4}}.$$

Using a, b and c, we obtain the hypergeometric equation expressed in standard form, i.e.

$$\zeta(1-\zeta)\frac{d^2f}{d\zeta^2} + \{c - (a + b + 1)\zeta\}\frac{df}{d\zeta} - abf = 0. \tag{9.22}$$

Expressing as $f(\zeta) = F(\mathrm{a, b, c}; \zeta)$, the solution of (9.22) is called a *hypergeometric function*.

The mesoscopic pseudospin variable $\sigma_\ell(x, \kappa)$ can then be expressed as

$$\sigma_\ell(x) = A\,\mathrm{sech}^\beta x F(\mathrm{a, b, c};\ \zeta). \tag{9.23}$$

However, we are only interested in extreme cases that can be specified by $\zeta \to 0$ and $\zeta \to 1$, in order to use it physically at distant points $x \to \pm\infty$, for which we expand $F(\mathrm{a, b, c}; \zeta)$ as

$$F(\mathrm{a, b, c}; \zeta) = 1 + \frac{\mathrm{ab}}{1!\mathrm{c}}\zeta + \frac{\mathrm{a(a+1)b(b+1)}}{2!\mathrm{c(c+1)}}\zeta^2 + \cdots. \tag{9.24}$$

Corresponding to $\zeta \to 0$, for $x \to +\infty$ we have

$$\sigma_\ell(x)_{x\to\infty} \to A2^\beta \exp(-\beta x),$$

whereas for $\zeta \to 1$, we consider $x \to -\infty$.

For calculating $\sigma_l(x)$ in these cases, it is convenient to use the following formula [4]:

$$\begin{aligned} F(\mathrm{a, b, c}; \zeta) &= \frac{\Gamma(\mathrm{c})\Gamma(\mathrm{c-a-b})}{\Gamma(\mathrm{c-a})\Gamma(\mathrm{c-b})} F(\mathrm{a, b, a+b-c+1}; \zeta) \\ &\quad + (1-\zeta)^{\mathrm{c-a-b}} \frac{\Gamma(\mathrm{c})\Gamma(\mathrm{a+b-c})}{\Gamma(\mathrm{a})\Gamma(\mathrm{b})} F(\mathrm{c-a, c-b, c-a-b+1}; 1-\zeta), \end{aligned}$$

where $\Gamma(\cdots)$ are *gamma functions.* It is noted that the first term dominates $F(\mathrm{a, b, c}; \zeta)$, if $\zeta \to 0$. On the other hand, if $\zeta \to 1$ we have $(1-\zeta)^{\mathrm{c-a-b}} \approx 2^\beta \exp(-\beta x)$ and $F(\ldots; 1-\zeta) \to 1$ in the second term, dominating $F(\mathrm{a, b, c}; \zeta)$. Considering a small β, $\mathrm{sech}^\beta x \approx 2^\beta \exp(+\beta x)$ and $F(\mathrm{a, b}, 1+\beta; \zeta) \to 1$ in the first term. In this case, writing $\beta = -ik$ to express propagating waves, we can write that

$$\sigma_\ell(x)_{x\to+\infty} \propto \frac{\Gamma(\mathrm{c})\Gamma(\mathrm{a+b-c})}{\Gamma(\mathrm{a})\Gamma(\mathrm{b})} \exp(+ikx) + \frac{\Gamma(\mathrm{c})\Gamma(\mathrm{c-a-b})}{\Gamma(\mathrm{c-a})\Gamma(\mathrm{c-b})} \exp(-ikx), \tag{9.25}$$

Using definitions of a, b and c, we notice that the numerators of these coefficients are $\Gamma(1+\beta)\Gamma(-\beta)$ and $\Gamma(1+\beta)\Gamma(\beta)$, respectively, whereas the denominators have singularities, depending on the value of v_0. In figure 9.3 is shown the curve of a *gamma function* $\Gamma(z)$ plotted against the variable z, where there are a number of poles at $z = 0, -1, -2, \ldots$. Hence, the first coefficient has singularities at

$$\mathrm{a, b} = \tfrac{1}{2} + \beta \pm \sqrt{v_0 + \tfrac{1}{4}} = -\ell, \tag{9.26}$$

whereas the second coefficient can be specified by singularities for the denominator to become infinity, i.e.

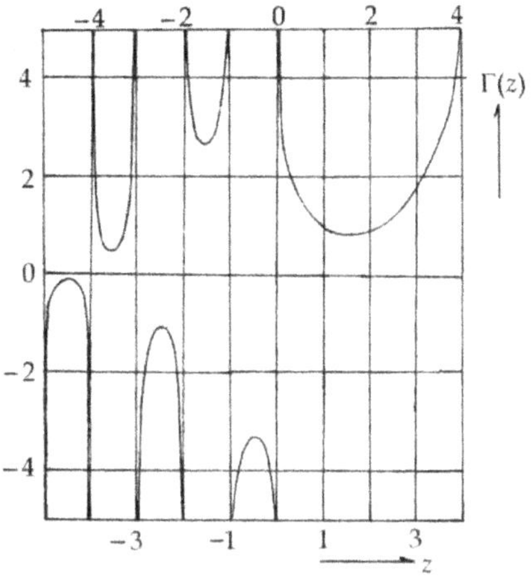

Figure 9.3. Gamma function $\Gamma(z)$.

$$\Gamma(c - \mathrm{a})\Gamma(c - \mathrm{b}) = \Gamma\left(\frac{1}{2} + \sqrt{v_o + \frac{1}{4}}\right)\Gamma\left(\frac{1}{2} - \sqrt{v_o + \frac{1}{4}}\right) = \frac{\pi}{\cos\left(\pi\sqrt{v_o + \frac{1}{4}}\right)} = \infty,$$

hence

$$\sqrt{v_o + \frac{1}{4}} = n + \frac{1}{2} \quad \text{or} \quad v_o = n(n + 1) \tag{9.27}$$

where n is any integer. Combining (9.26) and (9.27), from the constant a, we obtain

$$\beta = \left(n + \frac{1}{2}\right) - \ell - \frac{1}{2} = n - \ell, \tag{9.28}$$

where $l = 0,\ 1,\ 2,\ \ldots,\ n - 1$, corresponding to $T_0, T_1, T_2, \ldots, T_{n-1}$. Therefore the eigenvalues for $\beta > 0$ can be represented as

$$\beta_\ell = n - \ell = n,\quad n - 1, n - 2, \ldots \ldots, 1. \tag{9.29}$$

Imposing conditions for no transmission and reflection on (9.21), the pseudospin wave $\sigma_\ell(x)$ can be in-phase with the soliton potential $V_\ell(x)$. Therefore, by expressing the potential by eigenvalues n, the steady-state equation (9.14) for longitudinal wave functions $\sigma_\ell(x)$ can be written as

$$\frac{\mathrm{d}^2\sigma_\ell(x)}{\mathrm{d}x^2} + \left\{-\beta_\ell^2 + \ell(\ell + 1)\,\mathrm{sech}^2\,x\right\}\sigma_\ell(x) = 0. \tag{9.30}$$

Writing equation (9.30) for two adjacent levels at n and $n - 1$, we have

$$\frac{\mathrm{d}^2\sigma_n}{\mathrm{d}x^2} = \{-n^2 + n(n + 1)\,\mathrm{sech}^2\,x\}\sigma_n = 0$$

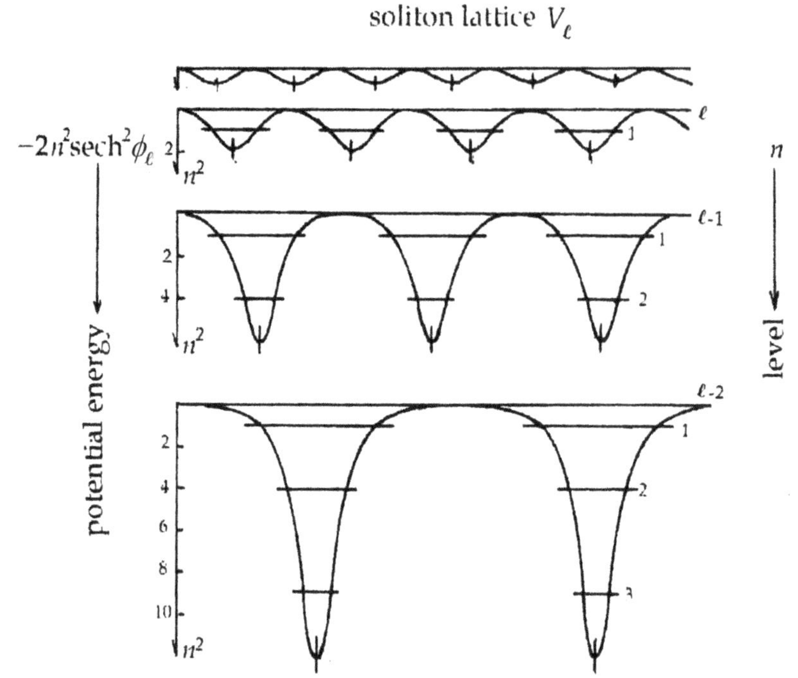

Figure 9.4. Soliton levels in cnoidal potentials.

and

$$\frac{d^2\sigma_{n-1}}{dx^2} + \left\{-(n-1)^2 + (n-1)n\ \mathrm{sech}^2 x\right\}\sigma_{n-1} = 0.$$

Hence, for a change $\sigma_{n-1} \rightarrow \sigma_n$ the potential value decreases by $\Delta V_{n-1,n} = -2n\mathrm{sech}^2 x$ that is a difference between discrete soliton potential energies. We consider that such transitions can take place as a thermal relaxation process with varying temperature. Although dynamically inaccessible, such transitions should be observed thermally as a consequence of phonon scatterings. Corresponding lattice displacements u_n should occur in a relaxation process between strain energies related to $V_{n-1}^2 \rightarrow V_n^2$, corresponding to $T_{n-1} \rightarrow T_n$ in virtue of the equipartition theorem.

Figure 9.4 shows a change of the periodic potential (9.20) with lowering temperature, where eigenvalues of the pseudospin density are indicated in each of the periodic Eckart potentials. Although not shown in the figure, the density levels are practically in a band structure whose widths are significantly broader at smaller n. In any case, stepwise thermal relaxation processes can logically be considered for transitions toward lowest level $-n(n+1)\mathrm{sech}^2 x$ in the potential well at a given temperature, while in the lattice counterpart, an entropy production occurs for the whole crystal to be stable at new symmetry. Therefore, the collective pseudospin σ_n and corresponding internal Weiss field together in-phase and combined to propagate can be called a *condensate* for mesoscopic order by analogy to classical condensation phenomena, as discussed in chapter 4.

9.6 Elemental solitons as Boson particles

Properties of the adiabatic potential represented by $\mathrm{sech}^2 x$ were investigated extensively by computer simulation of the Korteweg–deVries equation, producing convincing results that nonlinear waves describe collective motion of *quantized* pulses of *soliton particles* counted by an integral number, which is regarded as a thermodynamic variable $n = n(T, p)$. Due to a specific pulse-like sech^2–potential, which behaves like *free moving particles in the x–t plane*, this is named a *soliton* after its mathematical behavior. The *quantized solitons* behave like independent *quasi-particles* in a gas phase characterized by 'n' number of identical solitons, which are shown to obey the *Bose–Einstein statistics.*

In chapter 1, phonons were discussed like quantized particles of energy $\hbar\omega_k$ and momentum $\hbar \boldsymbol{k}$, where k is the wavenumber associated with the periodic lattice. It is noted that such quantization in field theory can not only be applied to phonons, but also to solitons moving through a periodic structure at a constant energy λ_n. *Creation* and *annihilation* operators, i.e. $\tilde{b}_n^\dagger$ and $\tilde{b}_n$, can therefore be applied for *soliton particles* to be governed by the commutation relations for *boson particles*, as explained in [5, 6].

Denoting the wavefunction of a collective mode of n solitons as σ_n, we have wave equation

$$\frac{d^2\sigma_n}{dx^2} + \{\lambda_n + V(n)\}\sigma_n = 0 \quad \text{where} \quad \sigma_n(x, t) = \exp(-i\tilde{\omega}_n t)\sigma_n(x).$$

For the eigenvalue λ_n and the corresponding frequency ω_n, we can define creation and annihilation operators at n by $\tilde{b}_n(t) = \tilde{b}_n(0)\exp(-i\tilde{\omega}_n t)$ and $\tilde{b}_n^\dagger(t) = \tilde{b}_n^\dagger(0)\exp(+i\tilde{\omega}_n t)$, as shown in chapter 3, thereby the wavefunction of the soliton field can be expressed as

$$\sigma(x, t) = \sum_{1\ldots n} \tilde{b}_n(t)\sigma_n(x) \quad \text{and} \quad \sigma^*(x, t) = \sum_{1\ldots n} \tilde{b}_n^\dagger(t)\sigma_n^*(x).$$

Accordingly, the soliton Hamiltonian H_n can be calculated with commutation relations

$$\left[H_n, \tilde{b}_n^\dagger \tilde{b}_n\right] = 0 \quad \text{and} \quad \left[\tilde{b}_{n'}, \tilde{b}_n^\dagger\right] = \delta_{n'n} \tag{9.31}$$

indicating that particles represented by $\tilde{b}_n^\dagger$ and $\tilde{b}_n$ are a *boson* characterized by elementary solitons of each energy λ_o, and $n = \sum_{1,\ldots,n} \tilde{b}_n^\dagger \tilde{b}_n$.

Hence, we have the expression $H_n = \lambda_o \tilde{b}_{n1}^\dagger \tilde{b}_{n2}$ composed of a wave packet of *two coherent* mobile particles signified by phases 1 and 2, when applied to a domain boundary, which is similar in form to $H_{\text{phonon}} = \sum_k \hbar\omega_k b_k^\dagger b_k$, where the lowest phonon energy $\frac{1}{2}\hbar\omega_o$ is considered at $k = 0$ and $T = 0$ K, if $p = \text{const}$.

These boson operators can thus be used for *coherent solitons* in favor of *symmetric boson properties*, to deal with stable lattice symmetry in an equilibrium Hamiltonian, which is therefore expressed as follows.

Theorem:

$$\langle \boldsymbol{H}_n \rangle_{\text{thermal}} = \mp V_n = \mp n\lambda_o \propto \mp \langle \sigma_{\perp 1} \, . \, \sigma_{\perp 2} \rangle_{\text{thermal}} = \mp \left\langle \sigma_\perp^2 \tilde{b}_{n1}^\dagger \tilde{b}_{n2} \right\rangle_{\text{thermal}} \tag{9.32}$$

for symmetric and antisymmetric soliton combinations, respectively. This relation implies that the soliton field in a thermodynamic environment is signified by a *singularity* due to symmetric and antisymmetric fluctuations of $\mp V_n$, which is responsible for transitions, as discussed in chapter 4.

Regarding the negative $-\langle \sigma_\perp^2 \rangle_{\text{thermal}}$, the negative potential $-V_n$ signifies the *stability of domain structure*, while the positive $+V_n$ can lead to *entropy production* with phonon interactions in the *condensate* of $\boldsymbol{\sigma}$ in the lattice. In other words, $\sigma_\perp^2$, which is given as the equilibrium solution for $\kappa = 1$ determines physically essential properties of the domain and surfaces in finite crystals.

For boundary walls between mesoscopic domains 1 and 2, as illustrated by figure 9.5, we need to consider *phase fluctuations* $\Delta\phi_\pm$ *of coherent* $\sigma_\perp(\phi)$ across each boundary plane, whereby anomalies arise from an energy gap $n\lambda_o$ between antisymmetric and symmetric fluctuations $\Delta\phi_+ - \Delta\phi_-$ and $\Delta\phi_+ + \Delta\phi_-$, respectively, at a given temperature T, which is due to the work

$$\left\langle \int_{-\pi/2}^{+\pi/2} \sigma_\perp(\phi) d\phi \right\rangle_{\text{thermal}} = \frac{1}{2} \left\langle \sigma_\perp^2 \right\rangle_{\text{thermal}}, \tag{9.33}$$

thereby the crystal volume V cannot be constant. Unlike critical anomalies discussed in chapter 4, the temperature T is arbitrary for *adiabatic processes* for $\Delta V \neq 0$ and $\Delta n \neq 0$. On the other hand, the conventional transition temperature T_c is specified by $\Delta n = 0$ but a large n associated with a seed *cluster, exhibiting the characteristic feature of order variables.*

Owing to *symmetric two-boson correlations in the present case*, the critical energy at T_c is always lowered from the conventional case at $\varepsilon = 0$, confirming the *structural stability* determined by *a negative hyperbolic potential* $-\text{sech}^2 \phi_c$, *which is in fact consistent with thermodynamic principles.*

For the pseudospin $\sigma_n(\phi)$ expressed by the phase ϕ, it should be noted that the classical displacement is not quantum-mechanically *acceptable* in the vicinity of $\phi = 0$, because of nonzero distributed $\lambda_o \neq 0$, corresponding to a small but finite correlation potential proportional to $\text{sech}^2 \phi_c$.

In contrast, the antisymmetric boson correlations give rise to phase fluctuations proportional to $\langle \Delta\phi_1 \Delta\phi_2 \rangle_{\text{thermal}}$ that is observed as critical anomalies at $\phi = 0$. Nevertheless, the thermal fluctuations determined by *nonlinear displacements*

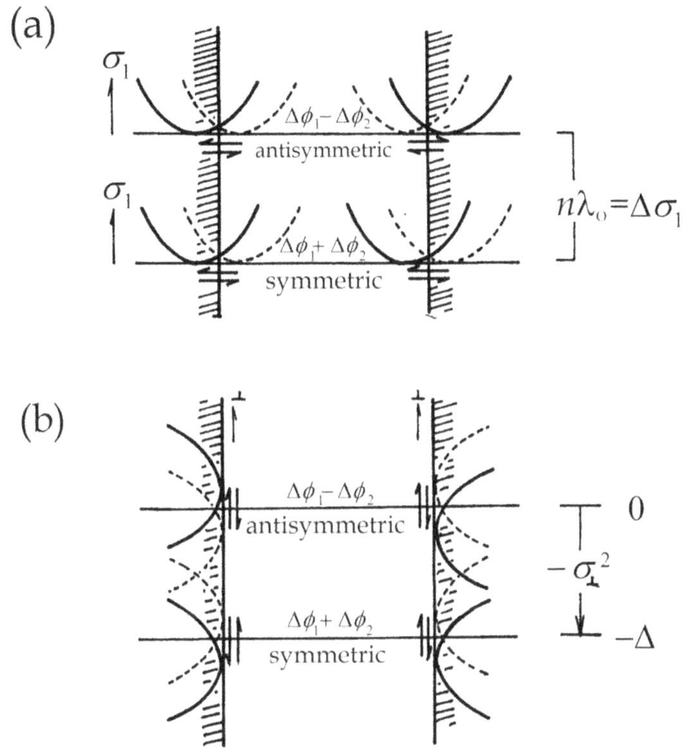

Figure 9.5. Critical anomalies at domain boundaries, where symmetric boson fluctuations give rise to lower energy states at a given temperature T_ℓ that is signified by the soliton number $n(T_\ell)$. (a) Longitudinal phase fluctuation $\Delta\phi$ of $\Delta\sigma_1$ across a domain boundary. (b) Transversal fluctuation energy proportional to $-\sigma_\perp^2$ assures always the structural stability of the domain wall, because of symmetrical two boson correlations.

represent the adiabatic potential at T_c, for which an inversion process $\boldsymbol{q} \rightleftarrows -\boldsymbol{q}$ of soliton energy

$$\mathcal{H} \to \mp\langle\mathcal{H}\rangle_{\text{thermal}} = \mp V_n \tag{9.34}$$

occurs in practice, characterized by $\kappa \to 1$ or $\boldsymbol{k} \to \boldsymbol{q}$. Expressed by the Born–Huang principle, the *negative* $-V_n$ gives the equilibrium potential determined by *symmetric combination* $\Delta\phi_+ + \Delta\phi_-$, whereas *positive* $+V_n \leqslant k_B T$ can make the crystal *unstable by antisymmetric combination* $\Delta\phi_1 - \Delta\phi_2$, leading to *entropy production* as will also be discussed in the next section.

Summarizing the above arguments, the following can be stated as follows.

Theorem: In mesoscopic equilibrium of a finite crystal, the coherent order variable $\sigma_n(\phi)$ is proportional to the Eckart potential $V_n(\phi)$, which is quantized as represented by the soliton number n.

9.7 Riccati's thermodynamic transitions

In section 9.5, we noticed that thermal or adiabatic transitions between soliton states take place in crystals as $V_{n-1}^2 \to V_n^2$ for $T_{n-1} \to T_n$, indicating the role played by *transverse components* σ_{n-1} and σ_n that need to be investigated to deal with these transitions, analogous to critical anomalies discussed in chapter 4, therefore called a Born–Huang transition. There is the mathematical *Riccati theorem* for such transitions to be due to *transversal components* signified by a soliton number n.

Taking two complex $\sigma_1(x)$ and $\sigma_2(x)$ at different temperatures T_1 and T_2, respectively, into consideration, we have corresponding transversal potentials $V_1(x)$ and $V_2(x)$ for their wave equations, which can be written as

$$\boldsymbol{D}^2\sigma_1(x) = \left\{-\beta^2 + V_1(x)\right\}\sigma_1(x) \text{ and } \boldsymbol{D}^2\sigma_2(x) = \left\{-\beta^2 + V_2(x)\right\}\sigma_2(x). \tag{9.35}$$

We assume that these order variables are connected by the relation

$$\sigma_2(x) = A(x, \beta)\sigma_1(x) + B\boldsymbol{D}\sigma_1(x), \tag{9.36}$$

which is adequate for the present analysis, while B is considered to be a small constant. Substituting (9.36) into (9.35), and setting the front factors of σ_1, ignoring $\boldsymbol{D}\sigma_1$ equal to zero, we obtain the relations

$$A_{xx} + V_{1x} + A(V_1 - V_2) = 0 \text{ and } 2A_x + (V_1 - V_2) = 0, \tag{9.37}$$

respectively. Eliminating $V_1 - V_2$ from (9.37), we derive the relation $A_{xx} - 2A_xA - V_{1x} = 0$, which is integrated as

$$A^2 - A_x - V_1 = -\tilde{\beta}^2,$$

where $-\tilde{\beta}^2$ is a constant of integration. Known as *Riccati's equation*, this expression can be linearized by transforming A to $\tilde{\sigma}(x)$ defined as

$$A = -\frac{\boldsymbol{D}\tilde{\sigma}}{\tilde{\sigma}} = -\boldsymbol{D}(\ln\tilde{\sigma}), \tag{9.38}$$

resulting in

$$\boldsymbol{D}^2\tilde{\sigma}(x) = \left\{-\tilde{\beta}^2 + V_1(x)\right\}\tilde{\sigma}(x).$$

This is identical to equation (9.35), if the function $\tilde{\sigma}(x)$ of (9.38) is obtained when $\tilde{\beta} = \beta$. In this case, using (9.38) in the second equation of (9.37), we can write

$$\boldsymbol{D}^2\sigma_2(x) = \left\{-\beta^2 + V_1(x) - 2(\ln\tilde{\sigma})\right\}\sigma_2(x) \tag{9.39}$$

and this leads to the following.

Theorem:

$$V_2(x) - V_1(x) = -2(\ln\tilde{\sigma}), \tag{9.40}$$

representing a significant property of the adiabatic transition between $V_1(x)$ and $V_2(x)$ observable as emission or absorption of a discrete energy $\mp 2 \ln \tilde{\sigma}$ to and from the modulated lattice, respectively.

In the above analysis, the variable x represents the phase of propagation, and the functions $\sigma_1(x)$ and $V_1(x)$ are *conformal* for the coordinate translation $x \rightarrow x - vt$, where v is constant, and t is an arbitrary temporal parameter. For normal thermodynamic applications, we consider for a *thermal process* $V_1(x) \rightleftarrows V_2(x)$ to be determined by the condensate for $T_1 \rightleftarrows T_2$ that can signify energy transfer to the lattice. Further significant is that the potential change $\Delta V = -2(\ln \tilde{\sigma})$ in (9.40) is temperature dependent, representing an adiabatic transfer of energy.

As the potential change $\Delta V(x - vt)$ occurs in the lattice, we have

$$\Delta V(x - \alpha t) = \left(\frac{\partial V}{\partial x}\right)_t \Delta x - v\left(\frac{\partial V}{\partial t}\right)_x \Delta t. \tag{9.41}$$

The first term on the right is expressed as

$$\left(\frac{\partial V}{\partial x}\right)_\tau \Delta x = 2\frac{\tilde{\sigma}'}{\tilde{\sigma}}\Delta x, \tag{9.42}$$

representing an *adiabatic change* if $\tilde{\sigma}'/\tilde{\sigma}$ is constant; whereas the second term

$$-v\left(\frac{\partial V}{\partial t}\right)_x \Delta t = -\frac{2v\tilde{\sigma}'}{\tilde{\sigma}}\Delta t, \tag{9.43}$$

is related to an *isothermal change* for $\Delta t =$ const. Considering $\Delta t \propto \Delta T$, like soft modes, the process (9.41) may be called *quasi-adiabatic* at constant volume $\Delta V = 0$, or *adiabatic* at constant pressure p.

In fact, Riccati's transformation (9.38) can be interpreted for the vector function $\tilde{\sigma}$ to rotate 2π in the complex plane, in other words inversion $\tilde{\sigma} \rightleftarrows -\tilde{\sigma}$, so that (9.42) and (9.43) are not independent, and the adiabatic process of (9.42) cannot be fully temperature independent.

When subjected to phonon scatterings for energy transfer to the lattice, the *positive potential change* $+\Delta V$ is related to a transversal $\sigma_\perp{}^2$, and can be calculated as proportional to

$$\sum_{\boldsymbol{K}'} \langle \boldsymbol{K} + \boldsymbol{q} | \sigma_\perp | \boldsymbol{K}' \rangle \langle \boldsymbol{K}' | \sigma_\perp^* | -\boldsymbol{K} + \boldsymbol{q} \rangle,$$

where $\boldsymbol{K}$, $\boldsymbol{K}'$ are *phonon wavevectors*, and $\boldsymbol{q}$ represents a wavevector for *lattice displacement*, which can be evaluated with *nonzero* phonon densities at $\boldsymbol{K} + \boldsymbol{q}$ and $-\boldsymbol{K} + \boldsymbol{q}$ for symmetric combination of $\sigma_\perp$ and $\sigma_\perp^*$ for possible entropy production. However, for critical fluctuations discussed in chapter 4, it is noted that $\pm\boldsymbol{q}$ do not contribute to $\sigma_\perp^2$, because of $\Delta n = 0$ specifically, otherwise there is significant probability for entropy production to lose energy as heat dissipation at domain boundaries. However, such a probability can generally exhibit changes in temperature and soliton number, if the thermodynamic state remains stable.

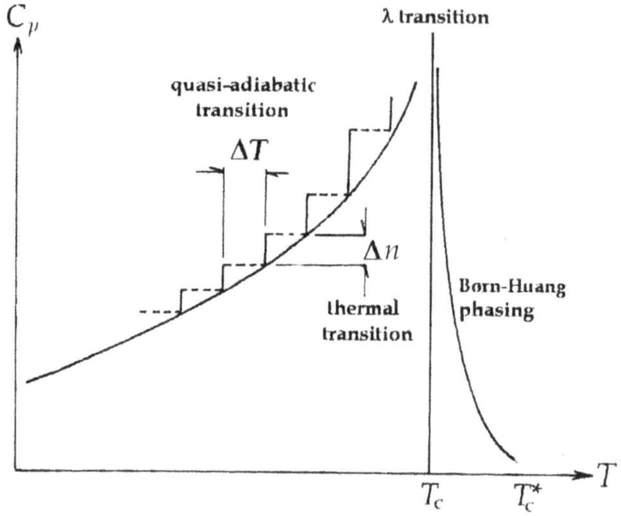

Figure 9.6. Schematic anomalies in the specific-heat curve $C_p - T$ of a λ transition. The steep rise for $T > T_c$ is due to Born–Huang phasing. The slow varying tail for $T < T_c$ consists effectively of a sequence of adiabatic and isothermal steps, characterized by Δn and ΔT, respectively. The adiabatic change is signified by a discontinuous change Δn in soliton number.

As applied to the conventional critical point T_c, the weak potential hump assumed at $\phi = 0$ may be considered the same as $\mathrm{sech}^2\,\phi$, which is, however, obscured in practice by uncertainties of microscopic phases.

On the other hand, the *negative* potential change $-\Delta V$ captures $-\Delta\sigma_\perp^2$ during a step for $\Delta n \neq 0$ that signifies a finite ΔT. Using the *equipartition theorem* for equilibrium states, $\left(\frac{\partial V}{\partial x}\right)_T$ and $\left(\frac{\partial V}{\partial t}\right)_x$ for phonon scatterings by the transversal $\sigma_\perp$ and $\sigma_\perp{}^*$ can be assigned, respectively, to $\Delta \boldsymbol{K} = 0$ and $\Delta \boldsymbol{K} = \boldsymbol{q}$ for $\Delta T = 0$ and for $\Delta n \neq 0$, $\Delta T = T_2 - T_1$ processes, respectively.

The former is for an isothermal process discussed in section 4.2.3 of chapter 4, whereas the latter process cannot be fully adiabatic, as involved in weak thermal relaxation, if $\Delta V \neq 0$. The expression (9.42) may therefore be called *quasi-adiabatic transitions* that is involved in a volume change Δ per ΔT, namely the quasi-entropy change $p\Delta / \Delta T = \Delta S^*$ can be produced in a crystal, where p is the external pressure.

Assuming that the transformation (9.38) considered for $\sigma_1 \to \sigma_2$ represents a Born–Huang transition, we can explain the corresponding experimental results. On the other hand, subject to phonon scatterings in crystals, a discontinuous change $\Delta\sigma_\perp = \sigma_{1\perp} - \sigma_{2\perp}$ shows singularity in the transversal lattice potential $\Delta V_\perp = V_1 - V_2$, corresponding to the temperature change ΔT.

The temperature-dependence in the tail of the specific heat anomaly below T_c, in so-called λ transitions, can therefore be analyzed as illustrated in figure 9.6. It is interesting to see that the tail for $T < T_c$ can be schematically interpreted as *stepwise thermal* and *quasi-adiabatic* changes, i.e.

$$\beta_\ell - \beta_{\ell+1} \propto T_\ell - T_{\ell+1} = \Delta T \quad \text{and} \quad \Delta V_{\ell,\ell+1} = 2n\text{sech}^2 x \propto \Delta n,$$

as indicated in the figure, respectively; which correspond to the *devil's staircase* in soliton theory.

However, if the critical fluctuations discussed in chapter 4 are interpreted as due to emission of elemental soliton particles, we have to consider change in strain energy of the lattice that creates uncertainties. Accordingly, the uncertainty for initial soliton emission is similar to quantum-mechanical *photon emissions from electronic systems*. In contrast, for coherent soliton emission $\sigma_{n+1} \to \sigma_n$ for $T < T_c$, there is no uncertainty in symmetric crystalline media.

In thermodynamics, the Gibbs function of a mesoscopic thermodynamic state can be expressed by

$$G(n_\sigma, n_\text{L}; p, T) = G_\sigma(n_\sigma; p, T) + G_\text{L}(n_\text{L}; p, T), \tag{9.44}$$

where n_σ and n_L are numbers of solitons in the order-variable system and the lattice, respectively, in thermodynamic equilibrium. At constant p but with varying T, we showed in the above that $n_\sigma = n_\sigma(T) \propto T$ and $\Delta n_\sigma \propto \Delta T$, hence $\Delta n_\text{L} \propto -\Delta T$. To express the long-range order for G_σ, we define the chemical potential μ_σ for the system of solitons with the conventional relation $\Delta G_\sigma = -\mu_\sigma \Delta n_\sigma \propto \Delta T$, thereby responsible for entropy production $\Delta S_\sigma(T) = (\mu_\sigma/T)\Delta n_\sigma(T)$ for $T < T_c$ [7].

Signified by equilibrium conditions, however, the Gibbs potential in an isothermal process is characterized by the cnoidal potential (9.14) expressed by a series of *negative Eckart's potentials*. According to section 9.6, each negative Eckart potential can *accommodate n* elemental solitons in equilibrium, while the *positive cnoidal potential* at a given T represents a lattice of positive Eckart potentials $+V_l(n)$, where the index n indicates the lattice site. Such a *soliton lattice* of $+V_l(n)$ is a significant concept for the lattice dynamics in later discussions.

In section 9.2, the Korteweg–deVries equation was derived in thermodynamic equilibrium, the solutions for which are restricted to $\kappa = 1$ for symmetrical crystals. On the other hand, in section 8.4 of chapter 8, we consider the Klein–Gordon equation for nonlinear waves in unspecified media. If soliton potentials can be found from the Klein–Gordon equation, those specific solutions can be accepted logically for non-equilibrium states to be stabilized by Born–Huang's principle. Accordingly, the complex potentials in equations (8.17*a*, *b*), i.e. $\pm u^2 \pm i\frac{\mathrm{d}u}{\mathrm{d}x}$, can be considered as consequent on the Korteweg–deVries equation. Namely, for λ-transitions shown in figure 9.5, the step Δn is associated with a complex potential $-\Delta n\left(\text{sech}^2 \phi \pm i\frac{\mathrm{d}\,\text{sech}\,\phi}{\mathrm{d}x}\right)$.

It is significant to realize that the soliton potential exists at domain boundaries and surfaces, but not at the critical point at T_c, because the energy gap 2Δ between $P-$ and $A-$modes of initial clusters cannot be determined as a solution of Korteweg–deVries' equation. The gap at T_c should be associated with the minimum soliton excitation $\lambda_0/2$, analogous to zero-point energy of phonons. Since there is no adiabatic potential at T_c, the positive energy state is unstable as characterized by antisymmetric boson correlations, which is consistent with Landau's postulate for critical anomalies discussed for (3.7*a*) in chapter 3. In contrast, mesoscopic domain boundaries formed with $\Delta n \neq 0$ at lower temperatures are stable in a negative Eckart's potential supported by symmetric displacements.

In the above arguments on soliton potential energies $\mp\Delta V_\ell$, we considered $-\Delta V_\ell$ and $+\Delta V_\ell$ for *pinning collective* σ_l and *entropy production at* T_ℓ, respectively, corresponding to the classical concept of a *condensate* composed of displacive σ_ℓ and lattice displacement, for which *Newton's action–reaction principle* holds as $-\Delta V_\ell + \Delta V_\ell = 0$. Consistent with the thermodynamic model of condensates, it is significant that the nature of the *internal Weiss field is clearly indicated by the soliton dynamics in the asymptotic approximation* (9.8*c*).

Exercises

1. Show that constant eigenvalues as specified by $\lambda_\tau = 0$ and $\kappa = 1$ are required for a crystal in equilibrium with surroundings. Hence, the adiabatic potential is always proportional to $\text{sech}^2 x$, where the coupling with the lattice is signified by a kink of $\tanh x$ at $x = 0$.
2. Review the mathematical argument on the Eckart potential characterized by discrete values with respect to the number n of elemental solitons. Discuss how and why the negative Eckart potential determines thermal stability.
3. Discuss the way that $\Delta G(n)/\Delta n$ determines the energy transfer rate to the lattice, leading to entropy production.

References

[1] Lamb G L Jr 1980 *Elements of Soliton Theory* (New York: Wiley)
Fujimoto M 2014 *Introduction to the Mathematical Physics of Nonlinear Waves* (Bristol: IOP Publishing)

[2] Fujimoto M 2010 *Thermodynamics of Crystalline States* (Berlin: Springer)

[3] Morse P M and Feshbach H 1953 *Methods of Theoretical Physics* (New York: McGraw-Hill) p 1651

[4] Abramowitz M and Stegan I A 1964 *Handbook of Mathematical Functions* (Gaithersburg, MA: National Bureau of Standards Applied Math Series), p 253

[5] Haken H 1973 *Quantenfeldtheorie des Festkörpers* (Stuttgart: B. G. Teubner)

[6] Henley E M and Thirring W 1962 *Elementary Quantum Field Theory* (New York: McGraw-Hill)

[7] Prigogine I 1955 *Introduction to Thermodynamics of Irreducible Processes* 2nd edn (New York: Wiley) ch 3

Chapter 10

Soliton mobility in dynamical phase space: time–temperature conversion for thermal processes

In experiments, soliton potentials have been detected as mobile solitary objects in crystals. Although limited to equilibrium and quasi-static phases, however, the soliton mobility with Galilean invariance from one site of space–time to another is confirmed in non-equilibrium phases as well, representing the space–time scale in crystals, which is conveniently expressed for the wavevector in a Brillouin zone.

Originating from lattice displacements, solitons are *quantized* objects by Reccati's theorem, as expressed by a number of pulses between discrete levels in Eckart's potentials, behaving like *free quasi-particles* in crystals. While such *soliton particles* are characterized by *boson* statistics in the field theoretical approximation, electron-pairs in a superconducting phase of a metal at very low temperatures travel as *charged fermion particles* in momentum space; exhibiting *Meissner effects*, as will be explained in chapter 14. In this chapter, the theoretical basis for soliton mobility in crystalline media under *time–temperature conversion* is discussed, prior to its application to metals and binary magnetic systems in chapters 14, 15 and 16, respectively. *The time–temperature conversion is necessary for the lattice dynamics to represent crystalline processes.*

10.1 Bargmann's theorem

In this section, Bargmann's mathematical theorem (1949) in nonlinear dynamics is discussed, showing that the steady Eckart's potential exhibits a wave in finite magnitude, behaving like an intense coherent pulse in the *dynamical phase space*. The theorem is indirectly in proof of the soft mode and related internal Weiss potentials in *conservative dynamics in crystals.*

doi:10.1088/978-0-7503-2572-1ch10

10.1.1 One-soliton solution

First, to a wave equation $\frac{d^2\sigma}{dx^2} + q^2\sigma = 0$ for a free running soft-mode wave $\sigma \sim \exp(\pm iqx)$ for an elemental soliton at T_l, we consider a potential V for the lowest-energy excitation at wavevector q, so that the propagation equation

$$\frac{d^2\sigma}{dx^2} + (q^2 - V)\sigma = 0 \tag{10.1}$$

is the same as the first equation of (8.17*c*) in chapter 8, providing, however, *amplitude-modulated waves* expressed by $\sigma = F(q, x)\exp iqx$ under equilibrium conditions, where the amplitude $F(q, x)$ is assumed as a polynomial of q.

In this case, writing

$$\sigma = \exp iqx\{2q + ia(x)\}, \tag{10.2}$$

the potential V can be zero for $q = 0$ for $a(x)$ to be a function of x, whose magnitude can increase with increasing q.

Differentiating σ_1 for (10.1), we have

$$\frac{da}{dx} = -V \quad \text{and} \quad \frac{d^2a}{dx^2} = Va. \tag{10.3}$$

Eliminating V_1 from these relations, we have $\frac{d^2a}{dx^2} = -a\frac{da}{dx}$, hence $\frac{1}{2}\frac{d}{dx}\left(\frac{da}{dx}\right)^2 = -\frac{1}{2}\frac{d(a^2)}{dx}$.

Integrating this, we obtain

$$\frac{da}{dx} + \frac{a^2}{2} = 2\mu^2,$$

where μ^2 can be considered as constant, *if assuming that the system is conservative.* Transforming $a(x)$ to another variable $w(x)$ by the relation $a = \frac{2}{w}\frac{dw}{dx}$, this equation becomes a linear equation

$$\frac{d^2w}{dx^2} - \mu^2 w = 0,$$

whose solution is given by $w = \alpha\exp(\mu x) + \beta\exp(-\mu x)$, where α and β are arbitrary. In this case, from (10.3) we obtain that

$$V = -2\frac{d^2(\ln w)}{dx^2} = -2\mu^2\,\text{sech}^2(\mu x - \theta), \quad \text{where } \theta = \frac{1}{2}\ln\frac{\beta}{\alpha}.$$

This is Eckart's potential obtained as a steady solution of Korteweg–deVries' equation, at $\kappa = 1$ and at a given temperature. Substituting this into equation (10.1), we see that $\theta = 4\mu^3 t$, the Eckart potential can be expressed as follows.

Theorem:

$$V = -2\mu^2 \operatorname{sech}^2(\mu x - 4\mu^3 t), \tag{10.4}$$

which agrees with the previous expression in chapter 9, if $\mu = \sqrt{v}/2$; where the phase is expressed by $\phi = \mu x - \theta = \frac{\sqrt{v}}{2}(x - vt)$ characterizing the steady solution by the phase $\phi(x - vt)$, in the vicinity of $\phi = 0$ and $\phi = \pm\pi/2$ in particular, which is *invariant* under the transformation $x - vt \rightarrow x'$ at a constant speed v. In this case, the soliton potential V is regarded as a mobile pulse in Galilean space-time (x, t).

On the other hand, if $\mu \neq \sqrt{v}/2$, the phase ϕ can also change in thermodynamic environment, shifting by time t that can represent a temperature change ΔT; corresponding to another significant feature of the soliton potential (10.1) in *thermal transitions* that are signified by *temporal transitions*.

Since (10.1) is also a specific solution of the Korteweg–deVries equation for $\kappa = 1$ in equilibrium crystals, Bargmann's postulate $\sigma(x) = F(x)\exp iqx$ at modified amplitude by a constant μ is a solution of conservative dynamics, restricted at different temperatures by different symmetries in a crystal, where these functions are oriented in a perpendicular to x direction as $\sigma_\perp(x) \propto V_\perp(x)$, where the former is pinned by the other.

Representing collective motion, as indicated by invariance of the wave vector $\boldsymbol{q} \rightarrow \boldsymbol{q} + \boldsymbol{G}$ in periodic lattice of soliton potentials (see chapter 11), the phase ϕ is specified by the Bloch theorem for Galilean mobility, while the amplitude $F(x)$ is determined by the number n of *coherent* elemental solitons. Owing to boson statistics, the coherent wave of n solitons propagates in-phase, where the phase ϕ is determined by the wave vector $\boldsymbol{q}$ in equilibrium crystals of *discrete structure*, as illustrated in figure 10.1.

Mathematically, Bargmann's result (10.1) is only a specific solution $\kappa = 1$ of the Korteweg–deVries equation; however, it has been physically confirmed that (10.1) is compatible with thermodynamic equilibrium. In contrast, the ellipsoidal solutions for $0 \leqslant \kappa < 1$ represent waves in *strained lattice structure*.

10.1.2 Two-soliton solution

Assuming that Bargmann's *amplitude function* $F(q, x)$ is expressed in the second-order of q, we can also show that the potential V satisfying the Korteweg–deVries equation can be composed of two independent Eckart's potentials, just by the superposition principle, allowing us to consider a *soliton gas* in adiabatic approximation. Here, we ignore all possible damping effect for simplicity.

Assuming

$$\sigma = \exp(iqx)\{4q^2 + 2ia(x) + b(x)\}, \tag{10.5}$$

we substitute this σ for σ in (10.1), and obtain

$$-\frac{\mathrm{d}a}{\mathrm{d}x} = V, \quad \frac{\mathrm{d}^2 a}{\mathrm{d}x^2} + \frac{\mathrm{d}b}{\mathrm{d}x} = Va \quad \text{and} \quad \frac{\mathrm{d}^2 b}{\mathrm{d}x^2} = Vb,$$

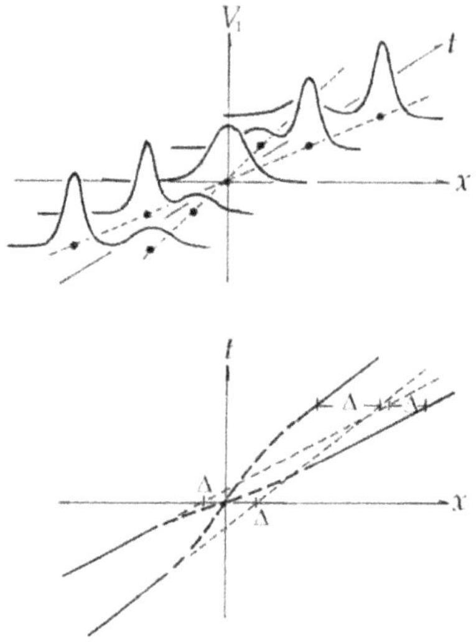

Figure 10.1. Numerical solutions of the Korteweg–deVries equation, assuming for the initial function as given by (10.4), and $\sqrt{\mu} = 0.022$. Curves A, B and C are obtained for $t = 0$, $1/\pi$ and $3.6/\pi$, respectively. Reproduced from [1].

which can be manipulated for expressions to be integrated.

Eliminating V from the above, we obtain a relation $\frac{d}{dx}\left(-\frac{1}{2}a^2 - \frac{da}{dx}\right) = \frac{db}{dx}$, which is integrated as

$$b + \frac{da}{dx} + \frac{1}{2}a^2 = 2c_1, \tag{10.6}$$

where c_1 is a constant.

Also, another expression, $b\left(\frac{d^2a}{dx^2} + \frac{db}{dx}\right) - a\left(\frac{d^2b}{dx^2}\right) = 0$, can be integrated as

$$\frac{1}{2}b^2 + b\frac{da}{dx} - a\frac{db}{dx} = 2c_2, \tag{10.7}$$

where c_2 is a constant.

Writing $a = \frac{2}{w}\frac{dw}{dx}$, we obtain from (10.6)

$$b = 2\left(c_1 - \frac{1}{w}\frac{d^2w}{dx^2}\right),$$

and from (10.7)

$$2\frac{dw}{dx}\frac{d^3w}{dx^3} - \left(\frac{d^2w}{dx^2}\right)^2 - 2c_1\left(\frac{dw}{dx}\right)^2 + w^2\left(c_1^2 - c_2\right) = 0.$$

The latter can be re-expressed as

$$\frac{d^4w}{dx^4} - 2c_1\frac{d^2w}{dx^2} + \left(c_1^2 - c_2\right)w = 0, \tag{10.8}$$

which can be solved in the form $w \approx \exp(\Omega x)$, where $\Omega^2 = c_1 \pm \sqrt{c_2}$. We can therefore express four roots of the integrated equation by $\Omega_1 = \pm\sqrt{c_1 + \sqrt{c_2}}$ and $\Omega_2 = \pm\sqrt{c_1 - \sqrt{c_2}}$, and (10.8) has a solution

$$w = \{\alpha_1 \exp(\Omega_1 x) + \beta_1 \exp(-\Omega_1 x)\} + \{\alpha_2 \exp(\Omega_2 x) + \beta_2 \exp(-\Omega_2 x)\}. \tag{10.9}$$

Here, these four constants are not independent, as they are related by $\alpha_1\beta_1\Omega_1^2 = \alpha_2\beta_2\Omega_2^2$. Accordingly, (10.9) can be expressed as

$$w = 2\Omega_2 \cosh(\Omega_1 x - \varphi_1) + 2\Omega_1 \cosh(\Omega_2 x - \varphi_2),$$

therefore,

$$a = \frac{2}{w}\frac{dw}{dx} = 2\Omega_1\Omega_2\frac{\sinh(\Omega_1 x - \varphi_1) + \sinh(\Omega_2 x - \varphi_2)}{\Omega_2 \cosh(\Omega_1 x - \varphi_1) + \Omega_1 \cosh(\Omega_2 x - \varphi_2)}.$$

Further writing that $\Omega_1 = q_1 + q_2$, $\Omega_2 = q_1 - q_2$ and $\varphi_1 = \theta + \chi$, $\varphi_2 = \theta - \chi$, we obtain $a = 2\left(q_1^2 - q_2^2\right)(q_1 \coth\phi_1 - q_2 \tanh\phi_2)^{-1}$, where $\phi_1 = q_1 x - \theta$ and $\phi_2 = q_2 x - \chi$.

Therefore, the potential for V in (10.5) can be expressed as

$$V = -\frac{da}{dx} = -2\left(q_1^2 - q_2^2\right)\frac{q_1^2 \operatorname{cosech}^2\phi_1 + q_2^2 \operatorname{sech}^2\phi_2}{(q_1 \coth\phi_1 - q_2 \tanh\phi_2)^2}. \tag{10.10}$$

Assuming that $|\phi_1| \sim 1$ and $|\phi_2| \sim 0$, (10.10) is approximated as

$$V \approx -\frac{2q_2^2\left(q_1^2 - q_2^2\right)\operatorname{sech}^2\phi_2}{(\pm q_1 - q_2 \tanh\phi_2)^2}, \tag{10.11a}$$

where $\pm P_1$ corresponds to the values for $\phi_1 \to \pm\infty$.

The resulting expressions (10.3) can be approximated as

$$V_\perp(q_1) = -2q_1^2 \operatorname{sech}^2(q_1 x - \theta \mp \Delta) \quad \text{and}$$
$$V_\perp(q_2) = -2q_2^2 \operatorname{sech}^2(q_2 x - \chi \mp \Delta), \tag{10.11b}$$

where $\Delta = \tanh\frac{q_2}{q_1}$, and the phase parameters θ and χ can be function of τ, being identical to the one-soliton potential (10.4), considering $\frac{d\theta}{d\tau} = 4q_1^3$ and $\frac{d\chi}{d\tau} = 4q_2^3$. Thus, Bargmann's potential V for σ_2 consists of *two independent solitons*, $V_\perp(q_1)$ and $V_\perp(q_2)$ at temperatures T_1 and T_2, respectively, where *many elemental solitons like to collide* at $\tau = 0$, or $T_1 = T_2$, as signified by the phase shift Δ. Figure 10.1 shows the result of numerical studies [1] on (10.11*b*), where two solitons, 1 and 2, exhibit virtually no deforming impact, but a phase shift. Equation (10.11*b*) for two independent 2sech^2-peaks were numerically simulated, as shown in the figure.

Notable is that two soliton potentials marked 1 and 2 overlap completely, as expressed by $V_\perp(x, 0) = -6\mathrm{sech}^2(qx \pm \Delta)$ at $\tau = 0$, exhibiting particle-like images for two *coherent soliton potentials* colliding as if isolated pulses, reflecting boson characters.

In crystalline states, solitons become in-phase at points of high structural symmetry, exhibiting intense waves, when approaching a macroscopically symmetric direction in a space–time phase diagram x versus t, as in figures 10.1 and 10.3, showing behavior of solitons that may be called *Galilean connection* in equilibrium crystals.

10.2 Riccati's theorem and the modified Korteweg–deVries equation

10.2.1 Riccati's theorem

Nonlinear waves are pinned by a soliton potential in-phase, as implied by the Korteweg–deVries equation. On the other hand, for soliton potentials, the Riccati theorem signifies the mobile pathway for the pinned energy to be transferred to the surrounding medium.

Assuming two values V', V'' of the potential V, the corresponding waves σ', σ'' can be linearly related as

$$\sigma'' = A(x, \lambda)\sigma' + B(x, \lambda)\frac{\mathrm{d}\sigma'}{\mathrm{d}x},$$

where

$$\frac{\mathrm{d}^2\sigma'}{\mathrm{d}x^2} = (\lambda + V')\sigma' \quad \text{and} \quad \frac{\mathrm{d}^2\sigma''}{\mathrm{d}x^2} + (\lambda + V'')\sigma'' = 0;$$

λ is the common eigenvalue. Assuming $B(x, \lambda)$ to be constant, the analysis can be made sufficiently simpler. As already discussed in chapter 9, from these relations we can derive Riccati's equation

$$A^2 - \frac{\mathrm{d}A}{\mathrm{d}x} - V' = \tilde{\lambda}, \tag{10.12}$$

where $\tilde{\lambda}$ is an integration constant. Equation (10.12) can be linearized by a transformation

$$A = -\frac{1}{\tilde{\sigma}}\frac{\mathrm{d}\tilde{\sigma}}{\mathrm{d}x}, \tag{10.13}$$

leading to

$$\frac{\mathrm{d}^2\tilde{\sigma}}{\mathrm{d}x^2} + (\tilde{\lambda} + V)\tilde{\sigma} = 0.$$

Note that this equation represents a wave equation for σ' at the eigenvalue λ, if $\tilde{\lambda} = \lambda$. In this case, writing $\tilde{\sigma} = \sigma'$ for $\tilde{\lambda} = \lambda$, otherwise we obtain

$$V' - V'' = 2\frac{\mathrm{d}^2(\ln \sigma')}{\mathrm{d}x^2} \quad \text{for} \quad \tilde{\lambda} \neq \lambda.$$

Hence, we have the equation

$$\frac{d^2\sigma''}{dx^2} + \left(\lambda + V' - 2\frac{d^2(\ln \sigma')}{dx^2}\right)\sigma'' = 0 \tag{10.14}$$

Equation (10.13) allows us to interpret (10.14) in physical terms such that there should be an energy transfer process between two states V' and V'' in the thermodynamic environment. Thus, the transformation $A = -\frac{d}{dx}\ln \sigma'$ gives rise to a change in the soliton potential between $V' \to V'' = V' - 2\frac{d^2 \ln \sigma'}{dx^2}$.

It is worth noting at this point that in crystals in a thermodynamic environment there should be sufficiently high phonon energies in the crystalline medium for $\hbar\omega_{\text{phonon}} = k_B(T' - T'')$ to induce the transition $V'(T') \to V''(T'')$. In chapter 9, we already discussed this kind of transition with respect to the number of elemental solitons, i.e. $\Delta V_{n,n-1} = -2n\,\text{sech}^2 x$ for $n \to n - 1$, indicating a thermal flow of solitons between different potentials. Assuming no phonon interactions available at a constant temperature T'. Such a transition as $\Delta V_{n,n-1}$ may be called *forbidden*, but it is *allowed* if the soliton number can change as $\Delta n = -1$. Adiabatic transition $\Delta V_{n,n\pm 1}$ occurs for $\Delta T_\pm = \pm(T' - T'')$ like radiative transitions, so is called a *quasi-adiabatic transition* thermodynamically.

10.2.2 Modified Korteweg–deVries equation

In section 10.1, we discussed the following.

Theorem: The two-component field $\sigma = \sigma' + i\sigma''$, for which the corresponding Schrödinger equation is signified by a complex potential $u^2 \pm i\frac{du}{dx}$, where $u(x) = -\frac{K(x)}{v}$.

Accordingly, the development equation with two components cannot be exactly compatible with the Korteweg–deVries equation for a complex potential $V(x)$ under practical conditions.

However, as required by eigenvalues independent of developing time t, i.e. $\frac{\partial \varepsilon_1}{\partial t} = 0$, we consider $V = u^2 \pm i\frac{du}{dx}$ to satisfy the equation

$$\frac{\partial V}{\partial t} - v\frac{\partial V}{\partial x} = 0, \tag{10.15}$$

and obtain

$$\frac{\partial V}{\partial t} - 6V\frac{\partial V}{\partial x} + \frac{\partial^3 V}{\partial x^3} = 0, \tag{10.16}$$

after combining with (10.15). Substituting the complex V to (10.15), we have

$$2V\left(\frac{\partial V}{\partial t} - v\frac{\partial V}{\partial x}\right) \pm i\left(\frac{\partial V}{\partial t} - v\frac{\partial V}{\partial x}\right) = 0,$$

which is nothing new.

On the other hand, substituting $u^2 \pm i\frac{\mathrm{d}u}{\mathrm{d}x}$ into (10.16), we obtain

$$2u\left(\frac{\partial u}{\partial t} - 6u^2\frac{\partial u}{\partial x} + \frac{\partial^3 u}{\partial x^3}\right) \pm i\left(\frac{\partial u}{\partial t} - 6u^2\frac{\partial u}{\partial x} + \frac{\partial^3 u}{\partial x^3}\right) = 0,$$

hence

$$\frac{\partial u}{\partial t} - 6u^2\frac{\partial u}{\partial x} + \frac{\partial^3 u}{\partial x^3} = 0, \tag{10.17}$$

which is called the *modified Korteweg–deVries equation* for the function u. Equation (10.17) is similar to (10.16), but not exactly the same as the original Korteweg–deVries equation. It is transformable from (10.16), sharing a common feature of propagation under a soliton potential expressed by the third-order derivative.

Letting $u = -i\tilde{u}$ with a real function $\tilde{u}$, (10.3) can be written as

$$\frac{\partial \tilde{u}}{\partial t} + 6\tilde{u}^2\frac{\partial \tilde{u}}{\partial x} + \frac{\partial^3 \tilde{u}}{\partial x^3} = 0. \tag{10.18}$$

By Riccati's transformation $\tilde{u} = -\frac{1}{\sigma}\frac{\mathrm{d}\sigma}{\mathrm{d}x}$, we can derive the complex potential

$$\tilde{V} = V_1 + iV_\perp = -\tilde{u}^2 - i\frac{\mathrm{d}\tilde{u}}{\mathrm{d}x} \tag{10.19}$$

for the wave equation of σ, which is written as

$$\frac{\mathrm{d}^2\sigma}{\mathrm{d}x^2} - V_\perp\sigma = 0 \tag{10.20}$$

in thermal equilibrium. This equation (10.20) can be modified for the change $V_\perp \to V_\perp + \lambda$, as

$$\frac{\mathrm{d}^2\sigma}{\mathrm{d}x^2} - (V_\perp + \lambda)\,\sigma = 0,$$

where the shift λ is independent of t.

This is invariant under the Galilean transformation $x \to x + 6\lambda t$. Here, the constant λ is the shift in eigenvalue that is consistent with the steady solution of soliton potential in sech^2-type. Therefore, the transformation process from (10.16) to (10.17) should be thermodynamically *irreversible between different eigenvalues at different temperatures*, thereby the Riccati's transition can explain *stepwise quasi-adiabatic thermodynamic changes*, as illustrated in figure 9.5 of chapter 9.

Comparing equation (10.16) with the modified version (10.17), the speed of propagation is shifted, facing different lattice symmetries, where the soliton potential $-\tilde{u}^2 \propto \mathrm{sech}^2(x - vt)$ remains invariant.

Also noticeable at this point is that such a soliton potential $-\tilde{u}^2$ is proportional to σ''^2, where $\sigma'' = \sigma_A$ and $\sigma' = 0$ at a critical temperature, consistent with section 4.2.3 and chapter 8, in relation to entropy production via Riccati's transitions under *time–temperature conversion*, which, however, remains implicit in computational studies in constant-temperature conditions.

10.3 Soliton mobility studied by computational analysis

The significance of the potentials (10.4) and (10.11*b*) as particle-like behavior in phase space of x and t was first demonstrated numerically by Zabusky and Kruskal (1965) [2], using computational analysis. Their numerical studies of the constants a and b in (10.6) have established the soliton image as mobile objects in crystalline media in pulse-shaped motion.

In deriving the Korteweg–deVries equation, we chose a specific value $a_3 = -4$ for mathematical convenience, resulting in the term $\frac{\partial^3 V}{\partial x^3}$ that is responsible for the dispersive nature of propagation. To deal with the initial situation, we can therefore specify the frequency as $\omega = v_0 k - \mu k^3$, where v_0 and μ are the initial phase velocity and dispersion parameter, respectively. In this case, the phase velocity at time t is determined by

$$v = v_0 + \mu^2 v_1 = v_0 + 2\mu^2 \mathrm{sech}^2\{\mu(x - v_0 t) - 4\mu^3 t\}, \tag{10.21}$$

where the second term on the right is discontinuous at $t = t_0$; $v = v_0$ and $\mu = 0$ at $t = t_0$.

On the other hand, it was found that equation (10.21) at $t = t_0$ could be replaced by the periodic function

$$v = v_0 + 2\mu^2 \cos(\mu x - \omega t), \tag{10.22}$$

whose nodal points correspond to the zero phase of (10.21) at many converging solitons. Symmetry axes in a crystal were considered as places where solitons are gathered in-phase. Also noted in (10.22) is that t represents a Galilean timescale different from real time, thereby μ is invariant with respect to the phase $\mu x - \omega t$, determined by symmetry of boson particles.

In their numerical analysis, Zabusky and Kruskal considered a sinusoidal wave

$$v = \cos \pi(x - t) \quad \text{for} \quad t < t_0, \tag{10.23}$$

allowing us to set the initial discontinuity at $x = 0$ and $\pi/2$ by $t = t_B = 1/\pi$.

Figure 10.2 shows the numerical results for a very weak dispersion determined by $\mu^{1/2} = 0.022$, where the amplitudes are plotted at $t = 0$, $1t_B$ and $3.6t_B$, where t_B is a development timescale. At these times, the emergence of soliton peaks is clearly seen in the simulated pattern, where they identified eight solitons at $t = 3.6t_B$, while just one peak was recognized at $t = t_B$.

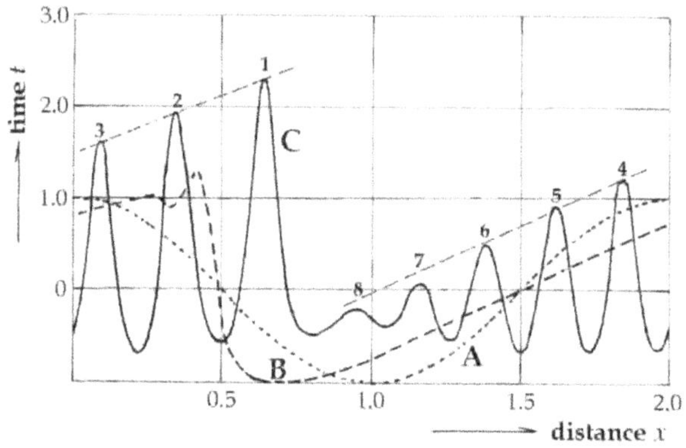

Figure 10.2. Eight solitons with varying time t: 1, 2, 3, …, 7, 8 as indicated in [1]. Reproduced with permission from [3].

Significant in this figure is that those soliton potentials are mobile with varying x and t, whose motion shows tracks in the x–t phase diagram, as sketched in figure 10.3. The discovery by these authors from graphical studies was that those soliton peaks in sech^2 shape move with varying time t with practically no change in lineshape, except at cross-points, and that they become amalgamated together at $t = \frac{1}{2}t_B$, showing a small number of peaks at higher amplitudes. In fact, the illustration in figure 10.3 for collision of two solitons, corresponds to crossing tracks of two peaks, implying that they become *coherent* in these quasi-static tracks, signifying that the nonlinearity in an equilibrium crystal is determined by the whole correlation energy.

We can interpret that (10.23) represents static equilibrium in the medium, in the sense that such soliton movement is simulated numerically, exhibiting particle-like behavior. In terms of physical preference, such bundled coherent solitons are considered to occur on symmetry planes of the medium. In crystalline media, we may therefore assume that a specific time $t = 0.5t_B$ can represent a critical temperature for a structural transition, suggesting symmetry planes are responsible for coherence in crystals.

It is important to realize that these computational studies showed the space–time mobility of soliton potentials of boson character in arbitrary directions in crystals, while the soliton was *restricted coherently more to symmetric directions than nonsymmetric directions*, providing *a stress map* in phase space of the system. It may be considered as the obvious consequence of a finite crystal in equilibrium with its surroundings, but was a new finding in Zabusky–Kruskal's computational studies of modulated lattices. Nevertheless, in equilibrium crystals numerical studies established soliton mobility through periodic structure. In particular, the two-component waves of mobile solitons propagate coherently in directions of $\boldsymbol{k}$-vectors in crystals, as characterized by longitudinal and transversal components.

Although somewhat deformed, the pulse-shaped $-n(n + 1)\mathrm{sech}^2\,\phi(x, t)$ potential is maintained in the phase space of (x, t), which is a significant feature of soliton particles, as demonstrated in figure 10.3. Although mobile in a restricted phase space by space group, solitons are not particles that are as free as phonons, but are *mobile*

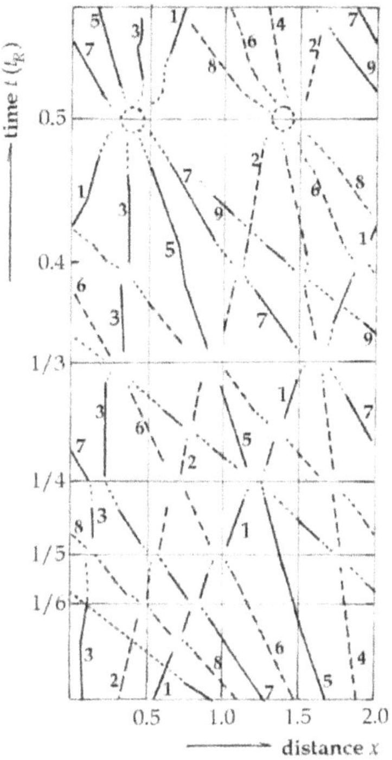

Figure 10.3. Loci of solitons numerically simulated in a x-t plane. Reproduced with permission from [3].

in Galilean space in finite crystals as determined by the lattice symmetry. In the advent of a large number n of positive coherent potentials along symmetric axes and surface area, the inside of crystals can be active for entropy production to separate domain structures in a thermodynamic environment.

In fact, there are a number of mathematical reference articles in literature for soliton mobility in general space, supporting the image of *soliton particles* discovered by computational studies [1]. However, we shall not discuss all of these computational results, as we only deal with the steady crystal space determined by lattice symmetry.

Exercises

1. Why are solitons mobile? Discuss the physical reason for the mobility.
2. Discuss the difference between the Korteweg–deVries equation and its modified version. Show that the difference can be attributed to a change in their eigenvalue values.

References

[1] Lamb G L Jr 1980 *Elements of Soliton Theory* (New York: Wiley)
[2] Zabusky N J and Kruskal M D 1965 *Phys. Rev. Lett.* **15** 240
[3] Taniuchi T and Nishihara K 1998 *Nonlinear Waves (Applied Mathematics Series)* (Tokyo: Iwanami) (in Japanese)

IOP Publishing

Solitons in Crystalline Processes (2nd Edition)
Irreversible thermodynamics of structural phase transitions and superconductivity
Minoru Fujimoto

Chapter 11

Toda's theorem of the soliton lattice

In chapter 2, a microscopic vector variable with a finite amplitude of displacement at site n was considered as represented by a pseudospin responsible for the modified lattice structure. However, such a model is inadequate for the inversion process in finite crystals, because of unavoidable uncertainties near zero displacement, disallowing quantum-mechanical observations. Therefore, the initial condition for nonlinear development cannot be analyzed with such a model, while critical correlations among pseudospins need to be evaluated. On the other hand, Toda [1] discovered new correlation potentials mathematically to avoid such uncertain initial problems.

In this chapter, introducing Toda's work, the nonlinearity of condensates originating from inversion can be discussed in matrix form to facilitate algebraic analysis of displacive transitions in crystals. Using the method called *inverse scatterings*, singularities in collective waves can be related analytically to irreversible processes in crystals with the soliton theory. Dealing with *binary transitions of displacive condensates* in crystals, the Toda theorem makes for *mesoscopic disorder* to be analyzable with the lattice, replacing pseudospin correlations by *exponential potentials*. Furthermore, with *exponential correlations*, soliton potentials emerge logically at singularities, so that Toda's theory of soliton lattice is a valid assumption for entropy production during binary transformations, as supplemented by the Korteweg–deVries equation. The soliton lattice, whose sites are signified by soliton numbers, represents properties of nonlinear lattice, exhibiting deformed structures during crystalline processes.

11.1 The Toda lattice

11.1.1 Theorem of dual chains for condensates

Assuming the partial displacement model in chapter 3, we consider a long chain of identical displacements in a one-dimensional structure. Such displacements can take place with respect to correlated ions at lattice sites, accompanied by counter-displacements in the latter.These pseudospins, defined as partial displacements, and the rest of the constituents should be mobile in-phase with crystals, which are therefore considered as classical *condensates*. Characterized by reduced masses, collective motion of pseudospins and the corresponding lattice can take place, which can be discussed with the concept of the *dual* model.

Figure 11.1 shows the model of dual chains, where one chain of mass m and spring constant κ_A is considered against another chain of mass M and spring constant κ_B, respectively. The relation can therefore be signified by momentum conservation, i.e. $m\dot{r}_n + M\dot{R}_n = 0$, where r_n and R_n are their displacements from the lattice point n.

Referring to these as a dual of chains A and B, the Hamiltonian can be written for the chain A as

$$\mathcal{H}_A = \frac{1}{2m}\sum_n p_n^2 + \sum_n \phi(r_n),$$

where the momentum can be determined by

$$p_n = \frac{\partial K_A}{\partial \dot{y}_n} = m\dot{y}_n$$

for the kinetic energy $K_A = \frac{1}{2}\sum_n m\dot{y}_n^2$.

Assuming $y_0 = 0$ for simplicity, we write

$$y_0 = 0,\ y_1 = r_0,\ y_2 = r_0 + r_1,\ \ldots;\ \dot{y}_0 = 0,\ \dot{y}_1 = \dot{r}_0,\ \dot{y}_2 = \dot{r}_0 + \dot{r}_1,\ \ldots$$

For the chain A of sites $n = 0, 1, \ldots, N-1$, the kinetic energy of collective motion is

$$K_A = \frac{m}{2}\sum_{n=0}^{N-1}(\dot{r}_0 + \dot{r}_1 + \cdots + \dot{r}_n)^2.$$

For each coordinate r_n, the conjugate momentum s_n can be defined as

$$s_n = \frac{\partial K_A}{\partial \dot{r}_n} = m\{(\dot{r}_0 + \dot{r}_1 + \cdots + \dot{r}_n) + (\dot{r}_0 + \dot{r}_n + \cdots + \dot{r}_{n+1}) + \cdots + (\dot{r}_0 + \dot{r}_1 + \cdots + \dot{r}_{N-1})\},$$

where $s_0 = 0$, and for momentum variables s_n we have relations

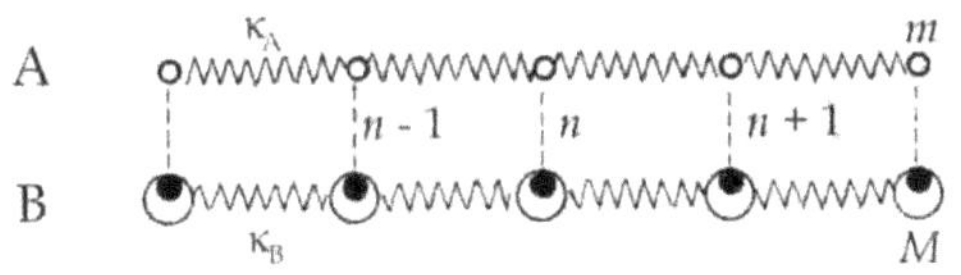

Figure 11.1. Dual lattice A of mass m and spring constant κ. Lattice B of mass M and spring constant κ'. Interacting particles $n-1$, n, $n+1$. Masses m are indicated by small open circles, and M are shown by large circles with holes.

$$s_{n-1} - s_n = m\dot{y}_n \quad \text{and} \quad s_N = 0.$$

The Hamiltonian of the chain A can be expressed as

$$\mathcal{H}_A = \frac{1}{2m}\sum_{n=0}^{N-1}(s_{n+1} - s_n)^2 + \sum_{n=0}^{N-1}\phi(r_n),$$

and the canonical equations of motion can be written as

$$\dot{r}_n = \frac{\partial \mathcal{H}_A}{\partial s_n} = -\frac{s_{n+1} - 2s_n + s_{n-1}}{m} \quad \text{and} \quad \dot{s}_n = -\frac{\partial \mathcal{H}_A}{\partial r_n} = -\phi'(r_n), \tag{11.1}$$

where the total potential $\sum_n \phi(r_n)$ is considered as acting on the lattice B.

Eliminating s_n from (11.1), we obtain the relation

$$m\ddot{r}_n = \phi'(r_{n+1}) - 2\phi'(r_n) + \phi'(r_{n-1}) \tag{11.2}$$

for the chain A, which is identical to

$$m\ddot{r}_n = \kappa_A(r_{n+1} - 2r_n + r_{n-1}), \tag{11.3}$$

if the potential $\phi(r_n)$ is given by a spring potential $\frac{1}{2}\kappa_A r_n^2$ of Hooke's law.

Assuming another harmonic chain for the dual chain B, we can further derive the equation

$$M\ddot{R}_n = \kappa_B(R_{n+1} - 2R_n + R_{n-1}), \tag{11.4}$$

using the relation $m\dot{r}_n = -M\dot{R}_n$. Equations (11.3) and (11.4) represent the dual, as illustrated in figure 11.1.

If the potential ϕ (r_n) is not in Hooke's type, the second equation in (11.1) can be solved as for the general relation

$$mr_n = -\chi(\dot{s}_n) \tag{11.5a}$$

to be revised by the equation of motion

$$\frac{d}{dt}\chi(\dot{s}_n) = s_{n+1} - 2s_n + s_{n-1}, \tag{11.5b}$$

allowing us to interpret the right side as a modified force. Defining the modified spring force by

$$f_n = -\phi'(r_n) = \dot{s}_n, \tag{11.5c}$$

the above can be re-expressed as

$$\frac{d^2\chi(f_n)}{dt^2} = f_{n+1} - 2f_n + f_{n-1} \tag{11.5d}$$

Defining a function $S_n = \int_{t_o}^{t} s_n\, dt$ adjusted by the integration constant at an appropriate time t_o, we can write

$$Y_n = \frac{1}{M}(S_{n-1} - S_n) \quad \text{and} \quad R_n = -\frac{1}{M}(S_{n+1} - 2S_n + S_{n+1}),$$

corresponding to y_n and r_n in harmonic lattice of Hooke's type, the equation of motion can be expressed as

$$\chi(\ddot{S}_n) = S_{n+1} - 2S_n + S_{n-1}. \tag{11.5e}$$

Paying attention to the momentum relation $m\dot{r}_n = -M\dot{R}_n$ for the dual, we can confirm that equations (11.5*a–e*) can be applied to dual lattices A and B in a similar manner, where s_n and S_n are functions of r_n and R_n, respectively, hence the system of (s_n, S_n) constitutes dual for the condensate (r_n, R_n). The dual relation can deal with two chains of m and M in thermodynamic applications.

11.1.2 Discovering the exponential potential

In this section, we discuss how Toda came up with the concept of the *exponential lattice*. Although a somewhat isolated mathematical topic, his derivation is instructive for those learning about process in particular. In any case, the adiabatic correlation potential is analyzable for lattices, if characterized by the Toda potential.

For general displacements, the variable $\dot{s}_n$ in (11.3) is related to $s_{n+1} - 2s_n + s_{n-1}$ among nearest neighbors connected by the *inverse function* of $\phi\ (r)$. Nevertheless, (11.3) is an asymptotic formula for s_n to be specified by s_{n+1} and s_{n-1}, for which $\phi(r_n)$ is a function of r_n. After examining various trial functions for these requirements, Toda found that elliptic functions are promising for analyzable lattices.

Assuming periodic nonlinear waves restricted by *the modulus* κ *of elliptic functions*, he found that sn- or cn-functions were not directly usable. However, using their *addition theorem* instead, he was able to express the potential $\phi(r)$ in analytical form, as outlined in the following.

The addition theorem of elliptic functions is written for two independent variables u and v in the formula

$$\mathrm{sn}^2(u+v) - \mathrm{sn}^2(u-v) = 2\frac{\mathrm{d}}{\mathrm{d}v}\frac{\mathrm{sn}\,u\,\mathrm{cn}\,u\,\mathrm{dn}\,u\,\mathrm{sn}^2\,v}{1 - k^2\,\mathrm{sn}^2\,u\,\mathrm{sn}^2\,v},$$

where we have $\mathrm{dn}^2\,u = 1 - \kappa^2\,\mathrm{sn}^2\,u$. However, he considered the function $\varepsilon(u) = \int_0^u \mathrm{dn}^2\,u\,\mathrm{d}u$, whose differential relations

$$\varepsilon'(u) = \mathrm{dn}^2\,u \quad \text{and} \quad \varepsilon''(u) = -2\kappa^2\,\mathrm{sn}\,u\,\mathrm{cn}\,u\,\mathrm{dn}\,u,$$

satisfy the formula

$$\varepsilon(u+v) + \varepsilon(u-v) - 2\varepsilon(u) = \frac{\varepsilon''(u)}{\frac{1}{\mathrm{sn}^2 u} - 1 + \varepsilon'(u)}.$$

Nevertheless, the function $\varepsilon(u)$ is not periodic, so a related *zeta-function*

$$Z(u) = \varepsilon(u) - (E/K)u \tag{11.6a}$$

was chosen for his analysis, because $\boldsymbol{Z}(u)$ is periodic in $2K$, where K and E are complete elliptic integrals of the first and second kinds, respectively. Accordingly, the following formula for zeta-functions was utilized for his calculation, namely

$$\boldsymbol{Z}(u+v)+\boldsymbol{Z}(u-v)-2\boldsymbol{Z}(u)=\frac{\mathrm{d}}{\mathrm{d}u}\ln\left\{1+\frac{\boldsymbol{Z}'(u)}{\frac{1}{\mathrm{sn}^2 v}-1+\frac{E}{K}}\right\}, \tag{11.6b}$$

where the variables u and v can be expressed specifically like a conventional phase, i.e.

$$u=2K\left(\nu t\pm\frac{n}{\lambda}\right) \quad \text{and} \quad \nu=\frac{2K}{\lambda},$$

where these parameters ν, n and λ are familiar quantities: frequency, refractive index and wavelength for the conventional phase of propagation, respectively. Therefore, the function $s_n(t)$ can be determined by the relation

$$s_n(t)=\left(\frac{2K\nu}{b/m}\right)\boldsymbol{Z}(u) \tag{11.6c}$$

Therefore, replacing $s_n(t)$ by a continuous variable $s(t)$ in crystals at a long wavelength, we can permit the function $\chi(\dot{s})=\frac{m}{b}\ln\frac{\frac{b\dot{s}}{m(2K\nu)^2}}{\frac{1}{\mathrm{sn}^2 v}-1+\frac{E}{K}}-m(\text{const.})$ to satisfy equation (11.3b), by using the relation

$$(2E\nu)^2=\frac{ab}{m}\left\{\frac{1}{\mathrm{sn}^2(2K/\lambda)}-1+\frac{E}{K}\right\}^{-1} \quad \text{and} \quad ab>0,$$

provided that suitable values can be found for these constants a and b. In this way, from the general relation $\dot{s}=-\phi'(r)$, Toda derived the relation

$$r=-\frac{1}{b}\ln\left(1+\frac{\dot{s}}{a}\right)+\sigma \tag{11.7a}$$

for the correlation potential $\phi(r)$, leaving a and b as adjustable parameters, and σ is another constant. The formula (11.5a) can then be expressed as

$$s=a\exp\{-b(r-\sigma)\}=-\phi'(r) \tag{11.7b}$$

and hence the following applies.

Theorem:

$$\phi(r)=\frac{a}{b}\exp\{-b(r-a)\}+ar+\text{const.}, \tag{11.7c}$$

which can represent the correlation potential as the *exponential lattice*, known as the *Toda lattice*. Although mathematical, the theorem plays a significant role for binary transitions to be analyzable with physical interpretation of structural phase transitions.

11.1.3 The Toda lattice

Setting a minimum of $\phi(r)$ at $r = 0$ and disregarding the added constant in (11.5*c*), Toda's potential can be expressed as

$$\phi(r) = \frac{a}{b}\exp(-br) + ar, \qquad \text{where} \qquad ab > 0. \tag{11.8a}$$

Figures 11.2(a) and (b) show $\phi(r)$ sketched for a, $b > 0$ and a, $b < 0$, respectively, and figure 11.2(c) illustrates a specific case of a large a and positive b.

Expanding $\phi(r)$ with respect to a small variation of r, we have

$$\phi(r) = \text{const.} + \frac{ab}{2}r^2 - \frac{2ab\alpha}{3}r^3 + \cdots,$$

where $\kappa = ab$ and $\alpha = -b/2$ are equivalent to a conventional *spring constant* and a nonlinearity parameter, respectively. In addition, for a large value of b, figure 11.2(c) represents the boundary considered as a rigid sphere of unspecified small radius, so that such a potential can be used to characterize the *displacive* pseudospin. At this stage, it should be remarked that *all ambiguities at the onset of microscopic displacements are embedded in the above definitions of a and b in a thermodynamic environment*, which are therefore constant of temperature and pressure.

Considering an external force f acting on the system, we can add a potential $f \cdot r$ to (11.8*a*) and write

$$\phi(r) = \frac{a}{b}\exp(-br) + (a + f)r. \tag{11.8b}$$

Although f is not necessarily a constant in this case, we realize that, if $\phi(r)$ satisfies the Korteweg–deVries equation, f is invariant for translation with respect to r, so that $f(r)$ can be assigned to intrinsic adiabatic potential for the Hamiltonian of a constant eigenvalue. Thus, the function f can be attributed to distant correlations in a crystal, namely the Weiss field, in the thermodynamic environment.

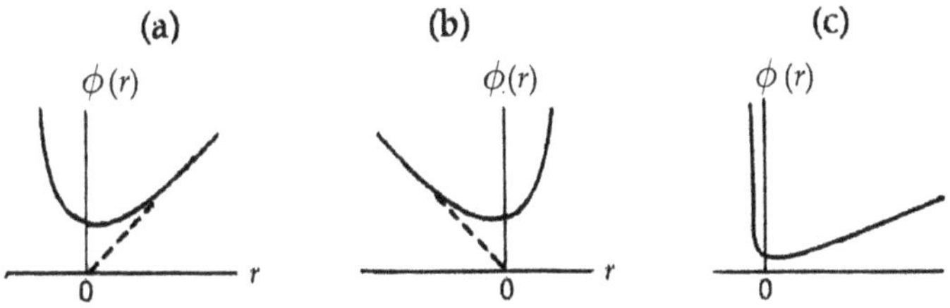

Figure 11.2. Toda's potential $\phi(r) = \frac{a}{b}\exp(-br) + ar$. (a) a, $b > 0$; (b) a, $b < 0$; (c) large b.

Assuming f is constant, however, (11.6b) can be expressed as

$$\phi(r) = \frac{a'}{b}\exp\{-b(r+\sigma')\} + a'r, \tag{11.8c}$$

where $a' = a + f$ and $a'\exp(-b\sigma') = a$, giving an identical potential to (11.8a). In any case, the equation of relative motion can be written as

$$m\frac{\mathrm{d}^2 r_n}{\mathrm{d}t^2} = a\{2\exp(-br_n) - \exp(-br_{n-1}) - \exp(-br_{n+1})\},$$

hence

$$\frac{\mathrm{d}}{\mathrm{d}t}\ln(a + \dot{s}_n) = \frac{b}{m}(s_{n-1} - 2s_n + s_{n+1}) \tag{11.9a}$$

for the dual lattice A, associated with the Toda's potential.

Differentiating the right side of (11.7a), we obtain $\frac{\ddot{s}_n}{a+\dot{s}_n} = \frac{b}{m}(s_{n-1} - 2s_n + s_{n+1})$, and further

$$\frac{\mathrm{d}^2}{\mathrm{d}t^2}\ln\left(1 + \frac{f_n}{a}\right) = \frac{b}{m}(f_{n-1} - 2f_n + f_{n+1}).$$

Integrating this, we have

$$\ln\left(1 + \frac{\ddot{S}_n}{a}\right) = \frac{b}{m}(S_{n-1} - 2S_n + S_{n+1}), \tag{11.9b}$$

which is the equation of motion of dual B, where the *nonlinear spring force* can be defined effectively as

$$f_n = a\{\exp(-br_n) - 1\} = \dot{s}_n \tag{11.9c}$$

It is noted that the Toda potential $\phi(r)$ in (11.7c) can replace Hooke's potential $\phi_{\mathrm{Hooke}}(r) = ar$, adding a finite $\phi(0) = a/b \neq 0$ at $r = 0$ due to the exponential term in the former, while $\phi_{\mathrm{Hooke}}(0) = 0$ in the latter case. Physically, however, $\phi(0) \neq 0$ is a logical assumption for lattice displacements to be determined by mutual correlations in finite crystals.

11.1.4 Nonlinear waves in finite crystals

The periodicity of nonlinear waves $s_n(t)$ is determined by the zeta function $\boldsymbol{Z}(u)$, as discussed previously, whose period can be determined by the expression

$$\exp(-br_n) - 1 = \frac{(2K\nu)^2}{ab/m}\left[\mathrm{dn}^2\left\{2\left(\frac{n}{\lambda} \pm \nu t\right) - \frac{E}{K}\right\}\right], \tag{11.10a}$$

where the wavelength λ and the frequency ν are related by the dispersion relation

$$2K\nu = \sqrt{\frac{ab}{m}}\left(\frac{1}{\mathrm{sn}^2(2K\lambda)} - 1 + \frac{E}{K}\right)^{-\frac{1}{2}}.$$

In these expressions, the complete elliptic integral K and the complete elliptic integral of the second kind E are both functions of the modulus κ, as expressed by

$$K = K(\kappa) = \int_0^{\pi/2} \frac{\mathrm{d}\theta}{\sqrt{1 - \kappa^2 \sin^2\theta}} = \int_0^1 \frac{\mathrm{d}x}{\sqrt{(1 - x^2)(1 - \kappa^2 x^2)}}$$

$$E = E(\kappa) = \int_0^{\pi/2} \sqrt{1 - \kappa^2 \sin^2\theta}\,\mathrm{d}\theta = \int_0^1 \sqrt{\frac{1 - \kappa^2 x^2}{1 - x^2}}\,\mathrm{d}x,$$

where $x = \sin\theta$.

For small values of the modulus κ, the elliptic dn^2-function in the above square brackets of (11.8*a*) is small, showing a wave similar to a trigonometric function, whereas for $\kappa \to 1$ it is characterized by peaked maximum and flat minimum, like in a surface wave on water. In a continuous medium, (11.8*a*) can equally be described by cn^2-function, satisfying the Korteweg–deVries equation; such a wave is traditionally called a cnoidal wave.

Because of $0 < \kappa \leqslant 1$ in general, we can assume $E/K \simeq 1 - \kappa^2/2$ in their definitions, we have a sinusoidal wave determined by

$$r_n \simeq -\frac{\omega^2\kappa^2}{8ab}\cos\left(\omega t \pm \frac{2\pi n}{\lambda}\right) \qquad \text{where} \qquad \omega = 2\pi\nu = \sqrt{ab/m}/\lambda.$$

On the other hand, the quantity in the square brackets in (11.8*a*) can be expanded into Fourier's series from the relation

$$\mathrm{dn}^2(2\pi K) - \frac{E}{K} = \frac{\pi^2}{K^2}\sum_{\ell=-\infty}^{\infty} \frac{\ell\cos(2\pi\ell x)}{\sinh(\pi\ell K'/K)},$$

where $K' = K(\kappa')$ and $\kappa' = \sqrt{1 - \kappa^2}$, to the expression

$$\mathrm{dn}^2(2\pi K) - \frac{E}{K} = \left(\frac{\pi}{2K'}\right)^2 \sum_{\ell=-\infty}^{\infty} \mathrm{sech}^2\left\{\frac{\pi K}{K'}(x - \ell)\right\} - \frac{\pi}{2KK'}, \tag{11.10b}$$

which is the same formula for finite crystals quoted in section 9.4 in chapter 9.

For such a *cnoidal* wave, (11.10*a*) can be expressed as

$$\exp(-br_n) - 1 = \frac{m}{ab}\left[\sum_{\ell=-\infty}^{\infty} \beta^2 \,\mathrm{sech}^2\{\alpha(n - \lambda\ell)\} - 2\beta\nu\right], \tag{11.10c}$$

where $\alpha = \frac{\pi K}{\lambda K'}$ and $\beta = \frac{\pi K\nu}{K'}$.

Equation (11.10*c*) indicates that the Toda potential $\phi_n(r)$ is equivalent to such sequential pulses of $\mathrm{sech}^2\{\alpha(n - \lambda l)\}$ lined-up as $-\infty \cdots < \ell - 1 < \ell < \ell + 1 < \cdots +\infty$ in one direction. As quoted in section 9.4, we have a cnoidal lattice at $\kappa = 1$ of

n-elemental solitons at sites ℓ as a Toda's lattice, where the potential $\phi(r_\ell)$ represents the correlations between nearest sech^2-peaks. It should be remarked that such a cnoidal lattice at $\kappa = 1$ is compatible with conservative systems, illustrating non-linear fields of soliton potentials (11.10*b*) in equilibrium crystals. Figure 11.3(a) shows such a cnoidal wave, whose dispersion relation is shown in figure 11.3(b). Thus the potential $\phi(r, \kappa)$ with the varying parameter $0 < \kappa < 1$ can be disregarded, if the medium is not in thermal equilibrium, considering the cnoidal wave for $\kappa = 1$ to represent nonlinear propagation of *n elemental solitons* in finite crystals in a thermodynamic environment.

For thermodynamics, the Fourier expansion (11.10*c*) can be considered for *finite crystals* at all temperatures in different equilibria, where each term specified by ℓ corresponds to the temperature T_ℓ, corresponding to the soliton number n_ℓ determined by $n_\ell - \lambda\ell = 0$; meaning that to each soliton lattice of n_ℓ we can assign a specific temperature T_ℓ. In this interpretation of finite crystals, we can consider a *cnoidal lattice of* sech^2*-type* specified by sites n at a given T_ℓ, which is consistent with the Korteweg–deVries equation (9.7) written for thermodynamic equilibrium.

Nevertheless, using elliptic ϑ-function related to $\boldsymbol{Z}$-function as

$$2K\boldsymbol{Z}(2xK) = \frac{\mathrm{d}}{\mathrm{d}x}\ln \vartheta_o(x),$$

the cnoidal wave and its momentum can by written as

$$s_n = \frac{m}{b}\frac{\mathrm{d}}{\mathrm{d}t}\ln \vartheta_0\left(\nu t \pm \frac{n}{\lambda}\right) \quad \text{and}$$
$$m\dot{y}_n = \frac{m}{b}\frac{\mathrm{d}}{\mathrm{d}t}\left\{\ln \vartheta_0\left(\nu t \pm \frac{n-1}{\lambda}\right) - \ln \vartheta_0\left(\nu t \pm \frac{n}{\lambda}\right)\right\}, \tag{11.10d}$$

respectively; these are signified by the modulus κ, but their stability cannot be determined mathematically.

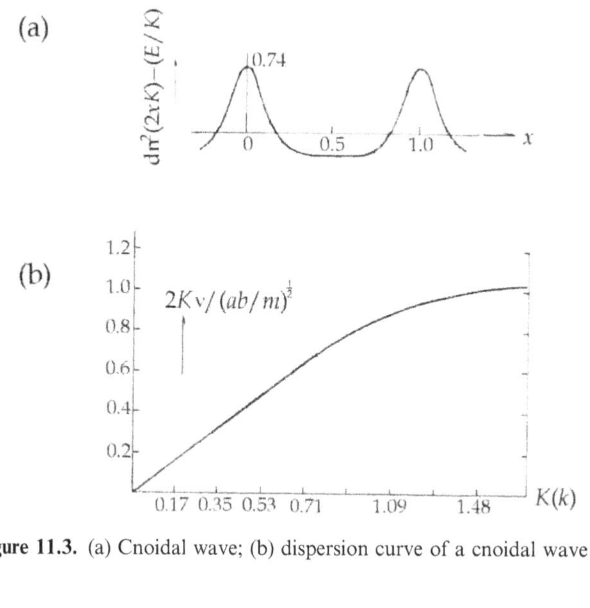

Figure 11.3. (a) Cnoidal wave; (b) dispersion curve of a cnoidal wave [1].

11.2 Developing nonlinearity with Toda's correlation potentials

The Toda potential (11.8*a*) is expressed in a convenient form to analyze binary ordering, in that the finite exponential term is positively added at the threshold transition. Flaschka [2] and Lax [3] showed that displacive order variables can all be composed by their matrix formulation of the exponential lattice, which is outlined in this section.

For an exponential lattice determined by the potential $\phi(r)$ in (11.8*a*), the displacive coordinate Q_n and its conjugate P_n can be defined to satisfy the equations of motion (11.9*a*).

$$\dot{Q}_n = P_n \quad \text{and} \quad \dot{P}_n = \exp\{-(Q_n - Q_{n-1})\} - \exp\{-(Q_{n+1} - Q_n)\}, \tag{11.11}$$

where

$$a_n = \frac{1}{2}\exp\left(-\frac{Q_{n-1} - Q_n}{2}\right) \quad \text{and} \quad b_n = \frac{1}{2}P_n. \tag{11.12}$$

Noted here is that the index n and signs of a_n, b_n are defined for relating (11.10*a*) to the Korteweg–deVries equation. In this case,

$$\dot{a}_n = a_n(b_n - b_{n+1}) \quad \text{and} \quad \dot{b}_n = 2\left(a_{n-1}{}^2 - a_n{}^2\right) \tag{11.13}$$

are considered for the equations of motion of a_n and b_n.

Noting that equations (11.11) are unchanged by reversing signs, we consider the recurrence relations for these variables, which are

$$a_{n+N} = a_n \quad \text{and} \quad b_{n+N} = b_n.$$

Following Lax [2], we write the Hamiltonian matrix for a_n and b_n in matrix form as follows.

Theorem: Lax' Hamiltonian

$$\boldsymbol{L} = \begin{pmatrix} b_1 & a_1 & & & & & & a_N \\ a_1 & b_2 & \ddots & & & & & \\ & \ddots & b_{n-1} & a_{n-1} & & & & \\ & & a_{n-1} & b_n & a_n & & & \\ & & & a_n & b_{n+1} & \ddots & & \\ & & & & \ddots & & b_{N-1} & a_{N-1} \\ a_N & & & & & & a_{N-1} & b_N \end{pmatrix}, \tag{11.14}$$

corresponding to equation (9.10) in chapter 9; and the *antisymmetric* matrix

$$\boldsymbol{B} = \begin{pmatrix} 0 & -a_1 & & & & & & & a_N \\ a_1 & 0 & & & & & & & \\ & & \ddots & & & & & & \\ & & & 0 & -a_{n-1} & & & & \\ & & & a_{n-1} & 0 & -a_n & & & \\ & & & & a_n & 0 & & & \\ & & & & & & \ddots & & \\ & & & & & & & 0 & -a_{N-1} \\ -a_N & & & & & & & a_{N-1} & 0 \end{pmatrix} \quad (11.15)$$

for the *developing operator of binary inversion* $a_n \rightleftarrows -a_n$ in the exponential lattice.

As discussed in chapter 9, the equation of motion for nonlinearity development can be written as

$$\frac{\mathrm{d}\boldsymbol{L}}{\mathrm{d}\tau} = \boldsymbol{BL} - \boldsymbol{LB}, \quad (11.16a)$$

where τ is a temporal parameter for developing nonlinearity, which is not necessarily the *real time.* Equation (11.16*a*) is known as Lax's development equation.

It is noted that the above formulation is an extended use of binary fluctuations discussed in section 4.2.2 in Toda's exponential lattice, where finite inversions $a_n \rightleftarrows -a_n$ are analyzable, despite the uncertainties for vector displacements.

Here the development operator $\boldsymbol{B}$ is *anti-symmetric* for binary variation, to which a unitary matrix $\boldsymbol{U}$ should be considered as

$$\frac{\mathrm{d}\boldsymbol{U}}{\mathrm{d}t} = \boldsymbol{BU}, \qquad \text{where} \qquad \boldsymbol{U}(0) = \mathbf{1}$$

and t is *real time* of the development along the antisymmetric axis.

By writing this relation in inverse form $\frac{\mathrm{d}\boldsymbol{U}^{-1}}{\mathrm{d}t} = -\boldsymbol{U}^{-1}\boldsymbol{B}$, we obtain the formula

$$\boldsymbol{UU}^{-1} = \boldsymbol{U}^{-1}\boldsymbol{U} = \mathbf{1} \quad \text{and} \quad \frac{\mathrm{d}}{\mathrm{d}t}(\boldsymbol{U}^{-1}\boldsymbol{LU}) = 0.$$

Accordingly, for the product $\boldsymbol{U}^{-1}\boldsymbol{L}(\tau)\boldsymbol{U}$ during development, we have the relation

$$\boldsymbol{L}(\tau) = \boldsymbol{U}(t)\boldsymbol{L}(0)\boldsymbol{U}(t)^{-1} \quad \text{or} \quad \boldsymbol{L}(0) = \boldsymbol{U}(t)^{-1}\boldsymbol{L}(\tau)\boldsymbol{U}(t), \quad (11.16b)$$

indicating that $\boldsymbol{L}(\tau)$ and $\boldsymbol{L}(0)$ are *unitarily equivalent*, regarding a general developing timescale τ and specific t for isothermal development, namely

$$\frac{\mathrm{d}}{\mathrm{d}t}\{\boldsymbol{U}^{-1}\boldsymbol{L}(\tau)\boldsymbol{U}\} = 0, \quad (11.16c)$$

signifying the conservative nature of the Hamiltonian $\boldsymbol{L}(\tau)$ during development.

In thermodynamic equilibrium, the wave equation $\boldsymbol{L}(0)\varphi(0) = \lambda(0)\varphi(0)$ for the order variable $\sigma(0) = \varphi^*(0)\varphi(0)$ defined at $t = 0$ is characterized by the eigenvalue $\lambda(0)$; therefore, for the developing process at τ, we can write

$$\boldsymbol{L}(\tau)\varphi(\tau) = \lambda(\tau)\varphi(\tau), \tag{11.16d}$$

where $\lambda(\tau)$ should be the same eigenvalue, keeping the relation $\lambda(\tau) = \lambda(0)$ in a given thermodynamic environment.

To translate from the real time t to the developing time τ, we assume a unitary transformation

$$\varphi(t) = \boldsymbol{U}\varphi(\tau) \quad \text{and} \quad \frac{\mathrm{d}\varphi(\tau)}{dt} = \boldsymbol{B}(\tau)\,\varphi(t), \tag{11.16e}$$

representing adiabatic development of the wave function $\varphi(\tau)$ under conditions of a constant eigenvalue $\lambda(\tau) = \lambda(0)$, signified by $\frac{\mathrm{d}\boldsymbol{B}(\tau)}{\mathrm{d}t} = 0$. The developing time τ is not necessarily the same as real time t; for example, instead of t a heat exchange time τ indicates stabilizing a deformed lattice in a thermodynamic environment.

It is notable that for $ab > 0$ in Toda's potential the operator $\boldsymbol{B}$ drives nonlinear developing $\boldsymbol{L}^+$ from left to right in one dimension, whereas for $ab < 0$ $\boldsymbol{B}$ drives development $\boldsymbol{L}^-$ from right to left. To express these actions mathematically, we define the operators

$$\boldsymbol{L}^{\pm} = \begin{pmatrix} b_1 & a_1 & & & \pm a_N \\ a_1 & b_2 & & & \\ & & \ddots & & \\ & & & b_{N-1} & a_{N-1} \\ \pm a_N & & & a_{N-1} & b_N \end{pmatrix}, \tag{11.16f}$$

whose eigenvalues are the same λ. As all matrix elements are real, we have the relations

$$\varphi^*\boldsymbol{L}\varphi = \varphi^*\boldsymbol{L}^{\pm}\varphi = \lambda^{\pm}\varphi^*\varphi, \quad \text{hence} \quad (\lambda^+ - \lambda^-)\varphi^*\varphi = 0.$$

Accordingly, $\lambda^+ = \lambda^- = \lambda$ is a real quantity, and hence the eigenfunction $\varphi^*\varphi$ for $\boldsymbol{L}$ expresses density flow, whereas wave function φ describes propagation of the nonlinear wave function.

The foregoing transformations are all *adiabatic* in a thermodynamic environment, where the wave function $\varphi(t)$ can be normalized with respect to its spatial distribution in a given system. At this stage, we consider that the development is in *real time*, but can be in any other development parameter such as temperature. With respect to real time, however, we should have $\lambda(t) \neq \lambda(0)$ for $t \neq 0$, which is associated with an *isothermal mechanism* in a thermodynamic environment.

Representing the total energy or total momentum of a conservative system, however, eigenvalues of $L(t)$ in (11.16*a*) can be obtained by solving the secular equation

$$\det|\lambda I - L| = 0, \tag{11.17}$$

associated with (11.6*a*), where I is a unit matrix. The eigenvalues λ_j can be obtained as real roots of the algebraic equation

$$\lambda_j^N + c_1\lambda_j^{N-1} + c_2\lambda_j^{N-2} + \cdots + c_{N-1}\lambda_j + c_N = 0.$$

Here, these coefficients c_j are functions of a_n and b_n, hence all determined by $\lambda_1, \lambda_2, \ldots, \lambda_N$. Equation (11.17) indicates that there are more than two conservative quantities related to a_n and b_n.

Nevertheless, the significance of these eigenvalues λ_j other than *total energy and momentum* has not been identified in computational studies, while Hénon and Heiles [4] suggested that λ_j should be related to stabile trajectories of the many-soliton system in dynamical phase space.

It is notable that in the Flaschka–Lax expressions of (11.11) and (11.12), the matrix formulation can be utilized for *any canonical variables* P_n and Q_n interacting mutually with correlation potential ϕ_n. Referring to the example of variables ρ_1', p_1' and v_1' in nonlinear propagation discussed together in section 9.1, we can postulate that at constant amplitudes, the *nonlinear phases* of order variables at sites $n-1$, n and $n+1$ constitute a *dual* in Toda's exponential lattice restricted to equilibrium states of crystals. Analyzing such correlations among *phases* in Toda' lattice by *phase solitons*, a nonlinear lattice in finite size will be discussed in chapter 12 *for domain structure and surface boundaries in equilibrium with surroundings.*

11.3 Infinite periodic lattice

Representing an exponential lattice, the operator $\boldsymbol{L}$ specifies the wave equation $\boldsymbol{L}\varphi = \lambda\varphi$, where the eigenvalue λ is invariant, independent of the developing time τ. The nonlinearity is determined by the development operator $\boldsymbol{B}(\tau)$ as determined by $\mathrm{d}\varphi/\mathrm{d}\tau = \boldsymbol{B}(\tau)\varphi(\tau)$. Accordingly, the equation of motion of $\boldsymbol{L}(\tau)$ can be obtained by differentiating $\boldsymbol{L}\varphi$, i.e. $\frac{\mathrm{d}\boldsymbol{L}}{\mathrm{d}\tau}\varphi + \boldsymbol{L}\frac{\mathrm{d}\varphi}{\mathrm{d}\tau} = \lambda\frac{\mathrm{d}\varphi}{\mathrm{d}\tau}$, from which the Lax wave equation

$$\left\{\frac{\mathrm{d}\boldsymbol{L}}{\mathrm{d}\tau} - (\boldsymbol{BL} - \boldsymbol{LB})\right\}\varphi = 0 \tag{11.18}$$

can be solved for the wave function φ.

While antisymmetric, the matrix $\boldsymbol{B}$ is accompanied in practice by an ambiguity $\Delta\boldsymbol{B}$ that is commutable with $\boldsymbol{L}$, namely $\Delta\boldsymbol{B}\cdot\boldsymbol{L} - \boldsymbol{L}\cdot\Delta\boldsymbol{B} = 0$. Furthermore, the corresponding wave function φ has space-time ambiguities in crystals, hence a state of φ is generally *unstable*, arising from not only spatial fluctuations, but also *unspecified boundaries* of the system. For such temporal fluctuations, however, we can consider *infinite crystals* for $\Delta\boldsymbol{B}$ to be corrected for observation in a thermodynamic environment.

Nevertheless, we consider an infinite lattice, where $n = 1, 2, \ldots, N$ is replaced by $n = -N/2, \ldots, N/2$, which can be extended to infinite lattice by letting $N \to \pm\infty$. Although the function φ cannot be normalized in the infinite lattice, the presence of a singularity allows us to define a finite *normalization factor*. However, such an infinite lattice is inadequate for finite crystals important in thermodynamics, for which we assume a long finite lattice instead, considering $n = -N/2$ to be identical to $n = +N/2$, where $-N/2$ and $+N/2$ represent the surfaces. This is a convenient assumption for finite periodic crystals, where both ends are regarded as surfaces for interacting with the surroundings. While hypothetical, it is such an acceptable model that does not contribute much to bulk properties of sufficiently large crystals.

For such an infinite lattice, Flaschka [2] wrote

$$L\varphi(n) = a_{n-1}\varphi(n-1) + b_n\varphi(n) + a_n\varphi(n+1) = \lambda\varphi(n) \tag{11.19a}$$

where

$$n = -N/2 \ldots, -2, -1, 0, 1, 2, \ldots +N/2 \quad \text{and} \quad \lambda = \text{const.},$$

expressing the Toda lattice in algebraic form. As noted in section 11.2, the function $\varphi(n)$ can represent any canonical variable related with σ_n at site n in finite lattice of total N sites that is regarded as infinite if $N \to \infty$, whose development equation is given by

$$\frac{\mathrm{d}\varphi(n)}{\mathrm{d}\tau} = B\varphi(n) = a_{n-1}\varphi(n-1) - a_n\varphi(n+1) \tag{11.19b}$$

Limiting interactions to the nearest neighbors, we can assume that $Q_{n+1} - Q_n = 0$ and $P_n = 0$, hence $a_n = \frac{1}{2}$ and $b_n = 0$, at long distance $|n| \gg 1$. Therefore, the asymptotic solution can be expressed in the form

$$\varphi(n) \sim \exp(\pm i\omega\tau \pm ikn) = \exp(\pm i\omega\tau)\{\exp(\pm ik)\}^n,$$

which should satisfy the relations

$$\frac{1}{2}\{\varphi(n-1) + \varphi(n+1)\} = \lambda\varphi(n) \quad \text{and} \quad \frac{\mathrm{d}\varphi(n)}{\mathrm{d}\tau} = \frac{1}{2}\{\varphi(n-1) - \varphi(n+1)\}.$$

Writing the above asymptotic solution as

$$\varphi(n) \sim \exp(\pm i\omega\,\tau)z^n, \qquad \text{where} \qquad z = \exp ik, \tag{11.20a}$$

we obtain

$$\lambda = \frac{1}{2}(z + z^{-1}) = \cos k \qquad \text{and} \qquad -\omega = \frac{1}{2i}(z - z^{-1}) = \sin k.$$

Here, z is a parameter for a singularity related to k, playing a dominant role in the following discussion. Also, it recognized that the parameter z determines the *nonlinear phase* as seen from the definition (11.20*a*), which is essential for the *phase solitons*, as will be introduced in chapter 12.

Since n is an integer, for the real quantity k we can set it in the range of $0 \leqslant k \leqslant 1$, in which the eigenvalue is restricted in the range $-1 \leqslant \lambda \leqslant +1$ where the corresponding ω may vary due to the ambiguity $\Delta\boldsymbol{B}$. On the other hand, for $|\lambda| > 1$ we have $|z| \neq 1$, hence the localized density of asymptotic $\varphi(0)$ implies *captured* elemental solitons.

In the active region, as generally $a_n \neq 1/2$ and $b_n \neq 0$, waves are scattered or captured by the potential depending on the value of z; that is why these mathematical processes are called *scatterings*. In fact, the potential can be evaluated from given asymptotic waves, which is the method of *inverse scattering*.

Equation (11.19*a*) can generally be expressed as

$$a_{n-1}\varphi(n-1) + b_n\varphi(n) + a_n\varphi(n+1) = \frac{1}{2}(z + z^{-1})\varphi(n), \tag{11.20b}$$

for which we consider an asymptotic formula

$$\varphi(n) = \phi(n, z)\exp(i\omega\tau), \qquad \text{where} \qquad \phi(n, z) = \sum_{n'=n}^{\infty} K(n, n')z^{n'}, \tag{11.20c}$$

replacing $\varphi(n) = z^n \exp(i\omega\tau)$ for $n \gg 1$.

This replacement is logical, assuming $K(n, n')$ is zero for $n \neq n'$. Further, using a given value of $K(n, n+1)$, we can determine $\phi(n-1, z)$ by $\phi(n+1, z)$ in (11.20*b*). In this way, $K(n-1, n')$ can be obtained for all other $n' \geqslant n-1$. Substituting (11.20*c*) into (11.20*b*), and comparing coefficients of z^{n-1}, z^n and z^{n+1} terms, we derive the following relations:

$$\begin{aligned}
&a_{n-1}K(n-1, n-1) = \frac{1}{2}K(n, n),\\
&a_{n-1}K(n-1, n) + b_nK(n, n) = \frac{1}{2}K(n, n+1),\\
&a_{n-1}K(n-1, n+1) + a_nK(n+1, n+1) + b_nK(n, n+1)\\
&\quad = \frac{1}{2}\{K(n, n) + K(n, n+2)\},\\
&a_{n-1}K(n-1, n+2) + a_nK(n+1, n+2) + b_nK(n, n+2)\\
&\quad = \frac{1}{2}\{K(n, n+1) + K(n, n+3)\},\\
&\dots\dots\dots\dots\dots
\end{aligned}$$

Solving the first two relations for a_{n-1} and b_n, we obtain

$$a_{n-1} = \frac{K(n, n)}{2K(n-1, n-1)} \quad \text{and} \quad b_n = \frac{K(n, n+1)}{2K(n, n)} - a_{n-1}\frac{K(n-1, n)}{K(n, n)},$$

hence

$$b_n = \frac{K(n, n+1)}{2K(n, n)} - \frac{K(n-1, n)}{2K(n-1, n-1)}.$$

Returning to their definitions, we have $\exp(Q_{n-1} - Q_n) = \left\{\frac{K(n, n)}{K(n-1, n-1)}\right\}^2$. Therefore, using the relation $P_n = s_{n-1} - s_n$, we derive the following equations:

$$s_n = -\frac{K(n, n+1)}{K(n, n)} \quad \text{and} \quad \dot{s}_n = \exp(Q_n - Q_{n+1}) - 1. \tag{11.21}$$

We can thereby confirm that the matrix method is compatible with the exponential potential in terms of the soliton number, and that the dual constitutes a soliton lattice composed of n-solitons at the site n; namely, the Toda lattice is a *soliton lattice*, as illustrated in figure 10.3 for a specific example.

11.4 Scattering and capture by singular soliton potentials

11.4.1 Reflection and transmission

In this section, the scattering of solitons by the adiabatic potential is discussed, defining reflection and transmission coefficients to visualize the processes [1, 2]. These coefficients are signified by *singularities*, leading to entropy productions for energy transfer and domain formation.

Equation (11.20*b*) is a linear equation, so that a general solution can always be given by a linear combination of specific solutions. Therefore, for the functions $\varphi(n, z)$ and related $\phi(n, z)$ in (11.20*c*), we take these asymptotic functions

$$\varphi(n, z) \to z^n, \ \varphi(n, z^{-1}) \to z^{-n} \qquad \text{for} \qquad n \to +\infty$$

and

$$\phi(n, z) \to z^{-n}, \ \phi(n, z^{-1}) \to z^n \qquad \text{for} \qquad n \to -\infty,$$

respectively, as specific solutions. Hence, their linear combinations

$$\varphi_+(n, z) = \alpha(z)\varphi(n, z^{-1}) + \beta(z)\varphi(n, z),$$
$$\varphi_-(n, z^{-1}) = \alpha(z^{-1})\varphi(n, z) + \beta(z^{-1})\varphi(n, z^{-1});$$

or conversely as

$$\phi_-(n, z) = \overline{\alpha}(z)\phi(n, z^{-1}) + \overline{\beta}(z)\phi(n, z),$$
$$\phi_-(n, z^{-1}) = \overline{\alpha}(z^{-1})\phi(n, z) + \overline{\beta}(z^{-1})\phi(n, z^{-1}),$$

are determined by these complex numbers α, β and their conjugates $\overline{\alpha}$, $\overline{\beta}$.

Assuming orthonormal relations among these asymptotic functions, these coefficients should hold relations listed below:

$$\alpha(z)\overline{\alpha}(z^{-1}) + \beta(z)\overline{\beta}(z) = 1, \qquad \alpha(z)\overline{\beta}(z^{-1}) + \beta(z)\overline{\alpha}(z) = 0;$$
$$\alpha(z^{-1})\overline{\alpha}(z) + \beta(z^{-1})\overline{\beta}(z^{-1}) = 1, \quad \alpha(z^{-1})\overline{\beta}(z) + \beta(z^{-1})\overline{\alpha}(z^{-1}) = 0;$$
$$\overline{\alpha}(z)\alpha(z^{-1}) + \overline{\beta}(z)\beta(z) = 1, \qquad \overline{\alpha}(z)\beta(z^{-1}) + \overline{\beta}(z)\alpha(z) = 0;$$
$$\overline{\alpha}(z^{-1})\alpha(z) + \overline{\beta}(z^{-1})\beta(z^{-1}) = 1, \quad \overline{\alpha}(z^{-1})\beta(z) + \overline{\beta}(z^{-1})\alpha(z^{-1}) = 0.$$

From these relations, we can derive

$$\overline{\alpha}(z) = \alpha(z), \ \beta(z) = -\beta(z^{-1}), \ \overline{\beta}(z^{-1}) = -\beta(z)$$

and

$$\alpha(z)\alpha(z^{-1}) = 1 + \beta(z)\beta(z^{-1}).$$

The last expression can be re-written particularly for $|z| = 1$ as

$$|\alpha(z)|^2 = 1 + |\beta(z)|^2. \tag{11.22}$$

Accordingly, to describe scatterings, it is convenient to define the function

$$S(n, z) = \frac{\varphi(n, z)}{\alpha(z)} = \varphi(n, z^{-1}) + R(z)\varphi(n, z), \qquad \text{where} \qquad R(z) = \frac{\beta(z)}{\alpha(z)}. \tag{11.23}$$

The asymptotic form of $S(n, z)$ can be determined by

$$S(n, z) \to z^{-n} + R(z)z^n \qquad \text{for} \qquad n \to +\infty$$

and

$$S(n, z) \to z^{-n}/\alpha(z) \qquad \text{for} \qquad n \to -\infty,$$

where $z^{-n} = \exp(-ikn)$ and $R(z)z^n = R(z)\exp ikn$ represent *incident* and *reflected* waves, respectively, moving to the left and to the right, and the coefficient $R(z)$ signifies the reflected amplitude. Accordingly, $R(z)$ is called the *reflection coefficient* and $T(z) = 1/\alpha(z)$ is the *transmission coefficient*; these coefficients are related as

$$|T(z)|^2 = 1 - |R(z)|^2 \tag{11.24}$$

11.4.2 Capture at singularities

Physically, it is important to deal with singularities of the transmission coefficient $T(z)$ for the energy flow to the surrounding medium to be specified. Here, we discuss the mathematical concept of singularity that can take place if the condition $\alpha(z) = 0$ is satisfied.

First writing

$$\varphi_+(n, z) = \alpha(z)\varphi(n, z^{-1}) + \beta(z)\varphi(n, z) \quad \text{and}$$
$$\varphi_+(n + 1, z) = \alpha(z)\varphi(n + 1, z^{-1}) + \beta(z)\varphi(n + 1, z),$$

we obtain

$$\alpha(z) = \frac{1}{W(n)}\{\varphi(n, z)\varphi(n + 1, z) - \varphi(n + 1, z)\varphi(n, z)\}, \tag{11.25a}$$

where

$$W(n) = \varphi(n, z^{-1})\varphi(n+1, z) - \varphi(n+1, z^{-1})\varphi(n, z)$$
$$= \varphi(n, z^{-1})\{\varphi(n+1, z) - \varphi(n, z)\} - \{\varphi(n+1, z^{-1}) - \varphi(n, z^{-1})\}\varphi(n, z)$$
$$= \begin{vmatrix} \varphi(n, z^{-1}) & \varphi(n+1, z^{-1}) \\ \varphi(n, z) & \varphi(n+1, z) \end{vmatrix}$$

is a *Wronskian determinant*.

Next, eliminating b_n from the relations

$$a_{n-1}\varphi(n-1, z^{-1}) + a_n\varphi(n+1, z^{-1}) + b_n\varphi(n, z^{-1}) = \frac{z + z^{-1}}{2}\varphi(n, z^{-1})$$

and

$$a_n\varphi(n-1, z) + a_n\varphi(n+1, z) + b_n\varphi(n, z) = \frac{z + z^{-1}}{2}\varphi(n, z),$$

we have

$$W(n-1)a_{n-1} = W(n)a_n.$$

Letting $n \to N$, this can be extended to N, as

$$W(n) = \frac{a_{n+1}}{a_n}W(n+1) = \frac{a_{n+1}}{a_n}\frac{a_{n+2}}{a_{n+1}}W(n+2) = \cdots = \frac{a_N}{a_n}W(N).$$

Considering the limit of $N \to \infty$,

$$\lim_{N\to\infty} W(N) = \lim_{N\to\infty}(z^{-N}z^{N+1} - z^{-N+1}z^{N}) = z - z^{-1} \quad \text{and} \quad \lim_{N\to\infty} a_N = \frac{1}{2},$$

so we have $a_n W(n) = \frac{1}{2}(z - z^{-1})$, which is independent of n. Equation (11.25*a*) can therefore be re-expressed as

$$\alpha(z) = \frac{2a_n}{z - z^{-1}}\{\varphi(n, z)\varphi(n+1, z) - \varphi(n+1, z)\varphi(n, z)\}. \tag{11.25b}$$

Assuming that $\alpha(z_j) = 0$, the parameter $z = z_j$ determines a specific value z_j for capturing in a given system. Therefore, the singularity can be characterized by $\beta(z_j)$ as

$$\varphi(n, z_j) = \beta(z_j)\phi(n, z_j). \tag{11.26}$$

Note that these asymptotic functions represent z^{-n} and z^n for $n \to -\infty$ and $n \to +\infty$, respectively, for $|z_j| < 1$ we expect that

$$\varphi(n, z_j) = \mu\phi(n, z_j) \to 0 \qquad \text{for} \qquad n \to \mp\infty,$$

where μ expresses the proportionality factor.

The singularity at $z = z_j$ due to $\lim_{z\to z_j}\frac{1}{\alpha(z)} \to \infty$ can be signified by the *residue*

$$\mathrm{Res}\left(\frac{1}{\alpha(z)}\right)_{z=z_j} = \left(1\Big/\frac{\mathrm{d}\alpha}{\mathrm{d}z}\right)_{z=z_j}, \tag{11.27a}$$

hence at this stage we study the behavior of the derivative $\mathrm{d}\varphi/\mathrm{d}z$ in the vicinity of $z = z_j$.

Differentiating the equation $a_{n-1}\varphi(n-1, z) + a_n\varphi(n+1, z) + b_n\varphi(n, z) = \lambda\varphi(n, z)$ with respect to z, we obtain

$$a_{n-1}\varphi'(n-1, z) + a_n\varphi'(n+1, z) + b_n\varphi'(n, z) = \lambda\varphi'(n, z) + \frac{\mathrm{d}\lambda}{\mathrm{d}z}\varphi(n, z).$$

Combining this with the equation

$$a_{n-1}\phi(n-1, z) + a_n\phi(n+1, z) + b_n\phi(n, z) = \lambda\phi(n, z),$$

we have

$$\frac{\mathrm{d}\lambda(z)}{\mathrm{d}z}\varphi(n, z)\phi(n, z) = a_{n-1}\{\varphi'(n-1, z)\phi(n, z) - \varphi'(n, z)\phi(n-1, z)\} \\ - a_n\{\varphi'(n, z)\phi(n+1, z) - \varphi'(n+1, z)\phi(n, z)\}.$$

Accordingly,

$$\frac{\mathrm{d}\lambda(z_j)}{\mathrm{d}z_j}\sum_{n'=-\infty}^{n}\varphi(n', z_j)\phi(n, z_j) = -a_n\{\varphi'(n, z_j)\phi(n+1, z_j) - \varphi'(n+1, z_j)\phi(n, z_j)\}$$

and

$$\frac{\mathrm{d}\lambda(z_j)}{\mathrm{d}z_j}\sum_{n'=n+1}^{\infty}\varphi(n', z_j)\phi(n, z_j) = a_n\{\varphi'(n, z_j)\phi(n+1, z_j) - \varphi'(n+1, z_j)\phi(n, z_j)\},$$

so that

$$\frac{\mathrm{d}\lambda(z_j)}{\mathrm{d}z_j}\sum_{n=-\infty}^{n=+\infty}\varphi(n, z_j)\varphi(n, z_j) = -a_n\left|\frac{\mathrm{d}}{\mathrm{d}z}\{\varphi(n, z)\phi(n+1, z) - \varphi(n+1, z)\phi(n, z)\}\right|_{z=z_j}$$

$$= -\left|\frac{\mathrm{d}}{\mathrm{d}z}\left\{\frac{z - z_j^{-1}}{2}\alpha(z)\right\}\right|_{z=z_j} = -\frac{1}{2}\left(z_j - z_j^{-1}\right)\left|\frac{\mathrm{d}\alpha(z)}{\mathrm{d}z}\right|_{z=z_j}$$

Since $\lambda = (z + z^{-1})/2$, $\frac{\mathrm{d}\lambda(z_j)}{\mathrm{d}z_j} = \frac{z_j - z_j^{-1}}{2z_j}$, therefore

$$\left|\frac{\mathrm{d}\alpha(z)}{\mathrm{d}z}\right|_{z=z_j} = -\frac{1}{z_j}\sum_{n=-\infty}^{n=+\infty}\varphi(n, z_j)\phi(n, z_j),$$

which can be written as

$$= -\frac{1}{z_j \mu^2 \beta(z_j)} \sum_{n=-\infty}^{n=+\infty} |\zeta(n, z_j)|^2,$$

where the function $\zeta(n, z_j)$ is normalized as $\sum_{n=-\infty}^{n=+\infty} |\zeta(n, z_j)|^2 = 1$. Asymptotically $\zeta(n, z_j) \to c_j z_j^n$ for $n \to \infty$, where $c_j = \mu\beta(z_j)$, hence we have

$$\left|\frac{\mathrm{d}\alpha(z)}{\mathrm{d}z}\right|_{z=z_j} = -\frac{\beta(z_j)}{z_j c_j^2} \quad \text{and} \quad \mathrm{Res}\left|\frac{1}{\alpha(z)}\right|_{z=z_j} = -\frac{z_j c_j^2}{\beta(z_j)}, \tag{11.27b}$$

which is determined by the value of z_j, as postulated in (11.25*b*).

11.5 The Gel'fand–Levitan–Marchenko theorem

For practical applications of the asymptotic theory, the most significant attempts are to obtain information on the expansion coefficients $K(n, n')$ in (11.20*c*). This task can be mathematically achieved by calculating the lattice potential from scattered waves given in asymptotic form, which is known as inverse scatterings. Notably, there exists a theorem that was worked out by Gel'fand and associates [5]. Known as the Gel'fand–Levitan–Marchenko (GLM) equation, their theorem constitutes the method of inverse scatterings for thermodynamic problems to be solved.

Considering z as a complex variable, we carry out integration of the function $S(n, z)z^{m-1}$ in the complex place, where $m \geqslant n$, along a closed circle C centered at the origin, as illustrated by figure 11.4.

Namely,

$$\frac{1}{2\pi i}\oint_C \frac{\varphi(z)}{\alpha(z)} z^{m-1}\mathrm{d}z = \frac{1}{2\pi i}\oint_C \{\phi(n, z^{-1}) + R(z)\phi(n, z)\} z^{m-1}\mathrm{d}z, \tag{11.28}$$

where these terms are Cauchy's integrals in the theory of complex variables. Using the relation

$$\phi(n, z) = \sum_{n'=n}^{\infty} K(n, n')z^{n}$$

in the integral on the right side of (11.28), the following expressions can be obtained from Cauchy's residue theorem. The first term on the right is

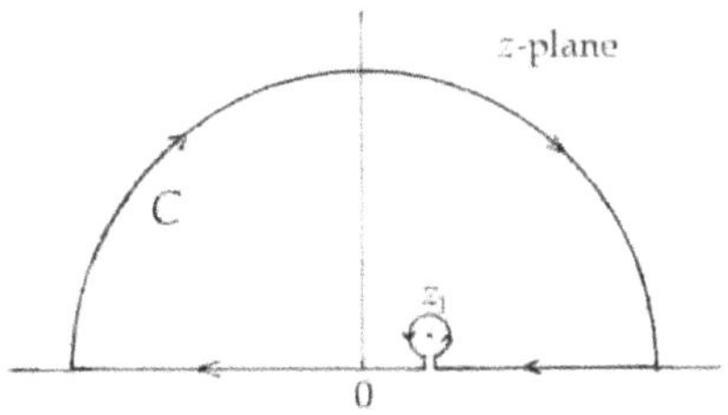

Figure 11.4. Cauchy's integral over a semi-circular path C. A singular point at $z = z_1$.

$$\frac{1}{2\pi i}\oint_C \phi(n, z^{-1})z^{m-1}\mathrm{d}z = \frac{1}{2\pi i}\sum_{n'=n}^{\infty} K(n, n')\oint_C z^{-n'+m-1}\mathrm{d}z = K(n, m) \qquad \text{for} \qquad m \geqslant n.$$

and the second term

$$\frac{1}{2\pi i}\oint_C R(z)\phi(n, z)z^{m-1}\mathrm{d}z = \frac{1}{2\pi i}\sum_{n'=n}^{\infty} K(n, n')\oint_C R(z)z^{n'+m-1}\,\mathrm{d}z.$$

Therefore, defining $F_c(m) = \frac{1}{2\pi i}\oint_C R(z)z^{m-1}\,\mathrm{d}z$, the right side of (11.28) can be expressed as

$$=K(n, m) + \sum_{n'=n}^{\infty} K(n, n')F_c(n' + m).$$

On the other hand, the integral on the left of (11.28) can be evaluated as

$$\frac{1}{2\pi i}\oint_C \frac{\varphi(n, z)}{\alpha(z)}z^{m-1}\,\mathrm{d}z = I_\alpha + I_o, \tag{11.29}$$

where I_α and I_o are residues at the poles of functions $1/\alpha(z)$ and $\varphi(n, z)z^{m-1}$, respectively. The former arises from the property of z_j to satisfy $\alpha(z_j) = 0$, whereas the latter I_o is involved in surfaces and domain boundaries for structural change of crystals. In this case, the origin of I_o must be interpreted as related to adiabatic internal changes in the thermodynamic phase. In practical systems, these singularities are evident in thermal experiments [6], while only a qualitative analysis has so far been carried out on structural changes. Nevertheless, such a singular behavior in the nonlinear dynamics is theoretically predicted, while the GLM theorem offers the method of inverse scatterings to determine the initial condition for nonlinearity development.

In (11.29), we have

$$I_\alpha = -\sum_j \varphi(n, z_j)z^{m-1}\,\mathrm{Res}\left|\frac{1}{\alpha(z)}\right|_{z=z_j} = -\sum_j \phi(n, z_j)z_j^m c_j^2 = -\sum_j c_j^2 \sum_{n'=n}^{\infty} K(n, n')z_j^{n'+m},$$

which is determined by the specific z_j.

In addition, the singular property of I_o can be evaluated from $\phi(n, z^{-1})$ in the vicinity of $z = 0$ within $S(n, z)$ in (11.22) given at this point. The function $S(n, z)$ satisfies the relation

$$a_{n-1}S(n-1, z) + a_n S(n+1, z) + b_n S(n, z) = \frac{z + z^{-1}}{2}S(n, z),$$

which can be approximated as

$$a_n S(n+1, z) \simeq \frac{1}{2z}S(n, z) \quad \text{for} \quad |z| \simeq 0.$$

Hence

$$S(n,\, z) \simeq 2za_n S(n+1,\, z) \simeq \cdots \simeq (2n)^N a_n a_{n+1} \cdots a_{n+N-1} S(n+N,\, z),$$

where

$$a_N = K(n+1,\, n+1)/2K(n,\, n),$$

thereby we have

$$S(N,\, z) \simeq z^{-N} \quad \text{at } z \simeq 0 \quad \text{for } N > 1.$$

In asymptotic expressions, we use $K(\infty,\, \infty) = 1$ in the limit of $N \to \infty$, therefore $S(n,\, z)$ can be evaluated by

$$S(n,\, z) \simeq \frac{1}{K(n,\, n)} z^{-n} \qquad \text{at} \qquad z \simeq 0.$$

Accordingly, letting $-n + m - 1 = -1$ in the above, for $n = m$, the function $S(n,\, z)$ is characterized by a pole near $z = 0$, and hence we can write

$$I_o = \frac{1}{K(n,\, n)} \delta(n,\, m) \qquad \text{for} \qquad m \geqslant n,$$

where $\delta(n,\, m)$ is a *Kronecker's delta* with respect to n and m.

Writing $F_o = \sum_j c_j^2 z_j^n$ with respect to the pole of I_o, the left side of (11.29) $= \frac{\delta(n,m)}{K(n,n)} - \sum_{n'=n}^{\infty} K(n,\, n') F_o(n' + m)$, which is equal to the right side of (11.29) $= K(n,\, m) + \sum_{n'=n}^{\infty} K(n,\, n') F(n' + m)$. As the result, we have

$$\frac{\delta(n,\, m)}{K(n,\, n)} = K(n,\, m) + \sum_{n'=n}^{\infty} K(n,\, n') F(n' + m), \tag{11.30}$$

where

$$F(m) = \frac{1}{2\pi i} \oint_C R(z) z^{m-1}\, dz + \sum_j c_j^2 z_j^m.$$

Equation (11.30) is a GLM equation generalized for discontinuous two poles of I_α and I_o. Basically, it can be used to determine $F(m)$ algebraically, which represents the lattice potential from scattered waves signified by given $\varphi(n,\, z)$.

Nevertheless, as practical calculations $K(n,\, m)$ are for $n \neq m$ as well as $n = m$, for which (11.30) can be modified by using another notation, $\kappa(n,\, m) = \frac{K(n,m)}{K(n,n)}$, where $m > n$. If $n \neq m$, equation (11.30) can be re-expressed as follows.

Theorem:

$$\kappa(n,\, m) + F(n+m) + \sum_{n'=n+1}^{\infty} \kappa(n,\, n') F(n' + m) = 0, \tag{11.31a}$$

which is a linear equation with respect to $\kappa(n, m)$ that is soluble *algebraically*. After solving for *off-diagonal* $\kappa(n, m)$, the diagonal $K(n, n)$ can be evaluated by

$$\frac{1}{K(n, n)^2} = 1 + F(2n) + \sum_{n'=n+1}^{\infty} \kappa(n, n')F(n' + m). \qquad (11.31b)$$

11.6 Entropy production at soliton singularities

11.6.1 Energy transfer at singularities

According to Flaschka's theory [2], the spatial profile of a nonlinear wave can be determined by (11.19*a*), while the development process can be described by (11.19*b*). The basic problem of nonlinear waves is the relation between the lattice potential and scattered waves, which can be obtained, in principle, by the GLM theorem. Accordingly, we could discuss the classical problem of initial conditions in nonlinear phenomena. Considering microscopically however, the initial condition is obscured by *space–time uncertainties* in quantum theory. Obscured by *bifurcation* or uncertainty in practice, the problem is nevertheless an interesting issue in the classical lattice dynamics. In any case, the development time in the above theory is not physically meaningful, if disregarding interactions at singularities with surrounding media.

Ignoring interactions however, the initial condition can be incorporated with the corresponding potential, as discussed in the following. Substituting the asymptotic expression for $n \to +\infty$

$$S(n, z, \tau) \to \{z^{-n} + R(n, \tau)n^{n}\}\exp(-i\omega\tau), \quad \text{where} \quad -i\omega = \frac{z - z^{-1}}{2},$$

into the development equation $\frac{\mathrm{d}}{\mathrm{d}\tau}S(n, z, \tau) = a_{n-1}S(n-1, z, \tau) - a_n S(n+1, z, \tau)$, we see that

$$\frac{\mathrm{d}}{\mathrm{d}\tau}S(n, z, \tau) \to \frac{1}{2}\{(z^{-n+1} - z^{-n-1}) + R(n, \tau)(z^{n-1} - z^{n+1})\}\exp(-i\omega\tau)$$
$$= \frac{z - z^{-1}}{2}\{z^{-n} - R(n, \tau)z^{n}\}\exp(-i\omega\tau) \quad \text{for} \quad n \to +\infty.$$

This must be identical to the direct differentiation of $S(n, z, \tau)$, i.e.

$$\frac{\mathrm{d}S(n, z, \tau)}{\mathrm{d}\tau} \to \frac{\mathrm{d}R(z, \tau)}{\mathrm{d}\tau}z^{n}\exp(-i\omega\tau) - i\omega\{z^{-n} + R(z, \tau)z^{n}\}\exp(-i\omega\tau).$$

Hence

$$\frac{\mathrm{d}R(z, \tau)}{\mathrm{d}\tau} = \left(\frac{z - z^{-1}}{2} + i\omega\right)R(z, \tau) = (z^{-1} - z)R(z, \tau)$$

and

$$R(z,\tau) = R(z,0)\exp(z^{-1} - z)\tau. \tag{11.32a}$$

Carrying out a calculation on the asymptotic wave function similar to the above, from the relation

$$\varphi(n,z,\tau) = \alpha(z,\tau)S(n,z,\tau) \to \{\alpha(z,\tau)z^{-n} + \beta(z,\tau)z^{n}\}\exp(-i\omega\tau)$$

we obtain

$$\alpha(z,\tau) = \alpha(z,0) \quad \text{and} \quad \beta(z,\tau) = \beta(z,0)\exp(z^{-1} - z)\tau. \tag{11.32b}$$

Finally, for a temporal variation of a pole characterized by $\left|\frac{d\alpha(z)}{dz}\right|_{z=z_1} = -\frac{1}{z_1\mu^2\beta(z_1)}\zeta(n,z_1)^2$ at z_1, we consider the asymptotic expression of $\zeta(n,z_1) \to c_1 z_1{}^n$ for $n \to +\infty$, where the temperature dependence of c_1 can be evaluated by

$$\dot{c}_1 = \frac{z_1{}^{-1} - z_1}{2}c_1 \quad \text{or} \quad c_1(\tau) = c_1(0)\exp\left(z_1{}^{-1} - z_1\right)\tau.$$

Experimentally, it is clear that these times associated with singularities I_α and I_o are not the same and distinct as related to a finite z_j and $z \approx 0$, respectively; the former is at *domain boundaries*, whereas the latter represents the *threshold of a phase transition*.

11.6.2 Soliton potentials at singularities

It is significant that the soliton propagation energy should all be absorbed by a singular potential if $R(z,\tau) = 0$, remaining the same regardless of τ due to the relation (11.32*a*). This means that all solitons are captured by the reflection-free singularity existing in the development process. In this section, such a singular pole of z in the complex plane is signified by a soliton potential of sech^2–type.

We assume that there is only one pole $z = z_1$ in the plane for simplicity, where we consider $|z_1| < 1$ for a real z_1. Writing $z = \pm\exp(-\gamma)$, where γ is an angle variable in the complex plane, the factor $c(t)$ defined in (11.27*b*) can be expressed as $c(t) = c(0)\exp\beta t$, where $\beta = \frac{z_1^{-1} - z_1}{2} = \pm\sinh\gamma$. Using these notations, the kernel of the integral equation (11.27*b*) can be written as $F(m) = c_1^2 z_1{}^m$, so that this GLM equation is

$$\kappa(n,m) + c_1^2 z_1^{n+m} + c_1^2 z_1{}^m \sum_{n'=n+1}^{\infty} \kappa(n,n')z_1{}^{n'} = 0.$$

Assuming $\kappa(n,m) = c_1 A^{(n)} z_1{}^m$, this algebraic equation can be solved with expressions

$$A^{(n)} = -\frac{c_1 z_1{}^n}{1 + \exp(2\delta)z_1^{2(n+1)}} \quad \text{and} \quad \exp\delta = \frac{c_1(t)}{\sqrt{1 - z_1^2}} = \exp(\delta_o + \beta t),$$

where $\exp\delta_0 = \frac{c_1(0)}{\sqrt{1-z_1^2}}$, hence we obtain

$$\{K(n, n)\}^{-2} = 1 + c_1^2 z_1^{2n} + c_1^3 A^{(n)} z_1^n \sum_{n'=n+1}^{\infty} z_1^{2n'} = \frac{1 + \exp(2\delta - 2n\gamma)}{1 + \exp\{2\delta - 2(n+1)\gamma\}}.$$

Accordingly, we have

$$\{K(n, n)\}^{-2} \to 1 \quad \text{for} \quad n \to +\infty; \quad \{K(n, n)\}^{-2} \to z_1^{-2} \quad \text{for} \quad n \to -\infty;$$

and

$$\begin{aligned} e^{-(Q_n - Q_{n-1})} - 1 &= \left\{\frac{K(n, n)}{K(n-1, n-1)}\right\}^2 - 1 \\ &= \frac{\{\exp\gamma - \exp(-\gamma)\}^2}{\{\exp(\gamma n - \delta) - \exp(-\gamma n + \delta)\}^2} \\ &= \beta^2 \operatorname{sech}^2(\gamma n - \beta t - \delta_0) = V(n, \phi), \end{aligned} \tag{11.33}$$

where

$$\phi = \gamma n - \beta t - \delta_0,$$

representing a soliton potential of n elemental solitons at site n, propagating to the right and left, depending on $\beta > 0$ and $\beta < 0$, respectively. Therefore, interpreting it as the *Weiss field*, the soliton potential has the peak at the singularity $z_1 \to 0$ at constituting *soliton lattice points n*, implying possible entropy production, depending on the number n in a thermodynamic environment, as discussed in chapters 4 and 9. With the lattice point signified by correlation potential energy, *the soliton lattice represents an adiabatic potential field that can be analyzed for the internal Weiss field with Born–von Kármán boundary conditions in finite crystals.*

11.7 The Toda lattice and the Korteweg–deVries equation

The Toda lattice is discussed with an *algebraically analyzable potential* for displacive variables, showing the same soliton solution as derived from the Korteweg–deVries equation. Dealing with lattice displacements, the soliton is an obvious consequence of Toda's theory. In this section, we show that the Toda lattice can be transformed to the Korteweg–deVries equation in an approximation process, keeping the integrability in the theory.

Writing the Toda lattice of variables V_n by

$$\frac{d^2}{dt^2}\ln(1 + V_n) = V_{n+1} + V_{n-1} - 2V_n, \tag{11.34}$$

we carry out a space–time translation $(x, t) \to (n, \tau)$, i.e. $t = \frac{\tau}{h^2}$ and $V_n(x, t) = h^2 u_n(\tau)$, where h is a positive parameter in the range $0 \leqslant h \leqslant 1$. In other words,

$$x = hn - \left(\frac{1}{h^2} - h^2\right)\tau,\ u(x, \tau) \rightarrow u_n(\tau)$$

and

$$\left\{\frac{\partial}{\partial\tau} - \left(\frac{1}{h^2} - h^2\right)\frac{\partial}{\partial x}\right\}^2 \ln\{1 + h^2u(x, \tau)\}$$
$$= \frac{1}{h^2}\{u(x + h) + u(x - h) - 2u(x, \tau)\}. \tag{11.35}$$

For $h = 1$ in this expression, (11.35) is the same as (11.34), while (11.35) represents a Toda's lattice, if $h \neq 1$. Considering a limiting case $h \rightarrow +0$, we have specifically $\frac{\partial}{\partial x}\left(-2\frac{\partial u}{\partial x} - \frac{1}{2}\frac{\partial u^2}{\partial x}\right) = \frac{1}{12}\frac{\partial^4 u}{\partial x^4}$ in (11.35), which can be integrated, if assuming $u \rightarrow 0$ for $x \rightarrow \pm\infty$, approaching a Korteweg–deVries equation

$$\frac{\partial u}{\partial\tau} + \frac{1}{2}u\frac{\partial u}{\partial x} + \frac{1}{24}\frac{\partial^3 u}{\partial x^3} = 0. \tag{11.36}$$

Thus, equation (11.31*a*, *b*) of Toda's lattice is consistent with the Korteweg–deVries equation (11.33) in the limit of $h \rightarrow 1$.

For a system of N solitons, we consider the Toda lattice (i) for $n = -\infty, \ldots, 0, 1, 2, \ldots, +\infty$, writing

$$\frac{\mathrm{d}}{\mathrm{d}t}\left\{\frac{1}{1 + V(n)}\frac{\mathrm{d}V_j(n)}{\mathrm{d}t}\right\} = V_j(n + 1) + V_j(n - 1) - 2V_j(n)$$

and

$$V(n) = \sum_{j=1}^{N} V_j(n) \tag{11.37}$$

where $V(n)$ is the adiabatic potential at the n-th particle of n coherent elemental solitons. $V_j(n)$ represents the mutual correlation potentials at one of the sites $j = 1, 2, \ldots, N$.

In fact, colliding solitons in the space–time diagram generally exhibit a distorted line shape that is a little too complex, however, it is relatively simpler to analyze with the Korteweg–deVries equation than with the Toda lattice.

Using the Korteweg–deVries equation for an individual function $u_j(n)$

$$\frac{\partial u_j}{\partial\tau} + 6u_j\frac{\partial u_j}{\partial x} + \frac{\partial^3 u_j}{\partial x^3} = 0$$

we obtain a relation for the potential $u(n)$ similar to the above $\boldsymbol{U}(n)$ in (11.33), which is written as

$$u(n) = \sum_{j=1}^{N} u_j(n), \tag{11.38}$$

as consistent with (11.36).

When a soliton is crossing over another, the distorted line-shape can be calculated simply by assuming that the overlapped area remains constant. Namely, $\int_{-\infty}^{+\infty} u_j \, \mathrm{d}x = 4\kappa_j$. In this case, as discussed in section 11.3, we consider the eigen-equation for the phase ϕ_j

$$\boldsymbol{L}\phi_j = -k_j^2\phi_j, \qquad \text{where} \qquad \boldsymbol{L} = -\frac{\partial^2}{\partial x^2} - u, \tag{11.39}$$

which can be modified by the development equation $\frac{\partial \phi_j}{\partial t} = \boldsymbol{B}\phi_j$, where $\boldsymbol{B} = -4\frac{\partial^3}{\partial x^3} - 6u\frac{\partial}{\partial x} - 3\frac{\partial u}{\partial x}$ from the Korteweg–deVries equation. Note that this wave function may represent angular phase ϕ_j of $\sigma = \sigma(\phi)$, as will be discussed in chapter 12, which is therefore focused on the ϕ in the following argument.

Considering the normalization condition to the eigen-function ϕ_j, i.e. $\int_{-\infty}^{+\infty} \phi_j^2 \, \mathrm{d}x = 1$, each soliton potential at j can be expressed as $u_j = 4k_j\phi_j^2$. For a collision of two solitons as shown in figures 10.1(a) and (b) in chapter 10, we take $N = 2$ and set linear combinations $\psi_j = c_1\phi_1 + c_2\phi_2$ for two independent solitons $j = 1, 2$ where $c_1^2 + c_2^2 = 1$, for the Lax commutator $\boldsymbol{BL} - \boldsymbol{LB}$ to be zero for conservative collision. Therefore, we obtain eigen-functions

$$\phi_1 = \frac{1}{\Delta}\begin{vmatrix} c_1 \exp \psi_1 & \dfrac{c_1c_2 \exp(\psi_1 + \psi_2)}{k_1 + k_2} \\ c_2 \exp \psi_2 & 1 + \dfrac{c_1c_2 \exp(2\psi_{2\kappa})}{k_1 + k_2} \end{vmatrix}$$

and

$$\phi_2 = \frac{1}{\Delta}\begin{vmatrix} 1 + \dfrac{c_1c_2 \exp(2\psi_1)}{k_1 + k_2} & c_1 \exp \psi_1 \\ \dfrac{c_1c_2 \exp(\psi_1 + \psi_2)}{k_1 + k_2} & c_2 \exp \psi_2 \end{vmatrix},$$

where

$$\Delta = det\left|\delta_{12} + \frac{c_1c_2 \exp(\psi_1 + \psi_2)}{k_1 + k_2}\right|, \psi_1 = k_1x_1 - 4k_1^3t_1 \quad \text{and} \quad \psi_2 = k_2x_2 - 4k_2^3t_2.$$

Calculating with eigen-functions ϕ_1 and ϕ_2 for the overlapped region, the colliding independent *phase solitons* ψ_1 and ψ_2 are in typical shape in the space–time (x, t). The foregoing algebraic calculation is a typical example of inverse scatterings, supporting for the Toda lattice to be thermodynamically consistent with nonlinear development with the Korteweg–deVries equation.

11.8 Topological strain mapping of mesoscopic Toda lattices

One-dimensional scatterings of a collective pseudospin σ_n by an Eckart potential, discussed in section 9.5, are found to have discrete magnitudes, referring to the parameter n. In section 9.2, for propagation of σ_n along the x direction, the driving potential $V(n, \phi)$ was found as a solution of the Korteweg–deVries equation, which is signified by the number of solitons n. In this model, the propagation is practically dispersive due to the third-order derivative $\boldsymbol{D}^3\sigma_n$ in the Korteweg–deVries equation, but becoming dissipative as well in the thermodynamic environment. Therefore, the adiabatic potential cannot be one-dimensional, as the transversal component $\sigma_\perp(\phi)$ exists that can interact with neighboring $\sigma_\perp(\phi)$ in the crystal. The potential $V(\phi)$ represents a *geometrical surface of strain distribution in crystal space*, expressed by the coordinates x, y and z. Considering coordinates y and z with one-dimensional x-axis, the three-dimensional Eckart potential $V(\boldsymbol{r})$ acts as a sink for a propagating wave $\boldsymbol{\sigma}(\boldsymbol{r})$. Such an expanded Eckart potential in three-dimensional crystals can therefore depict topological features of *the correlation field, representing distributed Weiss fields in a thermodynamic environment.*

Except for the critical region in the vicinity of T_c, the non-critical phase for $T < T_c$ can be represented by the adiabatic potential $V(n)$ of $\kappa = 1$, characterized as a function of soliton number n. The magnitude of $V(n)$ determines *long-range order* by means of n in a crystal. In idealized binary order, where the correlation field is primarily one-dimensional $V(n, x)$ in the x–y plane, consisting of a *trough* at $x = 0$ along the y-direction with the depth on the z-axis, forming a domain wall, as illustrated in figure 11.5. In a *hydrodynamic description*, these quantities $\sigma_l(n, x)$ and $V(n, y)$ can be simulated like fluid material, flowing through a crystal in thermodynamic equilibrium. Applying to condensates, we can map the mesoscopic state

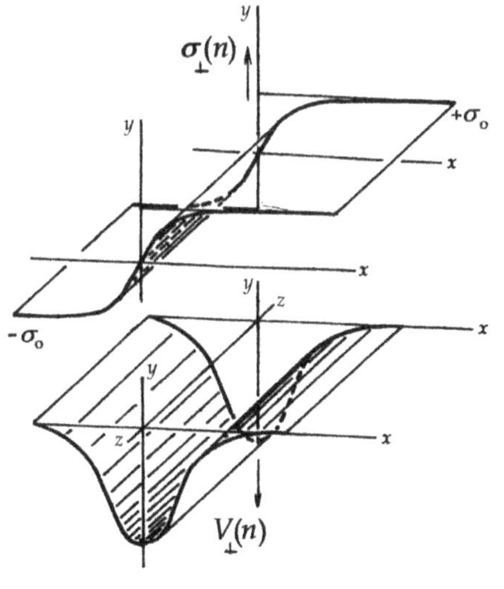

Figure 11.5. A three-dimensional view of a domain wall. (a) $\boldsymbol{\sigma}(n)$; (b) $V(n)$.

geometrically by surface $V(n, \pm L/2)$ for $\phi = \pm\pi/2$, where domain walls and other singularities can be illustrated in analogy to a geographical contour specified by altitudes. Such a domain pattern in a two-dimensional wall can be regarded as a specific case of Toda's soliton lattice, which is physically stabilized with *transverse correlations* across chains, as will similarly be discussed for a triple axis in chapter 12.

A basic theorem in the classical field theory suggests that a field consists of laminar (irrotational) and rotational component fields that are characterized by div $\boldsymbol{\sigma} \neq 0$ and curl $\boldsymbol{\sigma} \neq 0$, respectively. For example, the pattern shown previously in figure 8.5(a) represents a singularity in the laminar field; on the other hand, the rotational field can be specified by such an angular velocity as defined by $\boldsymbol{\omega} = \boldsymbol{r} \times \boldsymbol{\sigma}$, representing a field of vector potential $\boldsymbol{A} \propto \boldsymbol{\omega}$ related by the Lorentz condition div $\boldsymbol{A}(n) \propto \frac{\partial V(n)}{\partial t}$. Such a vector potential $\boldsymbol{A}(n)$ therefore represents a *vortex* in the potential field.

We notice that vortexes presumably occur at impurity or defective sites in crystals. Figure 11.6 shows a photograph of a TSCC crystal obtained under a polarizing microscope, which was taken from a crystal grown from aqueous solution at room temperature; exhibiting clearly distinguishable *ferroelastic domains* in hexagonal shape [7]. Domain boundaries are clearly distinguishable between shaded areas of the crystal, indicating the potential $V(n)$ in *trigonal* symmetry. The central line of the pattern parallel to the *a*-axis is terminated at some points, indicating a trigonal *vortex* around the singular *a*-axis.

Figure 8.6(a) shows vortexes in K_2ZnCl_4 crystal [8], where many *discommensuration lines* exhibit a clear indication of pseudopotentials of C_3 screw axis in this dark-field image of satellite reflections of electron beams from (100) planes, whereas points at which three lines join together can be interpreted as vortexes. Figure 8.6(b)

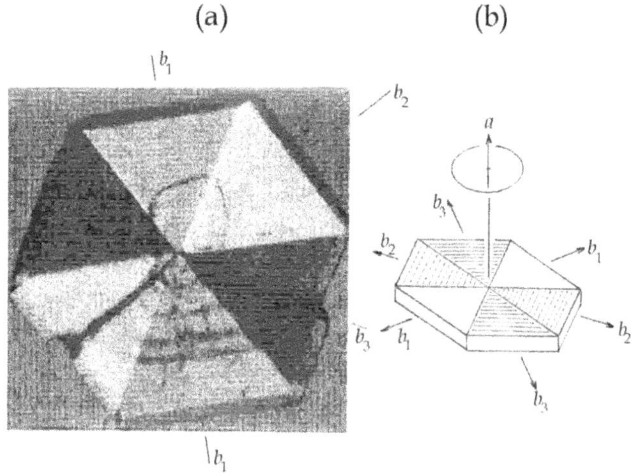

Figure 11.6. (a) A TSCC crystal photographed with a polarizing filter perpendicular to the *a*-axis [4]. (b) Schematically shows the presence of a vortex. v_1, v_2 and v_3 are directions of propagation in three elastic domains trigonally oriented.

shows a schematic interpretation of a vortex, which may occur at defective sites in the lattice.

Exercises

1. Compare the classical displacive lattice and the Toda lattice, and discuss the differences between their dynamical features.
2. Discuss the problem of energy transfer to thermodynamic environments in general terms. Can it be in relaxation or oscillatory? Discuss this problem as a matter of timescale of observation.
3. Using equation (11.26), we predicted that singularities of z may occur in two different modes I_α and I_0. Experimentally, these can be considered for domain boundaries and the initial problem. However, note that $\alpha(z_j) = 0$ is not necessarily for domain boundaries, it may occur in a domain that is not in equilibrium with surroundings, as discussed in section 9.7.

   Discuss this singularity with respect to the Born–Huang principle. In fact, it may be associated with a discontinuous change of Hénon–Heiles' conservative quantities in nonlinear systems.
4. Discuss the possibility of a creating vortex in the soliton field.
5. Toda's lattice is considered as the soliton lattice, where n elemental solitons constitute the lattice site n coherently. Is the coherency a conflicting statement regarding Heisenberg's uncertainties in quantum theory? Discuss the issue with respect to finite symmetry in crystals.
6. Toda's lattice is observable in principle, while only implicit in experiments on the order variable. Nevertheless, in section 11.8 we discussed it, representing the Weiss field. Clarify this issue by explaining why the potential $V(\phi, n)$ is visible in practical observations.

References

[1] Toda M 1987 *Lectures on Nonlinear Lattice Dynamics* (in Japanese) (Tokyo: Iwanami)
Toda M 1974 *Phys. Rep.* 18c
Toda M 1989 *Theory of Nonlinear Lattices* 2nd edn (Berlin: Springer)
[2] Flaschka H 1974 *Phys. Rev.* B **9** 1924
[3] Lax P D 1968 *Comm. Pure Appl. Math.* **21** 467
[4] Hénon M and Heiles C 1964 *Astro. J.* **69** 73
Hénon M 1974 *Phys. Rev.* B **9** 1921
[5] Gel'fand I M and Levitan B M 1955 *Am. Math. Soc. Transl.* **1** 253
[6] Fujimoto M 2013 *Thermodynamics of Crystalline States* 2nd edn (New York: Springer) p 129
[7] Fujimoto M and Jerzak S 1986 *Philos. Mag.* **B-53** 521
[8] Pan X and Unruh H-G 1990 *J. Phys. Cond. Matter.* **2** 323

Chapter 12

Phase solitons in adiabatic processes: topological correlations in the domain structure

A mesoscopic domain structure always exists in practical ordering processes. While associated with lattice imperfections in practice, the origin of domains can be attributed to an intrinsic potential specified by a *particular wavevector* $\boldsymbol{q}$ *that emerges adiabatically under critical conditions of structural changes*. In a cyclic transition process, collective pseudospins are characterized in the timescale of the whole crystal, linking all domains together. In thermal equilibrium, however, domains are not necessarily characterized by a single timescale, because of dissipative properties in different boundaries. Assuming a constant-pressure condition for an ordering process, the whole system is primarily in a single timescale in crystals. Domains are generally unstable in this sense and *disconnected* in non-equilibrium crystals, characterized in timescales of domain walls that interact elastically with the lattice.

A domain is primarily linked to another across the boundary zone, for which a *linear combination of domain fields*, called *Bäcklund's transformation*, can be postulated, where transversal correlations are significant, as described by the Toda potentials. The mesoscopic domain structure is described dynamically by Brillouin-zones of the collective order variables.

In addition to the *linear inversion* equivalent of 180°-rotation, *trigonal* 120°-rotation of C_3 symmetry is another example in this chapter, showing that the *phase soliton* plays a significant role across the domain boundaries; hence the observed structure can be determined at singularity of transversal waves of order variables. *Bäcklund's transformation* is a mathematical process to deal with phase correlations across domains, yet providing a general model of domains and crystal surfaces that is physically acceptable to explain *structural disorder*. In this chapter, we discuss the phase solitons arising from transversal correlations, leading to quasi-adiabatic transitions that can be responsible for domain separation.

The phase soliton theory describes a profile of ordering in terms of successive isothermal-adiabatic processes via soliton potentials, corresponding to Carnot's

doi:10.1088/978-0-7503-2572-1ch12

cycles in repetition, which is equivalent to statistical *long-range order*. Needless to say, in real crystals transversal correlations play a significant role in irreversible processes in modulated lattices.

12.1 The sine-Gordon equation

In chapter 11, using the Toda theory, we discussed nonlinear waves in a binary system, where the fluctuations are signified by an adiabatic soliton potential in thermodynamic environment. Also noted are symmetry changes in crystals and domain boundaries that occur with anisotropic phase-variation of the collective pseudospins.

Considering cases where the longitudinal wave is predominantly a function of angular variable ψ, $\sigma_1 = \sigma_1(\psi)$ is correlated in crystals by phase difference $\Delta\psi$ with the neighboring $\sigma_1' = \sigma_1'(\psi)$. At domain boundaries, the fluctuation $\Delta\psi$ arises from correlations between *transversal components* $\sigma_\perp(\psi)$ and $\sigma_\perp'(\psi')$ in the same transversal direction. For $\sigma_\perp(\psi + \Delta\psi)$ and $\sigma_\perp'(\psi + \Delta\,\psi')$, these $\Delta\psi$ and $\Delta\psi'$ are related with a *pseudopotential* $V(\psi)$, representing transversal nearest neighbor-correlations. It is realized that the variation of ψ is determined by the complex variable $\sigma = i\sigma_\perp$ in this case, where $\sigma_\perp$ propagates in a direction y perpendicular to the domain boundary.

Following section 8.4, we consider a complex potential for phase fluctuations, as described by a Klein–Gordon's equation

$$\frac{\partial^2\rho_\perp}{\partial t^2} - v^2\frac{\partial^2\rho_\perp}{\partial y^2} = -K^2(y,\,t)\,\rho_\perp$$

for the density function $\rho_\perp = \sigma_\perp{}^*\sigma_\perp$ to fluctuate along *transversal y-direction* at a speed v, where $K^2(y,\,t)$ is considered for a *real* interaction with the lattice that is responsible for a *pseudopotential* $v(y,\,t)$ to modulate the phase. The coordinate y should determine the variation of $\rho_\perp = \sigma_\perp^*\sigma_\perp$, specifying propagation $\pm y$ at the position $y = 0$ along perpendicular directions to σ_1, depending on the local symmetry.

In *domain boundary planes* perpendicular to σ_1, the amplitude of $\rho_\perp$ is primarily constant of temperature, while phase fluctuations are *adiabatic* between $\sigma_\perp$ and $\sigma_\perp'$. At a small amplitude σ_0, we can consider that approximately *sinusoidal* fluctuations $\sigma_\perp \sim \sigma_0 \exp i\Delta\psi$ and $\sigma_\perp' \sim \sigma_0 \exp i\Delta\psi'$ are correlated by Riccati's relation $\Delta\psi' = -\frac{1}{\Delta\psi}\frac{d\Delta\psi}{dy}$, so that $\sigma_\perp$ and $\sigma_\perp'$ are correlated via Toda's exponential potentials across the boundaries expressed as functions of coordinates y and t, characterized by $\sigma_\perp \rightleftarrows -\sigma_\perp'$. Signified by a finite value at zero distance, their *exponential correlation potentials* allow mathematical analysis of anomalies at domain boundaries, with respect to phase correlations.

Replacing $K(y,\,t)$ by $m = m(\psi)$ to deal with the discrete nature of $\Delta\psi$, we express the Klein–Gordon equation for the phase function $\psi\ (y,\,t)$ as follows.

Theorem:

$$\frac{\partial^2\psi}{\partial t^2} - v^2\frac{\partial^2\psi}{\partial y^2} + m^2\psi = 0. \tag{12.1}$$

Transforming space-time $(\pm y, t) \rightarrow (\xi, \eta)$ by $\xi = \frac{y - vt}{2}$ and $\eta = \frac{y + vt}{2}$ along the y-axis, equation (12.1) can be written as

$$\frac{\partial^2 \psi}{\partial \xi \partial \eta} = m^2 \psi. \tag{12.2}$$

Integrating this, for the *critical case*, we obtain

$$\frac{\partial \psi}{\partial \xi} = m^2 \int \psi \, \mathrm{d}\eta + r(\xi) \quad \text{and} \quad \frac{\partial \psi}{\partial \eta} = m^2 \int \psi \, \mathrm{d}\xi + s(\eta), \tag{12.3}$$

where $r(\xi)$ and $s(\eta)$ are arbitrary functions of ξ and η, respectively.

Defining $P(\psi) = \frac{\partial \psi}{\partial \xi}$ and $Q(\psi) = \frac{\partial \psi}{\partial \eta}$, equation (12.2) can be written as

$$Q\frac{\partial P}{\partial \psi} = P\frac{\partial Q}{\partial \psi} = m^2 \psi,$$

which can be integrated as the relation $Q = a^2 P$, where a is an arbitrary constant.

Integrating these relations further, we obtain $a^2 P^2 = m^2 \psi^2 + 2c$, where c is an integration constant. Assuming $c = 0$ for simplicity to consider $P = 0$ at $\psi = 0$, we can write

$$P(\psi) = \frac{\partial \psi}{\partial \xi} = \pm \frac{m}{a} \psi \quad \text{and} \quad Q(\psi) = \frac{\partial \psi}{\partial \eta} = \pm (ma)\psi, \tag{12.4}$$

allowing us to write for the phase function ψ to be expressed in the form

$$\psi = \exp(\pm i m \zeta + b) = A \exp\{\pm i m (qy - \Omega t)\}, \tag{12.5}$$

where

$$A = \exp b, \quad \zeta = \frac{\xi}{a} + a\eta, \quad q = \frac{1 + a^2}{2a} \quad \text{and} \quad \Omega = \frac{1 - a^2}{2a}.$$

Equation (12.5) manifests that the phase function ψ can be expressed by another phase $\zeta = qy - \omega t$, which can describe transversal fluctuations correlated in phase correlation along the y-direction perpendicular to the x-axis, assuming the *y-propagation is energetically in favor*. Writing $mq = k$ and $m\Omega = \omega$ for $a^2 < 1$ in this case, we obtain $\omega = vk$ and $\Omega = vq$, indicating that the wave (12.5) at q and Ω *correlating with neighboring waves* propagates at speed $v = \omega/k$ along the y-direction.

12.2 The Bäcklund transformation and domain boundaries

It is noted that binary inversion $\sigma_n(\omega) \rightleftarrows -\sigma_n(\omega)$ at a lattice site n generates a critical splitting similar to quantum-mechanical splitting. On the other hand, such domain inversion across domain walls is not exactly quantum-mechanical, but indicating for mathematical fluctuations to be expressed by a *linear combination* of two domain fields. Known as the *Bäcklund transformation* [1], such a linear combination is considered for the *correlated fields* to exhibit inversion symmetry across the domain boundaries.

Considering it as a binary transition between *opposite domains* denoted by superscripts (1) and (2), we assume linearly combined states of phase functions. That is as follows.

Theorem:

$$\psi_{\rm A} = (\psi^{(1)} + \psi^{(2)})/\sqrt{2} \quad \text{and} \quad \psi_{\rm P} = (\psi^{(1)} - \psi^{(2)})/\sqrt{2}$$

for the boundaries in analogy to A and P modes under critical conditions as previously discussed in chapter 4 with the complex vector $\psi_{\rm P} + i\psi_{\rm A}$. Writing that

$$\frac{\partial(\psi^{(1)} + \psi^{(2)})}{\partial\xi} = m^2\int(\psi^{(1)} + \psi^{(2)})\mathrm{d}\eta + r(\xi) = P(\psi^{(1)} + \psi^{(2)})$$

and

$$\frac{\partial(\psi^{(1)} - \psi^{(2)})}{\partial\eta} = m^2\int(\psi^{(1)} + \psi^{(2)})\mathrm{d}\xi + s(\eta) = Q(\psi^{(1)} - \psi^{(2)}),$$

we consider equation (12.4) to be extended to domains (1) and (2).

Differentiating these relations, we have

$$\frac{\partial\psi^{(1)}}{\partial\eta}\frac{\partial P}{\partial\psi^{(1)}} + \frac{\partial\psi^{(2)}}{\partial\eta}\frac{\partial P}{\partial\psi^{(2)}} = m^2(\psi^{(1)} + \psi^{(2)}) \quad \text{and}$$

$$\frac{\partial\psi^{(1)}}{\partial\xi}\frac{\partial Q}{\partial\psi^{(1)}} - \frac{\partial\psi^{(2)}}{\partial\xi}\frac{\partial Q}{\partial\psi^{(2)}} = m^2(\psi^{(1)} - \psi^{(2)}).$$

Manipulating derivatives, we arrange these results in the following equations:

$$P\frac{\partial Q}{\partial\psi^{(1)}} - Q\frac{\partial P}{\partial\psi^{(1)}} = -2m^2\psi^{(2)}, \quad P\frac{\partial Q}{\partial\psi^{(2)}} - Q\frac{\partial P}{\partial\psi^{(2)}} = 2m^2\psi^{(1)},$$
$$\frac{\partial Q}{\partial\psi^{(1)}} - \frac{\partial Q}{\partial\psi^{(2)}} = 0, \quad \text{and} \quad \frac{\partial P}{\partial\psi^{(1)}} + \frac{\partial P}{\partial\psi^{(2)}} = 0 \tag{12.6}$$

Using these expressions for the Klein–Gordon equation (12.2), we obtain the relations

$$P\frac{\partial^2 Q}{\partial\psi_{\rm A}^2} - Q\frac{\partial^2 P}{\partial\psi_{\rm P}^2} = 0,$$

hence

$$\frac{\partial^2 P}{\partial\psi_{\rm P}^2} + \kappa^2 P = 0 \quad \text{and} \quad \frac{\partial^2 Q}{\partial\psi_{\rm A}^2} + \kappa^2 Q = 0, \tag{12.7}$$

where κ^2 is an arbitrary parameter. Noting that the equations in (12.7) are identical, however, we set $\kappa^2 = -m^2$ or $\kappa = \pm im$ for $\kappa \neq 0$, thereby translating equation (12.7) into a similar form to (12.4), i.e.

$$P(\psi_A) = \pm\frac{m}{a}\psi_P \quad \text{and} \quad Q(\psi_P) = \pm(ma)\psi_A, \tag{12.8}$$

where

$$\psi_A = (\psi_1 + \psi_2)/\sqrt{2} = A\{\exp im\zeta + \exp(-im\zeta)\} = 2A\cos m\zeta,$$
$$\psi_P = (\psi_1 - \psi_2)/\sqrt{2} = A\{\exp im\zeta - \exp(-im\zeta)\} = 2iA\sin m\zeta\,.$$

Using the relations (12.6), we can derive wave equations for these ψ_A and ψ_P to satisfy:

$$\frac{\partial^2\psi_A}{\partial t^2} - v^2\frac{\partial^2\psi_A}{\partial y^2} = 0 \quad \text{and} \quad \frac{\partial^2\psi_P}{\partial t^2} - v^2\frac{\partial^2\psi_P}{\partial y^2} = -\sin\psi_P. \tag{12.9}$$

The first equation implies that phase modulation does not occur in ψ_A-mode, because of the absence of longitudinal P-potential on the left side. On the other hand, the second equation indicates that ψ_P-mode is modulated by $-\sin\psi_P$, suggesting that the second equation for a ψ_P state, called the *sine-Gordon equation*, corresponding to a modulated wave with lower eigenvalues.

Solving (12.7) for P and Q, we obtain

$$P = p\sin\kappa\psi_P \quad \text{and} \quad Q = q\sin(\kappa\psi_A + \theta),$$

where θ is an arbitrary phase difference between P and Q. However, choosing these constants in such a way that $\theta = 0$ and $pq = \pm 4$ for convenience, equation (12.8) can be found as in modified relations as

Theorem:

$$\frac{\partial\psi_A}{\partial\xi} = \frac{2}{a}\sin\frac{\psi_P}{2} \quad \text{and} \quad \frac{\partial\psi_P}{\partial\eta} = \pm 2a\sin\frac{\psi_A}{2}. \tag{12.10}$$

Equations (12.10) signify the presence of singular relations between ψ_P and ψ_A at domain boundaries, while domain volumes are considered as continuous quantities.

The sine-Gordon equation (12.9) for ψ_P can be solved in a simple way in the following. Here, using the new space–time variable $\pm\zeta = \pm(ky - \omega t)$ for ψ_P to be signified by propagation speed $v = \omega k$, the equation of motion can be expressed as

$$\frac{d^2\psi_P}{d\zeta^2} = \pm\frac{1}{1 - v^2}\sin\psi_P, \tag{12.11}$$

which is an equation for a classical pendulum [2], providing two distinct solutions expressed by

$$\psi_{\mathrm{P}}(+\zeta) = 4\tan^{-1}\left(\exp\frac{+\zeta}{\sqrt{1-v^2}}\right) \quad \text{and}$$
$$\psi_{\mathrm{P}}(-\zeta) = 4\tan^{-1}\left(\exp\frac{-\zeta}{\sqrt{1-v^2}}\right) \quad \text{for} \quad v^2 < 1. \tag{12.12}$$

It is clear in (12.12) that for the parameter $v \to \pm 1$, we have asymptotically the relation $\lim_{\zeta\to\zeta_0} \psi_{\mathrm{P}}(\pm\zeta) \to \psi_{\mathrm{P}}(\zeta_0) = \pm\,\pi/2$, resulting in the following.

Theorem:

$$\psi_{\mathrm{P}}(\zeta_0) - \psi_{\mathrm{P}}(-\zeta_0) = \Delta\psi_{\mathrm{P}}(\zeta_0) = \pi, \tag{12.13}$$

representing a *singular* character of angular gap between *kinks* of $\psi_{\mathrm{P}}(\pm\,\zeta_0)$ at domain boundaries specified $\pm\zeta_0$.

Using the first relation in (12.10), these $\psi_{\mathrm{P}}(\pm\zeta_0)$ can be replaced by $(\partial\psi_A/\partial\zeta)_{\pm\zeta_0}$ that is significant for a small ζ_0. Accordingly, relation (12.13) can be expressed as follows.

Theorem:

$$\left\{\left(\frac{\mathrm{d}\psi_{\mathrm{A}}}{\mathrm{d}\zeta}\right)_{+\zeta_0} - \left(\frac{\mathrm{d}\psi_{\mathrm{A}}}{\mathrm{d}\zeta}\right)_{-\zeta_0}\right\}\Delta\zeta \propto \frac{\pi}{a}\Delta\zeta, \tag{12.14}$$

where $\Delta\zeta = +\zeta_0 - (-\zeta_0) = 2\zeta_0$, corresponds to *hypothetical work* for inversion $\psi_{\mathrm{A}}(+\zeta_0) \to \psi_{\mathrm{A}}(-\zeta_0)$ across the width $\Delta\zeta$ of *domain wall* between $+\zeta_0$ and $-\zeta_0$. Writing $\psi_{\mathrm{A}}(+\zeta_0) - \psi_{\mathrm{A}}(-\zeta_0) = \tilde{\sigma}\Delta\zeta$ for convenience, equation (12.14) can be re-expressed as $\mathrm{d}\ln\tilde{\sigma}/\mathrm{d}\sigma = \text{const.}$, which is a Riccati's transformation discussed in section 9.7.

Corresponding to these kinks, the lattice should be modified by domain walls, for which we can consider an adiabatic imaginary potential proportional to $\left(i\frac{\mathrm{d}\psi_{\mathrm{A}}}{\mathrm{d}\zeta}\right)_{\pm\zeta_0}$ *perpendicular to the direction of* ζ. It is noted in (12.14) that the phase $\zeta_0 = K(y_0 - vt)$ is a function of time, taking an arbitrary direction on the boundary plane. Accordingly, $\left(\frac{\mathrm{d}\psi_{\mathrm{A}}}{\mathrm{d}\zeta}\right)_{\pm\zeta_0}$ are a pair of *opposite time-dependent shear stresses* in the lattice, so that the strain energy can be minimized in thermodynamic equilibrium (see chapter 8). Consequently, such complex potentials as $\pm\, i\frac{\mathrm{d}\psi_{\mathrm{A}}}{\mathrm{d}\zeta_0}$ are derived from the Klein–Gordon equation, allowing nonlinear propagation at another speed between $+\left(\frac{\mathrm{d}\psi_{\mathrm{A}}}{\mathrm{d}\zeta}\right)_{+\zeta_0}$ and $-\left(\frac{\mathrm{d}\psi_{\mathrm{A}}}{\mathrm{d}\zeta}\right)_{-\zeta_0}$, at an unspecified speed v to be determined by timescale of the boundary zone in between. The binary inversion visualized by 180°-rotation of $\psi_{\mathrm{P}}(\zeta)$ around $y = 0$ on the y-axis can then be discussed in analogy of the A- and P-mode phase fluctuations [2] in a critical region. The amplitude σ_0 of pseudospin mode is almost constant near critical points, where the vector wave can be expressed in complex form $\sigma = \sigma_1 \pm i\sigma_\perp = \sigma_1 \pm i\sigma_{\perp 0} f(\psi)$.

Further noticeable is that the first equation for ψ_A expressed by (12.9) lacks a driving potential, while the second equation indicates that ψ_P is driven in parallel to the potential $\pm 2a\ \sin(\psi_P/2)$. While vanishing in the critical region, the driving shear potentials appear as related to ψ_A in *parallel* to the domain wall, as indicated by (12.14).

Describing the behavior of ψ_P and ψ_A with equation (8.10*c*), the phase function

Theorem:

$$\sigma_\perp(\zeta_o) = \psi_A(\zeta_o) \propto \tanh\left(\zeta_o \pm \frac{\pi}{2}\right) = \pm\tanh\zeta_o \tag{12.15}$$

confirms the thermodynamic nature of *phase soliton*, carrying significance of transversal interactions, as will be discussed in section 12.7.1 for rotation $+\tanh\zeta_o \rightleftarrows -\tanh\zeta_o$. In addition, the potential energy proportional to $\sigma_\perp(\zeta_o)^2 \propto \text{sech}^2\,\zeta_o$ can be obtained for hypothetical work of inversion $\psi_A(+\zeta_o) \rightleftarrows \psi_A(-\zeta_o)$ that is *directional along the z-direction.*

Implied by the above argument, the binary inversion $+\boldsymbol{q} \rightleftarrows -\boldsymbol{q}$ across the boundaries in-between can be signified by the correlation energies $\pm J_{q,-q}\boldsymbol{\sigma}_q \cdot \boldsymbol{\sigma}_{-q}$, where $J_{q,-q} > 0$, providing entropy production from the symmetric combination $\sigma_A = (\boldsymbol{\sigma}_q + \boldsymbol{\sigma}_{-q})/\sqrt{2}$, and domain separation by antisymmetric $\boldsymbol{\sigma}_P = (\boldsymbol{\sigma}_q - \boldsymbol{\sigma}_{-q})/\sqrt{2}$, respectively.

12.3 Computational studies of Bäcklund transformation

The Bäcklund transformation provides a useful computational method for multi-domain disorder in equilibrium states. In the section, the method is discussed for a general numerical analysis, following Taniuchi and Nishihara [3].

We consider a system composed of multiple fields of domains, each one signified by different time-scale, although not required. Order variables in each domain are primarily driven as described by Klein–Gordon equations independently, as discussed in section 11.1. Hence, for the net field we write a *symmetric* combination

$$\psi = \sum_j \psi_j = \sum_{j=0}^{n} A_j \exp(\pm m\zeta_j), \tag{12.16}$$

where $\zeta_j = \frac{\xi}{a_j} + a_j\eta$ and a specific time-scale index $j = 1, 2, \ldots, n$ of domains, as shown in (12.5). Field amplitudes are related by

$$A_j = \gamma_j A_{j-1}, \quad \text{where } \gamma_j = \frac{a_{j-1}\beta - a_j\alpha}{a_{j-1} - a_j} \quad \text{for} \quad a_j \neq a_{j-1} \tag{12.17}$$

and α, β are mixing constants for A and P modes of transformation between $j-1$ and j. Accordingly, the coefficient γ_j can be either $+1$ or -1, corresponding to $\alpha = \beta$ and $\alpha = -\beta$, respectively.

We assume that $\psi_1 = A_1 \exp(\pm m\zeta)$ is symmetric, which are not an antisymmetric P-mode. On the other hand, all other fields j are correlated with $j-1$ by transformation. Such a transformation is two ways as signified by $\gamma_j = \pm 1$, where the notation $\gamma_j = -1$ is kept for the path of P-mode transforming $\psi_{j-1} \to \psi_j$. Bäcklund's transformations are described in this way, as expressed

$$\psi_1 = A_1 \exp(\pm im\zeta_1),$$
$$\psi_2 = A_2 \exp(\pm im\zeta_2) + \gamma_1 A_1 \exp(\pm im\zeta_1)$$

and

$$\psi_3 = \gamma_3 A_2 \exp(\pm im\zeta_2) + \gamma_3\gamma_2 A_1 \exp(\pm im\zeta_1).$$

We shall not proceed beyond ψ_3 in the present example of Bäcklund's transformation between ψ_1 and ψ_3, where ψ_2 is considered for a state in between. The state ψ_2 simulates a fictitious interaction between ψ_1 and ψ_3 for the computational analysis of the transition $\psi_1 \to \psi_3$. It is noted that the selection of ψ_2 is not unique. Assuming two choices ψ_2 and ψ_2', we take two paths expressed by

$$\psi_2 = B_2\psi_1 \quad \text{and} \quad \psi_2' = B_2'\psi_1. \tag{12.18}$$

Hence paths from ψ_1 to ψ_3 via these ψ_2, ψ_2' can be expressed as

$$\psi_3 = B_3\psi_2 = B_3B_2\psi_1 \quad \text{and} \quad \psi_3 = B_3\psi_2' = B_3B_2'\psi_1 \tag{12.19}$$

as illustrated in the *Lamb's diagram* in figure 12.1. Nevertheless, since these paths specified by arbitrary parameters A_2 and A_2' are adjustable mathematically for the general unequal relation $B_3B_2 \neq B_3B_2'$ to be the equality $B_3B_2 = B_3B_2'$, they make four hypothetical processes to be signified uniquely by two parameters a and a' for sine-Gordon equations, as indicated in the diagram.

Using a and a', the two paths can be replaced by the single transformation $\psi_1 \to \psi_2 \to \psi_3$, that is

$$\tan\frac{\psi_3}{4} = \frac{a + a'}{a - a'}\tan\frac{\psi_1 - \psi_2}{4}, \tag{12.20}$$

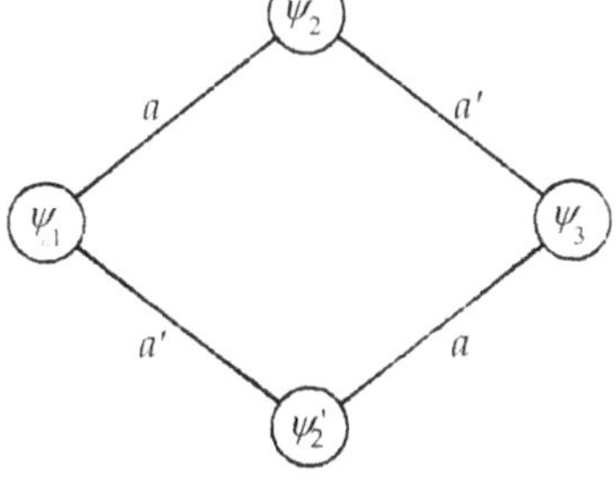

Figure 12.1. (a) Three ϕ-kinks of a phase soliton. (b) Three corresponding peaks of sech ϕ separated by a large $\Delta\phi$. (c) Three peaks with a small separation.

Known as Bäcklund's theorem, (12.20) is a convenient formula for computational work. From (12.20), we obtain

$$\psi_3 = 4\tan^{-1}\left\{\frac{a + a'}{a - a'}\frac{\sinh(\zeta_1 - \zeta_2)}{\sinh(\zeta_1 + \zeta_2)}\right\}. \tag{12.21}$$

Defining $\zeta_1 \pm \zeta_2 = \gamma_\pm(x - v_\pm t)$, where

$$\gamma_\pm = \gamma_1 \pm \gamma_2 \quad \text{and} \quad v_\pm = \frac{1 + v_1 v_2 \mp \sqrt{\left(1 - v_1^2\right)\left(1 - v_2^2\right)}}{v_1 + v_2},$$

We arrive at approximate expressions of ψ_3 from (12.21), namely

$$\psi_3 \approx 4\tan^{-1}[1 - \exp\{-\gamma_2(y - v_2 t - \Delta_2)\}] \quad \text{for} \quad y > v_+ t,$$
$$\psi_3 \approx 4\tan^{-1}[-\exp\{-\gamma_2(y - v_1 t + \Delta_1)\}] \quad \text{for} \quad v_+ t > y > v_- t,$$
$$\text{and} \quad \psi_3 \approx 4\tan^{-1}[\exp\{\gamma_2(y - v_2 t) + \Delta_2\}] \quad \text{for} \quad v_- t > y$$

where $\Delta_{1,2} = \gamma_{1,2}^{-1}\log|\frac{a + a'}{a - a'}|$.

Accordingly, in the limit of $t \to \pm\infty$, we can expect that the values of above ψ_3 are 0, $\pm 2\pi$ and 0 in the regions of $y > v_2 t + \Delta_2$, $v_+ t + \Delta_2 > y > v_- t - \Delta_1$ and $v_1 t - \Delta_1 > y$, respectively.

Figure 12.2 is the result of numerical simulation of ψ_3 performed by Taniuchi and Nishihara [3], showing *table-shaped curves* of $\psi_3(y)$ at specified times t. In the figure, the calculated derivatives $d\psi_3/dy$ of these curves are indicated, representing *antisymmetric transversal kinks* for the phase soliton potential significant for domain boundaries, which was discussed theoretically by (12.4).

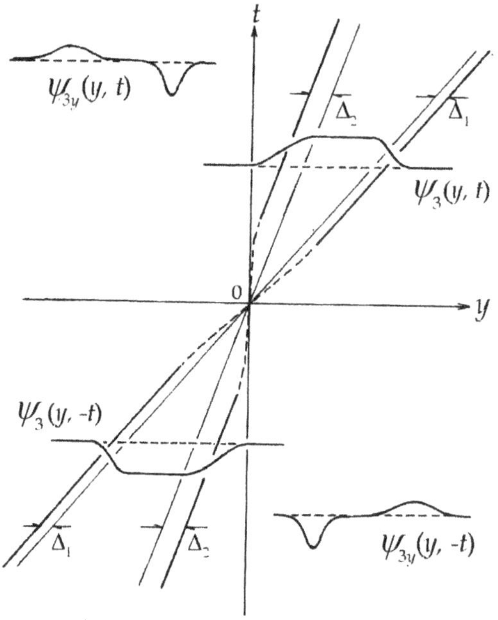

Figure 12.2. A computer simulation of phase solitons. Reproduced with permisson from [3].

12.4 Trigonal structural transitions

12.4.1 The sine-Gordon equation

Trigonal transitions have been observed in some crystals, where a pseudopotential in C_3-cyclic symmetry is regarded to be responsible. Most of these transitions have been characterized as incommensurate–commensurate transitions, which were also studied in charge-density waves, modulated lattice distortions, spin-density waves, helical spin order and some surface layer structure. We consider such *ferroelastic domains* in trigonal pattern, as shown in figure 11.6(a), which is a typical cyclic transition, and discussed in this section following Per Bak's general theory [4].

In tris-sarcosine calcium chloride (TSCC) crystals, trigonal displacements occur evidently with respect to the nearly *orthorhombic* a-axis with C_3 cyclic symmetry, which is transformed to slightly *monoclinic* by transition. In figure 11.6, we consider that displacements $\boldsymbol{u}_1$, $\boldsymbol{u}_2$ and $\boldsymbol{u}_3$ in these domains occur independently along directions b_1, b_2 and b_3. In the critical region however, they should interact angularly along a circle centered at the a-axis. Denoting the angular variable by ψ_i, where $i = 1, 2, 3$, such angular fluctuations can be expressed as

$$\psi_i = \psi_{i+}\exp(i\phi_{i+}) + \psi_{i-}\exp(-i\phi_{i-}) \quad \text{for} \quad 0 \leqslant \phi_i \leqslant 2\pi, \tag{12.22}$$

where ϕ_i is the phase of angular fluctuations around, say, $\phi_1 = 0^\circ$, $\phi_2 = 60^\circ$ and $\phi_3 = 120^\circ$. Following McMillan [5], the *phase fluctuation energy* can be expressed in expanded Gibbs potential, which is minimized with respect to the phase variation to obtain thermal equilibrium. He expressed the Gibbs function of the *phase density in terms of phase variables ψ_i correlated in C_3-symmetry, where $i = 1, 2, 3$*, i.e.

$$\begin{aligned}\Delta g = \sum_m^{\text{domain}} \Delta g_m &= \frac{1}{2}A\sum_{i=1}^{3}|\psi_i|^2 + \frac{1}{4}B\sum_{i=1}^{3}(|\psi_i|^2)^2 \\ &+ C\sum_{i=1}^{3}|\psi_i|^4 + D(\psi_1\psi_2\psi_3 + \psi_{-1}\psi_{-2}\psi_{-3}) \\ &+ D'\sum_{i=1}^{3}\left(\psi_i^3 + \psi_i^{-3}\right) + E\sum_{i=1}^{3}|(\nabla_i - i\delta_i)\psi_i|^2 + F\sum_{i=1}^{3}|(\boldsymbol{q}_i \times \nabla_i)\psi_i|^2\end{aligned} \tag{12.23}$$

representing the presence of phase correlations determined by lattice symmetry. Here, the corresponding wave vectors $\boldsymbol{q}_i$ are implicit, as included in the phase ϕ_i. Terms of A, B, C, D and D' are real responses from the lattice, development nonlinear propagation, among which C, D and D' contribute also to symmetry changes. The term of D' is a pseudopotential that is similar to (12.12) for C_3-screw axis, and the last terms of E and F compose complex potentials for solitons. Note that these terms are all additive, and the Δg represents condensates of the order variables ψ_i in crystals[note1].

[note1] The terms of D and D' are included in (12.14) phenomenologically. For the TSCC case, the origin may be attributed to the gravitational field where the sample is grown.

For each domain characterized by $\phi_1 \neq 0$ and $\phi_2 = \phi_3 = 0$, for instance, we can minimize Δg by calculating $(\partial \Delta g / \partial \phi_1)_{\phi_2 = \phi_3 = 0} = 0$, and obtain

$$\frac{1}{2}\left(\frac{\mathrm{d}\phi_1}{\mathrm{d}y} - \delta\right)^2 - \zeta(\cos 3\phi_1) - \frac{1}{2}\delta^2 + v_1 = 0,$$

where $\zeta = D'/E$, $v_1 = F\,|(\boldsymbol{q}_1 \times \nabla)\psi_1|^2$ and y is the spatial part of the phase ζ. This is the equation for the phase ϕ_1 determined by the potential $v(y)$, representing the adiabatic potential for the phase ϕ_1. Therefore, differentiating it once more to determine the center of fluctuation. In this way, we arrive at a sine-Gordon equation,

$$\frac{\mathrm{d}^2\phi_1}{\mathrm{d}y^2} - 3\zeta \sin(3\phi_1) = 0. \tag{12.24}$$

For ϕ_2 and ϕ_3, the same equation applies to determine stable fluctuations; and equation (12.13) is a sine-Gordon equation identical to screw-axis symmetry in chapter 8.

The solution of (12.24) for the trigonal case can be illustrated in figure 12.3(a), showing singularities at ϕ_1, ϕ_2 and ϕ_3, similar to figure 8.4. Corresponding to ϕ_i, three peaks of $\mathrm{sech}^2\,\phi_i$ are found in the perpendicular directions, which are separated by $\Delta\phi = 120°$, to the trigonal plane, as shown in figure 12.3(b) and (c), representing the *phase-soliton potential* at *strained domain boundaries*. It is notable that such an interpretation is valid for the trigonal plane with the soliton theory, while three phases ϕ_i may have different timescales for the corresponding domains to be in thermal equilibrium. Discontinuities at ϕ_i due to potentials $\mathrm{sech}^2\,\phi_i$ can be responsible for entropy production if the number of soliton exceeds a critical number n_c, as indicated in figure 12.3(c). A general three-dimensional view of the vector $\boldsymbol{\sigma}$ near the phase kink is sketched in figure 12.3(d).

Physically, it is significant that the correlation energies at singularities are transferred between the duals to maintain steady condensates in equilibrium with

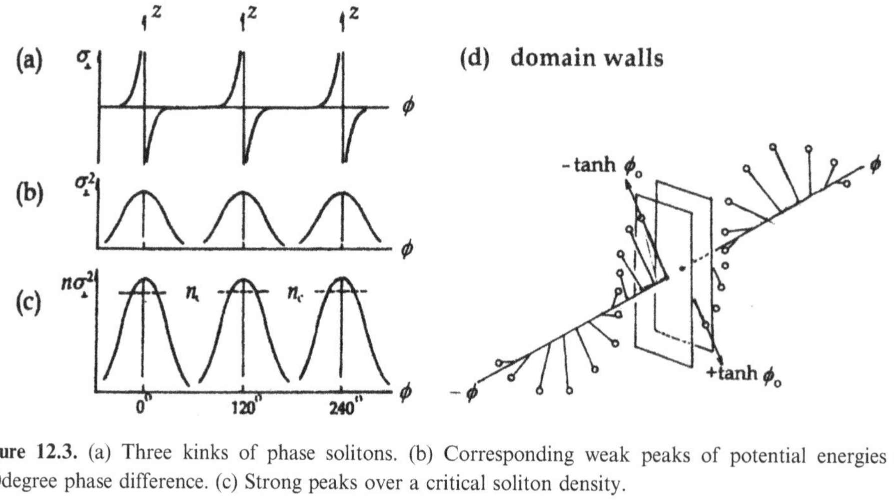

Figure 12.3. (a) Three kinks of phase solitons. (b) Corresponding weak peaks of potential energies for 120degree phase difference. (c) Strong peaks over a critical soliton density.

the phonon system in the surroundings, requiring energy exchange with the lattice. Such energy transfer occurs through internal interactions in a crystal, resulting in a modulated lattice in the direction of varying phase ϕ,[note2] provided that the soliton number is insufficient for energy exchange with phonons.

For the phase solitons observed as triple singularities, we realize there exist soliton transitions, causing fluctuations among singularities, which are generally called a *breezer* [1]. The phase soliton fluctuations discussed in section 12.3 can thus be analyzed as in the following.

Considering functions $\psi_1(\phi)$, $\psi_2(\phi)$ and $\psi_3(\phi)$, constituting a one-dimensional circular lattice, these can generally be re-defined as $\psi_{n-1}(\phi)$, $\psi_n(\phi)$ and $\psi_{n+1}(\phi)$ for Toda's lattice, where the angle ϕ should be an angular variable in the range $-\infty < \phi < +\infty$, where the lattice site n represents the number of solitons. Therefore, at these sites we can write wave equations for phase density functions $\sigma_{\perp n} \propto f(\psi_n)$ and $\sigma_{\perp n\pm 1} \propto f(\psi_{n\pm 1})$ characterized by real eigenvalues λ_n and $\lambda_{n\pm 1}$, respectively; for which we have wave equations

$$\frac{d^2\sigma_{\perp n}}{d\phi^2} + (\lambda_n - v_n)\sigma_{\perp n} = 0 \quad \text{and} \quad \frac{d^2\sigma_{\perp n\pm 1}}{d\phi^2} + (\lambda_{n\pm 1} - v_{n\pm 1})\sigma_{\perp n\pm 1} = 0. \tag{12.25}$$

For unequal eigenvalues $\lambda_n \neq \lambda_{n\pm 1}$, the transitions $n \rightleftarrows n \pm 1$ are determined by $\Delta\lambda_{n,n\pm 1} = \Delta v_{n,n\pm 1} = \mp 2n\, \mathrm{sec}\, h^2\phi$ as in section 10.2.1, both of which are thermodynamically detected, respectively, by adding or subtracting one elemental soliton under adiabatic or isothermal conditions with surrounding media, as discussed in chapter 9. Nonetheless, in such *breezer* fluctuations the *adiabatic soliton transition* by Δn should be dominant, and the corresponding entropy production $-\frac{\mu\Delta n(p_0 - p)}{T}$, where p_0 is constant, can be studied thermodynamically with varying external pressure p. In this case, $\Delta n(p_0 - p)$ can be regarded as an order parameter that is proportional to a parabolic $\sqrt{p_0 - p}$, if solitons are assumed to be *collision-free particles* in this near-critical region.

12.4.2 Observing adiabatic fluctuations

At this point, it is interesting to pay attention to the result of LEED experiments by Fain and Chinn [7] using krypton gas on a graphite surface that shows parabolic changes of the soliton lattice, as illustrated in figure 12.4, indicating adiabatic entropy production during the *irreversible* process in surface structure. Figure 12.5(a) shows schematically an energy diagram in C_3-symmetry for the variable Δn, showing adiabatic fluctuations, while $\Delta\phi$ in figure 12.5(b) is related to entropy production. Experimental work by Fain and Chinn has demonstrated that *adiabatic soft modes* are detected with varying pressure p externally.

Needless to say, such experiments should be carried out on modulated crystals as a whole, such that the sample specimen is of macroscopic size. Analogous to *polar or*

[note2] This is in fact a valid assumption that is correct in tetragonal lattice where $y = z$. However, for $y \simeq z$, Nakamura [6] showed that the phase soliton theory can be acceptable for orthorhombic symmetry as well.

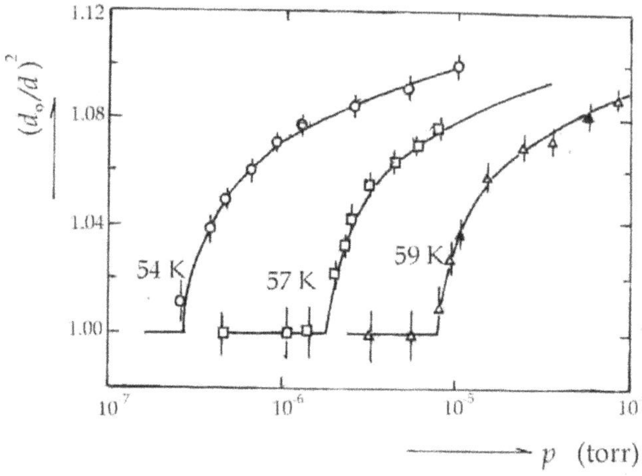

Figure 12.4. Adiabatic fluctuation processes with varying pressure. Reproduced with permission from [7].

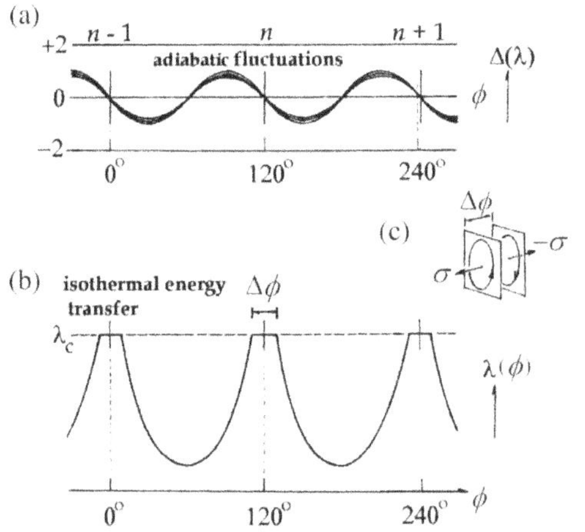

Figure 12.5. (a) Adiabatic fluctuation $\Delta n = \pm 1$, (b) isothermal energy transfers, (c) inversion $\boldsymbol{\sigma} \rightleftarrows -\boldsymbol{\sigma}$ across the domain wall.

magnetic materials, it is clear in classical physics that the displacive order parameter should be uniaxial with applied external stress uniformly, for which the host crystal should be in *ellipsoidal shape*. Considering uniaxial order in a crystal, the above view of the inside of a macroscopic crystal should be considered for domains, where surfaces are defined in acceptable accuracy, so that the external pressure can be regarded as uniaxial p_{int} with respect to the relation $-\boldsymbol{p} \cdot \mathrm{d}\boldsymbol{A} = -p\mathrm{d}A_\perp = p_{\text{int}}\mathrm{d}A_\perp$, where $\mathrm{d}\boldsymbol{A}$ and $\mathrm{d}A_\perp$ are differential elements on ellipsoidal and mathematical surfaces, respectively, as indicated in figure 12.6. In practice, the *isotropic external pressure p* is balanced effectively with the hypothetical internal pressure $\mp p_{\text{int}}$ on terminal surfaces $\pm A_\perp$ in equilibrium crystals related with $p_{\text{int}}\Delta v = \mu\Delta n$, where Δv is the volume change. Assuming the ideal-gas law $pv = nk_BT$ for the mobile soliton system of displacive order, we can derive the relation

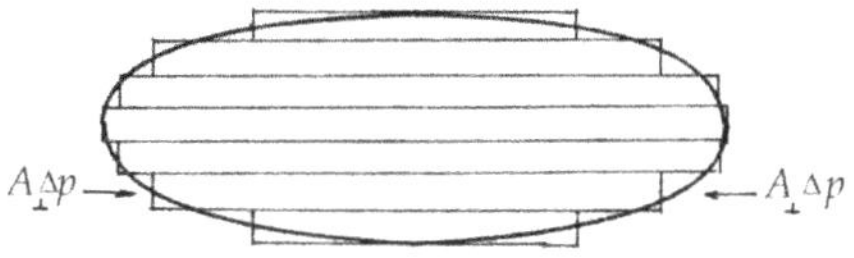

Figure 12.6. A macroscopic model of an ellipsoidal body for a finite lattice consisting of linear domains with surfaces. External pressure p and terminal surfaces $\pm dA_\perp$ are shown for external isotropic work $\mp p\, dA$ that is equivalent to uniaxial work $\pm A_\perp\, dp$. It is shown here that the crystal symmetry can be broken if *uniaxial pressure* is either $-dp$ or $+dp$, while it is invariant under *isotropic pressure* p.

$$p\mathrm{d}A_\perp = -A_\perp \mathrm{d}p \qquad \text{at} \qquad \Delta T = 0. \tag{12.26a}$$

for the adiabatic process in the lattice, we can study the *soft-mode response* from the pressure change $\mp\Delta p$. Shown by experimental curves, figure 12.4 illustrates that the pressure-dependent frequencies appear to be proportional to $\sqrt{\Delta p}$, which is characteristic soft-mode behavior. Theoretically, that can simply be verified with Landau's theory in chapter 4. Namely, writing a change in the Gibbs function as follows.

Theorem:

$$\Delta G(\Delta n, \Delta p) \approx \frac{1}{2}a(\Delta n)^2 \mp A_\perp \Delta p, \tag{12.26b}$$

the equilibrium condition $\Delta G = 0$ determines the parabolic relation $\Delta n \propto \sqrt{|\Delta p|}$ for adiabatic fluctuations.

In the experiments of Fain and Chinn [7], such adiabatic fluctuations were studied with applied pressure, where their results support nonlinear displacements on surfaces. Moreover, recent reports on superconductivity in a *metallic phase of hydrogen sulfide under high-pressure* [8] demonstrates that the transition temperature T_c to zero-resistivity depends on the applied pressure p, offering direct evidence for nonlinear lattice displacement, if the relation $T_c = T_c(p)$ can be verified with equation (12.17) for the presence of soft modes. (See chapter 15 for details.) It is noted that a more general discussion on adiabatic processes are discussed in Kittel's *Thermal Physics* [12]. At this point, we note that the significance of pressure p for thermodynamic systems was also emphasized in early theoretical work by Lee and Young [9].

12.5 Toda's theory of domain stability

We have so far defined the domain wall mathematically, but in practice domains are observed as space separating domains, which are normally composed of parallel crystal planes. In fact, Toda's theorem in chapter 11 is applicable to circularly correlated cases in crystals, where the correlation path can be assumed to be identified experimentally. The theory for a one-dimensional system can be modified for a circular system, where phase disorder is stabilized by the specific cyclic symmetry in the lattice. Analyzing such mesoscopic transitions with respect to the geometrical figure, we can find an important criterion for ordering in practical applications. In this section, Toda's theory for a multiple cyclic system is outlined for irreversible processes of entropy production, referring for detail to his Japanese textbook '*Nonlinear Lattice Dynamics*' [10].

Assuming that nearest-neighbor correlations in a soliton lattice are represented by a Toda potential signified by dynamical parameters a_n and b_n, which are actually characterized by magnitudes to scatter phonons with respect to the timescale of the thermodynamic environment, the nonlinearity is developed for the phase function as determined by the equation

$$\boldsymbol{L}\psi(n) = a_{n-1}\psi(n-1) + b_n\psi(n) + a_n\psi(n+1) = \lambda\psi(n),$$

where $-\infty < n < +\infty$. In such an infinite lattice, we have an eigen-equation for the function $\psi(n)$

$$\frac{\mathrm{d}^2\psi(n)}{\mathrm{d}\phi^2} + \{\lambda - u(\phi)\}\psi(n) = 0,$$

but by assuming

$$u(\phi) = u\left(\phi + \frac{2\pi}{N}\right), \tag{12.27}$$

we can convert the lattice to circular lattice for $0 \leqslant \phi \leqslant \frac{2\pi}{N}$. Here N can be any integer for infinite circular lattice, while $N = 3$ is of particular interest for C_3 symmetry in the present case.

If there is an additional mechanism between domains characterized by $N = 3$ and $N = 4$, we cannot use the exponential correlation potential, because the mechanism is concerned about different eigenvalues $\lambda(3)$ and $\lambda(4)$. However, we can assume that the Toda lattice in between is circularly uniform, specified by symmetries of different domains, but connected via Bäcklund's transformation.

For the boundary region between 3 and 4, Bäcklund's transformation can be expressed by a linear combination of wave functions ψ_1 and ψ_2 that represent domains 3 and 4, respectively. Hence, the transformation is determined by $\begin{pmatrix}\psi_1(N)\\ \psi_2(N)\end{pmatrix} = \boldsymbol{M}\begin{pmatrix}\psi_1(N+1)\\ \psi_2(N+1)\end{pmatrix}$, where the matrix $\boldsymbol{M}$ can be written as

$$\boldsymbol{M} = \begin{pmatrix}\psi_1(N) & \psi_1(N+1)\\ \psi_2(N) & \psi_2(N+1)\end{pmatrix}.$$

On the other hand, eliminating λ from the relations

$$a_n\psi_1(n+1) + b_n\psi_1(n) + a_{n-1}\psi_1(n-1) = \lambda\psi_1(n)$$
$$\text{and}\quad a_n\psi_2(n+1) + b_n\psi_2(n) + a_{n-1}\psi_2(n-1) = \lambda\psi_2(n),$$

we can derive determinantal *Wronskian* expressions

$$W = a_n \begin{vmatrix} \psi_1(n) & \psi_1(n+1) \\ \psi_2(n) & \psi_2(n+1) \end{vmatrix} = a_{n-1} \begin{vmatrix} \psi_1(n-1) & \psi_1(n) \\ \psi_2(n-1) & \psi_2(n) \end{vmatrix}.$$

Rising $n \to N$ and decreasing $n \to 1$, we obtain that

$$W = a_N \begin{vmatrix} \psi_1(N) & \psi_1(N+1) \\ \psi_2(N) & \psi_2(N+1) \end{vmatrix} = a_0 \begin{vmatrix} \psi_1(0) & \psi_1(1) \\ \psi_2(1) & \psi_2(0) \end{vmatrix} = a_0 \qquad (12.28)$$

Here, we have assumed that the last determinant (12.18) is given by $\begin{vmatrix} 1 & 0 \\ 0 & 1 \end{vmatrix} = 1$, if the *initial conditions* are given by

$$\psi_1(0) = \psi_2(1) = 1,\ \psi_1(1) = \psi_2(0) = 0 \quad \text{and} \quad \Delta(\lambda) = \psi_1(0) + \psi_2(1),$$

thereby keeping the domain symmetry via transit to another domain unchanged. In this case, $a_N = a_0$ because of the lattice periodicity, and we have

$$\det \boldsymbol{M} = \psi_1(N)\psi_2(N+1) - \psi_1(N+1)\psi_2(N) = 1,$$

On the other hand, if considering that the region between $(N, N+1)$ represents the boundary region (3, 4), we should write

$$c_1\psi_1(N) + c_2\psi_2(N) = \rho c_1\psi_1(0) = \rho c_1$$

and

$$c_1\psi_1(N+1) + c_2\psi_2(N+1) = \rho c_2\psi_2(1) = \rho c_2.$$

In this case, ρ should satisfy the equation

$$\rho^2 - \Delta(\lambda)\rho + 1 = 0,$$

where

$$\Delta(\lambda) = \psi_1(N) + \psi_2(N+1) = \text{trace}\, \boldsymbol{M}.$$

Solving these equations, we obtain the relation

$$\rho = \frac{1}{2}\left\{\Delta(\lambda) \pm \sqrt{\Delta(\lambda)^2 - 4}\right\}, \quad \text{where } -2 < \Delta(\lambda) < +2; \qquad (12.29)$$

hence in terms of ρ, the range $-1 < \rho < +1$ determines a *stable domain transformation characterized by C_3 symmetry*.

Using (12.17) for the boundary, we have $a_n = 1/2$ and $b_n = 0$ for equation $\boldsymbol{L}\psi(n) = \lambda\psi(n)$, which is written as $\frac{1}{2}\{\psi(n+1) + \psi(n-1)\} = \lambda\psi(n)$. Writing $\lambda = \cos\alpha$, this can be satisfied by $\psi_1 = -\frac{\sin\alpha(n-1)}{\sin\alpha}$, $\psi_2(n) = \frac{\sin\alpha n}{\sin\alpha}$ and $\Delta(\lambda) = 2\cos\alpha N$. Accordingly, the last relation of $\Delta(\lambda)$ versus N is characterized as periodic between $+2$ and -2. As illustrated in figure 12.5(a), transitions $n \rightleftarrows n \pm 1$ or $\Delta n = \pm 1$ are possible among $\psi(n)$ and $\psi(n \pm 1)$, indicating *steady fluctuations in the adiabatic processes for* $|\Delta(\lambda)| \leqslant 2$.

It is important to note at this point that the boundary wall between domains 1 and 2 can be specified by different orientation operators $\boldsymbol{u}_1$ and $\boldsymbol{u}_2$, for which we can write

$$\boldsymbol{L}_{11} = \boldsymbol{u}_1^{-1}\boldsymbol{L}\boldsymbol{u}_1, \qquad \boldsymbol{L}_{22} = \boldsymbol{u}_2^{-1}\boldsymbol{L}\boldsymbol{u}_2 \quad \text{and} \quad \boldsymbol{L}_{12} = \boldsymbol{u}_1^{-1}\boldsymbol{L}\boldsymbol{u}_2.$$

While $\boldsymbol{L}_{11} = \boldsymbol{L}_{22} = \boldsymbol{L}$ signify two domains 1 and 2 of the same symmetry group, $\boldsymbol{L}_{12}$ corresponds to the boundary characterized by a unitary relation $\psi_1{}^*\boldsymbol{L}_{12}\psi_2 = \psi_2{}^*\boldsymbol{L}_{21}\psi_1$ between ψ_1 and ψ_2, making density matrix elements $(\psi_*\psi)_{1,2}$ eigenvectors of $\boldsymbol{L}_{12}$ with real eigenvalues. Nevertheless, writing

$$\boldsymbol{L}_{12}\psi_1(N) = \lambda_1\psi_1(N) \quad \text{and} \quad \boldsymbol{L}_{12}\psi_2(N) = \lambda_2\psi_2(N),$$

we can derive the expression

$$\begin{aligned}(\lambda_1 - \lambda_2)\psi_1(n)\psi_2(n) &= \psi_2(n)\{a_n\psi_1(n+1) + b_n\psi_1(n) + a_{n-1}\psi_1(n-1)\}\\ &\quad - \psi_1(n)\{a_n\psi_2(n+1) + b_n\psi_2(n) + a_{n-1}\psi_2(n-1)\}\\ &= a_n\{\psi_1(n+1)\psi_2(n) - \psi_1(n)\psi_2(n+1)\}\\ &\quad - a_{n-1}\{\psi_1(n)\psi_2(n-1) - \psi_1(n-1)\psi_2(n)\}\end{aligned}.$$

Hence, we obtain

$$\begin{aligned}(\lambda_1 - \lambda_2)\sum_{n=1}^{N}\psi_1(n)\psi_2(n) &= a_N\{\psi_1(N+1)\psi_2(N) - \psi_1(N)\psi_2(N+1)\}\\ &\quad - a_0\{\psi_1(1)\psi_2(0) - \psi_1(0)\psi_2(1)\},\end{aligned}$$

which does not vanish for nonzero N, and hence we should have $\lambda_1 \neq \lambda_2$.

In this context, the domain wall region can have higher energies than stable domains, susceptible for adiabatic transitions to surrounding lattice media. Figure 12.5(b) illustrates that the soliton energy at these triple boundaries is transferred in a finite region $\Delta\phi$, becoming observable in isothermal experiments beyond the threshold indicated by λ_c in the figure. Here, $\Delta\phi \to 0$ takes place as $\lambda \to \lambda_c$, corresponding to singular soliton potentials v_n at a specific $n = n_c$; the energy transfer for $n < n_c$ is essentially *adiabatic* and confined within a deformed lattice in the limit λ_c, but becoming outwardly *isothermal* for $\lambda > \lambda_c$, if $n > n_c$.

However, note that these λ_1 and λ_2 in the above are perturbation results, which may not necessarily be compatible with the correlation potentials. Actually, Toda [10] derived a spectrum of compatible eigenvalues, called an *auxiliary spectrum*, thereby making all analyzable. However, his theory is not significantly innovative for energy transfer processes, remaining just for mathematical interest, so we shall not discuss it any further.

Discussed in section 12.2, a conventional classical model for such processes is *inversion* $\boldsymbol{\sigma} \rightleftarrows -\boldsymbol{\sigma}$ *across the boundary* as sketched in figure 12.5(c); although violating inside the boundary zone by deforming the structure, whose strain energy should be transferred to the surrounding, as confirmed by the Toda theory.

If n_c is sufficiently large, then a pair of oppositely twisting stresses exert a torque across the domain wall to separate domains adiabatically into two parts; bringing, however, the whole system into new equilibrium between them. We should note however that at a critical n_c, a quasi-adiabatic transition should take place, depending on the volume of domains, which is, however, undetermined, unless the size of a crystal and surface conditions are specified in practice.

Nevertheless, such adiabatic processes occur between different λ in surface layers under external compression, as reported in [7], showing parabolic plots as in figure 12.4 that represent the soft-mode character of pressure-dependent frequencies.

We discussed C_2 and C_3 symmetries as representative examples, however, the Toda lattice can theoretically be shown to connect C_3 and C_4 systems. The C_4 symmetry with 45° in a symmetric plane, is familiar from cubic Fe magnets. It is interesting that order variables in Toda's lattice are not necessarily in a straight-line geometrically, as long as the correlation potential $\phi(r)$ in (11.8*a*) can be adequately assumed, similar to triangular C_3 lattices. With such a model, it should be possible to discuss nonlinear propagations in helical chains in biological systems.

Applying Toda's theory to binary domains, we consider the interface between two domains C_2 and C_{-2}, for which we have $\Delta(\lambda) = \psi_0(1) + \psi_1(0) = 0$ in (12.29), indicating no adiabatic fluctuations in the boundary zone. Hence, the entropy production is entirely isothermal across normal domain boundaries, similar to the critical fluctuations discussed in section 4.2.3 of chapter 4.

12.6 Kac's theory of nonlinear development and domain boundaries

The domain boundary constitutes a transition zone between ordered domains, which can be stable after the stress energy is released to the surroundings. The crystal can otherwise be separated into independent domain structures. In a stable crystal, however, the properties of boundary zone are related to physically different domains. In the foregoing, they can be connected by Bäcklund's transformation, which is nonetheless a mathematical method to cover the transition area. Nonetheless, we have Kac's theory [11] supporting such a transformation, which is summarized in the following.

Considering Toda's potential (11.5*a*) of nearest neighbors in a one-dimensional chain, the nonlinear developing operator $\boldsymbol{B}$ was defined as (11.15). On the other hand, such an operator expressed in matrix form can be employed for inversion in a crystal. Kac and his coworkers extended initial interactions at n beyond $n \pm 1$, showing that Toda's potential is applicable for a steady condition with higher accuracy.

Assuming (11.28) is the simplest case, we write the matrix $\boldsymbol{B}_K = \boldsymbol{B} + \Delta\boldsymbol{B}_K$ in place of $\boldsymbol{B}$, where the index K specifies Kac's operators. The relation $\boldsymbol{B}_K\boldsymbol{L} - \boldsymbol{L}\boldsymbol{B}_K = 0$ should then be held for steady conditions against inversion, which is ascertained if $\Delta\boldsymbol{B}_K\,\boldsymbol{L} - \boldsymbol{L}\,\Delta\boldsymbol{B}_K = 0$. We assume

$$\Delta\boldsymbol{B}_K = \begin{pmatrix} 0 & \beta_1 & \gamma_1 & & \\ -\beta_1 & 0 & \beta_1 & \gamma_2 & \\ -\gamma_1 & -\beta_1 & 0 & & \\ & -\gamma_2 & & \ddots & \\ & & & & \end{pmatrix} \qquad (12.30a)$$

to express extended interactions that obey $\Delta \boldsymbol{B}_K \boldsymbol{L} - \boldsymbol{L}\,\Delta \boldsymbol{B}_K = 0$ under the restriction $\frac{\mathrm{d}L}{\mathrm{d}\tau} = 0$ for a steady state. Hence, off-diagonal elements

$$(n-1|\Delta \boldsymbol{B}_K \boldsymbol{L} - \boldsymbol{L}\Delta \boldsymbol{B}_K|n+2) \quad \text{and} \quad (n-1|\Delta \boldsymbol{B}_K \boldsymbol{L} - \boldsymbol{L}\Delta \boldsymbol{B}_K|n+1)$$

should be equal to zero for $\Delta \boldsymbol{B}_K$ to obey inversion symmetry. It is notable at this point that such $\Delta \boldsymbol{B}_K$ is a useful idea for geometrical details of the *domain wall to likely involve lattice dislocations and related defects.*

Elements of $\Delta \boldsymbol{B}_K$ as $\gamma_k \neq 0$ in the former can be written as $a_{k-1}\gamma_k = \gamma_{k-1}a_{k+1}$, and hence

$$\gamma_k = a_k a_{k+1}; \tag{12.30b}$$

and from the latter we obtain the relation

$$\beta_k = (b_k + b_{k+1})a_k. \tag{12.30c}$$

However, these Kac extension should be zero, if $\frac{\mathrm{d}L_K}{\mathrm{d}\tau} = 0$ where $\boldsymbol{L}_K = \boldsymbol{L} + \Delta \boldsymbol{L}_K$.

Kac's postulates are generally expressed as

$$\Delta \boldsymbol{L}_K = \begin{pmatrix} \ddots & & & & \\ & 0 & \alpha_{n-1} & & \\ & \alpha_{n-1} & 0 & \alpha_n & \\ & & \alpha_n & 0 & \alpha_{n+1} \\ & & & \alpha_{n+1} & 0 & \\ & & & & & \ddots \end{pmatrix}, \tag{12.31}$$

with (12.30*a*) rewritten as

$$\Delta \boldsymbol{B}_K = \begin{pmatrix} \ddots & & & & & \\ & 0 & 0 & \alpha_{n-2}\alpha_{n-1} & & \\ & 0 & 0 & 0 & \alpha_{n-1}\alpha_n & \\ & -\alpha_{n-2}\alpha_{n-1} & 0 & 0 & 0 & \alpha_n\alpha_{n+1} \\ & & -\alpha_{n-1}\alpha_n & 0 & 0 & 0 \\ & & & -\alpha_n\alpha_{n+1} & 0 & 0 \\ & & & & & & \ddots \end{pmatrix}, \tag{12.32}$$

and the developing equation is given as

$$\frac{\mathrm{d}\Delta \boldsymbol{L}_K}{\mathrm{d}\tau} = \Delta \boldsymbol{B}_K \Delta \boldsymbol{L}_K - \Delta \boldsymbol{L}_K \Delta \boldsymbol{B}_K. \tag{12.33}$$

Therefore, for these elements, we have the relation

$$\frac{d\alpha_n}{d\tau} = \alpha_n\left(\alpha_{n+1}^2 - \alpha_{n-1}^2\right),$$

which is analogous to equation (11.14).

Writing $\alpha_n = \frac{1}{2}\exp\frac{-\phi_n}{2}$, and defining timescale as τ instead of $\tau/2$ for convenience, the above relation is converted into the equation for ϕ_n, i.e.

$$\frac{d\phi_n}{d\tau} = \exp(-\phi_{n+1}) - \exp(-\phi_{n-1}),$$

allowing us to select ϕ_n for two independent transferable hypothetical lattices.

And both re-indexing ϕ_n by ϕ_{2n} to distinguish them, we write another expression

$$\frac{d\phi_{2n}}{d\tau} = \exp(-\phi_{2n-1}) - \exp(-\phi_{2n+1}). \tag{12.34}$$

Defining new coordinates Q_n and Q_n' by

$$\phi_{2n} = Q_n' - Q_n, \qquad \phi_{2n+1} = Q_{n+1} - Q_n' \quad \text{and} \quad \phi_{2n} + \phi_{2n+1} = Q_{n+1} - Q_n,$$

we have then

$$\frac{dQ_n}{d\tau} = \exp(-\phi_{2n-1}) + \exp(-\phi_{2n}) - c \quad \text{and}$$
$$\frac{dQ_n'}{d\tau} = \exp(-\phi_{2n}) + \exp(-\phi_{2n+1}) - c,$$

where c is a constant; both satisfy the relation (12.34). Differentiating once again, we have

$$\begin{aligned}\frac{d^2Q_n}{d\tau^2} &= -\frac{d\phi_{2n-1}}{d\tau}\exp(-\phi_{2n-1}) - \frac{d\phi_{2n}}{d\tau}\exp(-\phi_{2n}) \\ &= \exp\{-(\phi_{2n-2} + \phi_{2n-1})\} - \exp\{-(\phi_{2n} + \phi_{2n+1})\},\end{aligned}$$

that is

$$\frac{d^2Q_n}{d\tau^2} = \exp\{-(Q_n - Q_{n-1})\} - \exp\{-(Q_{n+1} - Q_n)\}. \tag{12.35}$$

Similarly

$$\frac{d^2Q_n'}{d\tau^2} = \exp\{-(Q_n' - Q_{n-1}')\} - \exp\{-(Q_{n+1}' - Q_n')\} \tag{12.36}$$

Thus, Kac's nonlinear development theory encompasses two independent exponential lattices that are connected via exponential potentials, supporting Toda's theory of domain walls composed of two lattice modes constituting so-called *auxiliary spectra of different eigenvalues in the boundary region.*

Resulting from the postulate (12.33), eigenvalues of $\Delta \boldsymbol{L}_K$ and $\Delta \boldsymbol{L}_K'$ can be kept constant with time τ and τ', respectively, in which case the transformation between Q_n and Q_n' is *canonical* because of their conservative nature, signifying the thermodynamic environment. Hence, Bäcklund's transformation can be regarded as canonical with respect to Toda's potentials, thereby *all states in a boundary zone are specified by their eigenvalues.* Assuming auxiliary spectra, Toda [10] elaborated a theory of a boundary zone, providing an analyzable interpretation of mesoscopic disorder. However, we shall not pursue his mathematical procedure, since published data in the present literature are not very informative about the boundary structure.

Figure 12.7 is a schematic view of these lattices, and figure 12.5(c) illustrates the nature of domain boundary. The transformation can be interpreted geometrically as rotations by $(\Delta \boldsymbol{L}_K)_{2n}$, $(\Delta \boldsymbol{L}_K)_{2n+1}$ and $(\Delta \boldsymbol{L}_K)_{2n+2}$, which are indicated by the corresponding angular variations $\Delta\phi_{2n}$, $\Delta\phi_{2n+1}$ and $\Delta\phi_{2n+2}$, respectively, indicating disoriented Q- and Q'-lattices in the figure. Related to the selection rules $\Delta n = \pm 1$, these rotations represent adiabatic fluctuations, as indicated in figure 12.5(c). Thus, Kac's theory allows us to also use the Toda theorem for non-straight lattices, whose applications are not specifically restricted to straight lattices.

Applying the theory to the C_3-transition of TSCC crystals, we consider two modified Toda lattices within each boundary region, linking canonical modes for connecting domains 1 and 2 as specified by *eigen-spectra* of L_1^+ and L_2^-. As calculated in section 12.6, the eigenvalues of ΔL_K and $\Delta L_K'$ are always additive to λ, being susceptible to heat transfer to the crystal, breaking the connection between domains if exceeding the limit. Figure 12.8(b) illustrates the domain configuration in TSCC in the plane specified by $L_i^\pm$, where $i = 1, 2$. It is noted that L_i^+ and L_i^- are not distinguishable experimentally, while in theory they are different with respect to the Toda potentials ϕ_+ and ϕ_- characterized by $ab > 0$ and $ab < 0$, respectively. Figure 12.8(a) shows kinks of $\sigma_\perp$ in the vicinity of domain boundaries 1, 2 and 3.

Reviewing the basic assumptions for Kac's lattice, transversal correlations are determined by the phase function, disregarding the direction of propagation, so that the lattice composed of mesoscopic domains is not necessarily in straight form as originally assumed. Accordingly, Kac's model can be applied to crystalline polymers, chained biological macro-molecules, etc, to discuss their transport problems. However, it is significant in the soliton theory that responsible transversal

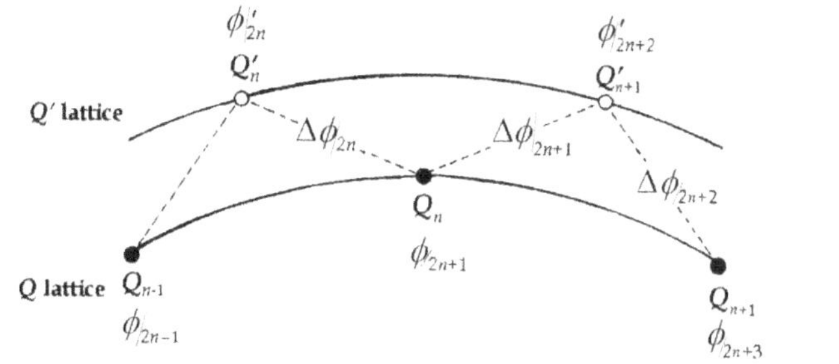

Figure 12.7. Schematic view of two Toda lattices in a Kac's system, which are convertible via Bäcklund's transformation, illustrating schematically that Q- and Q'-lattices are disoriented by rotation $\Delta\phi_{2n}$, $\Delta\phi_{2n+1}$ and $\Delta\phi_{\ 2n=2}$ in the boundary zone between two domains.

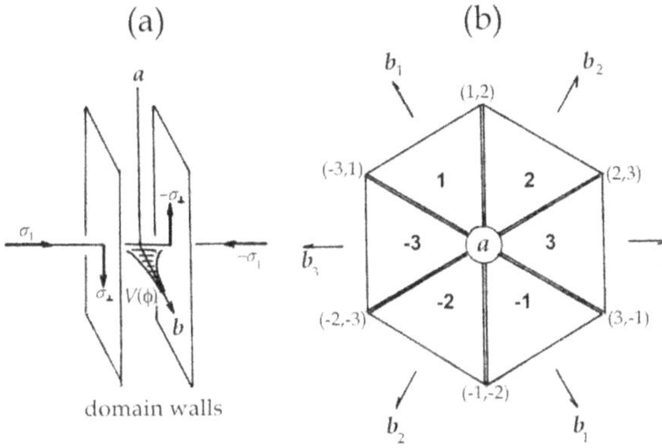

Figure 12.8. Domain structure of C_3 screw symmetry. Domain boundaries. $\sigma_{i\perp}(\phi)$ before entropy production and the potential $v_i(\phi)$ after energy transfer. Domain pattern in the bc plane. Domain boundaries are indicated by 1, 2 and 3. Each domain is characterized by $\boldsymbol{L}_i^{\pm}$, which is perpendicular to the a-axis.

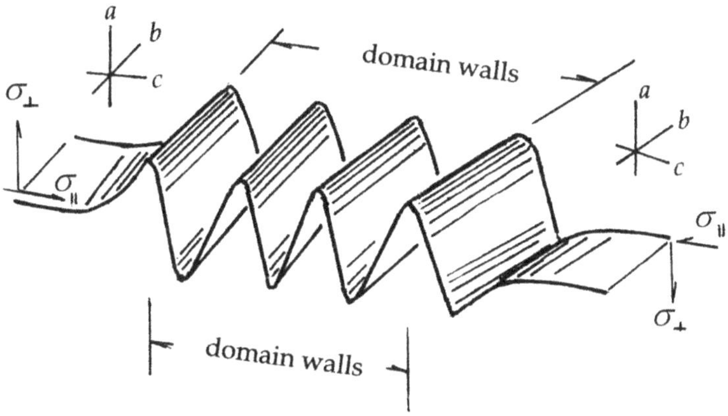

Figure 12.9. Simulated domain structure in terms of order variables ($\sigma_{\|}$, $\sigma_{\perp}$).

correlations should be determined by the host structure. Figure 12.9 shows a simulated profile of domain boundaries.

12.7 Domain separation: thermal and quasi-adiabatic transitions

12.7.1 Domain separation in finite crystals

Regarding the sine-Gordon equation (12.15) for trigonal transitions, the second term on the right $\zeta \cos 3\phi$ indicates the potential energy as given by $v(\phi) = v\left(\phi + \frac{2\pi}{3}\right)$, which can be re-expressed as

$$v(\phi) = \frac{1}{2}\alpha\phi(y)^2, \quad \text{where} \quad -\frac{\pi}{3} < \phi(y) < \frac{\pi}{3}$$

for a single domain.

On the other hand, an emerging adiabatic potential should cause *elastic strain* density $\eta(y)$, counteracting with a strain potential $\frac{1}{2}\beta\eta(y)^2$. In addition, we need to consider a coupling energy between the order variables and the lattice, which can be expressed by $\gamma\eta(y)\frac{\mathrm{d}\phi(y)}{\mathrm{d}y}$ following Per Bak [4].

Following Per Bak, we consider the free energy density $F(x)$ as defined by $F(T)=\frac{1}{L}\int_{-L/2}^{+L/2}F(y)\mathrm{d}y$, which is written as

$$F(y)=\frac{1}{2}\left\{\frac{\mathrm{d}\phi(y)}{\mathrm{d}y}-\delta\right\}^2+\frac{1}{2}\alpha\phi(y)^2+\gamma\eta(y)\frac{\mathrm{d}\phi(y)}{\mathrm{d}y}+\frac{1}{2}\beta\eta(y)^2+\frac{1}{2}\varepsilon\left\{\frac{\mathrm{d}\eta(y)}{\mathrm{d}y}\right\}^2,$$

where ε is another constant. Minimizing $F(y)$ with respect to $\phi(y)$ and $\eta(y)$, we obtain *Euler's equations*

$$\frac{\mathrm{d}^2\phi(y)}{\mathrm{d}y^2}-\alpha\phi(y)+\gamma\frac{\mathrm{d}\eta(y)}{\mathrm{d}y}=0 \quad \text{and} \quad \varepsilon\frac{\mathrm{d}^2\eta(y)}{\mathrm{d}y^2}-\beta\eta(y)-\gamma\frac{\mathrm{d}\phi(y)}{\mathrm{d}y}=0,$$

to be solved for equilibrium. These equations have the solutions

$$\phi(y)=A\sinh\kappa y,\quad -\frac{L}{2}\leqslant y\leqslant\frac{L}{2},\qquad \phi(y+L)=\phi(y)+\frac{2\pi}{3};$$

$$\text{and}\quad \eta(y)=B\cosh\kappa y\begin{pmatrix}1&0\\0&1\end{pmatrix},\quad -\frac{L}{2}\leqslant y\leqslant\frac{L}{2},\qquad \eta(y+L)=\eta(y).$$

Here, $\kappa^2=\frac{1}{2\varepsilon}\{\beta+\alpha\varepsilon-\gamma^2-\sqrt{(\beta+\alpha\varepsilon-\gamma^2)^2-4\alpha\varepsilon\beta}\}$, $A=\pi/3\sinh(\kappa L/2)$ and $B=A\gamma\kappa/(\varepsilon\kappa^2-\beta)$.

In this analysis, if the coupling γ is sufficiently strong in such a way that $(\sqrt{\alpha\varepsilon}+\gamma)^2>\beta$, we have $\kappa=0$, implying that there is no steady solution for isothermal process, meaning that phases are separated into two; otherwise nonzero κ keeps both $\phi(y)$ and $\eta(y)$ in fluctuation.

In the former case, the potential energy densities $\frac{1}{2}\alpha\phi\left(\frac{L}{2}\right)^2$ and $\frac{1}{2}\beta\eta\left(\frac{L}{2}\right)^2$ at $y=\frac{L}{2}$ can be expressed as

$$\frac{\alpha A^2}{2}\quad\text{and}\quad\frac{\beta B^2}{2}\mathrm{csch}^2\,\phi\left(\frac{L}{2}\right)=\frac{\beta B^2}{2}\left\{\coth^2\phi\left(\frac{L}{2}\right)-1\right\},$$

respectively, and the energy difference $\Delta E(\phi,\eta)$ between systems of $\phi(y)$ and $\eta(y)$ is given by

$$\Delta E(\phi,\eta)=\frac{\beta B^2}{2}\coth^2\phi\left(\frac{L}{2}\right).$$

Considering that these interactions are consequent on thermal processes, the energy exchange $\Delta E(\phi,\eta)$ should be characterized by the Boltzmann's average for equilibrium, namely probabilities for thermal averages $\langle\Delta E(\phi,\eta)\rangle$ and $\langle\Delta E(\eta,\phi)\rangle$ to be determined by factors $\exp\frac{-\langle\Delta E(\phi,\eta)\rangle}{k_BT}$ and $\exp\frac{\langle\Delta E(\eta,\phi)\rangle}{k_BT}$, respectively, because of the relation $\langle \Delta E(\phi, \eta)\rangle = -\langle \Delta E(\eta, \phi)\rangle$ in equilibrium. The above factor $\coth^2 \phi\left(\frac{L}{2}\right)$ can then be determined by thermal probability for $\langle \Delta E(\phi, \eta)\rangle$, so that $\langle \Delta E(\eta, \phi)\rangle$ is written as the reciprocal

$$\langle \Delta E(\eta, \phi)\rangle = \frac{\beta B^2}{2}\tanh^2 \phi\left(\frac{L}{2}\right) = \frac{\beta B^2}{2}\left(1 - \operatorname{sech}^2 \phi\left(\frac{L}{2}\right)\right);$$

hence

$$\langle \Delta E(\phi, \eta)\rangle = \frac{\beta B^2}{2}\left(-1 + \operatorname{sech}^2 \phi\left(\frac{L}{2}\right)\right).$$

Accordingly, we have the symmetric and antisymmetric relations

$$\langle \Delta E(\eta, \phi)\rangle + \langle \Delta E(\phi, \eta)\rangle = 0 \quad \text{and} \quad \langle \Delta E(\eta, \phi)\rangle - \langle \Delta E(\phi, \eta)\rangle = \pm 2\Delta v_{\perp}\left(\frac{L}{2}\right),$$

where the following applies.

Theorem:

$$\Delta v_{\perp}\left(\frac{L}{2}\right) = \mp\frac{1}{2}\beta B^2 \operatorname{sech}^2 \phi\left(\frac{L}{2}\right), \tag{12.37a}$$

which represents the *soliton potential energy* proportional to $\mp\operatorname{sech}^2\zeta$ for $\phi(\zeta)$ and $\eta(\zeta)$, respectively, at $y = L/2$ specified by the *phase* $\zeta = k(y - vt)$ *at a constant amplitude*; which is the Weiss potential, taking the same amplitude of longitudinal potential. As discussed in section 12.2, it is accessible for phonons to bring the whole system into thermal equilibrium. Therefore, the factor $\beta B^2/2$ or the number of solitons is sufficiently larger than the critical n_c, the interaction energy $\Delta v_{\perp}(\frac{L}{2})$ can be further transferred to the surroundings, confirming that *the soliton theory for symmetric crystals is compatible with the thermodynamic environment.* Namely, *the soliton potential energy* (12.37*a*) *exists always as an adiabatic Weiss field due to relative displacements in equilibrium crystals,* confirming the action–reaction principle for condensates.

For the function ϕ, the soliton potential energy is essential for $\phi\left(\frac{L}{2}\right)$ to be related for entropy production, because $\phi^2\left(\frac{L}{2}\right) \propto -\Delta v_{\perp}\left(\frac{L}{2}\right)$, which is essential for critical singularities in the angle ϕ in *the x–z* plane. In this sense, the soliton potential $\Delta v_{\perp}\left(\frac{L}{2}\right)$ can be assumed to lie in the z-direction. Corresponding to hypothetical work for the kinks to reverse the strain density expressed by $\tanh \eta(\zeta_0)$ at $L/2$, i.e. $\tanh \phi\left(\frac{L}{2}\right) \rightleftarrows -\tanh \phi\left(\frac{L}{2}\right)$, as illustrated in figure 12.3(c), being allowed by the Toda–Kac theorem. The quantity $2\Delta v_{\perp}\left(\frac{L}{2}\right)$ is the energy for inversion of $\left(\frac{d\psi_A}{d\zeta}\right)_{\zeta_0} \rightleftarrows -\left(\frac{d\psi_A}{d\zeta}\right)_{-\zeta_0}$ along the z-axis at T_c as indicated by the first relation of (12.10), so that we have the following.

Theorem:

$$\left(\frac{\mathrm{d}\psi_A}{\mathrm{d}\zeta}\right)_{\zeta_0} \approx \tanh \phi\left(\frac{L}{2}\right). \qquad (12.37b)$$

On the other boundary plane, we have the angular strains $+\eta(L/2)$ and $-\eta(L/2)$ as a *couple of shear rotations* around the y-axis across the domain boundary zone. Hypothetical work reversing the direction of $\phi(L/2)$ in the boundary layers at $y = L/2$ should be signified with transversal correlations between $\phi(\pm L/2)$ in the boundary zone. As noted in chapter 11 for Toda's theorem, the soliton theory can thus be applied to *longitudinal and transversal fluctuations together in a mesoscopic equilibrium state.*

Domain boundaries are stabilized in crystals by inversion of $+\tanh \phi(L/2) \rightleftarrows -\tanh \phi(L/2)$ in the z-direction creating either a negative energy $-\mathrm{sech}^2\,\phi(L/2)$, or otherwise the lattice is destabilized by positive $+\mathrm{sech}^2\,\phi(L/2)$. *Inverting local order*, the excess elastic energy in the latter case should be transferred to the surroundings to attain further stability, where the domains are separated to keep stability of the whole crystal at a given temperature.

At this point, we realize that the foregoing argument of energy transfer to the lattice is exactly the same as described in section 9.6, where creation and annihilation operators $\tilde{b}_n^\dagger$ and $\tilde{b}_n$ can explain the mechanism.

It is noted that the soliton theory is applicable to thermodynamic equilibrium conditions established within the symmetry group to keep a crystal in a lower energy state, confirming the presence of adiabatic potentials $\pm\mathrm{sech}^2\,\zeta_c$, known as Eckart's potentials (9.10), which is perpendicular to the *longitudinal nonlinear propagation* in crystals; hence we often express it as $v_\perp(\phi_c) = \pm v_0\,\mathrm{sech}^2\,\phi_c$.

In Per Bak's theory, equations (12.37*a*) and (12.37*b*) can be applied to sufficiently large L for stability of the domain structure. Indicated by the propagating parameter κ in the wave $\phi(y) = A \sinh \kappa y$, $\kappa = 0$ corresponds to the stability limit (12.29) of the phase correlation determined by a critical soliton number n_c for energy transfer to $\eta(y)$. In practice, n_c cannot be specified, unless volume, surface area and temperature of crystals are known. However, sufficiently intense shear stress will separate two domains facing across the boundary zone, constituting the basic mechanism of entropy production. Two independent domains have new surfaces hereafter that are the same boundary planes of the connected domains. Therefore, it is logical to consider that the waves $\phi(y)$ are pinned at $y = \pm L/2$, that can be applied to crystal surfaces as well.

12.7.2 Entropy production in isothermal and quasi-adiabatic transitions

In section 9.7, we discussed quasi-adiabatic transitions in the irreversible process of the specific heat for $T < T_c$, implying isothermal entropy production. While phonon scatterings by transverse components $\sigma_\perp$ is an obvious mechanism for energy loss, the problem of entropy production in lattice can be explained by the positive soliton potential energy $+\Delta v_\perp(\phi_c)$.

Representing pseudospin correlations by Toda's potentials, the pseudospin lattice is expressed by cnoidal series of Eckart potentials in the thermodynamic environment. In field-theoretical approximation, we discuss scatterings of n elemental solitons by the positive Eckart's potential $+v(n)$ specified by n at a given temperature T. While dealing specifically with phase correlations among $\boldsymbol{\sigma}_{\perp n}$, the phase soliton represents a specific case of constant amplitude, but sharing the feature specified by the soliton number, as shown by the Toda theorem.

During the transition process below T_c, the renormalized structure specified by $-L_n/4\pi \leqslant \phi \leqslant L_{n+1}/4\pi$ in succession, infinite one-dimensional lattice can be represented at given temperature by a series of reduced circular Brillouin zones of Toda's lattice. Referring to figures 9.4 and 9.5, figure 12.10(b) illustrates *changing soliton lattices in reduced Brillouin zones* with temperatures T_n signified by soliton numbers that are equivalent to *local order*, terminated by singular interfaces of widths $(\Delta\phi)_n$. In these gaps, the width is characterized by strain inversion between $\pm$ tanh $\phi(L_n/2)$, as shown in figure 12.10(a), which are responsible for entropy production in strained lattice under *thermodynamic conditions* exactly the same way as described in section 4.2.3. Here, circles indicate domain lengths with decreasing temperature. Figure 12.10(b,c) indicates corresponding soliton levels in adiabatic Eckart potentials along the perpendicular direction determined by $L_n/2$ and $L_{n+1}/2$, i.e. for $\Delta n = 1$, corresponding to temperature change $\Delta T_{n,n+1}$.

Here, the planes at $L_n/2$ and $L_{n+1}/2$ can be called boundaries between mesoscopic *mini-domains*

$$(-L_n/2, \ldots, +L_n/2) \quad \text{and} \quad (-L_{n+1}/2, \ldots, L_{n+1}/2); \tag{12.38}$$

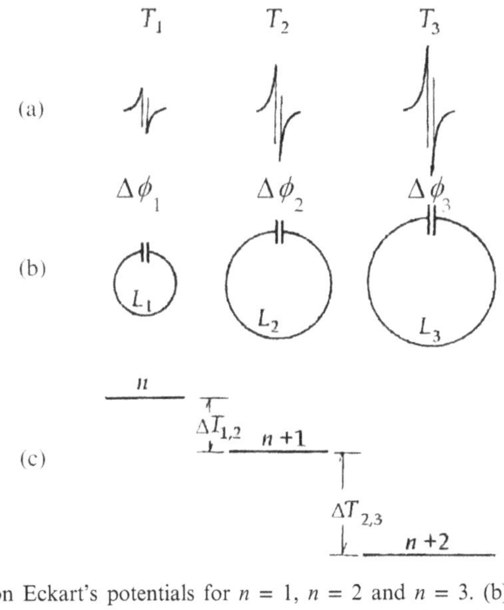

Figure 12.10. (a) Kinks on Eckart's potentials for $n = 1$, $n = 2$ and $n = 3$. (b) C_2-circular lattices for the corresponding Eckart's potentials. (c) Soliton levels for $n = 1$, $n = 2$ and $n = 3$.

and the left and right end-boundaries of mini-domains

$$(-L_1/2, \ldots, L_1/2) \quad \text{and} \quad (-L_N/2, \ldots, L_N/2) \tag{12.39}$$

represent respectively, the left and right surface elements of a crystal, $-\mathrm{d}A_{1\perp}$ and $+\mathrm{d}A_{\mathrm{N}\perp}$. It is notably significant that *all mini-domain boundaries and corresponding surface elements* (12.38) *and* (12.39) *are characterized with respect to the change in soliton number, i.e.* $\Delta n = \pm 1$, *during isothermal processes.*

On the other hand, under quasi-adiabatic conditions shown as a horizontal process for $\Delta n = 0$ in figure 9.5, a change in $\Delta\sigma_\perp^2$ is determined by an *off-diagonal element of inelastic phonon scatterings by positive potentials for* $T < T_c$, i.e. $\langle \boldsymbol{K}|\Delta\sigma_\perp|\boldsymbol{K}'\rangle_{1,2}^2 \propto \Delta T_{1,2}$ and $\langle \boldsymbol{K}|\Delta\sigma_\perp|\boldsymbol{K}'\rangle_{2,3}^2 \propto \Delta T_{2,3}$, etc, where the *equipartition theorem* can be applied statistically to phonon states and surroundings *for the lower-temperature phase.* In these cases, the temperature-dependence of $\Delta\sigma_\perp^2$ is proportional to related ΔT, if each step is independent of the lattice, giving rise to a parabolic curve $\Delta\sigma_\perp \propto \sqrt{\Delta T}$ in mean-field accuracy.

We remark here that the energy transfer at the critical temperature T_c of phase transitions occurs isothermally at $L = 0$ for $\Delta T_c = 0$, caused by finite value $\langle \boldsymbol{K}|\boldsymbol{\sigma}_o|\boldsymbol{K}'\rangle^2$ for the seed cluster to initiate mesoscopic processes that may not be parabolic in practice (see section 4.2.3).

Mesoscopic order in domain structure can thus be described in terms of *complex soliton potentials in a deformed lattice*, whose examples are reported for many structural transitions. Notably, the presence of soft modes is evident as found in *observed stepwise decay and rise* in the vicinity of T_c, describing the behavior of Δy_q between different soliton numbers $n = n(T)$, corresponding to the temporal variation of traditional *long-range order*.

At this stage, a further important remark can be made for the initial clustering process in phase transitions. As the nonlinear process can be discussed with respect to successive soliton potentials in the lattice for varying temperature, each adiabatic step in decreasing temperature can be assigned to Brillouin's zones of the soliton lattice at given $n^2(T)$, so that the initial clustering by Born–Huang's principle can be considered in the first domain structure at $n^2(T^*) = 1$ that determines the critical temperature T_c at $n^2(T_c) = 1$. The initial condition was discussed in chapters 4 and 5, for which this remark can be applied, while another significant example will be given later in chapter 14 for the Cooper pair to be analogous to the initial cluster.

It is notable that the critical temperature T_c is *independent of the volume* under constant pressure, while at a domain-separating temperature crystals exhibit *volume-dependent properties* ($\Delta v \neq 0$), unless the number $n(T)$ for coherent solitons can be specified with respect to practical sample crystals. With a sufficiently large number of $n(T)$, strong shear strains at T can exert forces that separate the crystal into two single crystals after the process of symmetry change.

Thermodynamically, the Toda lattice describes *isothermal processes* under constant-volume conditions $\Delta v = 0$, while in *adiabatic processes* the lattice response was not much studied experimentally so far. However high-pressure studies of graphite surfaces [8] and recently reported results from metallic hydrogen sulfide [9] showed

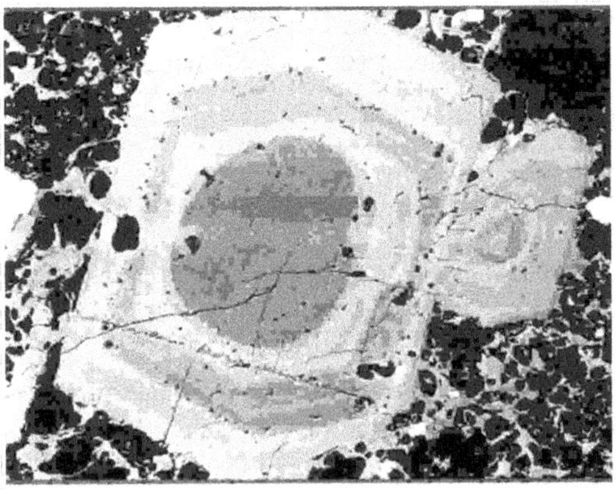

Figure 12.11. *Feldspar crystals* in hexagonal-plate shape found in volcanic rock. These crystals can be responsible for a volcanic explosion, if existing in large quantity at critical conditions for temperature and pressure, similar to domain separation in single crystals. (Courtesy of Professor J D Blundy, University of Bristol.)

adiabatic processes against applied pressure Δp. In these works, while the lattice is assumed to be *homogenous*, and experimentally crystals are restricted to *ellipsoidal samples* to facilitate a uniform field, as shown in figure 12.6.

Further remarks should be made at this stage that those domains defined above can be thermodynamically stable, if the strain energy is transferred to the surroundings; on the other hand, after the transfer, the domain boundaries in thermal equilibrium are characterized by $\Delta n \neq 0$, as indicated by Kac's theory in section 12.6. Although theoretically domain structure is due to finite size, it is a natural fact that single crystalst can be sensitive in producing entropy at specific T and p, as seen from an example of a volcanic explosion shown in figure 12.11.

Exercises

1. Screw-axis symmetry C_3 in crystals is composed of triple-axial rotation and translation along the axis. In this case, in addition to phase solitons arising from the C_3 rotation, as discussed in section 12.3, we expect an amplitude soliton along the screw-axis. Discuss the axial soliton, which may explain the shape of TSCC crystals in flat-disk shape, as an example.
2. Assume different space-times for the fields ψ_1 and ψ_3 in the Bäcklund transformation discussed in section 12.4. If considering two component wave $\boldsymbol{\psi}_3 = \psi_3(\mathrm{A}) \pm i\psi_3(\mathrm{P})$ in this case, we have such a complex soliton potential as proportional to $\mathrm{sech}^2\,\phi \pm i\,\mathrm{sech}(\phi - \Delta\phi)$, where $\Delta\phi$ represents a phase difference corresponding to different time scales. Accordingly, we may consider that $\Delta\phi$ signifies thickness of the domain wall. Evaluate this interpretation, referring to kink reversal of transversal waves discussed in section 8.2.

3. The soliton transitions $n \rightleftarrows n \pm 1$ are thermodynamically adiabatic. Discuss how such processes can be observed, referring to the experimental work by Fain Jr. and Chinn [7].

References

[1] Lamb G L Jr 1963 *Elements of Soliton Theory* (New York: Wiley)
[2] Fujimoto M 2013 *Thermodynamics of Crystalline States* Chap. 6 (New York: Springer)
[3] Taniuchi T and Nishihara K 1998 Nonlinear waves *Applied Mathematics Series* (in Japanese) (Tokyo: Iwanami)
[4] Bak P 1978 Solitons in Incommensurate Systems in Solitons and Condensed Matter Physics ed A R Bishop and T Schneider (Berlin: Springer)
Janner A and Janssen T 1977 *Phys. Rev.* **B15** 643
Bak P and Janssen T 1978 *Phys. Rev.* **B17** 436
[5] McMillan W L 1976 *Phys. Rev.* B **14** 1496
[6] Nakamura A 1983 *J. Phys. Soc. Jpn.* **52** 380
[7] Fain S C Jr and Chinn M D 1977 *J. Phys.* **38** C4-99; 1978 *Phys. Rev. Lett.* 39 146
[8] Drozdov A P, Eremet M I, Ksenofontov I A and Shylin S I 2015 *Nature* **525** 7567
[9] Yang C N and Yang C P 1969 *J. Math. Phys.* **10** 1115
[10] Toda M 1997 *Nonlinear Lattice Dynamics. Applied Mathematics Series* (Tokyo: Iwanami) (in Japanese) ch 4
[11] Kac M and van Moerbecke P 1975 *Adv. Math.* **16** 160
[12] Kittel C 1969 *Thermal Physics* Chap. 5 (New York: Wiley)

Part IV

Superconducting, magnetic, polymer and liquid crystals

The soliton theory can be applied to these systems of solid states, while soft matters need to be discussed with their method of observation. Particularly significant is however, the well-known Fröhlich mechanism as interpreted by the soliton theory can be applied to superconductors of many types, where the lattice distortion is responsible for most superconductivity in common.

IOP Publishing

Solitons in Crystalline Processes (2nd Edition)
Irreversible thermodynamics of structural phase transitions and superconductivity
Minoru Fujimoto

Chapter 13

Phonons, solitons and electrons in modulated lattices

When discovered by Kamerlingh Onnes, the *superconductivity* in metals was a peculiar phenomenon, which is now considered as related to a thermodynamic phase transition in the conduction–electron system in a periodic lattice in finite size at very low temperatures. Assisted by lattice displacements in finite magnitude interacting with electrons, such an electronic phase transition is a basic phenomenon in *multi-electron systems* in crystals. In this chapter, we summarize statistical properties of phonons, solitons and electrons in the momentum space, prior to discussing the spontaneous superconductivity.

At very low temperatures, *condensed electrons* in the lattice are mutually correlated in Fermi–Dirac statistics, which is entirely different from phonons and solitons. Described by the *Pauli principle*, the electron–electron correlations are characterized by spin–spin interactions proportional to the scalar product of spins $s_i \cdot s_j$. In the normal conducting states, only statistical properties of conduction electrons are discussed, however, in the superconducting states both phonons and solitons are also responsible for structural stability.

The statistical properties in *conservative dynamics* of those particles are fundamental for thermodynamic quantities in metals, where solitons play an essential role for equilibrium conditions with the lattice. Time-dependent interactions of electrons and solitons in the modulated lattice are particularly significant for superconducting phenomena, as will be discussed in chapter 14.

13.1 Phonon statistics in metallic states

In chapter 1, phonons were considered as *quasi-particles* in quantum theory of lattices characterized by the energy $\hbar\omega_K$ and momentum $\hbar\boldsymbol{K}$, where harmonic lattice vibrations are quantized with respect to ω_K and $\boldsymbol{K}$. Quantized vibrations can thus be equivalent to a large number of phonons in the normal phase, behaving as

doi:10.1088/978-0-7503-2572-1ch13

independent quasi-particles; phonons are considered as collision-free particles, keeping the lattice symmetry invariant in the thermodynamic environment. This is a valid assumption for sufficiently large crystals, as the thermal interaction takes place on surfaces, and the surroundings can be discussed separately in thermodynamics.

Classical normal modes of lattice vibrations can be indexed by the wave vector $\boldsymbol{K}$ for the Hamiltonian $\mathcal{H}_K$, while characterized quantum-mechanically by creation and annihilation operators b_K and $b_K^\dagger$ discussed in chapter 1. The Hamiltonian and total number of phonons are then expressed by

$$\mathcal{H} = \sum_K \mathcal{H}_K = \sum_K \hbar\omega_0\left(b_K^\dagger b_K + \frac{1}{2}\right) \quad \text{and} \quad N = \sum_K b_K^\dagger b_K, \tag{13.1}$$

respectively.

Thermodynamic properties of phonon gas can therefore be determined by a system in normal modes specified by $\boldsymbol{K}$ with energies $\varepsilon_K = n_K \hbar\omega_K$ consisting of $n_K = b_K^\dagger b_K$ phonons at a given temperature T, moving freely in all directions of $\boldsymbol{K}$. Note that ω_K and $\boldsymbol{K}$ are the characteristic frequencies and wavenumber of the modes, the number of phonons n_K can be arbitrary, and the total $\sum_K n_K = N$ is left unknown, which however can be determined with thermodynamic conditions.

As defined in section 1.2, the operators $b_K^\dagger$ and b_K are defined for changing energies ε_K plus and minus one phonon energy $\hbar\omega_K$, hence called the creation and annihilation operators, respectively. Taking the relation $[b_{K'}, b_K^\dagger] = \delta_{K',K}$, for instance, we note that these operators are commutable, i.e. $b_{K'} b_K^\dagger = b_{K'}^\dagger b_K$ for unequal $K' \neq K$, signifying that the lattice is invariant for exchanging phonons $(\hbar\omega_K, \boldsymbol{K})$ and $(\hbar\omega_{K'}, \hbar\boldsymbol{K}')$. These two phonons are *unidentifiable quasi-particles* and $\hbar\omega_K = n_K \hbar\omega_0$ at $|\boldsymbol{K}| \neq 0$, while each phonon is characterized by the rest energy $\hbar\omega_0$ at $|\boldsymbol{K}| = 0$. Considering a large number of identical element $\hbar\omega_0$, the vibration field is equivalent to a gaseous material of n_K particles of $\hbar\omega_0$ characterized by Bose–Einstein statistics.

For a small $|\boldsymbol{K}|$ compared with reciprocal lattice spacing, the lattice can be considered as a continuous medium. In the field theory, harmonic displacements $q_i(\boldsymbol{r}, t)$ and the corresponding conjugate momenta $p_i(\boldsymbol{r}, t)$ can be regarded as continuous variables $q(\boldsymbol{r}, t)$ and $p(\boldsymbol{r}, t)$, propagating in the direction of $\boldsymbol{K}$. In a classical field, these variables can be expanded into Fourier series with respect to the *phase* $\boldsymbol{K} \cdot \boldsymbol{r} - \omega_K t$, where $\boldsymbol{r}$ and t represent arbitrary space-time in a crystal.

On the other hand, in the quantized field expressed by the wave function $\psi(\boldsymbol{r}, t)$ and the conjugate momentum $\pi(\boldsymbol{r}, t) = \rho\frac{\partial\psi(\boldsymbol{r}, t)}{\partial t}$, where ρ is an effective density. Normalizing in a periodic structure of volume Ω, these field variables can be expressed as

$$\psi(\boldsymbol{r}, t) = \frac{1}{\sqrt{\Omega}} \sum_K \sqrt{\frac{\hbar}{2\rho\omega_K}} \left\{ b_K \exp i(\boldsymbol{K} \cdot \boldsymbol{r} - \omega_K t) + b_K^\dagger \exp i(-\boldsymbol{K} \cdot \boldsymbol{r} - \omega_K t) \right\} \tag{13.2a}$$

and its generalized momentum

$$\pi(\boldsymbol{r}, t) = \frac{i}{\sqrt{\Omega}} \sum_{\boldsymbol{K}} \sqrt{\frac{\hbar\rho\omega_k}{2}} \{ -b_{\boldsymbol{K}} \exp i(\boldsymbol{K}\cdot\boldsymbol{r} - \omega_{\boldsymbol{K}} t) + b_{\boldsymbol{K}}^{\dagger} \exp i(-\boldsymbol{K}\cdot\boldsymbol{r} - \omega_{\boldsymbol{K}} t)\}, \tag{13.2b}$$

where $b_{\boldsymbol{K}}^{\dagger} = b_{-\boldsymbol{K}}$. Using the normalizing condition, we have

$$\int_{\Omega} \psi(\boldsymbol{r}, t)\delta(\boldsymbol{r} - \boldsymbol{r}')d\Omega = \psi(\boldsymbol{r}', t)$$

for the field $\psi(\boldsymbol{r}, t)$, we can show the following commutation relations for these field variables ψ and π, i.e.

$$\pi(\boldsymbol{r}, t)\psi(\boldsymbol{r}', t) - \psi(\boldsymbol{r}', t)\pi(\boldsymbol{r}, t) = \frac{\hbar}{i}\delta(\boldsymbol{r} - \boldsymbol{r}'), \tag{13.3a}$$

$$\psi(\boldsymbol{r}, t)\psi(\boldsymbol{r}', t) - \psi(\boldsymbol{r}', t)\psi(\boldsymbol{r}, t) = 0 \tag{13.3b}$$

and

$$\pi(\boldsymbol{r}, t)\pi(\boldsymbol{r}', t) - \pi(\boldsymbol{r}', t)\pi(\boldsymbol{r}, t) = 0. \tag{13.3c}$$

Writing the Hamiltonian density as $\mathcal{H} = \frac{\pi^2}{2\rho} + \frac{\kappa}{2}(\text{grad}\,\psi)^2$ where κ is a restoring constant, (13.3*a*–*c*) are solutions of the wave equation $-i\hbar\frac{\partial\psi(\boldsymbol{r}, t)}{\partial t} = \mathcal{H}\psi(\boldsymbol{r}, t)$, as derived in the *Lagrangean formalism*. Nevertheless, we shall not discuss the formal theory here, while interested readers are referred to a standard textbook, e.g. *Quantenfeldtheorie des Festkörpers* by Haken [1].

For thermodynamics, we have to consider that energy levels $n_k\hbar\omega_o, \ldots$ represent excited levels that are occupied by n_k phonons at a given temperature T. Considering the correlations among independent phonons, the isothermal probability $\exp\left(-\frac{n_K\hbar\omega_o}{k_BT}\right)$ for the state $\varepsilon_K = n_K\hbar\omega_o$ should be multiplied by the statistical weight λ^{n_K}, expressing *quantum-mechanically correlated phonons*, where λ is called the *adiabatic probability* for adding one phonon to this level in the thermodynamic environment.

Owing to such phonon–phonon correlations, phonon numbers n_K can vary, as if driven by an adiabatic potential that is expressed equivalently by λ. In fact, λ is related to the chemical potential μ, if writing as $\lambda = \exp\frac{\mu}{k_BT}$ in open *thermodynamic systems* [2]. The phonon gas is *open* to the correlated system, hence for equilibrium crystals we can write the partition function of $\mathcal{H}_K$ as

$$Z_K = \sum_{n_K} \lambda^{n_K} \exp\left(-\frac{n_K\hbar\omega_o}{k_BT}\right) = \sum_{n_K} \left\{\lambda \exp\left(-\frac{\hbar\omega_o}{k_BT}\right)\right\}^{n_K}.$$

This expression is an infinite power series of a small quantity $x = \lambda\exp\left(-\frac{\hbar\omega_o}{k_BT}\right) < 1$, so that Z_k can be written simply as $\frac{1}{1-x}$, summing over $n_K = 0, 1, 2, \ldots, \infty$; thus we obtain

$$Z_K = \frac{1}{1 - \lambda \exp\left(-\frac{\hbar\omega_0}{k_B T}\right)}.$$

The partition function for the whole crystal is therefore determined by the product $Z = \Pi_K Z_K$, with which the thermal average of the number of phonons can be calculated as

$$\langle n \rangle = \frac{1}{Z}\sum_K n_K \lambda^{n_K} \exp\left(-\frac{n_K \hbar\omega_0}{k_B T}\right) = \lambda \frac{\partial}{\partial \lambda} \ln Z = \lambda \frac{\partial}{\partial \lambda} \sum_K \ln Z_K.$$

Writing $\langle n \rangle = \sum_K \langle n_K \rangle$, we then obtain

$$\langle n_K \rangle = \lambda \frac{\partial}{\partial \lambda} \ln Z_K = -x \frac{d}{dx} \ln(1 - x) = \frac{x}{1 - x} = \frac{1}{\frac{1}{\lambda}\exp\frac{\hbar\omega_0}{k_B T} - 1}.$$

Returning to μ, the average number of phonons $\langle n_K \rangle$ in the $\boldsymbol{K}$-state is therefore expressed by

$$\langle n_K \rangle = \frac{1}{\exp\frac{\hbar\omega_0 - \mu}{k_B T} - 1}. \tag{13.4}$$

In the limit of $T \rightarrow \frac{\hbar\omega_0 - \mu}{k_B} \approx 0$, we obtain $\langle n_K \rangle \rightarrow \infty$, meaning that if $\mu = \hbar\omega_0$, almost all phonons occupy the first level $\varepsilon_K = \hbar\omega_K$ at the limiting temperature T, which is therefore called the *Bose–Einstein condensation*, in contrast with the classical condensation that never takes place for $\mu = 0$, at which phonons are independent particles free from correlation.

13.2 Solitons in modulated metals

Practical crystal lattices of pure metals are by no means in a *perfectly symmetrical structure*. At elevated temperatures, lattice excitations are dominated by vibrating ions in equilibrium crystals, independent of conduction electrons. On the other hand, at very low temperatures electrons can interact with the lattice of displaced ions from regular sites, which is significant for the superconducting phase in metals, as postulated by Fröhlich (see chapter 14).

We assume that such correlated displacements at very low temperatures are statistically different from random anharmonic vibrations, resulting in deformed structure due to classical displacements of reduced masses. Such a lattice deformation does not occur in the normal phase, but is caused *spontaneously at a specific condition* as a transition to another modulated structure. At this stage, we make a brief remark on statistical properties of displacements representing solitons, in order to avoid possible confusion with vibrational displacements.

In terms of mobility, solitons behave in crystals in a different manner from phonons, while both represent *different excitations in the same lattice*. As discussed in section 9.6, solitons are characterized by boson creation and annihilation operators, $\tilde{b}_n \exp(-in\tilde{\omega}_s t)$ and $\tilde{b}_n^\dagger \exp(in\,\tilde{\omega}_s t)$, where ω_s is a corresponding frequency of the displacement mode, determined by the soliton energy $\lambda_\nu = n\lambda_o$ [1].

Therefore, *soliton particles* should obey the same boson statistics as phonons in field theory, but in a different manner from phonons in terms of discrete soliton energies of excitation. For solitons the energy and momentum are λ_n and $\approx n\boldsymbol{q}$, respectively, whereas for phonons these are $\hbar\omega_n$ and $\hbar\boldsymbol{K}$. In this case, solitons can be described like *quantized displacements* with respect to the lattice structure, which is by no means the same as randomly quantized lattice vibrations with respect to the Planck constant $\hbar$, however, sharing the same statistical feature of bosons with phonons.

Nevertheless, phonons and solitons are both characterized by the wave vectors at a given temperature T, their numbers at very low temperatures can be estimated by the order of *boson correlation energy determined by the Debye limit* $k_B\Theta_D$. Therefore, in spite of the difference, conduction electrons at Fermi's level in metals are modulated by $\lambda_n \approx k_B\Theta_D$ at the critical temperature T_c, as will be discussed in chapter 14.

In any case, soliton particles provide an image of classical lattice interaction together with lattice displacements and phonons. While the interactions can be either elastic or inelastic scatterings in general, both phonons and solitons can be employed in the lattice to deal with properties of superconductors in the field-theoretical approximation.

13.3 Conduction electrons in normal metals

In the normal phase of a metallic superconductor, the system of multiple electrons is independent of the lattice. While phonons and solitons represent the lattice, electrons in crystals are not quite independent from the lattice, and the lattice has to be subjected to the Bloch theorem of periodic structure. Nevertheless, correlated electrons as a group should be characterized by the *Fermi–Dirac statistics*.

13.3.1 The Pauli principle for electrons

Characterized by the charge and mass, an electron is a classical particle, however, in metal it is a quantum particle signified by mutual correlations with other electrons. Electrons, as *quantum-mechanically unidentifiable particles*, must obey Pauli's principle at closer mutual distances, permitting only two electrons to correlate with anti-parallel spins in the orbital states together. An electron is signified by an internal degree of freedom called spin $\boldsymbol{s}$, specified by two opposite directions $s_z = \pm\frac{1}{2}$, thereby determining lower energy for two electrons to be coupled as proportional to $-\boldsymbol{s}_1 \cdot \boldsymbol{s}_2$.

Two electrons correlated via *spins* are combined in a different manner from their electrostatic Coulomb's repulsion forces between the *two charges*. Although the Coulomb interaction cannot be ignored in principle, the electron–spin correlations are regarded as dominant at close distances.

A state of two unidentifiable electrons can be described by a wavefunction $\psi(\boldsymbol{r}_1, \sigma_1; \boldsymbol{r}_2, \sigma_2)$, where $\boldsymbol{r}_1$, $\boldsymbol{r}_2$ and σ_1, σ_2 express their positions and intrinsic spins, respectively. Defining an operator $\boldsymbol{P}$ to exchange particles 1 and 2 in such a two-particle state [3], we write

$$\boldsymbol{P}\psi(\boldsymbol{r}_1, \sigma_1; \boldsymbol{r}_2, \sigma_2) = \psi(\boldsymbol{r}_2, \sigma_2; \boldsymbol{r}_1, \sigma_1).$$

Operating $\boldsymbol{P}$ once again on both sides, the state can return to the original state, therefore we have

$$\boldsymbol{P}^2\psi(\boldsymbol{r}_1, \sigma_1; \boldsymbol{r}_2, \sigma_2) = \boldsymbol{P}\,\psi(\boldsymbol{r}_2, \sigma_2; \boldsymbol{r}_1, \sigma_1) = \psi(\boldsymbol{r}_1, \sigma_1; \boldsymbol{r}_2, \sigma_2),$$

from which we obtain

$$\boldsymbol{P}^2 = 1 \quad \text{or} \quad \boldsymbol{P} = \pm 1, \tag{13.5}$$

thereby the wave function of two unidentifiable electrons can be either a *symmetric* or *antisymmetric* combination of wave functions with respect to their exchange. Hence, the operator $\boldsymbol{P}$ is called the *exchange operator*.

For a pair of electrons, $\boldsymbol{P}$ and the lattice Hamiltonian $\mathcal{H}$ are commutable as $[\mathcal{H}, \boldsymbol{P}] = 0$, so that eigenfunctions of $\boldsymbol{P}$ can be determined by linear combinations of $\psi(\boldsymbol{r}_1, \sigma_1; \boldsymbol{r}_2, \sigma_2)$ and $\psi(\boldsymbol{r}_2, \sigma_2; \boldsymbol{r}_1, \sigma_1)$, where the anti-symmetric combination determines the lower energy es expressed by

$$\{\psi(\boldsymbol{r}_1, \sigma_1; \boldsymbol{r}_2, \sigma_2) - \psi(\boldsymbol{r}_2, \sigma_2; \boldsymbol{r}_1, \sigma_1)\}/\sqrt{2},$$

corresponding to the spin–spin interaction $\propto -\, \boldsymbol{s}_1 \cdot \boldsymbol{s}_2$.

In contrast, phonons and solitons are *spinless particles*, for which the wave function for two particles is signified by $\boldsymbol{P} = +1$ for a symmetric combination as expressed by

$$\{\psi(\boldsymbol{r}_1, \boldsymbol{r}_2) + \psi(\boldsymbol{r}_2, \boldsymbol{r}_1)\}/\sqrt{2},$$

as already discussed in previous chapters in terms of *phase coherence*.

Experimentally, phonons and electrons provide typical examples of different exchange operators, $\boldsymbol{P} = +1$ and $\boldsymbol{P} = -1$, respectively; accordingly, expressing *intrinsic* properties of these particles with spins. *Phonons* in crystals and electromagnetic *photons* are both characterized by zero spins and $\boldsymbol{P} = +1$, whereas *electrons and protons* are particles of spin $\pm\frac{1}{2}$ and $\boldsymbol{P} = -1$. In general, elementary particles are characterized by spins $s_z = 1n$, where n is either *even* or *odd* integers, corresponding to the exchange operator of $+1$ or -1, respectively; which can be applied to *permutable constituents in idealized pure crystals.*

In normal metals, we consider electrons in nearly free motion, where electron–electron correlations required by Pauli's principle are significant at very low temperatures. Disregarding correlations, for a metallic crystal of cubic volume $V = L^3$ with periodic boundary conditions, we assume a *single-particle state* for each electron as the first approximation, which is expressed by a plane wave

$$\psi_{k,\sigma} = (\exp i\boldsymbol{k} \cdot \boldsymbol{r})\chi(s) \tag{13.6}$$

with energy eigenvalues $\varepsilon(\boldsymbol{k}) = \frac{\hbar^2 \boldsymbol{k}^2}{2m^*}$, where $\boldsymbol{k}$, m^* are the wave vector and effective mass, respectively;, and $\chi(s)$ is the spin function. Due to Pauli's principle, each of these energy states specified by $\boldsymbol{k}$ and $\boldsymbol{s}$ can be occupied by a single electron and a pair of electrons with antiparallel spins; these energy levels in a metal are thereby filled up to an energy level specified by $\boldsymbol{k} = \boldsymbol{k}_\text{F}$, called the *Fermi level*. If the number of electrons is odd, only the top level $\varepsilon(\boldsymbol{k}_\text{F})$ is occupied by one electron with an arbitrary spin, while all other levels $\varepsilon(\boldsymbol{k})$ for $k < k_\text{F}$ are filled with two electrons with antiparallel spins at $T = 0$ K. Electrons near the Fermi level $\varepsilon(\boldsymbol{k}_\text{F})$ can then be excited by an applied electric field $\boldsymbol{E}$, generating an electrical current; while all other electrons stay intact at levels $\varepsilon(\boldsymbol{k})$ below $\varepsilon(\boldsymbol{k}_\text{F})$. In this *Sommerfeld model*, the total number of electronic states can be determined by the spherical volume of radius k_F in the reciprocal space, multiplied by two, i.e. $\frac{4\pi k_\text{F}^3}{3} \times 2$, where the factor 2 is the spin degeneracy. On the other hand, a small cube of volume $\left(\frac{2\pi}{L}\right)^3$ in the reciprocal space is occupied by one orbital state only. Denoting the total number of electrons per volume by N, we have the relation $N = 2 \times \frac{4\pi k_\text{F}^3}{3} / \left(\frac{2\pi}{L}\right)^3$. Letting $L^3 = V$ the volume of a crystal, $\rho_\text{o} = \frac{N}{V}$ is the number density of electrons, the Fermi energy can be expressed as

$$\varepsilon_\text{F} = \varepsilon(\boldsymbol{k}_\text{F}) = \frac{\hbar^2}{2m^*}\left(3\pi^2 \rho_\text{o}\right)^{\frac{2}{3}}. \tag{13.7}$$

13.3.2 The Coulomb interaction of electrons in metals

In the Sommerfeld model, Coulomb interactions between electrons was ignored, while justifiable as the matter of approximation. In addition, normal crystals are electrically neutralized by ionic charges, hence Coulomb's force is not explicit in macroscopic scale. Nevertheless, Thomas and Fermi [4] showed that the static interactions are primarily insignificant in a multi-electron system, as verified in their theory by the screening effect.

Placing an additional charge $-e$ near each origin of lattice sites, there should be no significant electrostatic effect. With a static potential energy $V_\text{o} = \frac{1}{4\pi\varepsilon o}\frac{e^2}{r}$ at each site, where ε_o is the dielectric constant of the metal, the kinetic energy $\varepsilon(\boldsymbol{k}_\text{F})$ of an electron is perturbed as $\varepsilon(\boldsymbol{k}_\text{F}') = \varepsilon(\boldsymbol{k}_\text{F}) + V_\text{o}$, resulting in the additional charge density $\rho' = \frac{1}{3\pi^2\hbar^3}\left(2m^* V_\text{o} + p_\text{F}^2\right)^{\frac{3}{2}}$ in the crystal, where $p_\text{F} = \hbar k_\text{F}$ is the momentum at Fermi level. Using this density ρ', we can write the Poisson equation $\nabla^2 V' = -\frac{\rho' - \rho_\text{O}}{\varepsilon_\text{O}}$, while the Coulomb potential V_o should satisfy the Laplace equation $\nabla^2 V_\text{o} = 0$. We can therefore calculate the density difference given by

$$\rho' - \rho_o = \frac{(2m^*)^{\frac{3}{2}}}{3\pi^2\hbar^3\varepsilon_o}\left\{\left(V_o + \frac{p_F^2}{2m^*}\right)^{\frac{3}{2}} - \left(\frac{p_F^2}{2m^*}\right)^{\frac{3}{2}}\right\}$$

$$\simeq \frac{p_F^3}{3\pi^2\hbar^3\varepsilon_o}\left\{\left(1 + V_o/p_F^2/2m^*\right)^{\frac{3}{2}} - 1\right\}$$

$$\simeq \frac{4m^* p_F}{\pi\hbar^3} V_o\left\{1 + \mathcal{O}\left(V_o/p_F^2/2m^*\right)\right\}.$$

Assuming that $V' = V_o\{1 + \mathcal{O}(V_o/p_F^2/2m^*)\}$ for a small ratio $V_o/p_F^2/2m^* = \frac{V_o}{\varepsilon_F}$, the Poisson equation can be reduced to

$$\frac{1}{r^2}\frac{d}{dr}\left(r^2\frac{dV'}{dr}\right) = -\kappa V',$$

where $\kappa = \frac{4m^* p_F}{\pi\hbar^3\varepsilon_o}$. The solution of this equation can be expressed by

$$V' \sim \frac{e^2}{r}\exp(-\kappa r^2),$$

implying that the charge $-e$ placed at $r = 0$ is *screened* by the exponential factor in effective length given by $\kappa^{-\frac{1}{2}}$. Such a screening length depends on the value of $\kappa \gtrsim k_F$, which is estimated to be numerically shorter than the nearest neighbor distance in a typical metal. In the crude assumption of $V_o < \varepsilon_F$, Coulomb's interactions are ignorable in metals, supporting Sommerfield's *free electron model* in normal metals.

13.3.3 The Bloch theorem for single electrons in periodic structure

In a stable metallic crystal at a given p–T condition, conduction electrons cannot be entirely free from the periodic lattice potential,

$$V(\boldsymbol{r}) = V(\boldsymbol{r} + n_1\boldsymbol{a}_1 + n_2\boldsymbol{a}_2 + n_3\boldsymbol{a}_3),$$

where (n_1, n_2, n_3) are integers along the symmetry axes $\boldsymbol{a}_1$, $\boldsymbol{a}_2$ and $\boldsymbol{a}_3$. The Fourier transform of $V(\boldsymbol{r})$ can be defined by

$$V(\boldsymbol{r}) = \sum_{\boldsymbol{G}} V_{\boldsymbol{G}} \exp i\boldsymbol{G}\cdot\boldsymbol{r}, \tag{13.8a}$$

where $\boldsymbol{G} = h\boldsymbol{a}_1^* + k\boldsymbol{a}_2^* + l\boldsymbol{a}_3^*$ is a translation vector in the reciprocal lattice, corresponding to the lattice translation $\boldsymbol{R} = n_1\boldsymbol{a}_1 + n_2\boldsymbol{a}_2 + n_3\boldsymbol{a}_3$.

In adiabatic approximation, the electronic motion is perturbed by the lattice potential, resulting in a modulated wavefunction

$$\psi_k(\boldsymbol{r}, \boldsymbol{s}) = u_k(\boldsymbol{r}) \exp i\boldsymbol{k}\cdot\boldsymbol{r}\,\chi(\boldsymbol{s}), \tag{13.8b}$$

where the amplitude $u_k(\boldsymbol{r})$ is a periodic function in the lattice. However, the Bloch function cannot be uniquely determined, as seen from the relation

$$\psi_{k,\sigma}(\boldsymbol{r}, s) = \{u_k(\boldsymbol{r}) \exp i(\pm \boldsymbol{G} \cdot \boldsymbol{r})\} \exp i\{(\boldsymbol{k} \mp \boldsymbol{G}) \cdot \boldsymbol{r}\}\chi(s), \tag{13.8c}$$

which is held at any lattice point $\boldsymbol{r} = \boldsymbol{R}_n$ for any $\boldsymbol{k}$ satisfying the relation

$$\boldsymbol{k} \mp \boldsymbol{G} = \boldsymbol{k}; \tag{13.8d}$$

therefore (13.8*c*) is another Bloch's function. Along the normal direction of a *crystal plane* (*h, k, l*), (13.9) can be interpreted as a Bragg diffraction of the plane wave $\exp i\boldsymbol{k} \cdot \boldsymbol{r}$ by crystal planes of $\pm\boldsymbol{G}$, reflecting and interfering in phase for a constructive diffraction pattern. Such diffraction is originated from *elastic* collisions of electrons and periodic lattice. Therefore, the Bloch wave in the lattice behaves like a free wave elastically reflected from Brillouin-zone boundaries. Corresponding to (13.9), the electron energy is specified by the equation $\varepsilon(\boldsymbol{k} \mp \boldsymbol{G}) = \varepsilon(\boldsymbol{k})$, from which we obtain $\pm 2\boldsymbol{k} \cdot \boldsymbol{G} + \boldsymbol{G}^2 = 0$, hence $\boldsymbol{k} = \boldsymbol{G}/2$ indicates zone boundaries. Therefore, for the Bloch waves, the reciprocal space can be divided into many of these first zones surrounded by planes $\pm\boldsymbol{G}/2$. It is usually sufficient to have one zone to deal with the energy of an independent electron in a periodic lattice. Figure 13.1 shows schematically the effect of a periodic lattice in one dimension.

In fact, at zone boundaries, two waves $\exp i\left(\pm\frac{G}{2} \cdot \boldsymbol{r}\right)$ are not independent of each other, as perturbed by the lattice potential, resulting in splitting of the degenerate energy into $\varepsilon\left(\frac{G}{2}\right) \pm V_{G/2}$. Therefore, the perturbed energy $\varepsilon(\boldsymbol{k})$ forms a band structure characterized by forbidden energy gap $2V_{G/2}$, as shown in the figure. In addition, the energy band theory shows that the mass for a Bloch electron should be calculated as an effective mass $m^* = \frac{1}{\hbar^2}\frac{\mathrm{d}^2\varepsilon}{\mathrm{d}k^2}$, and

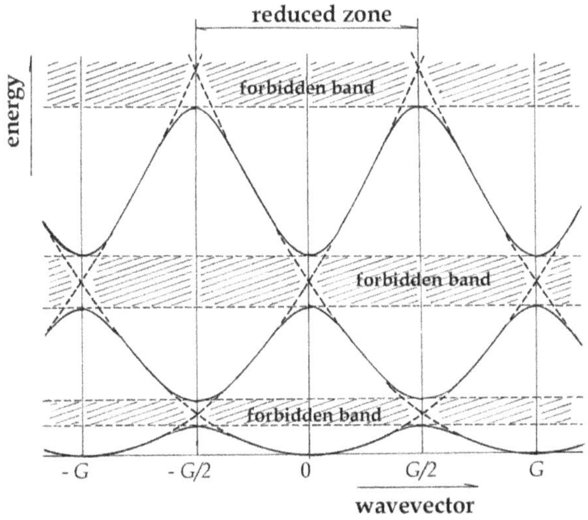

Figure 13.1. Energy band structure for a Bloch electron in one dimension.

$$\varepsilon(\boldsymbol{k}) = \sum_{i,j} \frac{\hbar^2}{m^*} k_i k_j + \text{const.} \tag{13.9}$$

Therefore, electrons in a metal can be described primarily as free particles, although modulated by the periodic lattice.

13.4 The multi-electron system

Although described as nearly free in metals at very low temperature, the spin–spin correlations among conduction electrons are significant as characterized by parity −1 governed by the Pauli principle. Due to such correlations, thermal properties of conduction electrons cannot be determined by the Boltzmann statistics.

We consider a system of N electrons that are primarily independent free particles. The wavefunction Ψ of the system can be constructed with *one-electron wave function* $\psi_{k,\sigma}$ in (13.6) with energies $\varepsilon(k) = \frac{\hbar^2 k^2}{2m^*}$. Counting these electrons by 1, 2, …, N, the wavefunction of N electrons can be constructed with a linear combination of products of $\psi_{k,\sigma}$ in the form of Slater's determinant, i.e.

$$\Psi(1, 2, \ldots, N) = \begin{vmatrix} \psi_{k_1,\sigma}(1) & \psi_{k_2,\sigma}(2) & \psi_{k_3,\sigma}(3) & \cdots & \psi_{k_N,\sigma}(N) \\ \psi_{k_1,\sigma}(2) & \psi_{k_2,\sigma}(3) & \psi_{k_3,\sigma}(4) & \cdots & \psi_{k_N,\sigma}(N-1) \\ \psi_{k_1,\sigma}(3) & \psi_{k_2,\sigma}(4) & \psi_{k_3,\sigma}(5) & \cdots & \psi_{k_N,\sigma}(N-2) \\ \vdots & \vdots & \vdots & \vdots & \vdots \\ \psi_{k_1,\sigma}(N) & \psi_{k_2,\sigma}(N-1) & \psi_{k_3,\sigma}(N-2) & \cdots & \psi_{k_N,\sigma}(1) \end{vmatrix},$$

for the collective wavefunction Ψ to satisfy Pauli's principle.

In the presence of a lattice potential $V(n)$, the eigenstate of one electron should be determined by the Schrödinger equation

$$-\frac{\hbar^2}{2m^*}\Delta\psi_{k',\sigma} + V\psi_{k',\sigma} = \varepsilon(k').$$

In this approximation, $\Psi(1, 2, \ldots, N)$ is not the eigenfunction of the whole system, but the above determinant composed of the perturbed one-electron functions $\psi_{k,\sigma}(n_k)$ can represent the perturbed state of N electrons.

For indistinguishable electrons, the number $n_k = 1, 2, \ldots, N$ in the determinant $\Psi(1, 2, \ldots, N)$ can be signified, if an electron is *present or absent* in each of the one-electron states $\psi_{k,\sigma}(n_k)$, by letting $n_k = 1$ or $n_k = 0$, respectively [5]. For example, by $\Psi(0_k, 1_k, \ldots, 1_k)$ we mean that the first state is empty, and all others are occupied by one electron each.

Creation and annihilation operators $a_k^\dagger$ and a_K for single electrons can be defined for electrons to express the following properties:

$$a_k^\dagger \psi_{k,\sigma}(0_k) = \psi_{k,\sigma}(1_k), \tag{13.10a}$$

$$a_k^\dagger \psi_{k,\sigma}(1_k) = 0, \tag{13.10b}$$

$$a_k \psi_{k,\sigma}(0_k) = 0, \tag{13.10c}$$

and

$$a_k \psi_{k,\sigma}(1_k) = \psi_{k,\sigma}(0_k). \tag{13.10d}$$

Among these, particularly (13.10*b*) manifests Pauli's principle; namely, no more than one particle can be in any state specified by k and σ. It follows from (13.10*b*) and (13.10*c*) that

$$\left(a_k^\dagger\right)^2 = (a_k)^2 = 0.$$

To express specifically *paired electrons* in the system, which will be discussed as the *Cooper pair* in the next chapter 14, we can use two single-creation operators $a_k^\dagger$ and $a_{k'}^\dagger$, and write

$$\begin{aligned} &a_k^\dagger a_{k'}^\dagger \{\psi_1(0_1)\psi_2(0_2)\cdots\psi_{k,\sigma}(0_k)\cdots\psi_{k',\sigma'}(0_{k'})\cdots\} \\ &= \psi_1(0)\psi_2(0)\cdots\psi_{k,\sigma}(1_k)\cdots\psi_{k',\sigma'}(1_{k'})\cdots. \end{aligned} \tag{13.11a}$$

The antisymmetric nature of the Cooper pair introduced in chapter 14 for example can then be expressed as

$$a_k^\dagger a_{k'}^\dagger \Psi(\ldots, 0_k, \ldots, 0_{k'}, \ldots) = -a_{k'}^\dagger a_k^\dagger \Psi(\ldots, 0_{k'}, \ldots, 0_k, \ldots). \tag{13.11b}$$

Therefore, exchanging two single-creation operators is equivalent to exchanging $\psi_{k,\sigma}$ and $\psi_{k',\sigma'}$ together in the Slater determinant. It is noted that (13.10) and (13.11) are satisfied, if these operators obey the commutation rules

$$\begin{aligned} &\left[a_k, a_{k'}^\dagger\right]_+ = a_k a_{k'}^\dagger + a_{k'}^\dagger a_k = \delta_{k,k'}, \\ &\left[a_k^\dagger, a_{k'}^\dagger\right]_+ = 0, \quad \text{and} \quad [a_k, a_{k'}]_+ = 0. \end{aligned} \tag{13.12}$$

Similar to phonons, the number operator in the state $\psi_{k,\sigma}$ can be defined by

$$n_k = a_k^\dagger a_k, \tag{13.13}$$

so that the total number of electrons is expressed as $N = \sum_k n_k$. We can confirm from the commutator relations in (13.12) that

$$n_k \psi_{k,\sigma}(0_k) = a_k^\dagger a_k \psi_{k,\sigma}(0_k) = 0$$

and

$$n_k \psi_{k,\sigma}(1_k) = a_k^\dagger \{a_k \psi_{k,\sigma}(1_k)\} = a_k^\dagger \psi_{k,\sigma}(0_k) = \psi_{k,\sigma}(1_k),$$

indicating that eigenvalues of the number operator n_k are 0 and 1, respectively.

The Hamiltonian for multi-electrons can be expressed as

$$\mathcal{H} = \sum_k \varepsilon(\boldsymbol{k}) n_k = \sum_k \varepsilon(\boldsymbol{k}) a_k^\dagger a_k, \tag{13.14}$$

where k in the above one-dimensional theory can be replaced by a vector $\boldsymbol{k}$, for which $\varepsilon(\boldsymbol{k})$ is the eigenvalue of the one-electron Hamiltonian in three dimensions. While some other perturbations can be considered in general, (13.13) and (13.14) can be used for the multi-electron system, specifying the one-electron eigenvalue by $\varepsilon(\boldsymbol{k})$.

The exchange operator of single electrons $\boldsymbol{P} = -1$ is scalar, and the correlation energy between two electrons at $\boldsymbol{r}_i$ and $\boldsymbol{r}_j$ can be described generally by a scalar potential $V(\boldsymbol{r}_i - \boldsymbol{r}_j)$, representing spatial correlations. The total correlation energy can generally be expressed as

$$\mathcal{H}_{\text{int}} = \frac{1}{2}\sum_{i \neq j} V(|\boldsymbol{r}_i - \boldsymbol{r}_j|), \tag{13.15}$$

which can be determined by matrix elements between two one-electron states. For nearly free single electrons, such $\mathcal{H}_{\text{int}}$ may have non-zero matrix elements in metals, if the potential $V(|\boldsymbol{r}_i - \boldsymbol{r}_j|)$ is a periodic function associated with lattice periodicity. Assuming that

$$\begin{aligned} V(|\boldsymbol{r}_i - \boldsymbol{r}_j|) &= \sum_{\pm \boldsymbol{q}} V(\boldsymbol{q}) \exp\{-i\boldsymbol{q}.\, |\boldsymbol{r}_i - \boldsymbol{r}_j|\} \quad \text{or} \\ V(\boldsymbol{q}) &= \frac{1}{\Omega}\int_\Omega \mathrm{d}^3|\boldsymbol{r}_i - \boldsymbol{r}_j| \exp i\boldsymbol{q}.\, |\boldsymbol{r}_i - \boldsymbol{r}_j|, \end{aligned} \tag{13.16}$$

where $\boldsymbol{q}$ represents a lattice vector of soliton, the matrix elements of $\mathcal{H}_{\text{int}}$ can be written as

$$\langle \boldsymbol{k}, \boldsymbol{k}'|\mathcal{H}_{\text{int}}|\boldsymbol{k}'', \boldsymbol{k}'''\rangle = \frac{1}{2\Omega}\iint_\Omega \mathrm{d}^3\boldsymbol{r}_i\, \mathrm{d}^3\boldsymbol{r}_j V(|\boldsymbol{r}_i - \boldsymbol{r}_j|) \exp i\{(\boldsymbol{k} - \boldsymbol{k}'')\cdot \boldsymbol{r}_i + (\boldsymbol{k}' - \boldsymbol{k}''')\cdot \boldsymbol{r}_j\},$$

where these integrals are calculated over a sufficiently large volume Ω. Therefore, $V(\boldsymbol{q})$ in (13.16) related to the element of $\mathcal{H}_{\text{int}}$ determine a secular perturbation for the relations

$$\boldsymbol{k} - \boldsymbol{k}' = \boldsymbol{q} \quad \text{and} \quad \boldsymbol{k}'' - \boldsymbol{k}''' = -\boldsymbol{q} \tag{13.17a}$$

or

$$\boldsymbol{k} - \boldsymbol{k}' = \boldsymbol{k}''' - \boldsymbol{k}''. \tag{13.17b}$$

This implies that in the collision process of two electrons, a soliton $n\lambda_q$ is emitted or absorbed by the lattice, so that (13.16) simulates an *inelastic scattering*. For such a process, the matrix element of $\mathcal{H}_{\text{int}}$ can be expressed as

$$\langle \boldsymbol{k}, +\boldsymbol{q}|\mathcal{H}_{\text{int}}|\boldsymbol{k}'', -\boldsymbol{q}\rangle = \frac{1}{2\Omega}\sum_{\boldsymbol{k},\boldsymbol{k}'',\pm\boldsymbol{q}} V(\boldsymbol{q})a^{\dagger}_{\boldsymbol{k}+\boldsymbol{q}}a_{\boldsymbol{k}}a^{\dagger}_{\boldsymbol{k}''-\boldsymbol{q}}a_{\boldsymbol{k}''}. \tag{13.18}$$

In a static periodic lattice, we noticed that electrons can be *scattered* elastically as Bragg's diffraction, while similar collisions can also be considered for electron–lattice interaction in modulated crystals. As will be discussed in chapter 14, Fröhlich proposed such phonon–electron interactions at very low temperatures.

13.5 The Fermi–Dirac statistics

In a metallic crystal, energy levels $\varepsilon(\boldsymbol{k})$ of one conduction electron are practically continuous, which makes them thermally accessible in the thermodynamic environment. Obeying the Pauli principle, the one-electron state ψ_k is either vacant or occupied by one electron in a specified spin state, i.e. $\psi_k(0)$ or $\psi_k(1)$ with thermal probabilities 1 or $\exp\left(-\frac{\varepsilon(k)}{\mathrm{k_B}T}\right)$, corresponding to excitation energy zero or ε, respectively. Considering the spin–spin correlations among electrons, the latter should be modified by multiplying the factor λ as in the case of phonons; we can write the Gibbs sum

$$Z_k = 1 + \lambda \exp\left(-\frac{\varepsilon(\boldsymbol{k})}{\mathrm{k_B}T}\right).$$

For electrons in typical metals, the energy $\varepsilon(\boldsymbol{k})$ can be occupied by only one electron in an arbitrary spin state, so that $n_k = 1$. Therefore, electrons participating in conductivity are restricted in the highest energy $\varepsilon_{\mathrm{F}}(\boldsymbol{k})$ called *Fermi level.* The thermodynamic probability for a one-particle state in $\varepsilon_{\mathrm{F}}(\boldsymbol{k})$ to be occupied by 1 electron is expressed as

$$\boldsymbol{P}(\varepsilon, 1) = \frac{1}{\exp\frac{\varepsilon-\mu}{\mathrm{k_B}T} + 1}. \tag{13.19}$$

Electrons obey the probability (13.8) at a given temperature T, which is called the *Fermi–Dirac distribution.*

Exercises

1. Phonons and solitons are differently quantized particles, although arising from the same crystal lattice; quantized quantum-mechanically and with respect to anharmonic displacements, respectively. Discuss this difference, as it is a confusing issue in field theory, if disregarding their physical origins.
2. Bosons and fermions are essentially different particles. However, at sufficiently high temperatures, both behave like classical particles. Discuss this issue, regarding their distribution functions.

References

[1] Haken H 1973 *Quantenfeldtheorie des Festkörpers* (Stuttgart: B. G. Teubner)

[2] Kittel C and Kroemer H 1980 *Thermal Physics* (San Francisco, CA: Freeman)
[3] Kurşunoglu B 1962 *Modern Quantum Theory* (San Francisco, CA: Freeman)
[4] Landau L D and Lifshitz E M 1958 *Quantum Mechanics, Theoretical Physics* vol 3 (London: Pergamon)
[5] Kittel C 1963 *Quantum Theory of Solids* (New York: Wiley)

IOP Publishing

Solitons in Crystalline Processes (2nd Edition)

Irreversible thermodynamics of structural phase transitions and superconductivity

Minoru Fujimoto

Chapter 14

Soliton theory of superconducting transitions

While multi-electron systems are essential for normal metals, electron–lattice interactions are the significant mechanism for superconductivity. Based on Fröhlich's proposal, Cooper elaborated the electron-pair model for superconducting charge carriers formed in the modulated lattice, thereby the theory of superconducting metals was established by Bardeen, Cooper and Schrieffer. Interpreting Fröhlich's interactions with the soliton theory, however, the persistent current emerges logically from the mobile Cooper pair in momentum space, in addition to the traditional Bardeen–Cooper–Schrieffer (BCS) theory. Moreover, the soliton theory allows the BCS theory to be applicable to high-T_c superconductivity that arises from the modulated lattice. The superconducting transition can therefore be analyzed with mobile Cooper-pairs of charged particles, analogous to pseudospin clusters in structural phase transitions.

On the other hand, while using the soliton theory, it was found Fröhlich's original proposal was hypothetical, but with the related *Meissner's effect* justified as a logical initial condition for the superconducting mechanism for clusters, as explained below and in chapter 12 of Kittel's textbook 6th edition [1] in detail. In this chapter, Cooper's electron pairs originally considered for metals are discussed as the mechanism in modulated lattices, but applied to such charged carrier-particles as protons as well, for the origin of superconductivity.

14.1 The Fröhlich condensate and the Meissner effect

In contrast to normal metals, in a strained lattice where lattice points can be displaced, inelastic interactions of charge carriers allow energy-momentum exchanges with the hosting lattice. At extreme low temperatures, lattice vibrations are so slow that free electrons at the Fermi level in metals or those in specific conduction bands in high-T_c materials can be in synchronous motion with *lattice displacements*, as if carrier-particles are *attached* to ionize all lattice sites.

doi:10.1088/978-0-7503-2572-1ch14

Fröhlich [2] assumed that conduction electrons can be attached to ions at lattice sites at very low temperatures, located at displaced positions with a reduced mass from the normal ionic sites. Denoting such a displacement mechanism from a regular lattice point $\boldsymbol{r}_\text{o}$ by $\Delta\boldsymbol{r}$, the displaced lattice coordinate is $\boldsymbol{r} = \boldsymbol{r}_\text{o} + \Delta\boldsymbol{r}$; thereby the lattice is spontaneously *polarized with an energy* $eV(\boldsymbol{r})$ that can be expressed as follows.

Fröhlich's proposal:

$$eV(\boldsymbol{r}) = eV(\boldsymbol{r}_\text{o}) + e\Delta\boldsymbol{r}\cdot\nabla_{\boldsymbol{r}_\text{o}}V(\boldsymbol{r}_\text{o}) \quad \text{on the surface,} \tag{14.1a}$$

where $V(\boldsymbol{r}_\text{o})$ represents the *lattice correlation voltage* characterized by the space group. It is noted that such charge attachments should induce an *electrical polarization* $e\Delta\boldsymbol{r}$ on surface-sites, in order for a test magnet to float upon the surface by an *image force*, as illustrated in figure 14.1(a) and (b). In fact, it is notable that such a mechanism of *negative* charge attachment can arise from *Meissner's diamagnetism* for the thermodynamic phase transition to a superconducting phase, transformed by an applied magnetic field $\boldsymbol{H}$, as will be discussed in chapter 16, assuming that *such a dipole* $e\Delta\boldsymbol{r}$ *on the surface* is equivalent to *Fröhlich's displacement* due to *soliton excitations* for superconductivity.

In contrast, in the *periodic structure deep inside the specimen*, it is fundamental that such a displaced position of attached electron by $\pm\Delta\boldsymbol{r}$ should at all sites be expressed by *symmetrical Fourier series* as follows.

Theorem:

$$\Delta\boldsymbol{r} = \sum_{\boldsymbol{q}}\Delta\boldsymbol{r}_{\pm\boldsymbol{q}}\exp i(\pm\boldsymbol{q}\cdot\boldsymbol{r} - \bar{\omega}t) \tag{14.1b}$$

Figure 14.1. Meissner's effect of a small test magnet: (a) a small magnet floating above the superconductor $YBa_2Cu_3O_7$(YBCO) cooled by liquid nitrogen floating on the surface by the image force at $T = T_c$, effective magnetic densities are distributed over the shaded surface layer. Reproduced from http://en.wikipedia.org/wiki/High-temperature_superconductivity. (b) For $T < T_c$, the sample crystal as a whole is binary fluctuating, while the magnet floats at T_c as Fröhlich proposed.

at particular wave vectors $\pm\boldsymbol{q}$ and time t in the direction determined by lattice symmetry in equilibrium crystals due to the *inversion energy* ΔU_0 due to (14.1*a*) and (14.1*b*) for $\boldsymbol{q} \rightleftarrows -\boldsymbol{q}$, while obscured by *bifurcation* due to multi-phase onset of mutual correlations among all displacements. Macroscopically, equation (14.1*b*) can be justified for a sample crystal in ellipsoidal shape.

Nevertheless, unlike such specific electron attachments on the surface at minimum excitation, inside the specimen it is fundamental for a negative $-\Delta U_0$ to have *symmetrical lattice modulation* at $\pm\boldsymbol{q}$ and corresponding frequencies $\mp\tilde{\omega}$ below critical temperatures, characterized by a mobile *dipole moment* $\pm e\Delta\boldsymbol{r}_q$, exhibiting modulated axial displacements in a whole crystal, which decay as $\langle \boldsymbol{q} \rightleftarrows -\boldsymbol{q}\rangle_{\text{thermal}} \rightarrow 0$, characterizing the inside of a crystal by $\boldsymbol{B} = 0$.

In contrast, Fröhlich's displacement waves can take place to represent a *surface magnetic density on the metallic sample sheet* perpendicular to the surface under experimental conditions, so that $\Delta\boldsymbol{r}$ in (14.1*a*) can be represented by *asymmetric* $+\Delta z$ perpendicular to the flat surface of the x–y-plane[note1].

Equation (14.1*b*) is due to a law of nature that all finite lattices created naturally or synthetically are invariant of *geometrical symmetry* except for Earth's gravity, as resulting from *Bose–Einstein's statistics for solitons*. Therefore, we must assume the soliton inversion $\boldsymbol{q} \rightleftarrows -\boldsymbol{q}$ characterizes all sites inside the metallic solid. Nevertheless, Fröhlich's formula (14.1*a*) *on surfaces* is related only with $+\boldsymbol{q}$ to represent the *magnetic image* of a test magnet, which is noted to be consistent with Meissner's diamagnetism along the direction of $\boldsymbol{M} = -\mu_0\boldsymbol{H}$. (See chapter 16.) That relation must correspond to Fröhlich's *specific charge–lattice interaction of symmetrically coherent displacements for superconductivity in general.* Figure 14.1(a) illustrates how the sample crystal is deformed by $\boldsymbol{q} \rightleftarrows -\boldsymbol{q}$ on the surface for $T < T_c$, when a small test magnet is floated on the surface by an image force at T_c, as shown in figure 14.1(b), where *Fröhlich's distortion* exists only on the surface.

The above arguments on Fröhlich' proposal can be subjected to the critical uncertainty, attributing to the symmetry changes, where solitons play an essential role for lattice deformation. In the *phase soliton theory*, we consider that the gradient $\nabla_{r_0} V(\boldsymbol{r}_0)$ in (14.1*a*) generates lattice density waves inside the crystal *induced by the Klein–Gordon equation*

$$\left(\frac{\partial^2}{\partial t^2} - v^2\nabla_{z_0}^2\right)\rho(\Delta z) = -K^2\rho(\Delta z)$$

for the density $\rho(\Delta z)$ of displacements existing inside the specimen, where $K = -\frac{\Delta z}{v}$ represents the source of waves arising from a *pseudopotential energy* K^2, and $v = \tilde{\omega}/q$ is the phase velocity of propagation symmetrically along $\pm z$ directions.

[note1] This is a reasonable assumption, while the flat large sheet can be considered approximated as a specific case of spheroidal shape, where the surface charge from $+\boldsymbol{q}$, whereas at all other sides the inversion $\boldsymbol{q} \rightleftarrows -\boldsymbol{q}$ characterizes the inside. Therefore, Fröhlich's proposal (14.1*a*) can be considered dynamically as related to the inside specified by (14.1*b*), which, however, needs to be justified by logical thermodynamic principles.

As discussed in chapters 8 and 12, arising from *modulated dipolar lattice in the z-axis*, the propagating gradient $\nabla_{z_o} V(z_o)$ is signified by inversions $\boldsymbol{q} \rightleftarrows -\boldsymbol{q}$ with their energies expressed by a complex soliton potential in the form $-u^2 \pm i(\mathrm{d}u/\mathrm{d}x)$, where u is a function of the phase $\boldsymbol{q} \cdot \boldsymbol{r} - \tilde{\omega}t$. Fröhlich's electron–lattice interaction can thus be attributed to the *modulated structure* pinned on surfaces in the lattice.

Assuming that the lattice is composed of constituent ions of mass M, with a restoring constant K associated with the Fourier transform $(\Delta z)_{\pm \boldsymbol{q}}$ of displacement that can be represented by *boson creation and annihilation operators* $\tilde{b}^{\dagger}_{\pm \boldsymbol{q}}$ and $\tilde{b}_{\pm \boldsymbol{q}}$ in the field theory. It is significant that these operators represent *finite lattice displacements that are actually different from phonon modes at infinitely small amplitudes* [3]. For such finite-small displacements to occur *initially*, we can write the Hamiltonian $\mathcal{H}_{\pm \boldsymbol{q}}$ for these displacement operators for n_s solitons as

$$\mathcal{H}_{\pm \boldsymbol{q}} = \lambda_o \left(\tilde{b}^{\dagger}_{\pm \boldsymbol{q}} \tilde{b}_{\pm \boldsymbol{q}} + \frac{1}{2} \right), \qquad \text{where} \qquad \tilde{b}^{\dagger}_{\pm \boldsymbol{q}} \tilde{b}_{\pm \boldsymbol{q}} = n_s, \tag{14.1c}$$

where the eigenvalue of the displacement wave is expressed as $\lambda_{\boldsymbol{q}} = n_s \lambda_o$, where λ_o and n_s represent the elementary soliton energy and its number, respectively, analogous to the zero-point energy of phonon system. Here, the soliton number is written as n_s for superconductors, indicating the number of carrier particles per volume, and the superconducting transition temperature T_c is determined initially by $\Delta n_{s,\mathrm{ini}} = \pm 1$, for the Cooper pair.

On the other hand, the perturbed Hamiltonian of *electrons* in their displacing lattice can generally be written as

$$\mathcal{H} = \mathcal{H}_o + \mathcal{H}_{\pm \boldsymbol{q}} + \mathcal{H}_{\mathrm{int}} + \mathcal{H}_{\mathrm{coul}}, \tag{14.2a}$$

where $\mathcal{H}_o$ is for single electrons, i.e.

$$\mathcal{H}_o = \sum_{\pm \boldsymbol{k}} \varepsilon(\boldsymbol{k}) a^{\dagger}_{\boldsymbol{k}} a_{\boldsymbol{k}}, \tag{14.2b}$$

which is composed of two *one-electron modes* expressed by creation and annihilation operators $a^{\dagger}_k$ and a_k specified by the wavevector $\boldsymbol{k}$ for kinetic energy $\varepsilon(\boldsymbol{k}) = \frac{\hbar^2 k^2}{2m^*}$, where m^* is the effective electron mass. The lattice Hamiltonian for displacements is specified by

$$\mathcal{H}_{\pm \boldsymbol{q}} = n_s \lambda_o \tilde{b}^{\dagger}_{\pm \boldsymbol{q}} \tilde{b}_{\pm \boldsymbol{q}} + \text{const.}, \tag{14.2c}$$

where the magnitude $|\boldsymbol{q}|$ specifies the soliton number n_s at a given temperature.

The Coulomb interaction term $\mathcal{H}_{\mathrm{coul}}$ is formally added to (14.2*a*), although ignorable if charges in metals are shielded out at mutual distance, as described in section 13.3.2, where each electron behaves like a free particle in the first-order approximation. Accordingly, we write *one-electron wavefunction* in the first order approximation as $\psi_k(\boldsymbol{r}) = \psi_o \exp i(\boldsymbol{k} \cdot \boldsymbol{r} + \varphi_k)$ for each electron at a wave vector $\boldsymbol{k}$ and energy $\varepsilon(\boldsymbol{k})$, where the phase constants φ_k are randomly distributed over the $\boldsymbol{k}$-space.

The interaction term $\mathcal{H}_{\text{int}}$ arising from the term of (14.1*a*) is thermodynamically signified by a matrix element of $\Delta z \cdot \nabla V(z_0)$, i.e. a thermal average

$$\left\langle \Delta z_{\pm q} \left| \int_\Omega \psi_k(+\boldsymbol{q})\Delta z \cdot \nabla_{z_0} V(z_0)\psi_k(-\boldsymbol{q})\mathrm{d}^3 z \right| \Delta z_{\mp q} \right\rangle_{\text{thermal}} \tag{14.3}$$

where the bra $\langle \Delta z_{+q}|$ and ket $|\Delta z_{-q}\rangle$ are wave functions of the displacements before and after electron scatterings in the crystal, respectively.

Because of sinusoidal electronic functions at the initial stage of transition, the integral in (14.3) does not vanish if $\boldsymbol{k}' - \boldsymbol{k} = \pm \boldsymbol{q}$, which is significant in equilibrium at low frequencies, if observed at z_0 in long timescale of the thermodynamic environment. In this case, $\mathcal{H}_{\text{int}}$ can be proportional to a *delta function* $\delta(\boldsymbol{k}' - \boldsymbol{k} \mp \boldsymbol{q} + \boldsymbol{G})$ in equilibrium crystals, where $\boldsymbol{G}$ can be *any reciprocal vector* of the soliton lattice. Nevertheless, we select only the first Brillouin zone $\boldsymbol{G} = 0$ for (14.2*a*), disregarding $\boldsymbol{G} \neq 0$ for all other zones.

The matrix elements (14.3) have undetermined $\Delta\varphi_k$ in the factor $\exp i\Delta\varphi_k$ arising from the phase difference during scattering processes, and are maximized at $\Delta \boldsymbol{k} = \pm \boldsymbol{q}$, if assuming a phasing process $\Delta\varphi_k \to 0$ for minimum structural strains due to the Born–Huang principle. In equilibrium, $\mathcal{H}_{\text{int}}$ can thus represent Fröhlich's *charged condensates* in phase with lattice excitations, characterized by $\Delta \boldsymbol{k} = \pm \boldsymbol{q}$ and $\Delta\varphi_k \to 0$.

With respect to these specific displacements, off-diagonal matrix elements do not vanish, i.e. $\langle \Delta z_{+q}|\cdots|\Delta z_{-q}\rangle = \langle \Delta z_{+n_s}|\cdots|\Delta z_{-n_s}\rangle \neq 0$, where the latter is expressed by soliton numbers $+n_s$ and $-n_s$, describing that the time-dependent $\mathcal{H}_{\text{int}}$ can thermally relax to equilibrium by phasing $\Delta\varphi_k \to 0$, forming stable condensates in the modified lattice. In a specific heat curve, an abrupt rise at critical temperature T_c can be expected, when the phasing is completed at a singularity in the displacement mode.

In Fröhlich's postulate, $\mathcal{H}_{\text{int}}$ at a given time can therefore be expressed in the field theory as

$$\mathcal{H}_{\text{int}} = -e\int_\Omega \mathrm{d}^3 z\{\rho(z)\Delta z \cdot \nabla_{z_0} V(z_0)\} \tag{14.4}$$

where $\rho = \sum_i \rho_q \exp(-iqz_i)$ is the modulated electron density to interact with Δz, and $\mathcal{H}_{\text{int}}$ showing the average integral over unit volume of the Toda lattice.

The Fourier transform $\rho_q = \sum_i \rho(z_i)\exp(iqz_i)$ can be written as $\rho(z)\exp(iqz)$ for a small continuous value of single $|\boldsymbol{q}|$, so that

$$\rho_q = \sum_k a_{k+q}^\dagger a_k \tag{14.5}$$

can be used in field theory. Therefore $\mathcal{H}_{\text{int}}$ can be expressed by the field variables ρ_q and $\tilde{b}_q$ as follows.

Theorem:

$$\mathcal{H}_{\text{int}} = i\sum_{q} eD_q\left(\rho_q\tilde{b}_q^{\dagger} - \rho_q^{\dagger}\tilde{b}_q\right), \tag{14.6}$$

where D_q is a factor proportional to the gradient $-\nabla_z\Delta z_q$, so that the factor eD_q is *time-dependent*, as related to the phase factor in $\Delta z_q \sim \exp i\phi_q$, where $\phi_q = \boldsymbol{q}\cdot\boldsymbol{r} - \bar{\omega}t + \varphi_q$. Such a time-dependent D_q will be utilized in later discussions as $D_q = D_q{}'\exp i\phi_q$, where $D_q{}'$ is constant of the phase ϕ_q.

It is also significant that in (14.6) $\mathcal{H}_{\text{int}}$ signifies inversion processes $eD_q \rightleftarrows -eD_{-q}$ or $D_q \rightleftarrows -D_{-q}$ at T_c *independent of the charge e*, corresponding to *uniaxial shear strain inversion* in the crystal expressed by $\tanh\phi_q \rightleftarrows \tanh\phi_{-q}$, where the *discontinuity* arises from a change of the initial soliton potentials for $\Delta n_{\text{s,ini}} = \pm 1$, as discussed in section 12.7.1; which is also confirmed for high-T_c superconductors of 3*d-electronic* and *protonic charges* in cuprate layers and in the H3S phase of hydrogen sulfide, respectively, as will be explained in chapter 15.

At this stage, we assume that the Hamiltonian $\mathcal{H}_\text{o}$ represents dynamics of *two single-electrons* in the first order, and $\mathcal{H}_{\text{int}}$ is considered as a perturbation. *Fröhlich's condensates* charged by electron attachments in metals are thus signified by displacements $\Delta z_{\pm q}$ in finite magnitudes.

14.2 The Cooper pair and superconducting transition

Due to *off-diagonal* elements of the displacement matrix with distributed phases $\Delta\phi_q$, Fröhlich's condensates depend significantly on the lattice, related to correlations among charged particles in the multi-electron system at low temperatures.

We consider a process for *two charged condensates* to bind together signified by inversion $\boldsymbol{q} \rightleftarrows -\boldsymbol{q}$ across the *mini-domain boundary* for $\Delta n_{\text{s,ini}} = 0 \rightarrow \pm 1$ *to be induced by* $\tilde{b}_q^{\dagger}$ and $\tilde{b}_q$ in the *initial Weiss field for phase transition*, where $n_{\text{s,ini}}$ is the initial soliton number that should be sufficiently large at the critical point T_c. Denoting two Fröhlich condensates by 1 and 2 on each side of the *boundary walls*, Cooper [4] assumed the critical vectors $\boldsymbol{q}_{1,2}$ and $-\boldsymbol{q}_{1,2}$ to be specified by $\Delta\phi_1 = \Delta\phi_2$, while $\Delta\phi_1 = -\Delta\phi_2$ for electrons at domain boundaries. In fact, two charges are synchronized with $\Delta z_{q_{1,2}}$ and $\Delta z_{-q_{1,2}}$ specified for $\boldsymbol{q}_{c1}\|\boldsymbol{q}_{c2}$ to form a stable condensate with $\Delta\phi_1 = \Delta\phi_2$.

Owing to the time-dependent exponential factor of $eD_{q_{1,2}}$ in (14.6), such *parallel phasing* actually occurs for the lattice in the thermal process that is the property of boson displacements, providing *symmetric* $eD_{q_{c1,2}}$ and $\Delta\phi_{c1,2}$ for two condensates 1 and 2 to form a stable pair in thermal equilibrium at T_c. We actually have the relation $\boldsymbol{q}_{c1,2}(T_c) \perp \boldsymbol{q}_{1,2}(T > T_c)$, thereby the pair model can logically be sketched as follows, supporting Cooper's hypothesis on paired electrons.

We write the Hamiltonian for such a pair of condensates 1 and 2 as $\tilde{\mathcal{H}}$, which should be different from (14.2*a*), while $\tilde{\mathcal{H}}$ and $\mathcal{H}$ are related by a canonical transformation with an action variable S associated with a *thermal process* to the thermodynamic environment. We therefore write

$$\mathcal{H} = \mathcal{H}_o + \alpha\mathcal{H}', \tag{14.7}$$

where $\mathcal{H}_o$ represents kinetic energies of two single-electrons given by (14.2*b*); and the perturbation $\mathcal{H}_{int}$ of (14.6) is replaced by $\alpha\mathcal{H}'$, where the specific relation $\alpha \to eD_{q_c}$ confirms for the Hamiltonian $\tilde{\mathcal{H}}$ to be in *thermal equilibrium*.

A canonical transformation

$$\tilde{\mathcal{H}} = \exp(-S)\mathcal{H}\exp S \tag{14.8}$$

is assumed for converting $\mathcal{H}$ to be equilibrium $\tilde{\mathcal{H}}$, where the *action variable* S is chosen for the lattice to be in minimal strains in the thermodynamic environment, expressing essentially the Born–Huang principle for clustering by analogy.

Expanding $\tilde{\mathcal{H}}$ into a power series, we have

$$\tilde{\mathcal{H}} = \mathcal{H} + [\mathcal{H}, S] + \frac{1}{2}[[\mathcal{H}, S], S] + \cdots.$$

Using (10.7) for $\mathcal{H}$, the above expression can be written for a small S as

$$\tilde{\mathcal{H}} = \mathcal{H}_o + \alpha\mathcal{H}' + [\mathcal{H}_o, S] + [\alpha\mathcal{H}', S],$$

ignoring higher-order terms than S^2. If S is determined in such a way that

$$\alpha\mathcal{H}' + [\mathcal{H}_o, S] = 0, \tag{14.9}$$

we obtain

$$\tilde{\mathcal{H}} = \mathcal{H}_o + [\alpha\mathcal{H}', S]. \tag{14.10}$$

Here, $\tilde{\mathcal{H}}$ can be characterized by the second term on the right for the system to be in equilibrium at $\alpha = eD_{q_c}$, hence the term $eD_{q_c}[\mathcal{H}', S]$ expresses the interaction energy of *coupled condensates in equilibrium*, while $\mathcal{H}$ of (14.7) is the Hamiltonian unrelated with the equilibrium condition.

Writing $\mathcal{H}'$ as a diagonal matrix $\begin{pmatrix} +\lambda_q & 0 \\ 0 & -\lambda_q \end{pmatrix}$, the eigenvalues $\pm\lambda_q$ correspond to symmetric and antisymmetric displacements

$$\Delta z_{qA} = (\Delta z_{+q} + \Delta z_{-q})/\sqrt{2} \quad \text{and} \quad \Delta z_{qP} = (\Delta z_{+q} - \Delta z_{-q})/\sqrt{2},$$

respectively, corresponding to $\Delta n_{s,ini} = 0$ and $\Delta n_{s,ini} = 2$, analogous to A- and P-modes in binary fluctuations discussed in chapter 4. From (14.10), we consider off-diagonal elements of the matrix S, i.e.

$$\langle\Delta z_{q\mathrm{A}}|S|\Delta z_{q\mathrm{P}}\rangle = -\, ieD_q\sum_k \frac{a^{\dagger}_{k-q}a_k}{\varepsilon_k - \varepsilon_{k-q} - \lambda_q} \quad \text{and}$$
$$\langle\Delta z_{q\mathrm{P}}|S|\Delta z_{q\mathrm{A}}\rangle = ieD_q\sum_{k'} \frac{a^{\dagger}_{k'+q}a_{k'}}{\varepsilon_{k'} - \varepsilon_{k'+q} + \lambda_q}, \tag{14.11}$$

perturbing $\pm\lambda_q$ by $\begin{pmatrix} 0 & -eD_q\ldots \\ eD_q\ldots & 0 \end{pmatrix}$, thereby the perturbed $\tilde{\mathcal{H}}$ can be expressed as

$$\tilde{\mathcal{H}} = \mathcal{H}_\mathrm{o} + \sum_q \frac{e^2D_q^2}{2}\sum_{k,k'} a^{\dagger}_{k'+q}a_{k'}a^{\dagger}_{k-q}a_k\left(\frac{1}{\varepsilon_k - \varepsilon_{k-q} - \lambda_q} - \frac{1}{\varepsilon_{k'} - \varepsilon_{k'+q} + \lambda_q}\right). \tag{14.12}$$

In (14.12), scatterings $\boldsymbol{k} \rightarrow \boldsymbol{k} - \boldsymbol{q}$ and $\boldsymbol{k}' + \boldsymbol{q} \rightarrow \boldsymbol{k}'$, cause *emission* and *absorption* of a soliton particle $(\lambda_\mathrm{n}(q), \boldsymbol{q})$ between uniaxial displacements Δz_q and Δz_{-q}, respectively.

In equation (14.12), the two terms in the brackets indicate singular behaviors at $|\Delta\boldsymbol{k}_\mathrm{c}| = |\boldsymbol{q}_\mathrm{c}|$ and $\Delta\boldsymbol{k}_\mathrm{c} = \Delta\boldsymbol{k}_\mathrm{c}'$ with $\Delta z_{q_{\mathrm{c}1}}$ and $\Delta z_{q_{\mathrm{c}2}}$ on both sides of T_c. Since $\boldsymbol{q}_{\mathrm{c}1}$ and $\boldsymbol{q}_{\mathrm{c}2}$ at these singularities are *in-phase*, the *parallel relation* $\pm\boldsymbol{q}_{\mathrm{c}1}\|\pm\boldsymbol{q}_{\mathrm{c}2}$, *perpendicular to non-critical* $\pm\boldsymbol{q}$, can be considered for the thermal process to stabilize a pair with *symmetric* displacements, $\Delta z_\mathrm{A}(\phi_\mathrm{c}) = (\Delta z(\phi_{\mathrm{c}1}) + \Delta z(\phi_{\mathrm{c}2}))/\sqrt{2}$, owing to *boson* displacements, as discussed in section 9.6. This is accompanied with *antisymmetric* scatterings of attached electrons $\Delta\boldsymbol{k}_\mathrm{c} = -\Delta\boldsymbol{k}_\mathrm{c}'$ for zero energy at T_c. Stabilizing a pair of symmetric polarized displacements in this case is essentially a pinning mechanism by the *negative Weiss' potential* $\langle -\mathrm{sech}^2\,\phi_{\mathrm{c}1} - \mathrm{sech}^2\,\phi_{\mathrm{c}2}\rangle_\mathrm{thermal} = -2\mathrm{sech}^2\,\phi_\mathrm{c}$.

In this case, the modulated electronic charge by lattice should be symmetric as induced by solitons; namely, the processes as

$$\Delta\boldsymbol{k}_\mathrm{c}(1, 2) = -\Delta\boldsymbol{k}_\mathrm{c}'(1, 2) \quad \text{and} \quad \pm\,\boldsymbol{q}_{c,1}\|\pm\boldsymbol{q}_{c,2}, \tag{14.13a}$$

providing thermal pathway $\Delta n_\mathrm{s} = 2$ for leading to equilibrium at T_c, where the electron pair interacts between *conservative* condensates at T_c. The above mechanism for a pair of parallel $(\boldsymbol{q}_{\mathrm{c}1}, \boldsymbol{q}_{\mathrm{c}2})$ in-phase accompanies *antisymmetric electron scatterings in Fermi–Dirac statistics*, at a critical point T_c. For *polar solitons as scatterers* however, antisymmetric scatterings of electrons can be considered for electrons to *fluctuate* by $\Delta\boldsymbol{k}_\mathrm{c}$, as indicated by the first relation in (14.11), indicating that these electrons participate as a combined pair particle.

Figure 14.2(a) shows a scattering geometry diagram for $\pm\boldsymbol{q}$ in non-critical condition, whereas figure 14.2(b) illustrates that the corresponding $\boldsymbol{q}_{\mathrm{c}1}$ and $\boldsymbol{q}_{\mathrm{c}2}$ are in parallel in *all perpendicular directions*, i.e. $\pm\boldsymbol{q} \perp (\boldsymbol{q}_{\mathrm{c}1}, \boldsymbol{q}_{\mathrm{c}2})$, creating a *persistent circular current* in momentum space, as consistent to

$$\langle\boldsymbol{\sigma}_{q_{\mathrm{c}1}}\cdot\boldsymbol{\sigma}_{q_{\mathrm{c}2}}\rangle_{T_c} = \langle\boldsymbol{\sigma}_{+q_\mathrm{c}}\cdot\boldsymbol{\sigma}_{-q_\mathrm{c}}\rangle_{T_c} = +1 \tag{14.13b}$$

in equation (3.6) in chapter 3.

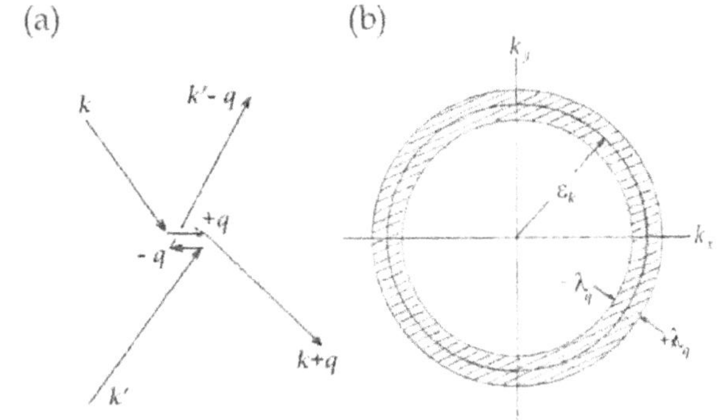

Figure 14.2. (a) Soliton geometry for the Cooper pair for non-critical wavevector $\boldsymbol{q}$. (b) One-electron energy surface ε_k in two dimensions, modulated by soliton energies $\pm\lambda_q$.

The off-diagonal elements $a^{\dagger}_{k-q}a_k$ and $a^{\dagger}_{k'+q}a_{k'}$ of the electron density matrices are associated with the coupling process, so that $\tilde{\mathcal{H}}$ represents two electrons as a bound object, for which the momentum conservation law requires a coupling for specific electron scatterings with fluctuations $\Delta\boldsymbol{k}_c$. Therefore, as inferred from figures 14.2(a) and (b), the binding energy of two electrons can be determined by (14.12) as averaged thermally. Namely,

$$\langle\tilde{\mathcal{H}} - \mathcal{H}_o\rangle_{T_c} = e^2 D^2_{q_c} \sum_{q_c, k_c} \frac{\lambda_{q_c}}{(\varepsilon_{k_c+q_c} - \varepsilon_{k_c})^2 - \lambda^2_{q_c}} a^{\dagger}_{k_c-q_c} a_{k_c} a^{\dagger}_{-k_c+q_c} a_{-k_c}, \tag{14.14a}$$

which can be re-expressed for bound electrons 1 and 2 as

$$\langle\tilde{\mathcal{H}} - \mathcal{H}_o\rangle_{T_c} = -\sum_{q_c}^{1,2} V(2\boldsymbol{q}_c)_{T_c} a^{\dagger}_{k_c-q_c} a_{k_c} a^{\dagger}_{-k_c+q_c} a_{-k_c}, \tag{14.14b}$$

where the following applies.

Theorem:

$$-V(2\boldsymbol{q}_c) = e^2 D^2_{q_c} \frac{\lambda_{q_c}}{(\varepsilon_{k_c+q_c} - \varepsilon_{k_c})^2 - \lambda^2_{q_c}}. \tag{14.14c}$$

The interaction $\tilde{\mathcal{H}}$ is *attractive* if $V(2\boldsymbol{q}_c) > 0$; implying that such an electron pair can carry a combined charge $2e$ stabilized by *the boson statistics of two Fröhlich's lattice distortions* $2\boldsymbol{q}_c$ *in the superconducting state*. Therefore, from (14.14*c*), we should have

$$|\varepsilon_{k\pm q_c} - \varepsilon_k| < \lambda_{q_c}, \tag{14.15}$$

otherwise $V(2\boldsymbol{q}_c) < 0$, we have $\langle\tilde{\mathcal{H}} - \mathcal{H}_o\rangle_{T_c} > 0$ that represents a *repulsive interaction*.

It is noted from (14.8) that $V(2\boldsymbol{q}_c)$ is related to the potential energy for inversion $2\boldsymbol{q}_c \rightleftarrows -2\boldsymbol{q}_c$, corresponding to eigenvalues of Lax's operator $\boldsymbol{L}$ at T_c. Hence, the soliton potential energy is proportional to $V(2\boldsymbol{q}_c) \propto \Delta z_{\boldsymbol{q}_c}^2$, where two *symmetric Fröhlich's displacements* $\Delta z_{\boldsymbol{q}_c} = (\Delta z_{\boldsymbol{q}_{c1}} + \Delta z_{\boldsymbol{q}_{c2}})/\sqrt{2}$ are stabilized by *two symmetric bosons* associated with $\Delta z_{\boldsymbol{q}_{c1}}$ and $\Delta z_{\boldsymbol{q}_{c2}}$.

Analogous to *binary transitions*, the relation $\lambda_{\pm \boldsymbol{q}_c} \propto u_c \sim \pm \tanh \phi_c$ at T_c exhibits the potential difference between $\pm i(\mathrm{d}u/\mathrm{d}\phi)_c$ in the lattice, corresponding to *uniaxial inversion* $\phi_c \rightleftarrows -\phi_c$. However, the *transversal potential energy* for this particular inversion is proportional to $-2u_c^2$ at $T = T_c$ corresponding to a negative energy $-2u_c^2 \propto -2\,\mathrm{sech}^2\,\phi_c$ for both $\boldsymbol{q}_{c1} \| \boldsymbol{q}_{c2}$ in parallel to form symmetric $2\boldsymbol{q}_c \perp \pm \boldsymbol{q}$, representing internal elastic work in crystals to stabilize parallel inversion $\boldsymbol{q}_{c1,2} \rightleftarrows -\boldsymbol{q}_{c1,2}$. We therefore write

$$-V(2\boldsymbol{q}_c) \propto e^2 D_{\boldsymbol{q}_c}^2 = e^2 n_{\mathrm{s,ini}}^2 \lambda_{\boldsymbol{q}_c}^2 \propto -2u_o^2\,\mathrm{sech}^2\,\phi_c$$

for entropy production. As warranted by the *negative* soliton potential $-2\mathrm{sech}^2\,\phi_c$, the charge $2e$ is *thermodynamically stable* as determined by negative $-V(2\boldsymbol{q}_c)$. It is noted that the adiabatic work for uniaxial inversions $\pm\boldsymbol{q}_c \rightleftarrows \mp\boldsymbol{q}_c$ is sufficiently small for $\Delta n_{\mathrm{s,ini}} = \pm 1$ to be in the category of equation (12.18) to obtain a stable pair. It is notable that a *Cooper's pair* [4] is composed of two electrons via soliton potentials, which has, however, nothing to do with *the antisymmetric spin–spin correlations of fermion particles.*

With respect to such singular inversion, Cooper's interaction produces a *pseudo-particle* of charge $e' = 2e$, playing the role of charge carriers in superconducting states. Such a *pseudo-particle* emerges at T_c as consequent on the canonical transformation by the action S defined by (14.6), which is analogous to *pseudospin clusters* in binary transitions.

The potential $V(2\boldsymbol{q}_c)$ in (14.7) represents the *binding energy* of a Cooper's pair, so that the superconducting phase transition is characterized by the thermal relation per periodic translational unit via positive $+V(2\boldsymbol{q}_c)$ as

$$\mathrm{k_B} T_c = V(2\boldsymbol{q}_c) = e^2 D_{\boldsymbol{q}_c}^2 = e^2 (n_{\mathrm{s,ini}} \lambda_{\boldsymbol{q}_c})^2, \qquad (14.16a)$$

where the initial soliton density $n_{\mathrm{s,ini}}$ at T_c signifies the strength of thermal average $+V(2\boldsymbol{q}_c)$ in the $\boldsymbol{k}$-space. Hence, *the transition temperature can be characterized as proportional to* $n_{\mathrm{s,ini}}$, as follows.

Theorem:

$$T_c \propto e^2 n_{\mathrm{s,ini}}^2 \lambda_{\boldsymbol{q}_c}^2, \qquad (14.16b)$$

where the critical energy (14.9*a*) is transferred to the lattice via positive $+2n_{s,ini}\,\mathrm{sech}^2\,\phi_c$, if $n_{s,ini}$ is sufficiently large at T_c. Accordingly, the soliton mechanism describes the same as seed formation in classical theory of phase transitions.

14.3 Persistent supercurrents

Further significant in soliton theory is that the Cooper pair supports not only the double charge $e' = 2e$ effectively but also the *Meissner effect*, characterizing dynamic superconducting transition. We notice that the direction of critical displacements Δz_{q_c} can be *specified in crystal space, while left as arbitrary in the momentum space.*

Characterized by the spatial phase $\phi_c = q_c z$, the inversion of $eD_{q_c} \propto |\Delta z_{q_c}|$ at T_c should take place *in perpendicular* directions to noncritical vectors $\pm\boldsymbol{q}$ at $\phi_c = 0$, i.e. $2\boldsymbol{q}_c \perp \pm\boldsymbol{q}$ in the A-mode σ_A at T_c, whose inversion energy is specified by the potential $V(2\boldsymbol{q}_c) \propto e^2 D_{q_c}^2$ oriented along $2\boldsymbol{q}_c$, pointing in *any direction perpendicular to the non-critical* $\boldsymbol{q}$.

Hence, in the momentum space, the charge $2e$ carried by the potential $V(2\boldsymbol{q}_c)$ in parallel to $2\boldsymbol{q}_c$ exhibits *vortex-like rotation* by the angle $2\phi_c$ in the range $0 \leqslant \phi_c \leqslant 2\pi$ in the perpendicular plane. Using the explicit relation $D_{q_c} = D'_{q_c} \exp i\phi_c$, the *complex function*

$$V(2\boldsymbol{q}_c) = |V(2\boldsymbol{q}_c)| \exp i(2\phi_c) \tag{14.17a}$$

should be defined in momentum space as the time-dependent Cooper's pair.

Along a circular trajectory, a superconducting charge generates a ring-current density $\boldsymbol{j}_s$ that can be given by the time-derivative of the complex potentials as follows.

Theorem:

$$j_s = e\frac{\mathrm{d}}{\mathrm{d}t}\{|V(2\boldsymbol{q}_c)| \exp 2i\phi_c(\boldsymbol{r}, t)\} = 2e|V(2\boldsymbol{q}_c)|\dot{\phi}_c \exp i(2\phi_c + \pi),$$

which is expressed as $|\boldsymbol{j}_s| = 2er_s\dot{\phi}_c \exp i(2\phi_c + \pi)$, where $r_s = |V(2\boldsymbol{q}_c)|$ represents the radius of trajectory. Defining the angular speed $v_s = r_s\dot{\phi}_c$, the *persistent current* can be expressed conventionally as $j_s = 2ev_s \exp i(2\phi_c + \pi)$ for the Cooper pair to circulate in the current density

$$j_s = 2ev_s, \qquad m'\frac{\mathrm{d}v_s}{\mathrm{d}t} = -2e\boldsymbol{E}_s \quad \text{and} \quad \frac{\partial j_s}{\partial t} \propto \boldsymbol{E}_s \tag{14.17b}$$

where m' and $\boldsymbol{E}_s$ are, respectively, the effective mass and the electric field responsible for $\boldsymbol{j}_s$ to satisfy *the charge–current continuity for Cooper's pair* [5]. Figure 14.3(b) shows such a ring current density $\boldsymbol{j}_s = \boldsymbol{j}_s(\phi_c)$ at $\boldsymbol{q} = \boldsymbol{q}_c$ with respect to the z-axis for

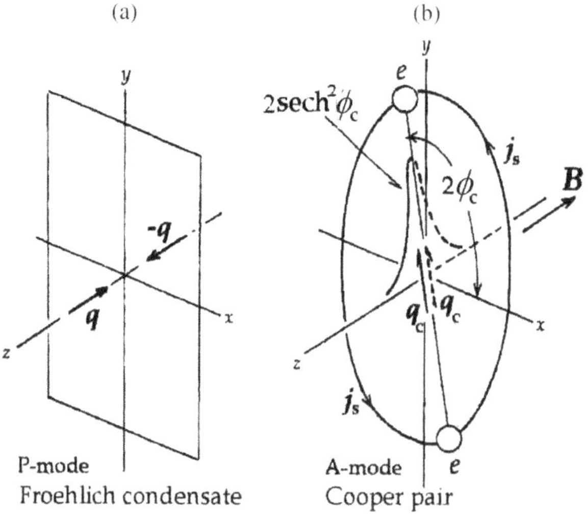

Figure 14.3. (a) Antisymmetric displacements in P-mode of a Fröhlich condensate at non-critical temperatures. (b) Cooper's pair $2e$ and persistent ring current j_s at the critical q_c in A-mode, producing a local magnetic field as $\text{curl}\boldsymbol{j}_s = \boldsymbol{B}$. The soliton potential energy $\text{sech}^2\,\phi_c$ associated with parallel $\boldsymbol{q}_c$ is directional and circularly rotating in momentum space.

the non-critical $\boldsymbol{q}$. Figure 14.3(a) illustrates for comparison the non-critical $\boldsymbol{q} \neq \boldsymbol{q}_c$, accompanied by no persistent current.

A magnetic field $\boldsymbol{B}$ can then be determined as indicated in figure 14.3(b) for the non-vanishing vector $\text{curl}\boldsymbol{j}_s$, where $\boldsymbol{j}_s$ is a circular persistent current density $\boldsymbol{j}_s = e'\boldsymbol{v}_s$, as described by the *Biot–Savart law* familiar in electromagnetic theory [4]. Since $\boldsymbol{B} = \text{curl}\,\boldsymbol{A}$ for the vector potential $\boldsymbol{A}$, the above relation between $\boldsymbol{B}$ and $\text{curl}\boldsymbol{j}_s$ can be replaced by $\boldsymbol{A} \propto \boldsymbol{j}_s$, leading to the *London equation* that was derived before the Bardeen–Cooper–Schrieffer theory. In this way, the soliton mobility of Cooper's pair $e' = 2e$ can explain the *Meissner effect* in terms of the London theory, which will be discussed in chapter 15. Notably, it is an experimental result that the persistent current is idealized by *observed long life*, however the *infinite life* is actually predicted by the soliton theory, *assuming no interactions with surroundings.*

Illustrated in figure 14.3(a), the persistent current density $\boldsymbol{j}_s$ of a Cooper pair is considered as charged pseudo-particles e' multiplied by the number $n_{s,\text{ini}}$ for mesoscopic expression $\boldsymbol{J}_s = n_{s,\text{ini}}\boldsymbol{j}_s$, based on which Bardeen, Cooper and Schrieffer [6] elaborated the theory of superconducting transitions.

A Cooper pair can be treated as a single charged quasi-particle of charge $e' = 2e$ and mass $m' = 2m_{\text{eff}}$, whose motion in momentum space at a given temperature is determined by the time-dependent *phase soliton voltage* $V(2\boldsymbol{q}_c) \propto n_{s,\text{ini}}^2 \exp i(2\phi_c + \pi)$. The symmetric Cooper's pair $e' = 2e$ circulates by rotating soliton potential energy $-2\,\text{sech}^2\,\phi_c$ in momentum space, satisfying the continuity equation of charge and current; accordingly, the Cooper pair is classified as a quantum-mechanically *unidentifiable fermion particle* in the superconducting phase.

Sharing a similar but opposite mechanism at domain boundaries, it is notable that the superconductivity is developed with a *quasi-adiabatic process under constant temperature*. Contrary to an adiabatic superconducting transition, *conventional domain separation* can occur in an *isothermal process in the lattice* to transfer strain energy to the surroundings, while entropy productions can be signified by the positive soliton potential $+2n_s \operatorname{sech}^2 \phi_c$ with a large soliton number n_s.

Further, according to section 12.7.1, the mesoscopic domain structure exists at all temperatures T below T_c, the soliton mechanism should be valid *at any temperatures for* $T < T_c$ *to be signified by* $n_s(T)$. Therefore, the supercurrent can be observed at all temperatures in the superconducting state stable at all temperatures $T = \lambda_q^2 e^2 n_s^2$ for $T < T_c$, as will be discussed in chapter 15. Thermodynamically, the supercurrent is observed as *adiabatic processes for* $\Delta V \neq 0$ below T_c; for that matter, Anderson [7] defined a *binary pseudospin* $\boldsymbol{\sigma}(\phi)$ associated with the superconducting charge $2e$ and current $\boldsymbol{j}_s$, supporting the BCS theory.

14.4 Critical energy gap and the superconducting ground state

14.4.1 Energy gap between normal- and superconducting states

As characterized by scatterings $\Delta \boldsymbol{k}_c = -\Delta \boldsymbol{k}_c'$, the Cooper pair can be described with respect to the center-of-mass coordinate system, where $\boldsymbol{k}_c' + \boldsymbol{k}_c = 0$. However, the center, $\boldsymbol{k}_c' + \boldsymbol{k}_c$ can actually be uncertain by a non-zero $\Delta \boldsymbol{k}_c$ that is equivalent to adiabatic potential $V(2\boldsymbol{q}_c)$ in equilibrium at T_c. Considering such pairs emerge at the threshold of a superconducting phase, the transition can be characterized by inversion symmetry $\boldsymbol{q}_c \rightleftarrows -\boldsymbol{q}_c$, while being distributed with respect to $\Delta \boldsymbol{k}_c$.

The transition is initiated by the potential $-\nabla V(z_0)$ emerging at T_c, where critical anomalies appear as related to these fluctuations. It is noted that the Cooper pair is associated with symmetric fluctuations $(\Delta z_{q_{c1}} + \Delta z_{-q_{c2}})/\sqrt{2}$ in lower energy for the negatively charged quasi-particle $e' = 2e$, while we have the antisymmetric mode $(\Delta z_{q_{c1}} - \Delta z_{-q_{c2}})/\sqrt{2}$ for two *uncoupled condensates* at higher energy. The stable Cooper pair is symmetrically modulated by lattice waves $\Delta z_{\pm q_c} \sim \exp(\pm i q_c z)$, so that the modulated wave function is given by $\psi_k(\Delta z_{q_c} + \Delta z_{-q_c})/\sqrt{2}$. Therefore, the corresponding kinetic energy is modulation energy between ε_{k+q_c} and ε_{k-q_c}, can be written as

$$\varepsilon_{k\pm q_c} - \varepsilon_k = \pm \frac{\hbar^2 k q_c}{m}.$$

Consequently, the Fermi level specified by $k = k_F$ is split into two, showing an *energy gap* at T_c that is given by

$$2\frac{\hbar^2 k_F q_c}{m} = E_g, \tag{14.18a}$$

where the critical wave vector q_c at T_c can be replaced by the *coherence length* $\xi_0 = 1/2q_c$ for some applications, that is

$$\xi_{0} = \frac{\hbar^2 k_F}{mE_g}. \tag{14.18b}$$

The critical singularity is due to the discontinuity in forming Cooper pairs, similar to creating clusters in structural transitions.

For a Cooper pair, we consider scattering processes signified by $\boldsymbol{k} \to \boldsymbol{k}' = \boldsymbol{k} + \boldsymbol{q}_c$, where $-\boldsymbol{k} \to -\boldsymbol{k}' = -\boldsymbol{k} - \boldsymbol{q}_c$ with respect to the center-of-mass coordinate. Therefore, in the relative coordinate frame of reference, the electronic wave functions $\psi_k(z)$ and $\psi_{k'}(z)$ must be functions of the relative coordinate $z = z_1 - z_2$, which can be set as $z = 0$ at the center of coordinates. We can therefore write Schrödinger equation for the pair at z, as

$$\left\{\frac{\hbar^2}{2m}\left(k^2 + \sum_{k'} k'^2\right) + \tilde{\mathcal{H}}\right\}\psi_{k,k'}(z) = \varepsilon\psi_{k,k'}(z),$$

where ε is the eigenvalue, and $\psi_{k,k'}(z) = \sum_k \alpha_k \exp i\boldsymbol{k}.\boldsymbol{z}$. Denoting the unperturbed energy by E_k, we have the secular equation

$$(E_k - \varepsilon)\alpha_k + \sum_{k'} \alpha_{k'}\langle \boldsymbol{k}, -\boldsymbol{k}|\tilde{\mathcal{H}}|\boldsymbol{k}', -\boldsymbol{k}'\rangle = 0,$$

where $\boldsymbol{k} = \boldsymbol{k}' - \boldsymbol{q}_c$ and $-\boldsymbol{k} = -\boldsymbol{k}' + \boldsymbol{q}_c$. The perturbation $\tilde{\mathcal{H}}$ in (14.8) is related to distributed $\boldsymbol{k}'$ deviating from $\boldsymbol{k}_F$ in a small limited range. In this case, assuming the second term on the left as a constant C in the first approximation, we obtain $\alpha_k = \frac{C}{E_k - \varepsilon}$. In the second approximation, the summation is replaced by an integral over the corresponding energy $E_{k'}$, i.e.

$$(E_k - \varepsilon)\alpha_k = -V\int_{E_F}^{E_k} \alpha_{k'}\rho_{k'}\,\mathrm{d}E_{k'}$$

where $E_{k'} - E_F = \Delta$. Assuming $\rho_{k'} \approx \rho_F$, we obtain

$$\frac{1}{\rho_F V} = \int_{E_F}^{E_{k'}} \frac{\mathrm{d}E_{k'}}{E_{k'} - \varepsilon} = \ln\frac{E_{k'} - \varepsilon}{E_F - \varepsilon} = \ln\frac{E_{k'} - E_F + \Delta}{\Delta}.$$

With the relation $E_{k'} - E_F = \lambda_q$, we have the following expression.

Theorem:

$$\Delta = \frac{2\lambda_{q_c}}{\exp\frac{1}{\rho_F V} - 1}, \tag{14.19}$$

representing the binding energy of a Cooper pair with respect to the Fermi level, which should be consistent with (14.8).

It is notable that using Toda's theorem in the soliton theory, we can interpret order variable waves of the Cooper pair for the superconducting phase transitions in metals, for all of which the *nonlinear lattice deformation* proposed by Fröhlich is responsible.

14.4.2 Anderson's pseudospins for the Cooper pair

Following Anderson [7, 8], in field theoretical approximation, we write the Hamiltonian for interacting electrons with a Cooper pair as expressed by

$$\mathcal{H} = \sum_k \left(\varepsilon_k a_k^\dagger a_k + \varepsilon_{-k} a_{-k}^\dagger a_{-k}\right) - \sum_{k,-k} V(\boldsymbol{k}, -\boldsymbol{k}; 2\boldsymbol{q}_c)\, a_k^\dagger a_{-k}^\dagger a_{-k} a_k \tag{14.20}$$

where $\varepsilon_k = \varepsilon_{-k}$ are the eigenvalues of a single electron at $\boldsymbol{k}$ and $-\boldsymbol{k}$ on the Fermi surface. Here one-electron energy in metals is considered at degenerate ε_k at the Fermi level. The above Hamiltonian can be re-expressed by

$$\mathcal{H} = \sum_{k,k'} \varepsilon_k a_k^\dagger a_k - \sum_{k,k';q} V(\boldsymbol{k}, \boldsymbol{k}'; 2\boldsymbol{q}_c) a_{k'+q}^\dagger a_{k-q}^\dagger a_{k'} a_k. \tag{14.21}$$

On the other hand, the wave function can be expressed as $\psi(\ldots, n_k, \ldots, n_{k'}, \ldots)$, where n_k, $n_{k'}$ take values either 1 or 0, signifying that one-particle states $\boldsymbol{k}$ and $\boldsymbol{k}'$ are either occupied or unoccupied, respectively. On the other hand, in BCS theory, the Cooper pair is described by a two-particle wavefunction $\psi(n_k, n_{k'})$, and hence using number operators $n_k = a_k^\dagger a_k$ and $n_{-k}^\dagger = a_{-k}^\dagger a_{-k}$, we can re-write (14.21) for the BCS Hamiltonian $\mathcal{H}_{\rm BCS}$ as

$$\mathcal{H}_{\rm BCS} = -\sum_k (1 - n_k - n_{-k})\varepsilon_k - V(\boldsymbol{k}, -\boldsymbol{k}; 2\boldsymbol{q}_c) \sum_{k,-k} a_k^\dagger a_{-k}^\dagger a_{-k} a_k, \tag{14.22}$$

becoming a constant in thermal equilibrium. Applying (14.22) to $\psi(n_k, n_{-k})$ in the two-particle sub-space, we have paired states for $n_k = n_{-k}$, corresponding to $\boldsymbol{k}$ and $-\boldsymbol{k}$, where each of the one-electron states are occupied or unoccupied, and hence

$$(1 - n_k - n_{-k})\psi(1_k, 1_{-k}) = -\psi(1_k, 1_{-k}) \quad \text{and}$$
$$(1 - n_k - n_{-k})\psi(0_k, 0_{-k}) = \psi(0_k, 0_{-k}).$$

Therefore, expressing these paired states by column matrices $\begin{pmatrix}0\\1\end{pmatrix}$ and $\begin{pmatrix}1\\0\end{pmatrix}$, respectively, the operator $1 - n_k - n_{-k}$ can be expressed by a matrix

$$1 - n_k - n_{-k} = \begin{pmatrix}1 & 0\\ 0 & -1\end{pmatrix} = \sigma_{k,z},$$

which is similar to the z-component of Pauli's matrix $\boldsymbol{\sigma}$. Further noting that

$$a_k^\dagger a_{-k}^\dagger \psi(1_k, 1_{-k}) = 0 \quad \text{and} \quad a_k^\dagger a_{-k}^\dagger \psi(0_k, 0_{-k}) = \psi(1_k, 1_{-k}),$$

the operator $a_k^\dagger a_{-k}^\dagger$ can be assigned to the x- and y-components of Pauli's spin $\boldsymbol{\sigma}$. Considering that

$$\sigma_{k,x} = \begin{pmatrix} 0 & 1 \\ 1 & 0 \end{pmatrix} \quad \text{and} \quad \sigma_{k,y} = \begin{pmatrix} 0 & -i \\ i & 0 \end{pmatrix},$$

we can define $\sigma_k^+ = \sigma_{k,x} + i\sigma_{k,y} = \begin{pmatrix} 0 & 2 \\ 0 & 0 \end{pmatrix}$ and $\sigma_k^- = \sigma_{k,x} - i\sigma_{k,y} = \begin{pmatrix} 0 & 0 \\ 2 & 0 \end{pmatrix}$. Then, we obtain the relations

$$a_k^\dagger a_{-k}^\dagger = \frac{1}{2}\sigma_k^- \quad \text{and} \quad a_{-k}a_k = \frac{1}{2}\sigma_k^+,$$

representing density matrices of *two single electrons of Cooper pairs*, as defined by (13.11*a*) and (13.11*b*), showing *no theoretical conflict with the soliton theory*.

Using these results, the collective vector mode $\boldsymbol{\sigma}_k$ can be described by the Hamiltonian

$$\begin{aligned} \langle \mathcal{H}_{\mathrm{BCS}} \rangle_{\mathrm{thermal}} &= -\sum_k \varepsilon_k \sigma_{k,z} - \frac{V}{4}\sum_{k',k} \sigma_{k'}^- \sigma_k^+ \\ &= -\sum_k \varepsilon_k \sigma_{k,z} - \frac{V}{4}\sum_{k',k} (\sigma_{k',x}\sigma_{k,x} + \sigma_{k',y}\sigma_{k,y}). \end{aligned} \tag{14.23}$$

In this argument, $\sigma_{k,z}$ is an operator for creating or annihilating *paired states* $\pm\boldsymbol{k}$. However, writing the second term of (14.23) as a scalar product of vectors $\boldsymbol{\sigma}_k$ and a field $\boldsymbol{F}_k$ defined by a vector

$$\boldsymbol{F}_k = \left(-\frac{V}{2}\sum_{k'} \sigma_{k',x}, \; -\frac{V}{2}\sum_{k'} \sigma_{k',y}, \; \varepsilon_k \right),$$

we have the following.

Theorem:

$$\langle \mathcal{H}_{\mathrm{BCS}} \rangle_{\mathrm{thermal}} = -\sum_k \boldsymbol{\sigma}_k \cdot \boldsymbol{F}_k. \tag{14.24}$$

This expression represents the interaction energy of classical vectors $\boldsymbol{\sigma}_k$ in the Weiss field $\boldsymbol{F}_k$, corresponding to an internal Weiss field originating from other $\boldsymbol{\sigma}_{k'}$ in the system. If $\boldsymbol{\sigma}_k \| \boldsymbol{F}_k$, $\langle \mathcal{H}_{\mathrm{BCS}} \rangle_{\mathrm{thermal}}$ is minimum in equilibrium, characterizing these vectors in-phase.

In (14.24), the direction of such classical vectors $\boldsymbol{\sigma}_k$ can be expressed by an angle θ_k from the z-axis, representing the effective sinusoidal phase. Assuming that $\boldsymbol{\sigma}_k$ is in the x–z-plane for simplicity, we write the relation $\boldsymbol{\sigma}_k \| \boldsymbol{F}_k$ as

$$\frac{F_{k,x}}{F_{k,z}} = \frac{\sigma_{k,x}}{\sigma_{k,z}} = \frac{\frac{V}{2}\sum_{k'} \sigma_{k',x}}{\varepsilon_k} = \tan\theta_k. \tag{14.25}$$

In fact, V is a function of $\boldsymbol{k}'$ and $\boldsymbol{k}$, depending on the distance between $\boldsymbol{\sigma}_k$ and $\boldsymbol{\sigma}_{k'}$, which can, however, be considered as constant in the critical region determined by a small $|\boldsymbol{k}|$ and ε_k. In addition, $V(\boldsymbol{k}', \boldsymbol{k}; 2\boldsymbol{q}_c)$ in (14.21) has a well-defined singularity at $\theta_k = 0$, where we can write $\sigma_{k',x} = \sigma_{k'_0} \sin\theta_{k'}$. Noticing that the amplitude $\sigma_{k'0}$ depends only on $|\boldsymbol{k}'|$, we can write $V_k' = \sum_{k'} V(\boldsymbol{k}', \boldsymbol{k}; 2\boldsymbol{q})\sigma_{|k'|_0}$, and obtain

$$\tan\theta_k = \frac{V_k'}{2\varepsilon_k}\sum_{k'} \sin\theta_{k'}. \tag{14.26}$$

Such a phase angle θ_k as determined by (14.26) expresses the mesoscopic wave of a superconducting cluster. Nevertheless, we pay attention to the singular behavior of $\tan\theta_k$ at $\theta_k = 0$, which represents a boundary wall between $\sigma_{k0} = +1$ and -1, like a domain-wall in magnetic ordering. Therefore, by writing $\tan\theta_k = \frac{\Delta_k}{\varepsilon_k}$, the singularity at $\theta_k = 0$ can be attributed to the parameter Δ_k, which satisfies the relation

$$\Delta_k = \frac{V_k'}{2}\sum_{k'}\frac{\Delta_k}{\sqrt{\Delta_k^2 + \varepsilon_{k'}^2}},$$

or

$$1 = \frac{V_k'}{2}\sum_{k'}\frac{1}{\sqrt{\Delta_k^2 + \varepsilon_{k'}^2}}. \tag{14.27}$$

Here, replacing the summation $\sum_{k'}$...... by integration over the distributed ε_k in a region between $-\lambda_q$ and $+\lambda_q$, this relation can be calculated as

$$1 = \frac{V'\rho_F}{2}\int_{-\lambda_{q_c}}^{+\lambda_{q_c}}\frac{d\lambda}{\sqrt{\Delta^2 + \lambda^2}} = V'\rho_F \sinh^{-1}\frac{\lambda_{q_c}}{\Delta},$$

omitting indexes k and k' for simplicity, where ρ_F is the density of states at the Fermi level $\varepsilon = \varepsilon_F$. For $V'\rho_F \ll 1$, we can derive the expression

$$\Delta = \frac{\lambda_{q_c}}{\sinh\frac{1}{V'\rho_F}} \cong 2\lambda_{q_c}\exp\left(-\frac{1}{V'\rho_F}\right), \tag{14.28}$$

showing that the energy gap Δ is positive, if $V' > 0$. In this case, the Weiss field $\boldsymbol{F}$ is by no means static, while interpretable as a field for inverting the spin $\boldsymbol{\sigma}_k$ between $\sigma_{kz} = \pm 1$, for which the work required is $2|\boldsymbol{F}| = 2\sqrt{\varepsilon_k^2 + \Delta_k^2}$. The minimum work is 2Δ at the Fermi level that is obtained in the limit of $\varepsilon_k \to 0$, representing the energy gap in a superconductor at $T \leqslant T_c$.

In the BCS theory that will be discussed in chapter 15, the Hamiltonian $\mathcal{H}_{BCS}$ is assumed for generating Cooper pairs at the Fermi level, so that the ground state of a superconductor can be specified by (14.24) plus the energy for creating electron pairs. Assuming $\boldsymbol{\sigma}_{ky} = 0$, the ground state can be characterized by

$$E_o = -\sum_k \varepsilon_k \sigma_{ko} \cos\theta_k - \frac{V}{4}\sum_{k',k}\sigma_{k'x}\sigma_{kx} + \sum_{\text{pair}} 2\varepsilon_k$$

$$= -\sum_k \varepsilon_k \sigma_{ko}\left(\cos\theta_k + \frac{1}{2}\sin\theta_k \tan\theta_k\right) + \sum_{\text{pair}} 2\varepsilon_k,$$

where $\sum_{\text{pair}} 2\varepsilon_k$ is added for the pair to raise the Fermi energy E_o and the second term can be simplified by (14.27) as

$$\sum_k \varepsilon_k \sin\theta_k \tan\theta_k = \sum_k \frac{\Delta^2}{\sqrt{\varepsilon_k^2 + \Delta^2}} = \frac{2\Delta^2}{V'}.$$

Replacing the sum $\sum_k$...... by integration, we obtain

$$E_o - \sum_{\text{pair}} 2\varepsilon_k = 2\rho_F \int_0^{\lambda_{q_c}} \left(\lambda - \frac{\lambda^2}{\sqrt{\lambda^2 + \Delta^2}}\right) d\lambda.$$

Assuming $\Delta \ll \lambda_{q_c} \approx \hbar\omega_D$, we can evaluate E_o as

$$E_o - \sum_{\text{pair}} 2\varepsilon_k = \rho_F (\lambda_{q_c})^2 \left(1 - \sqrt{1 + \left(\frac{\Delta}{\lambda_{q_c}}\right)^2}\right) \approx -\frac{1}{2}\rho_F \Delta^2$$

In thermodynamics, we consider for the energy gap Δ to be a function of temperature. Near the critical temperature T_c, the interaction potential $V_k' = \sum_{k'} V(\boldsymbol{k}', \boldsymbol{k};\ 2\boldsymbol{q}_c)\sigma_{|k'|o}$ can be calculated as a statistical average $V\sum_{k'}\langle\sigma_{k'z}\rangle$, where

$$\langle\sigma_{k'z}\rangle = \frac{\exp\frac{(+1)F_{k'z}}{k_B T} - \exp\frac{(-1)F_{k'z}}{k_B T}}{\exp\frac{(+1)F_{k'z}}{k_B T} + \exp\frac{(-1)F_{k'z}}{k_B T}} = \tanh\frac{F_{k'z}}{k_B T}.$$

Hence, the relation (14.26) can be written as

$$\tan\theta_k = \frac{V}{2\varepsilon_k}\sum_{k'} \tanh\frac{F_{k'z}}{k_B T}\sin\theta_{k'} = \frac{\Delta}{\varepsilon_k},$$

where $F_{k'z}$ is in energy units, so we replaced it by $\sqrt{\varepsilon_{k'}^2 + \Delta_{k'}^2}$. If the transition threshold is characterized by $\Delta = 0$, this equation can be utilized to specify the critical temperature T_c as

$$1 = V\sum_{k'}\frac{1}{2\varepsilon_{k'}}\tanh\frac{\varepsilon_{k'}}{k_BT}.$$

Replacing $\sum_{k'}\ldots$ by integration again, we obtain

$$\frac{2}{V\rho_F} = \int_{-\lambda_{q_c}}^{+\lambda_{q_c}}\frac{d\lambda}{\lambda}\tanh\frac{\lambda}{2k_BT} = 2\int_0^{\frac{\lambda_{q_c}}{2k_BT_c}}\frac{\tanh\xi}{\xi}d\xi, \tag{14.29}$$

which is the BCS formula for T_c. By graphical integration, these authors showed that T_c is given as

$$k_BT_c = 1,\ 14\lambda_{q_c}\exp\left(-\frac{1}{V\rho_F}\right), \tag{14.30}$$

where λ_{q_c} was originally estimated by Debye's relation $2\lambda_{q_c} \approx \hbar\omega_D \approx k_B\Theta_D$,[note2] which, however, is considered as a reasonable assumption.

Figure 14.4 shows the theoretical BCS curve of $\frac{E_g}{k_BT_C}$ versus $\frac{T}{T_C}$, compared with experimental results on some metallic superconductors [8]. Combined with (14.28),

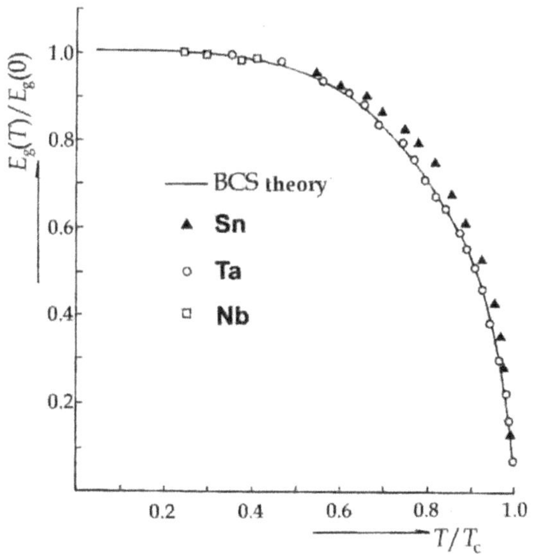

Figure 14.4. Energy gap E_g as a function of T/T_c. The BCS formula (14.30) compared with experimental results. Reproduced with permission from [1].

[note2] Phonons are quantized field of lattice vibrations, whereas solitons are quantized lattice displacements. It is significant to distinguish these excitations as different modes, although signified by the same order of magnitudes at a given temperature in crystals, because they are both boson particles. See chapter 13.

the energy gap is related to T_c as $2E_g = 2\Delta = 3.5\ T_c$, while the experimental values of $2E_g/T_c$ obtained from Sn, Al, Pb and Cd were 3.5, 3.4, 4.1, 3.3, respectively, all of which fit with the BCS theory approximately. The results show that $2\Delta/T_c$ is almost independent of nuclear mass M, but quite reasonable for the BCS theory to be amended with the soliton concept. However, both Δ and T_c depend on M in principle, as verified by small discrepancies in the figure, whose numerical analysis showed some isotopic effect existing as the relation $M^{0.5}T_c$ = const., confirming the BCS results (14.30), related to the Debye frequency ω_D that is proportional to $M^{-0.5}$.

Exercise

1. Fröhlich's condensate for electron attachment is hypothetical, but acceptable as a model of electron–lattice interaction at extreme low temperatures. Discuss it physically to see if it is an acceptable model of initial condensate.
2. Discuss the physical origin of the Cooper pair. Why does the Fröhlich's condensate need to be a pair?
3. Compare the Cooper pair with the spin–spin coupling, and discuss how they are different. Why can we not apply the Pauli principle to the Cooper pair?
4. Review the reason why the Cooper pair is formed in the momentum space, and generating a persistent current. Why do we consider solitons in favor of harmonic phonons?
5. Discuss the reason for the superconducting transition to be analogous to forming mesoscopic binary domains. Also, discuss the basic difference from the conventional binary transition with respect to the soliton theory.

References

[1] Kittel C 1986 *Introduction to Solid State Physics* 6th edn (New York: Wiley) ch 12

[2] Fröhlich H 1950 *Phys. Rev.* **79** 845
Fröhlich H 1952 *Proc. R. Soc.* **A215** 291

[3] Haken H 1973 *Quantenfeldtheorie des Festkörpers* (Stuttgart: Teubner)

[4] Cooper L N 1956 *Phys. Rev.* **104** 1189
Cooper L N 1961 *Phys. Rev. Lett.* **6** 689

[5] Fujimoto M 2007 *Physics of Classical Electromagnetism* (New York: Springer)

[6] Bardeen J, Cooper L N and Schrieffer J R 1957 *Phys. Rev.* **108** 1175

[7] Anderson P W 1958 *Phys. Rev.* **112** 900

[8] Kittel C 1963 *Quantum Theory of Solids* (New York: Wiley) ch 8

IOP Publishing

Solitons in Crystalline Processes (2nd Edition)
Irreversible thermodynamics of structural phase transitions and superconductivity
Minoru Fujimoto

Chapter 15

High-T_c superconductors

Superconductivity was once considered as a phenomenon specific in metals at very low temperatures, but in 1986 it was also discovered in various layered oxides, other compounds, and more recently in a hydrogen-bonding system in the range of 30–135 K. In higher temperature ranges, these materials have since been extensively studied for practical applications, constituting a new subject of investigation on high-temperature superconductivity (HTS). Shortly after the first discovery of high-T_c superconductivity, Anderson [1] worked out a theoretical description of these materials, postulating the *valence-bond theory*, which is still developing today [2, 3]. At present, the origin of HTS is still unclear, while experimentally it is now believed to arise from *specifically modulated lattice structures*, supporting Fröhlich's mechanism of the charged particles–lattice interaction. Incorporated with the soliton concept for lattice modulation as discussed in chapter 14, however, Anderson's collective valence electrons can explain the Cooper pair, so that no significant revision of the responsible mechanism is necessary for the original BCS theory.

In addition, recent studies on metallic hydrogen sulfide suggest that *protonic charge carriers* occur similarly as in metals, so that the BCS theory can be applied to superconductors at all temperatures, with displaced charges as initially suggested by Fröhlich. Besides, the soliton theory signifies distorted lattice at phase transitions.

15.1 Superconducting transitions under isothermal conditions

15.1.1 Layer structure of YBaCuO superconductors: cuprates

It is of utmost importance for superconductivity to identify specific charge carriers within modulated crystalline structure. Instead of free electrons in metals, *unpaired electrons* of the iron-group elements in crystals can constitute the mechanism for superconducting phenomena. Copper oxides known as *cuprates* are in a specific layer structure of CuO_2 planes, where the superconductivity can take place in space between layers, originating from *unpaired electrons* of Cu^{2+} ions. The more layers of

doi:10.1088/978-0-7503-2572-1ch15

CuO_2, the higher critical temperature T_c; exhibiting highly anisotropic conductivity and superconductivity for $T > T_c$ and $T < T_c$, respectively.

Figure 15.1 illustrates a view of the unit-cell structure of yttrium barium copper oxide ($YBa_2Cu_3O_{7-x}$), which is a typical high-temperature superconductor, where the proportion of the three different metals Y:Ba:Cu in the structure is in molar ratio 1:2:3, respectively, referred to as the 123 superconductors [2].

Shown in the figure, unit cells of perovskite type with Ba at the center are stacked up with two CuO_2 planes as layer structure with an Y ion in between, which is perpendicular to the *c*-axis.

Theorem: These interfacing double-layers look like domain-boundary planes in conventional ordering systems, as discussed in chapter 12. Originating from unpaired 3d-electrons in Cu^{2+} ions, there should be a *conduction band* between modulated planes, which can be analyzed as *deformed interfacing layers* by soliton dynamics with the crystal field theory [3].

15.1.2 The Cooper pair in high-T_c superconductors of cuprate layers

In the cuprate layer structure illustrated in figure 15.2, the unpaired electron of Cu^{2+} is in atomic orbital $3d(x^2 - y^2)$ for $T > T_c$, and there clearly exists an internal potential $V(z)$ to modulate the lattice, which may be considered as a *pseudopotential* in the periodic structure. Combining with another unpaired electron across the layers, a symmetric electron-pair can be formed with the energy $-eV(z)$, thereby propagating together in collective motion for $T < T_c$ along with the layer structure.

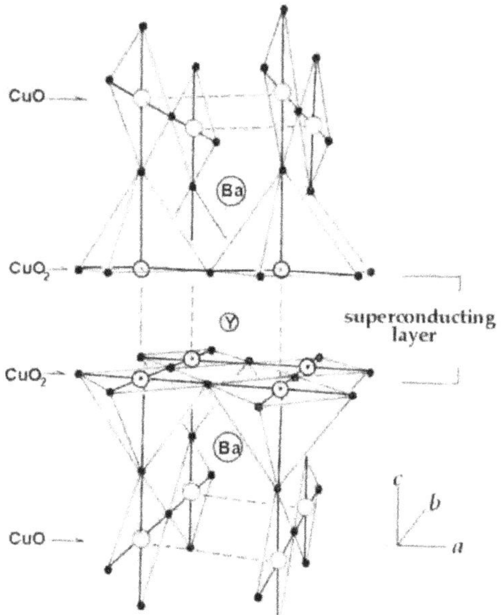

Figure 15.1. The cuprate layer structure with Y ions in-between of $YBa_2Cu_3O_{7-x}$ lattice. Reproduced with permission from [2].

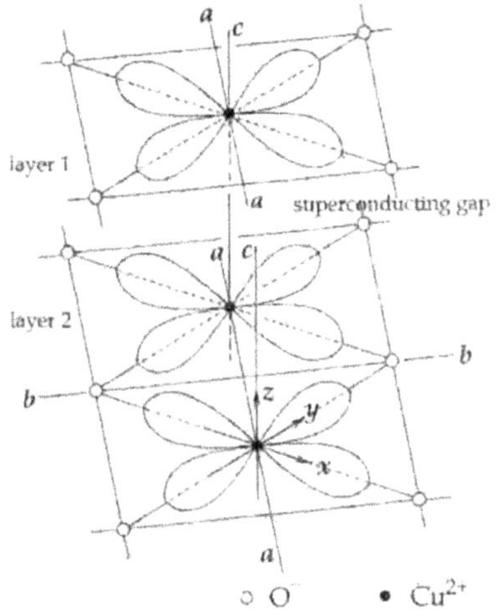

Figure 15.2. Configuration of spin-unpaired orbitals of Cu^{2+} in cuprate layers 1 and 2, showing the superconducting gap in between. Crystallographic axes are indicated by a, b and c. Symmetry of a $3d(x^2 - y^2)$ orbital in Cu^{2+} ion is shown as x-, y- and z-directions.

Such collective motion should therefore be susceptible to a structural change by $-eV(z)$, constituting the electron–lattice interaction energy, similar to Fröhlich's mechanism in metals at very low temperatures. Accordingly, the superconducting transition can be signified by the corresponding displacements $\Delta z_\pm$ arising from derivatives of the pseudopotential $-eV(z)$ that represents the correlation energy for these units in a crystal. Considering an orthorhombic-to-tetragonal symmetry change with respect to the c-axis, $\Delta z_\pm$ is perpendicular to the a–b-plane, allowing us to consider the classical *uniaxial displacements* $e\Delta z_\pm = (e\Delta z_0)f(\pm qz - \tilde{\omega}t)$ on both sides of the potential $-eV(z)$, exhibiting *rotation* of the orbital 3d $(x^2 - y^2)$ around the z-axis. The uniaxial displacement $\Delta z_\pm$ can then be associated with rotating angles $\varphi_\pm = \pm qz - \tilde{\omega}t$ of the 3d orbital $x^2 - y^2$, which are considered as the phase function of potential $V(z, t)$ that is in-phase with $\varphi_\pm(z, t)$, namely

$$V(z, t) = V_0(x, y)\exp i\varphi_\pm = \pm V_0(x, y)\exp i(qz \mp \tilde{\omega}t)$$

that is for the layers to be modulated in space–time at q and $\tilde{\omega}$. Comparing with normal metals, the Fröhlich attachment in metals at very low temperatures is exactly the same mechanism for 3d-electrons in cuprates, with respect to types of lattice modulation.

Using the soliton theory, the uniaxial collective variable $\varphi_\pm(z, t)$ propagating in the x–y-plane between the layers can be specified by a complex potential $u^2(z) \pm i du(z)/dz$, which is invariant of translation $z \to z \mp \tilde{\omega}t/q$. Superconducting transition occurs with a two-boson combination of inversion $+q_{1,2} \rightleftarrows -q_{1,2}$ at T_c, which is energetically favorable for a Cooper pair to be stabilized by *symmetric fluctuation* $(\Delta z_{c1} + \Delta z_{c2})/\sqrt{2}$ due to boson characters of displacements Δz_{c1} and Δz_{c2}, which is the same as in metals at very low temperatures.

Such a symmetric displacement as $\Delta z_A = (\Delta z_{c1} + \Delta z_{c2})/\sqrt{2}$ signified by boson operators $\tilde{b}_q$ and $\tilde{b}_q^\dagger$ in crystal, the Fröhlich's lattice–electron interaction can be described in the field-theoretical application as

$$\mathcal{H}_{\text{int}} = -i\sum_q eD_q\left(\rho_q\tilde{b}_q^\dagger - \rho_q^\dagger\tilde{b}_q\right), \tag{15.1}$$

where $\rho_q = \sum_k a_{k+q}^\dagger a_k$ is the density matrix of electrons, and $eD_q \propto -e\nabla V(z)\cdot(\Delta z)_o$, representing strength of the lattice interaction in uniaxial displacements. Corresponding to (15.1), the elastic lattice responds with the shear strain inversion, as discussed in chapter 12.

On the other hand, the 3d electrons in the gap space are subjected to a periodic lattice potential in the x–y-plane, which can be represented in the crystal field between layers as free particles of *effective mass* m^*, whose single-particle Hamiltonian is expressed as $\mathcal{H}_o = \frac{\hbar^2k^2}{2m^*}$, corresponding to the Fermi level of a metallic conductor. It is notable that the above describes the Fröhlich mechanism exactly in the same manner as in superconducting metals at very low temperatures.

Repeating the same argument for metals to be modified by (15.1), we can derive the Cooper pair of two charges and persistent current in cuprate layers in a thermodynamic environment. Limited to a specific wave vector $\boldsymbol{q}$ of modulation, we write the perturbation as $\mathcal{H}_{\text{int}} = eD_q\mathcal{H}'$, and calculate perturbed Hamiltonian $\tilde{\mathcal{H}} = \mathcal{H}_o + eD_q\mathcal{H}'$ for two polarized condensates on each side of the domain boundary, and $\mathcal{H}_o = \sum_k \varepsilon_k a_k^\dagger a_k$, where $\varepsilon_k = \hbar^2\boldsymbol{k}^2/2m^*$, is the unperturbed Hamiltonian for two 3d electrons. Considering a canonical transformation $\tilde{\mathcal{H}} = S^{-1}\mathcal{H}S$ from Fröhlich's condensates to the system of Copper pairs, matrix elements of the action S can be defined as

$$\langle\Delta z_A|S|\Delta z_P\rangle = \begin{pmatrix} 0 & -ieD_q\sum_k \dfrac{a_{k-q}^\dagger a_k}{\varepsilon_k - \varepsilon_{k-q} - \lambda_q} \\ ieD_q\sum_{k'} \dfrac{a_{k'+q}^\dagger a_{k'}}{\varepsilon_{k'} - \varepsilon_{k'} + \lambda_q} & 0 \end{pmatrix}$$

for symmetric and antisymmetric displacements, $\Delta z_A = (\Delta z_+ + \Delta z_-)/\sqrt{2}$ and $\Delta z_P = (\Delta z_+ - \Delta z_-)/\sqrt{2}$, respectively. Therefore, the Hamiltonian $\tilde{\mathcal{H}}$ of the electron-pair can be written as

$$\tilde{\mathcal{H}} = \mathcal{H}_o + \frac{e^2D_{q_{1,2}}^2}{2}\sum_{k',k} a_{k'-q_1}^\dagger a_{k'} a_{k+q_2}^\dagger a_k\left(\frac{1}{\varepsilon_{k'} - \varepsilon_{k'-q_1} - \lambda_{q_1}} - \frac{1}{\varepsilon_k - \varepsilon_{k+q_2} + \lambda_{q_2}}\right).$$

Here, electron scatterings $\boldsymbol{k}' \to \boldsymbol{k}' + \boldsymbol{q}_1$ and $\boldsymbol{k} \to \boldsymbol{k} - \boldsymbol{q}_2$ are considered to be *emission* and *absorption* for elemental soliton states $(\lambda_{q_1}, \boldsymbol{q}_1)$ and $(\lambda_{q_2}, \boldsymbol{q}_2)$, respectively. Corresponding to such scatterings as $\Delta\boldsymbol{k}' = +\boldsymbol{q}_1$ and $\Delta\boldsymbol{k} = -\boldsymbol{q}_2$, as illustrated in figure 14.2(a), two electrons of Fröhlich's condensates are bound for antisymmetric scatterings $\Delta\boldsymbol{k} = -\Delta\boldsymbol{k}'$ to form a single quasi-particle of $-2e$, if $\boldsymbol{q}_{c1}$ and $\boldsymbol{q}_{c2}$ are in parallel in equilibrium at T_c. In this case, the thermal average

$$\langle\tilde{\mathcal{H}} - \mathcal{H}_o\rangle_{T_c} = \sum_{\boldsymbol{q}_{c1}}^{\boldsymbol{q}_{c2}} e^2 D_{\boldsymbol{q}_c}^2 \sum_{\pm\boldsymbol{k}_c} \frac{\lambda_{\boldsymbol{q}_c}}{(\varepsilon_{\boldsymbol{k}_c\pm\boldsymbol{q}_c} - \varepsilon_{\boldsymbol{k}_c})^2 - \lambda_{\boldsymbol{q}_c}^2} a_{\boldsymbol{k}_c-\boldsymbol{q}_c}^{\dagger} a_{\boldsymbol{k}_c} a_{-\boldsymbol{k}_c+\boldsymbol{q}_c}^{\dagger} a_{-\boldsymbol{k}_c},$$

which is re-expressed as

$$= -V(2\boldsymbol{q}_c) a_{\boldsymbol{k}_c-\boldsymbol{q}_c}^{\dagger} a_{\boldsymbol{k}_c} a_{-\boldsymbol{k}_c+\boldsymbol{q}_c}^{\dagger} a_{-\boldsymbol{k}_c},$$

where

$$-V(2\boldsymbol{q}_c) = e^2 D_{\boldsymbol{q}_c}^2 \sum_{\pm\boldsymbol{k}_c} \frac{\lambda_{\boldsymbol{q}_c}}{(\varepsilon_{\pm\boldsymbol{k}_c\mp\boldsymbol{q}_c} - \varepsilon_{\pm\boldsymbol{k}_c})^2 - \lambda_{\boldsymbol{q}_c}^2}$$

can be considered for a Cooper pair. Unlike in metals at very low temperatures, the double-layer structure in cuprates is equivalent to a mini-domain-wall for $\Delta n_{\text{s,ini}} = \pm 1$ of soliton number $n_{\text{s,ini}}$ in a binary transition, discussed in chapter 12.

Here, at T_c on each side of the singularity the vector $\boldsymbol{q}_{c1}$ on one layer should be parallel to $\boldsymbol{q}_{c2}$ on the other, so that $\boldsymbol{q}_{c1} + \boldsymbol{q}_{c2} = 2\boldsymbol{q}_c$ determines the negative energy $-V(2\boldsymbol{q}_c)$. The quantity $V(2\boldsymbol{q}_c)$ represents the binding energy of charge $-2e$ of a Cooper pair stabilized in *symmetrical* uniaxial displacements $(\Delta\boldsymbol{r}_{\text{A}})_{\boldsymbol{q}_c}$ in boson states, trapping the pair in the soliton potential $-2\text{sech}^2\,\phi_c$, which corresponds to cluster formation in binary transitions, which is consistent with Fröhlich's model for surface modulation at T_c.

Accordingly, the critical temperature T_c should be determined by the relation to $+V(2\boldsymbol{q}_c)$, i.e.

$$\text{k}_\text{B}T_c = V(2\boldsymbol{q}_c) \quad \text{or} \quad \text{k}_\text{B}T_c \propto e^2 n_{\text{s,ini}}^2 \lambda_{\boldsymbol{q}_c}^2, \tag{15.2}$$

where $n_{\text{s,ini}}$ is the soliton density at T_c, implying that $T_c \propto n_{\text{s,ini}}^2$, which seems to be supported by the published reports for T_c of HTS. Confirmed in (15.2), the proportionality $T_c \propto n_{\text{s,ini}}^2$ *is a valid postulate in general, as derived from the soliton theory.*

Noting further that the potential $V(2\boldsymbol{q}_c)$ is oriented along the c-axis, the perturbed $\tilde{\mathcal{H}} - \mathcal{H}_o$ indicates that it is in arbitrary orientation of $2\boldsymbol{q}_c \| \boldsymbol{c}$ perpendicular to the *tetragonal* $q_a q_b$-plane. Therefore, in the momentum space it is the time-dependent potential

$$V(2\boldsymbol{q}_c) = |V(\Delta\boldsymbol{k}_c;\ 2\boldsymbol{q}_c)| \exp 2i(\boldsymbol{q}_c \cdot \boldsymbol{r} - \tilde{\omega}_{\boldsymbol{q}_c} t),$$

leading to a vector potential $\boldsymbol{A}$ by the Lorentz condition in electromagnetic theory, so that we can write

$$\boldsymbol{A} \propto \boldsymbol{j}_\text{s} = (-2e)\, v_\text{s} \tag{15.3}$$

for the *persistent current* $\boldsymbol{J}_\text{s} = n_\text{s}\, \boldsymbol{j}_\text{s}$ carried by charges $-2en_\text{s}$, leading to the London equation.

As in a metal superconductor, the vector j_s represents a circular persistent current in the x–y-plane perpendicular to the displacement $\Delta z_\pm$, whose propagation can be terminated at the superconducting surface, as will be explained in section 16.2, whereas inside the cuprate structure the vector $\boldsymbol{M} \propto \mathrm{curl}\boldsymbol{j}_s$ represents Meissner's diamagnetism. In practice, the superconducting current $\boldsymbol{j}_s$ detected in HTS is persistently circular in the a–b-plane, as explained in the popular demonstration [2] of a floating small magnet over the surface at $T < T_c$.

Theoretically, high-T_c superconductivity in cuprate structure can thus arise from the same mechanism in metals at very low temperatures, while the conductivity and Meissner results for high T_c superconductors have been substantiated [3, 5]. Different T_c arising from nonlinear interactions of the structure is still not clearly understood in general for HTS at the present time particularly in another group of HTS known as *iron-based superconductors* [2]. However, in both cases of conventional and high-T_c superconductors, soft modes characterized by the wave vector $2\boldsymbol{q}_c$ in parallel lattice are considered responsible for the interaction with lattice.

It is particularly interesting that Tarnawski [4] published experimental work on HTS $RBa_2Cu_3O_{7-x}$ where R represents rare-earth elements, interpreting effects of applied magnetic field to specific heat measurements with respect to lattice distortion. While known to be a logical interpretation, it was the first attempt to use that principle on superconducting transitions, verified by Eremets and his group [5] on the H3S phase of *hydrogen sulfide*, as will be discussed in the next section.

15.1.3 Layer structure in YBaCuO and other cuprate superconductors

These cuprate superconductors are listed in table 15.1 showing critical temperatures observed with various numbers of stacked CuO layers either in orthogonal or tetragonal symmetry. As indicated in the table, the layer structure can be considered as analogous to domain walls in normal lattice consisting of soliton layers.

Table 15.1. Critical temperatures of cuprate layer structures[a].

Formula (Notation)	T_c (K)	Number of cuprate planes in unit cell
$YBa_2Cu_3O_7$ (123)	92	2
$Bi_2Sr_2CuO_6$ (Bi2201)	20	1
$Bi_2Sr_2CaCu_2O_8$ (Bi–2212)	85	2
$Bi_2Sr_2Ca_2Cu_3O_6$ (Bi–2223)	110	3
$Tl_2Ba_2CuO_6$(Tl–2201)	80	1
$Tl_2Ba_2CaCu_2O_8$(Tl–2212)	108	2
$Tl_2Ba_2Ca_2Cu_3O_{10}$(Tl–2223)	125	3
$HgBa_2CuO_4$(Hg–1201)	94	1
$HgBa_2CaCu_2O_6$(Hg–1212)	128	2
$HgBa_2Ca_2Cu_3O_8$(Hg–1223)	134	3

[a] Data from http:////en.wikipwdia.org//wiki//High-temperature_superconductivity.

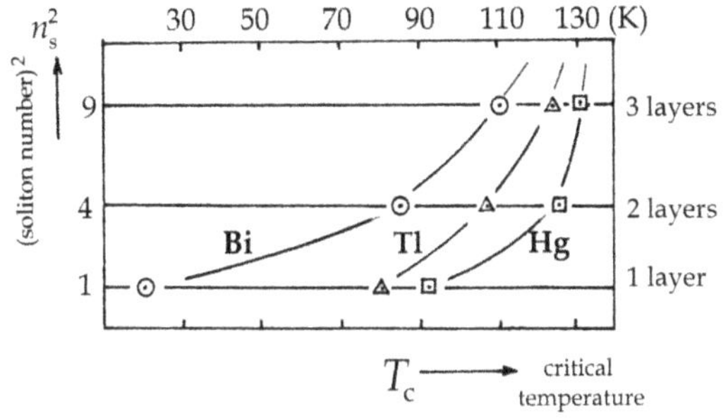

Figure 15.3. Veryfying the soliton formula $T_c \propto n_s^2$ with experimental results in table 15.1.

Figure 15.3 shows plots of T_c against the number of cuprate layers in unit cell, exhibiting parabolic lines for varying solution potential energies, like domain wall structure in figure 12.9 of chapter 12, signified by the distribute soliton potential energies. It offers clear evidence for solitons to be involved in the superconducting mechanism of Cooper's pairs, representing the elastic lattice. It is interesting that the result is found to support the phase transition formula $T_c \propto n_{s,\text{ini}}^2$ exhibiting parabolic curves predicted for volume changes under isothermal conditions. Figure 15.3 should be due to distributed lattice singularities in cuprate layer-structure, which is similar to figure 12.4 for graphite surface under pressure studied by Fain and Chinn [6]. However, there is no immediate relation found in-between, while both are supporting evidence for the soliton dynamics to be interpreted similar to figure 16.4 in chapter 16.

15.1.4 Layer structure of high T_c superconductors

Iron-based superconductors contain layers of iron and pnictogen, such as arsenic, phosphorous or chalcogen. This group exhibits superconductors at higher temperatures T_c in wide range, in addition to the cuprate system [2].

Although speculated, it is likely due to $3d(x^2 - y^2)$ unpaired electrons of ionic irons in layer structure exhibiting a conduction band, like in cuprate superconductors. Observed superconductivity from any layer structure urges one to identify the detail. According to the soliton theory, the critical temperature should be *proportional to the squared soliton number*, which can be studied experimentally under applied high pressure *p*or magnetic field H, but no such data are available as yet in the present literature.

15.2 Protonic superconducting transitions under high pressure conditions

A phase named H3S of condensed hydrogen sulfide is the only high-T_c superconductor reported in the literature, which is a unique protonic conductor discovered to show superconducting transition at around −70 °C. It is unique in terms of proton conductors, but classified as a high-T_c superconductor at the present time of our knowledge.

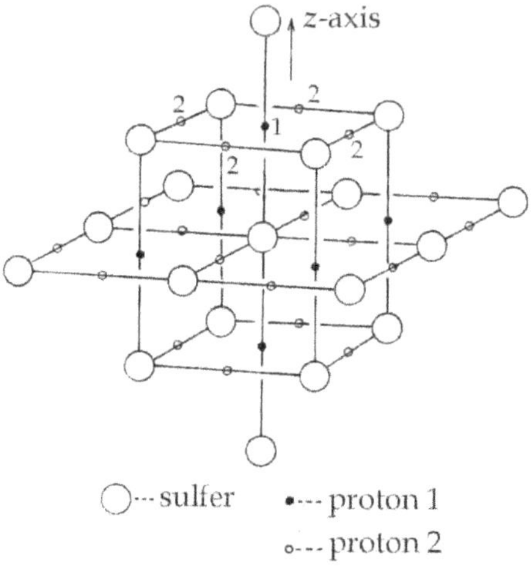

Figure 15.4. Lattice of the H3S phase in metallic hydrogen sulfide. At high pressure around 150 GPa along the z-axis, hydrogen-bonding protons 1 are displaced to be mobile by soliton potentials on the axis. Reproduced with permission from [9].

15.2.1 Metallic hydrogen sulfide

The metallic phase of condensed hydrogen sulfide, characterized by the H3S phase (H2S plus H), exhibits superconductivity at around −70 °C under applied high-pressure conditions at 150 GPa, providing a rare example of pressure-dependent zero-resistance phenomena [7].

The crystal structure of the H3S phase has recently been confirmed by Shimizu [8], as illustrated in figure 15.4, where the unit cell on top of the figure can be considered as an active element in deformed structure along the z-axis.

We consider the fluctuating proton 1 along the z-axis in the figure is a charged particle responsible for uniaxial modulation that is similar to Fröhlich's electron–lattice interaction expressed by a *pseudopotential induced by external pressure p on surfaces.* In the absence of conduction electrons in H3S, *moving protons* should be charge carriers in the compressed phase of hydrogen-bonding crystals. Figure 15.5(a) shows *protonic zero-resistance curves* for superconducting transitions in H3S reported in [4], constituting the first example of *phase transitions related to proton–lattice interaction.*

Unlike in ordinary superconductors, the protonic current of hydrogen-bonding is expected for the superconducting current in hydrogen sulfide, which was confirmed by measurements on deuterated D3S crystals [4], as illustrated in figure 15.5(b). In practice, this can be substantiated by observing Meissner's uniaxial diamagnetization $\boldsymbol{M}_{2\mathrm{p}}$ arising from two protons with the relation

$$\boldsymbol{M}_{2\mathrm{p}} \approx (0.5 \times 10^{-3})\boldsymbol{M}_{2\mathrm{e}},$$

where $\boldsymbol{M}_{2\mathrm{e}}$ is hypothetically a magnetization of n_s electron-pairs in other equivalent superconductors. While the magnitude $|\boldsymbol{M}_{2\mathrm{p}}|$ cannot be evaluated with unknown value of n_s, the corresponding persistent protonic current $\boldsymbol{j}_s$ should be observed in

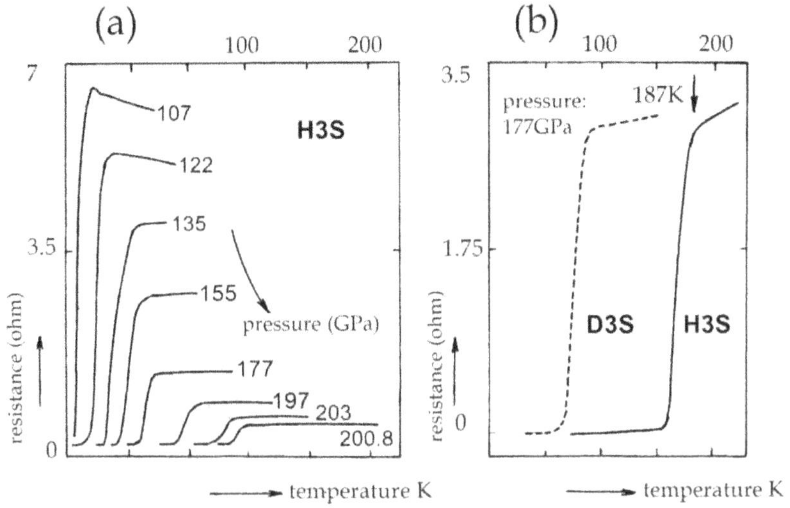

Figure 15.5. (a) Zero-resistance curves in H3S observed at various external pressures and temperatures. (b) Comparison of zero-resistance phenomena between H3S and D3S at 177 GPa and 185 K. Data from [4].

the tetragonal *x–y*-plane. On the other hand, the protonic current is evident from zero-resistance experiments with deuterated D3S crystals in exactly the same way as in H3S superconductors [4]. Accordingly, the *modulated structure* is responsible for supercurrents of protons or deuterons in the H3S or D3S phase, respectively, of condensed hydrogenated or deuterated sulfide.

The Cooper pair of protons can therefore be proposed for superconductivity in hydrogen sulfide, allowing for the BCS theory to describe the mechanism of superconductivity. Signified by the zero-resistance data under high pressure conditions [4, 7], the *protonic persistent current* $\boldsymbol{j}_s$ or Meissner's *protonic diamagnetism* should be observed in the hydrogen-bonding structure of hydrogen sulfide H3S.

With the permission of Dr Eremets and associates, we attempted to analyze their data [4] regarding the soliton theory, which is described as sketched in the following.

The normal protonic current $\boldsymbol{j}_n$ caused by an applied field $\boldsymbol{E}_{ext}$ for $T > T_c(p)$ has been confirmed experimentally to obey the Ohm law as in $\boldsymbol{j}_n \propto \boldsymbol{E}_{ext}$, whereas a pressure-dependent change of the electrical resistivity to virtual zero was observed at $T_c(p)$. In contrast, the non-zero current $\boldsymbol{J}_n \neq 0$ and zero current $\boldsymbol{J}_n = 0$ for $T < T_c$ and $T > T_c(p)$, respectively, characterize the protonic current clearly in figure 15.5(a). In fact, regarding these figures, observed $T_c(p)$ can be plotted against p to show a parabolic relation as shown in figure 15.6(a), demonstrating that the pressure-dependence of $p_0 \approx T_c(p)$ is equivalent to $T_c(p_0)$ *determined as proportional to the parabolic soliton density* $n_s(p_0)$, as indicated by equation (15.2). Observed pressure-dependent temperatures $T_c(p)$ in *soft-mode behavior are also undeniable evidence for an internal Weiss field* to participate in superconducting phenomena in H3S phase, exhibiting, however, the *characteristic continuity of protonic normal-super currents.*

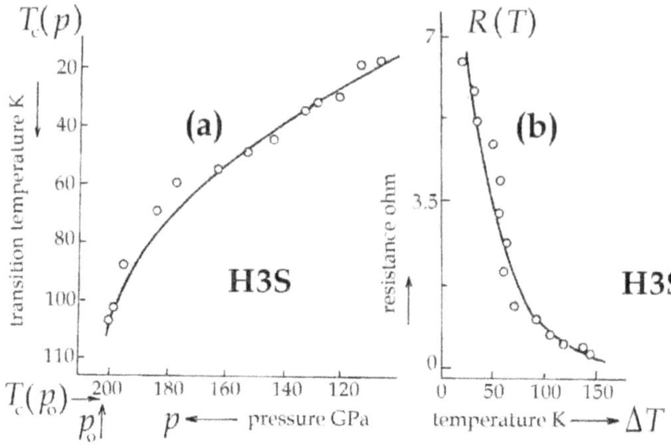

Figure 15.6. $T_c(p)$ plotted as a function of applied pressure p for H3S, representing peaks of resistance near critical pressure p_0. (a) The curve shows that $T_c(p) \propto n_S(p)$, the soliton density, which is proportional to $\sqrt{p_0 - p}$ in the mean-field approximation. (b) The curve for the resistance $R(p)$ is signified by $n_c(p) \propto (p - p_0)^{-1}$ for in isothermal processes for $p_0 > p$. (c) The resistance $R(T)$ versus $\Delta T_c(p)$. Data from [9].

15.2.2 Order variables in hydrogen sulfide

Superconducting currents occur in the H3S phase under high hydrostatic pressure, so that the *lattice stress* by external pressure p should be considered for displacing lattice points in the z-direction for Cooper pairs of protonic charges +2e to create *modulated Fröhlich's mechanism.* This mechanism should be similar to electron attachments in order for straining metallic structure to be described by the *modulated structure* originated from nonlinear displacements.

Defining the order parameter with *Anderson's pseudospin vector* $\boldsymbol{\sigma}_k$ for the protonic Cooper pair, as discussed in chapter 14, the BCS Hamiltonian of proton pairs can be expressed as

$$\langle \mathcal{H}_{\text{BCS}} \rangle_{\text{meanf}} = -\sum_k \varepsilon_k \sigma_{kz} - \frac{V}{4} \sum_k (\sigma_{kx}\sigma_{ky} + \sigma_{ky}\sigma_{kx})$$

in the momentum space of proton pairs in energy states ε_k, where the *nonlinear pseudospin* $\boldsymbol{\sigma}_k$ is regarded as proportional to the soliton density n_k, like free particles in the superconducting state. Therefore, we can consider that n_k plays the role of order parameter.

Expressing the second term on the right by the *mean-field average* $F_{kz} = \frac{V}{4}\langle \sigma_{kx}\sigma_{ky} + \sigma_{ky}\sigma_{kx} \rangle$, the mean-field average of $\mathcal{H}_{\text{BCS}}$ can be expressed as

$$\langle \mathcal{H}_{\text{BCS}} \rangle_{\text{meanf}} = -\langle \varepsilon_{\boldsymbol{k}} \rangle \langle \sigma_{kz} \rangle - \langle \sigma_{kz} \rangle F_{kz}.$$

The thermal average $\langle \mathcal{H}_{\text{BCS}} \rangle_{\text{thermal}}$ can then be written with respect to the average $\langle \sigma_{kz} \rangle_{\text{thermal}} = \sigma_z$, i.e.

$$\langle \mathcal{H}_{\text{BCS}} \rangle_{\text{thermal}} = -\langle \varepsilon_{\boldsymbol{k}} \rangle \sigma_z - F_z \sigma_z,$$

where $F_z = \langle F_{kz}\rangle_{\text{thermal}}$ is the Weiss field in the mean-field approximation, determining T_c approximately, giving no information on the deformed lattice.

Under an applied high pressure p on the crystal surface, the energy of proton pairs cannot be a constant at a given temperature, because boundary conditions are changed by applied pressure. In a compressed flat-plate specimen, the eigenvalue $\langle \varepsilon_k\rangle$ of proton pairs is not so widely spread so that their variation along the z-axis can be expressed as

$$\Delta\langle \varepsilon_k\rangle = \frac{\partial\langle \varepsilon_k\rangle}{\partial z}\Delta z = -pA_\perp \Delta z.$$

Thermodynamically, $\Delta\langle \varepsilon_k\rangle$ is compensated by applied work $-pA_\perp\,\Delta z$ on the crystal surface, where $A_\perp$ is the *effective area*, considering the Theorem for a macroscopically homogeneous superconductor in figure 12.6.

This formula allows for the compressed sample crystal to be analyzed with displacements Δz under high pressure p at a given $T_c(p)$ with respect to Δp. Assuming for the lattice strain energy to be distributed uniformly in crystals at $T_c(p)$, the displacement Δz can be converted to the pressure difference Δp, as expressed by $-pA_\perp\,\Delta z = -zA_\perp\,\Delta p$ at a given temperature, assuming the gaseous soliton is described by the ideal gas law $pV = n_s k_B T$ where $V = A_\perp z$.

Writing the change in the above Gibbs function density against the pressure change Δp, the relation

$$\Delta G(\sigma_z;\ \Delta p) = -\langle \varepsilon_k\rangle\sigma_z - F_z\sigma_z + \sigma_z(A_\perp z)\,\Delta p$$

is obtained, indicating that $\langle \varepsilon_k\rangle$ varies with Δp, explaining pressure-dependent transition temperature $T_c(p)$. The function $\Delta G(\sigma_z;\ \Delta p)$ suggests *pressure-dependent soft-mode analyses* of σ_z. For the sample crystal to be in ellipsoidal shape, we can determine the nonlinearity development with respect to a time-dependent pressure-change Δp in the H3S phase of hydrogen sulfide.

Here the last term $\sigma_z(A_\perp z)\Delta p$ on the right is equivalent approximately to the second-order term $\frac{1}{2}n_s{}^2$ in Landau's expansion of the Gibbs function, where the soliton density can be considered as the order parameter of a parabolic feature $n_s(p) \propto \sqrt{|\Delta p|}$, as shown in figure 15.5(a). For H3S, the relation $T_c(p) \propto n_s(p)$ is a valid expression given by (15.1), indicating that the adiabatic transition temperature $T_c(p)$ can be expressed by a parabolic pressure-dependence $\sqrt{p_0 - p}$ of the order parameter in the noncritical region, where p_0 is a constant and $p < p_0$ with respect to initial protonic conduction.

In fact, the external magnetic field H could have been equivalent to the external pressure p as an adiabatic stress that penetrates onto the H3S phase through the experimental chamber, however, such an experiment was not particularly performed in their work, as a technical matter of convenience.

It is noted that curves in figure 15.4(a) and (b) were obtained from resistance measurements for the *normal proton current* $\boldsymbol{J}_n = n_s\,\boldsymbol{j}_n$, for which *the resistance* $R(T)$ *can depend on* $n_s(T)$, which is signified by Curie–Weiss behavior. In figure 15.5(b), the resistance curve of $\boldsymbol{J}_n$ shows an averaged soliton density

$$\langle n_s(p)\rangle \propto (p_0 - p)^{-1} \qquad \text{for} \qquad p < p_0,$$

where $p_o = T_c(p)$, and $\langle n_n(p_o)\rangle = 0$ hence $\langle \boldsymbol{J}_n(p_o)\rangle = 0$. In this figure, Currie–Weiss behavior of the conduction current for $p > p_o$ is evident near the transition point p_o, as shown in figure 15.5(c), co-existing with some supercurrent that is undetectable by resistance measurements during the transition process broadened by $n_s(T)$.

In contrast, Meissner's diamagnetism for $\boldsymbol{M} = \mathrm{curl}\, \boldsymbol{J}_s$ should take over the superconducting state in superconducting phase H3S for $T_c(p) \leqslant T_c(p_o)$, while unconfirmed by their high-pressure studies. However, it is appreciable that the *normal proton current* for $T_c(p) > T_c(p_o)$ was confirmed in hydrogen-bonding systems for the first time.

Moreover, despite unconfirmed Meissner's effects in [4, 7], equation (15.2), figures 15.3 and 15.5(a) provide sufficient evidence for the soliton theory to be valid for displacive transitions in general, as substantiated by applied pressure in H3S crystals.

In the examples discussed in the sections above, the superconductivity in relation to *mobile Cooper pairs of protons* cannot be studied by resistivity measurements. Nevertheless, the lattice in H3S is *modulated at a different temperature or pressure as in other types of superconductors*, while the supercurrent phenomenon related to polarized uniaxial proton displacements should be further investigated from electromagnetic properties of Cooper pairs for $T \leqslant T_c$.

Summarizing chapters 14 and 15, it is notably significant that the superconducting phase can arise from not only free electrons but also free protons in conduction band in deformed crystals, as predicted by Fröhlich. Experimentally, that should be attributed to Meissner's effect, leading to Cooper pairs that are analogous to clusters for structural transitions. That should be a *common mechanism* in binary distorted lattices, and the critical point can be determined as related in general with the *specific critical soliton density*.

Exercises

1. Fröhlich's interaction of electron (or proton) and lattice is regarded for initiating superconductivity, for which a gradient of the lattice potential is essential. Review this mechanism for generating solitons in general. Discuss if such process would lead to terminated superconductivity in modulated crystals, and determine how it could be done.
2. Superconductivity never occurs for $T > T_c$. Explain why in terms of charge carriers against Cooper pairs. Could it be a better criterion for critical temperature T_c?
3. Strictly speaking, the BCS theory can be applied to a superconductor at constant volume. Modify the theory for a thermodynamic condition under constant pressure p to show the critical temperature is a function of p, as in hydrogen sulfide.

References

[1] Anderson P W 1987 *Science* **235** 4793

[2] http://en.wikipedia.org/wiki/High-temperature_superconductivity

[3] Abragam A and Bleaney B 1970 *Electron Paramagnetic Resonance of Transition Ions* (Oxford: Clarendon) ch 7
[4] Drozdov A P, Eremets M J and Troyan I A 2014 *Conventional superconductivity at 190 K at high pressures* arXiv: 1412.0460
[5] 1997 High Temperature Superconductivity *Mol. Phys. Rep.* vol. 15/16 (Poznan: Polish Academy of Sciences)
[6] Fain S C Jr and Chinn M D 1977 *J. Phys.* **38** C4–99
Fain S C Jr and Chinn M D 1978 *Phys. Rev. Lett.* **39** 146
[7] Drozdov A P, Eremets M I, Ksenofontov I A and Shylin S I 2015 *Nature* **525** 7567
[8] Shimizu K 2015 www.asahi.com (private communication)
[9] Einaga M 2016 *Nat. Phys.* **12** 835

IOP Publishing

Solitons in Crystalline Processes (2nd Edition)
Irreversible thermodynamics of structural phase transitions and superconductivity
Minoru Fujimoto

Chapter 16

Superconducting phases in metallic crystals

Now that the nature of the Cooper pair is clarified by soliton theory, the supercurrent arises from zero resistivity to satisfy the continuity of charge–current, as a natural consequence of binary phase transition in crystals. Therefore, the properties of the superconducting state can be described logically for displacive crystals in general as a characteristic phenomenon. However, experimental results in the literature are presently limited to low-temperature metals only, while plausible models of Cooper pairs in high-T_c superconductors are still under investigation. Accordingly, Meissner's diamagnetism in these compounds has not yet been confirmed in detail.

Cooper's electron pairs constitute a multiple particle system assisted by phase solitons in metallic lattices, providing the model of *superconducting* charge carriers in momentum space. The electromagnetic phenomena discovered in early physics can be analyzed with Cooper pairs responsible for *persistent currents*, which are now considered for high-T_c superconductors as well.

In this chapter, we review thermodynamic and electromagnetic properties of persistent currents in metals for $T < T_c$ as interpreted with the soliton theory, thereby showing that traditional Ginzburg–Landau and Bardeen–Cooper–Schrieffer theories can all be legitimate with no significant revision of the original theories other than the dynamic properties of Cooper pairs. Mobile phase solitons in momentum space, however, will delineate the earlier hypothesis for the continuity of the supercharge-persistent current.

16.1 Meissner's diamagnetism

Historically, superconducting metals were first characterized by resistance-free electrical currents, and later by their *perfect diamagnetism*. Meissner's diamagnetism related to *persistent* supercurrents was unfamiliar at the time of discovery, but is now explained with respect to their soliton properties. Thermodynamic analyses of the superconducting phase were also carried out as a phase transition with an external

doi:10.1088/978-0-7503-2572-1ch16

magnetic field, showing evidence for solitons to participate in the responsible mechanism.

16.1.1 The Meissner effect

Based on experiments with a small magnet floating on a superconducting material below T_c, Meissner proposed that the magnetic induction field inside the metal should be zero, i.e. $\boldsymbol{B} = 0$, characterizing superconducting states. On the other hand, Kamerlingh Onnes first discovered *zero-resistivity* of electric currents, leading to the relation $\dot{\boldsymbol{B}} = 0$ with respect to electromagnetic principles at the time of discovery. At that time, the zero resistivity was not easily related to Meissner's proposal $\boldsymbol{B} = 0$ for superconducting phenomena.

Figure 16.1(a) sketches a uniform magnetic induction $\boldsymbol{B}_o = \mu_o \boldsymbol{H}$ in an applied magnetic field $\boldsymbol{H}$ to a conductor in elliptic shape at a temperature above T_c, where $\mu_o = 4\pi \times 10^{-7}$ volts(A–m)$^{-1}$ is a constant of the vacuum inside and outside the conductor. On the other hand, for $T < T_c$ the magnetic flux of $\boldsymbol{B}$ are all expelled to outside the conductor, as shown in figure 16.1(b) [1], while inside the conductor is characterized by $\boldsymbol{B} = 0$.

From the relation $\boldsymbol{B} = \mu_o \boldsymbol{H} + \boldsymbol{M} = 0$ for inside metals, the magnetization vector $\boldsymbol{M}$ must be assumed to exist, and is expressed as

$$\boldsymbol{M} = -\mu_o \boldsymbol{H}, \tag{16.1}$$

which was referred to as *perfect diamagnetization*. It was an interesting finding that the superconductor is magnetized to $\boldsymbol{M}$, which remains in the crystal, even after the applied field is removed, when such a *trapped magnetization* $\boldsymbol{M}$ was recognized experimentally, as illustrated in figure 16.1(c).

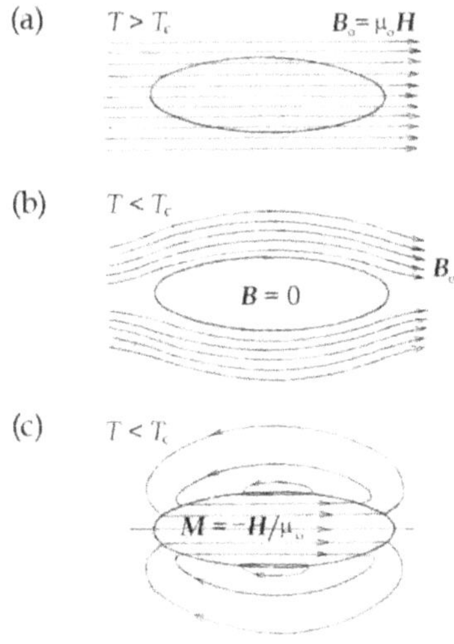

Figure 16.1. Meissner effect. (a) Normal conductor $T > T_c$ in an applied field $B_o = \mu_o H$. (b) Superconducting state $T < T_c$: all flux of B_o are expelled to outside, $B = 0$ inside. (c) Superconducting state $T < T_c$: B_o is removed. Inside is diamagnetic $M = -\mu_o H$.

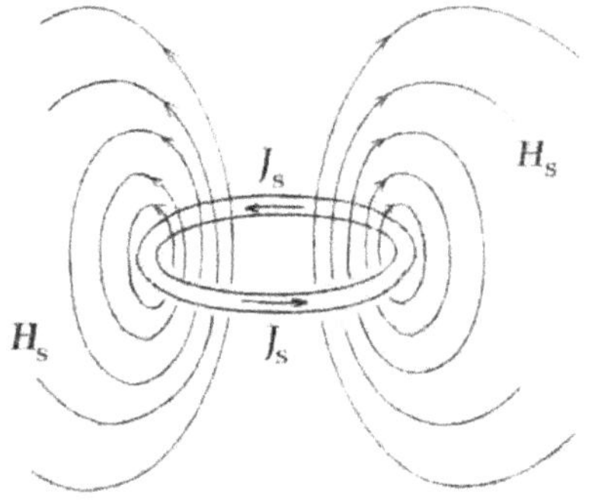

Figure 16.2. A superconducting ring $T < T_C$: after removing B_0, persistent current remains as curl $\boldsymbol{J}_s = \boldsymbol{B}_s$.

Later, Collins carried out an experiment on a superconducting ring, whose result reported in [1] indicates the presence of a *persistent current* $\boldsymbol{J}_s$, lasting as long as 2.5 years, as illustrated in figure 16.2. Noticing that the ring is a *multiply-connected* body different mathematically from Meissner's *singly-connected* ellipsoidal conductor, we can consider that the diamagnetic $-\boldsymbol{M}$ is equivalent to curl $\boldsymbol{J}_s$, as supported by the soliton theory.

Taking a current element J_s d$\boldsymbol{s}$ on the ring, the magnetic field in the vicinity can be described by the Biot–Savart law $\boldsymbol{B}(\boldsymbol{r}-\boldsymbol{r}_0) \approx J_s \oint_{\text{circle}} \frac{\mathrm{d}\boldsymbol{s}\times(\boldsymbol{r}-\boldsymbol{r}_0)}{|\boldsymbol{r}-\boldsymbol{r}_0|^3}$ in outside space, showing magnitudes at a distance $\boldsymbol{r}-\boldsymbol{r}_0$ from the ring center. Using the standard definition $\boldsymbol{B}$ = curl $\boldsymbol{A}$, the *vector potential* $\boldsymbol{A}$ can then be obtained as proportional to J_s, leading to the *London equation of supercurrent*, as illustrated in figure 14.3. In this figure, however, persistent flow density $\boldsymbol{j}_s$ of charge $2e$ is shown as signified by the fast *soliton velocity*, while Collins' persistent $\boldsymbol{J}_s$ is macroscopic.

In these descriptions, the *trapped diamagnetism* $-\boldsymbol{M}$ signified at $\boldsymbol{H} = 0$, permits us to consider the relation $-\boldsymbol{M} \propto$ curl $\boldsymbol{J}_s$ for figure 16.1(c), indicating that the superconductivity can be characterized either by diamagnetization $-\boldsymbol{M}$ or a curl of persistent current $\boldsymbol{J}_s$, which is in fact consistent with London's equation, as will be explained in section 16.2. Nonetheless, this relation was not derivable directly from *standard electromagnetic theory for conduction currents*, and hence was considered as *unusual*[note1] at the time of discovery.

Why then does the Cooper pair generate a persistent current $\boldsymbol{j}_s$? Theoretically, their relation can be attributed to the *time-dependent soliton potential* of superconducting carrier $2e$ that acts as *a time-dependent electric field* $\boldsymbol{E}_s$, holding the

[note1] It was unusual macroscopically at the time of discovery. However, what was unusual was the persistent current, although orbiting electrons around atomic nuclei defined a magnetic moment, consistent with the electromagnetic theory. Nevertheless, the pairing of two charges as a single $2e$ was indeed unusual, together with zero resistance and perfect diamagnetism. See further comments in section 17.1 on atomic magnetic moment supported by the angular momentum. Persistent currents for curl $\boldsymbol{j}_s \neq 0$ are therefore consistent with Maxwell's electromagnetism. We realize that Meissner's diamagnetism and supercurrent are properties of a mesoscopic lattice, which makes the latter essentially different from Ohm's current associated with transport phenomena.

relation $\frac{\partial j_s}{\partial t} \propto \boldsymbol{E}_s$, explaining the relation $\boldsymbol{j}_s = 2e\boldsymbol{v}_s$ (14.10), as considered for the continuity relation between $2e$ and $\boldsymbol{j}_s$; thus the persistent $\boldsymbol{j}_s$ belongs to a category of $\mathrm{curl}\,\boldsymbol{j}_s \neq 0$ that is entirely different from the *normal conduction current* $\boldsymbol{j}_n$ characterized $\mathrm{curl}\,\boldsymbol{j}_n = 0$ for Ohm's law $\boldsymbol{j}_n \propto \boldsymbol{E}_{ext}$. Nonetheless, by Biot–Savart's law, $\mathrm{curl}\,\boldsymbol{j}_s$ can be defined as a *magnetic moment* leading to macroscopic magnetization $\boldsymbol{M}$. According to section 14.2, *the supercurrent occurs with domain structure at any temperature below T_c, as determined thermodynamically by the soliton density at all temperatures for $T < T_c$.*

16.1.2 Specific-heat anomalies of superconducting transitions

Superconducting phase transitions are signified by specific-heat anomalies exhibited by a crystal of the electrons and lattice. In the first approximation, the specific heat can be discussed with Sommerfeld's free electrons and Debye's phonons. In this model, the specific heat C_V is expressed by

$$C_V = C_{el} + C_{ph}, \quad \text{where } C_{el} = \gamma T \text{ and } C_{ph} \propto T^3$$

at low temperatures, where γ is a constant of electron gas in Sommerfeld's model [2] and C_{ph} obeys the T^3–law for $T < \Theta_D$ (see chapter 1). The transition anomaly is then described by $\Delta C_{el} = C_V - C_{ph}$ detected as a function of temperature T in the vicinity of T_c.

Experimentally, Debye's formula $C_{ph} \propto T^3$ was found to work very well in metals for $T \leqslant T_c$, while a deviation ΔC_{el} from Sommerfeld's formula is evident, as expressed empirically in the form

$$\Delta C_{el} \sim \exp\frac{-b}{T}, \tag{16.2}$$

indicating that the constant b predicts the presence of an energy gap Δ between normal and superconducting states, as given by equation (14.10) determined with Boltzmann statistics for A and P modes for $T \leqslant T_c$.

16.1.3 Thermodynamic analysis

As shown by (16.1), we have the formula $\boldsymbol{M} = -\mu_0 \boldsymbol{H}$ for a superconducting state. However, it was further discovered that Meissner's diamagnetization $-\boldsymbol{M}$ decreases to 0 by increasing $\boldsymbol{H}$ to $\boldsymbol{H}_c$ for $T < T_c$, showing a critical behavior at $\boldsymbol{H}_c$. The superconducting transition can thus be signified by two critical parameters T_c and H_c, which is particularly significant for *experimental investigation* of phase transitions.

Figure 16.3(a) shows entropy curves in normal and superconducting phases of Al metals as a function of temperature, where $S_S < S_N$ for $T < T_c$, while only S_N is observed with magnetic fields $H > H_c$. Figure 16.3(b) [2] shows experimental results of C_V/T versus T^2 analysis on Ga metals, where the superconducting phase is absent, if the applied field is above $H_c = 500$ G; while exhibiting typical superconductivity at $H_c = 0$. For these results, the author of [2] used the formula

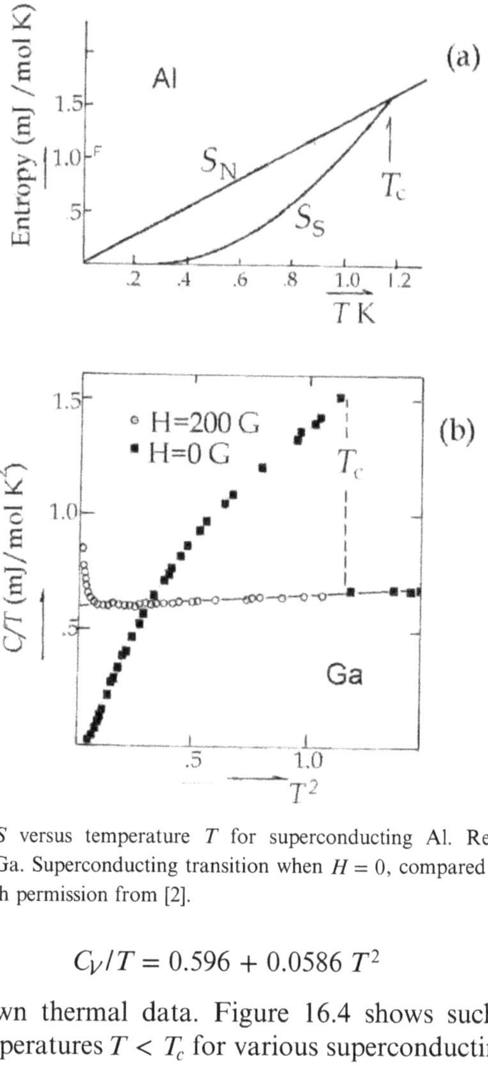

Figure 16.3. (a) Entropy S versus temperature T for superconducting Al. Reproduced from [1]. (b) C_V/T versus T^2 curves for Ga. Superconducting transition when $H = 0$, compared with no transition when $H = 200\text{G}$. Reproduced with permission from [2].

$$C_V/T = 0.596 + 0.0586\,T^2$$

obtained from his own thermal data. Figure 16.4 shows such critical fields H_c measured against temperatures $T < T_c$ for various superconducting metals, showing parabolic curves between super- and normal-conducting phases.

For a bulk of superconductor in ellipsoidal shape, Meissner demonstrated that the body has a magnetic energy density $\boldsymbol{M} \cdot \boldsymbol{H}$ per volume in a uniform magnetic field $\boldsymbol{H}$, which represents the work to remove Meissner's magnetization $\boldsymbol{M}$ from a given field $\boldsymbol{H}$. Therefore, the work to increase the magnetic energy can be expressed by $\boldsymbol{H} \cdot \mathrm{d}\boldsymbol{M}$ per volume. Accordingly, for such a superconducting body, the first law of thermodynamics can be expressed as

$$\mathrm{d}U = T\,\mathrm{d}S - p\,\mathrm{d}V + \boldsymbol{H} \cdot \mathrm{d}\boldsymbol{M}. \tag{16.3a}$$

Defining the Gibbs free energy as $G = U - T\,\mathrm{d}S + p\,\mathrm{d}V - H\,\mathrm{d}M$, the equilibrium condition can be specified by $\mathrm{d}G \geqslant 0$, minimizing the value of G. In equilibrium between superconducting and normal states, we can then set the conditions

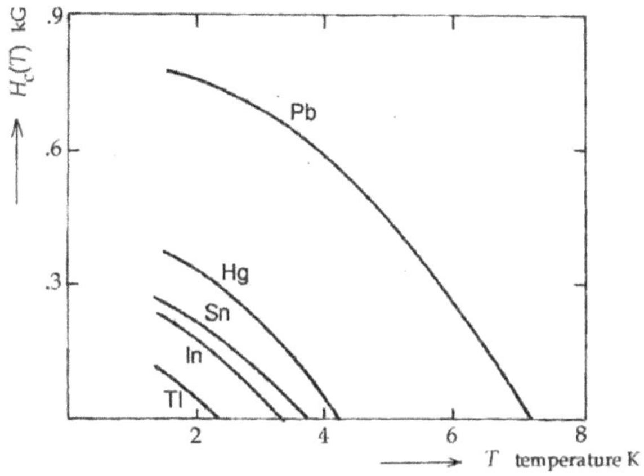

Figure 16.4. H_c versus T diagrams for representative superconducting metals. Reproduced with permission from [2].

$$G_s(H_c) = G_n(H_c) \quad \text{where } G_n(H_c) = G_n(0).$$

Consequently, we have

$$G_s(H_c) = G_n(0) - \mu_o \int_0^{H_c} H \cdot dH = G_n(0) - \frac{\mu_o}{2} H_c^2,$$

from which differences in entropies of these phases can be calculated as

$$S_n - S_s = -\frac{d}{dT}(G_n - G_s) = -\mu_o H_c \frac{dH_c}{dT},$$

and the latent heat at the transition is defined as

$$L = T(S_n - S_s) = -\mu_o T H_c \frac{dH_c}{dT}.$$

Observed curves of H_c versus T shown in figure 16.4 for representative superconducting metals indicate that the derivative dH_c/dT is negative, hence $L < 0$ or $S_n < S_s$, implying that the transition to superconducting state is signified by a larger entropy or negative latent heat transferred to the lattice.

The specific anomaly at $T \leqslant T_c$ can also be calculated as

$$\Delta C_V = \mu_o T \frac{\partial}{\partial T}(S_n - S_s) = \mu_o T \left\{ H_c \frac{d^2 H_c}{dT^2} + \left(\frac{dH_c}{dT} \right)^2 \right\}.$$

As $H_c = 0$ at $T = T_c$, we have specifically

$$(\Delta C_V)_{T_c} = \mu_o T_c \left(\frac{dH_c}{dT} \right)_{T_c}^2. \qquad (16.3b)$$

For $T \to 0$ particularly, the third law of thermodynamics suggests $S_n = S_s$, which could have been analyzed under investigation for the soliton theory [3]. Therefore, as indicated in figure 15.4(a), we have the relation

$$\lim_{T\to 0} \frac{dH_c}{dT} \to 0,$$

which, however, cannot be confirmed experimentally, as prohibited by the third law of thermodynamics.

Nevertheless, the forgoing thermodynamic analysis has confirmed for Meissner's diamagnetism to be a correct magnetic variable for superconducting current density $\boldsymbol{j}_s$ that is characterized by $\mathrm{curl}\,\boldsymbol{j}_s \neq 0$.

However, the experimental curves in figure 16.4 were observed as linear as $H_c \propto T_c - T$ for the region of $T < T_c$, implying a parabolic temperature-dependence of

$$H_c(T) \approx \sqrt{T_c - T} \tag{16.3c}$$

In fact, the relation

Theorem:

$$(M_c H_c)_{T_c} \propto T_c \text{ at the critical point} \tag{16.3d}$$

was empirically confirmed with the formula (16.3*c*) with respect to the soliton theory, as consequent on the soliton analysis, as indicated by the critical soliton number $(n_s)_{\text{critical}} = n_s(T_c,\ p_0)$ as in *classical ideal gas* characterized by T and H.

In fact, it is *significant in experimental studies on superconductivity* that the applied magnetic field acts on the superconducting phase from outside the apparatus, like applied pressure at T_c, as confirmed by Eremets *et al* (see [5, 7] in chapter 15). Such experiments could have been attempted for a *volume change in adiabatic transitions*, but not attributed effectively to external pressure, as shown by Tarnowski's experiments on high-T_c superconductor $RBa_2Cu_3O_{7-x}$ [3], which have not verified the relations (16.3*e*, *d*) from the soliton theory.

16.2 Electromagnetic properties of superconductors

Idealized in early physics by *infinite electrical conductivity*, however, the superconductivity was revised later to exhibit *perfect diamagnetism* after Meissner's discovery. These experimental results were first assumed to be consistent with the Maxwell theory of electromagnetism, but later recognized as being otherwise, for which the Cooper pair were considered responsible.

Normal and superconducting phases in metals were separated by an externally *applied magnetic field* $\boldsymbol{H}$, where the former obeyed the conventional Ohm law, while the latter exhibits the Meissner effect. Superconductivity was then considered to arise from a phase transition to generate Cooper pairs at the critical point.

A normal current density $\boldsymbol{J}_n$ is determined by an *applied electric field* $\boldsymbol{E}_{ext}$ as

$$\boldsymbol{J}_n = \sigma \boldsymbol{E}_{ext}, \qquad (16.4a)$$

where σ is conventional conductivity. On the other hand, owing to equation (14.8), the equation

$$\Lambda \frac{\partial \boldsymbol{J}_s}{\partial t} = \boldsymbol{E}_s \qquad (16.4b)$$

can be considered for supercurrent $\boldsymbol{J}_s$ for $T \leqslant T_c$ to be resistance-free, where Λ is a constant of metal, because the electric field $\boldsymbol{E}_s$ in (16.4*b*) originates from the *soliton potential.*

Equation (16.4*a*) is the Ohm law for normal conduction electrons, while (16.4*b*) describes current density expressed as $\boldsymbol{J}_s = n_s e' \boldsymbol{v}_s$ of n_s superconducting particles carrying charge e' that are in accelerating motion, i.e.

$$m' \frac{\mathrm{d}\boldsymbol{v}_s}{\mathrm{d}t} = e' \boldsymbol{E}_s,$$

as mentioned in equation (14.8) of chapter 14. Here, superconducting charge carriers are considered to be particles with unknown charge e' and mass m', and n_s is the number of soliton potentials per volume, according to the soliton theory. Hence, the parameter Λ in (16.4*b*) can be expressed as

$$\Lambda = \frac{m'}{n_s e'^2} \qquad (16.5)$$

For the Ohm law, the electric field is an *irrotational vector*, i.e. curl $\boldsymbol{E}_{ext} = 0$, whereas the field $\boldsymbol{E}_s$ in (16.4*b*) is *rotational*, characterized by curl $\boldsymbol{E}_s \neq 0$. Equation (16.4*b*) suggests that $\boldsymbol{J}_s$ should be originated from such an *inductive* field $\boldsymbol{E}_s$, and the Maxwell equations for *supercurrent density* $\boldsymbol{J}_s$ can be written as

$$\text{curl}\, \boldsymbol{E}_s = -\frac{\partial \boldsymbol{B}}{\partial t} \quad \text{and} \quad \text{curl}\, \boldsymbol{B} = \mu_0 \left(\boldsymbol{J}_s + \varepsilon_0 \frac{\partial \boldsymbol{E}_s}{\partial t} \right).$$

Combining the first equation with (16.4*b*), we obtain

$$\text{curl}\left(\Lambda \frac{\partial \boldsymbol{J}_s}{\partial t} \right) = -\frac{\partial \boldsymbol{B}}{\partial t}. \qquad (16.6)$$

Assuming $\frac{\partial \boldsymbol{E}_s}{\partial t} = 0$ for a steady current $\boldsymbol{J}_s$, the Maxwell equation can be reduced to curl $\boldsymbol{B} = \mu_0 \boldsymbol{J}_s$, indicating that $\boldsymbol{B}$ is distributed over the superconductor. Combining this result with (16.6), we can write

$$\frac{\partial \boldsymbol{B}}{\partial t} = -\text{curl}\left(\frac{\Lambda}{\mu_o} \text{curl} \frac{\partial \boldsymbol{B}}{\partial t} \right) = -\lambda^2 \,\text{curl}\,\text{curl} \frac{\partial \boldsymbol{B}}{\partial t} = \lambda^2 \nabla^2 \frac{\partial \boldsymbol{B}}{\partial t},$$

where the factor λ is a constant defined as

$$\lambda^2 = \frac{\Lambda}{\mu_o}.$$

The time derivatives $\dot{\boldsymbol{J}}_s$ and $\dot{\boldsymbol{B}}_s$ represent local properties of the field, not for the bulk body that exhibits the Meissner effect. We notice that (16.4*b*) is for $\dot{\boldsymbol{J}}_s$, and the corresponding $\dot{\boldsymbol{E}}_s$ should determine $\dot{\boldsymbol{B}}_s$ that penetrates to a finite depth λ from the superconducting surface. Therefore, writing the above equations for the component $\dot{B}_z$ penetrating along the normal z direction, for simplicity, perpendicular to the surface, we obtain

$$\lambda_d^2 \nabla^2 \dot{B}_z - \dot{B}_z = 0, \tag{16.7}$$

whose solution is given by $\dot{B}_z = \dot{B}_o \exp\left(-\frac{z}{\lambda_d}\right)$, where $z = \lambda_d$ represents the effective penetration depth of $\dot{B}_z$. Hence, we have the relation $\dot{B}_z = 0$ for $z > \lambda_d$.

However, we should have $B_z = \text{const.}$ for $0 < z < \lambda_d$, and $B_z = 0$ for $z > \lambda_d$. Here, the relation $B_z = \text{const.}$ is by no means the same as $B_z = 0$, so that the Meissner effect cannot be explained by the Maxwell theory.

To obtain the Meissner effect, London revised (16.6) and (16.7) as

$$\text{curl}(\Lambda \boldsymbol{J}_s) = -\boldsymbol{B} \tag{16.8}$$

and

$$\boldsymbol{B} = \lambda_d{}^2 \nabla^2 \boldsymbol{B}. \tag{16.9}$$

Equation (16.8) is consistent with the Cooper pair, because the same equation as (16.8) is derived from the definition of the Cooper pair in chapter 14, while the constant λ_d in (16.9) describes the penetration depth of $\boldsymbol{B}$.

Based on these London's equations, we can describe that the Meissner effect is *incomplete*, limited within surface layer of thickness λ_d, while the effect $\boldsymbol{B} = 0$ is *complete* beyond the depth $z = \lambda_d$. Equation (16.9) may be oversimplified for practical Meissner effect, giving, however, the correct order magnitude of λ_d that is estimated as about 10^{-5} cm.

The superconducting state is characterized by the presence of persistent ring current density described by curl $\boldsymbol{j}_s$ with respect to the z–axis, signifying the intrinsic magnetic field $\boldsymbol{H}_s(z)$. In Meissner's experiment, illustrated in figure 16.1(c), the field in the penetrating region is composed of intrinsic and applied fields, i.e. $\boldsymbol{B}_z = \mu_o(\boldsymbol{H} + \boldsymbol{H}_s(z))$, where the intrinsic field is partially cancelled. On the other hand, inside the conductor $\boldsymbol{B}_z = 0$, exhibiting *perfect diamagnetism* $\mu_o \boldsymbol{H}_s = \boldsymbol{M}$, which is consistent evidence for the Cooper pair.

London's equation (16.8) for expressing the relation with a Cooper pair can be modified by the vector potential $\boldsymbol{A}$ defined by $\boldsymbol{B} = \text{curl}\,\boldsymbol{A}$ derived from the basic relation div $\boldsymbol{B} = 0$, corresponding to curl $\boldsymbol{E}_s \neq 0$. Therefore, (16.8) is written as

$$\text{curl}(\Lambda \boldsymbol{J}_s + \boldsymbol{A}) = 0, \tag{16.10a}$$

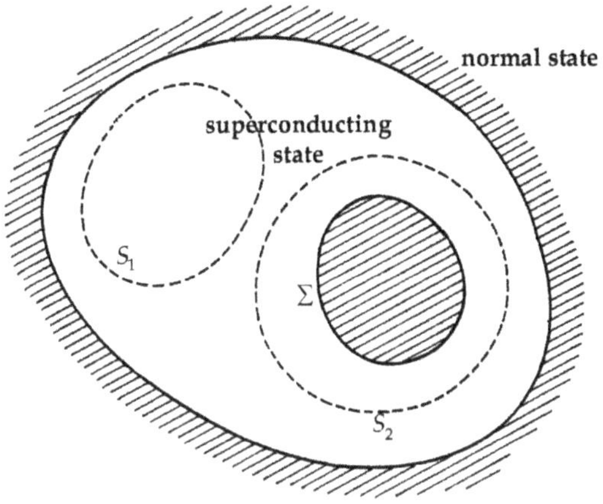

Figure 16.5. A superconducting body surrounded by a normal conductor. Closed surface S_1 is filled by superconducting medium completely, whereas the surface S_2 encloses a void or normal medium marked Σ. S_1 and S_2 are singly- and multiply-connected.

expressing an exact relation with the Cooper pair associated with a persistent current that is supported by fast soliton mobility.

Macroscopically, we assume for simplicity that a superconducting body is *simply-connected*, as illustrated by the curve C_1 in figure 16.5, and write

$$\Lambda \boldsymbol{J}_s + \boldsymbol{A} = \nabla \chi', \tag{16.10b}$$

where χ' is an arbitrary scalar function that satisfies $\nabla^2 \chi' = 0$, thereby determining unique values of $\boldsymbol{J}_s$ and $\boldsymbol{A}$ at any point inside C_1.

On the other hand, in a *multiply-connected* body, (16.10*a*) should be re-expressed in integral form to obtain a relation between the *persistent supercurrent* on surfaces and the vector potential $\boldsymbol{A}$. However, leaving the multiple case aside, in a simply-connected body we have chosen the *gauge* χ' for div$\boldsymbol{A} = 0$ of a *time-independent* case of the Lorentz condition. Nevertheless, this particular χ' is called the *London gauge* to simplify equation (16.10*b*) as

$$\boldsymbol{J}_s = -\frac{1}{\Lambda}\boldsymbol{A} = -\frac{1}{\mu_o \lambda_d^2}\boldsymbol{A}, \tag{16.11}$$

which is the relation between the Cooper pair and persistent current discussed in section 14.2. Besides, we note that (16.11) explains the persistent current in Collins' experiment [1].

With *London's* equations (16.8) or (16.10*a*), the Meissner effect can be explained for a simply-connected metal. Equation (16.8) indicates that each of n_s superconducting carriers gains a momentum by $\frac{m'}{e'}\boldsymbol{j}_s$, where $n_s \boldsymbol{j}_s = \boldsymbol{J}_s$ is the total supercurrent density.

For a system of independent superconducting particles, we can write the Hamiltonian

$$\mathcal{H}_{\text{system}} = \sum_i \frac{1}{2m'} \{\boldsymbol{p}_s(\boldsymbol{r}_i) - e'\boldsymbol{A}(\boldsymbol{r}_i)\}^2, \tag{16.12}$$

where the momentum $\boldsymbol{p}_s(\boldsymbol{r}_i)$ and position $\boldsymbol{r}_i$ constitute dynamical field variables that can interact with the electromagnetic field. Assuming the field of $(\boldsymbol{p}_s(\boldsymbol{r}_i), \boldsymbol{r}_i)$ as continuous variables, we can therefore define the Hamiltonian density $\mathcal{H}$ by writing $\mathcal{H}_{\text{system}} = \int_{\text{volume}} \mathcal{H}\, d^3\boldsymbol{r}$, i.e.

$$\mathcal{H} = \frac{1}{2m'} \{\boldsymbol{p}_s(\boldsymbol{r}) - e'\boldsymbol{A}(\boldsymbol{r})\}^2,$$

which is a *one-particle Hamiltonian* of the superconducting particles associated with one soliton particle.

The vector potential $\boldsymbol{A}(\boldsymbol{r})$ is invariant under a gauge transformation

$$\boldsymbol{A}(\boldsymbol{r}') = \boldsymbol{A}(\boldsymbol{r}) + \nabla\chi_s(\boldsymbol{r}),$$

where the scalar function $\chi_s(\boldsymbol{r})$ can be an arbitrary gauge selected from the relation $\nabla^2\chi_s(\boldsymbol{r}) = 0$. Here, the gauge $\chi_s(\boldsymbol{r})$ is specified by the suffix s, because it is related to the time-dependent displacement $\Delta\boldsymbol{r}_q(t)$, as will be explained later. The gauge invariance of $\mathcal{H}$ in this case must be fabricated for the gauge $\chi_s(\boldsymbol{r})$ to be consistent with the invariance of time-dependent electromagnetic field.

These carrier particles of electron pairs are quantized objects in the momentum space, so that $\boldsymbol{p}_s(\boldsymbol{r})$ can be expressed by a differential operator $-\boldsymbol{i}\boldsymbol{n}\hbar\nabla_r$ in equilibrium crystals. On the other hand, as applied to the lattice to the condensate system, the London gauge should be characterized in mesoscopic timescale, so we express it specifically as $\chi_s(\boldsymbol{r})$ for the following discussion.

Accordingly, the Schrödinger equation $\mathcal{H}\psi(\boldsymbol{r}) = E\psi(\boldsymbol{r})$ for the energy eigenvalue E can be written with $\boldsymbol{A}(\boldsymbol{r})$ invariant under gauge transformation. We postulate that the equation

$$\frac{1}{2m'}\{-i\hbar\nabla_{r'} - e'\boldsymbol{A}(\boldsymbol{r}')\}^2\psi(\boldsymbol{r}') = E\psi(\boldsymbol{r}')$$

is transformed to

$$\frac{1}{2m'}\{-i\hbar\nabla_r - e'\boldsymbol{A}(\boldsymbol{r}) - e'\nabla_r\chi_s(\boldsymbol{r})\}^2\psi(\boldsymbol{r}) = E\psi(\boldsymbol{r}).$$

We notice that the latter equation can be satisfied with a transformation

$$\psi(\boldsymbol{r}') \rightarrow \exp\frac{ie'\chi_s(\boldsymbol{r})}{\hbar}\psi(\boldsymbol{r}),$$

where the function $\psi(\boldsymbol{r}')$ is characterized by a phase shift $\frac{e'\chi_s(\boldsymbol{r})}{\hbar}$.

On the other hand, the momentum $\boldsymbol{p}_s(\boldsymbol{r}) = -i\hbar\nabla_r$ is a quantum field operator related to the speed $v(\boldsymbol{r})$ of superconducting carrier particles, hence we can write $m'v(\boldsymbol{r}) = \boldsymbol{p}_s(\boldsymbol{r}) - e'\boldsymbol{A}(\boldsymbol{r})$. The total momentum $\boldsymbol{P}_s(\boldsymbol{r}) = n_s\boldsymbol{p}_s(\boldsymbol{r})$ can then be written as

$$P_s(r) = e'\Lambda J_s(r) + e'A(r) \tag{16.13}$$

Accordingly, equations (16.7*a*) and (16.7*b*) become

$$\text{curl } P_s(r) = 0 \quad \text{and} \quad P_s(r) = n_s \nabla\chi_s(r),$$

where $\chi_s(r)$ is a scalar function in a simply-connected conductor. Considering $-n_s\chi_s(r)$ as the transformation gauge, the London equation (16.10*a*) is equivalent to

$$P_s(r) = 0. \tag{16.14}$$

Thus, equation (16.14) implies that total long-range ordered momenta associated with superconducting charge carriers remain zero in equilibrium states. Nevertheless, it is significant to notice that in the above theory the gauge function $\nabla_r\chi_s$ of the vector potential A is playing the role of adiabatic potential for mobile solitons in the momentum space of condensates. Therefore, the gauge invariance for $\nabla_r^2\chi_s(r) = 0$ is consequent on $P_s(r) = 0$, as stated by (16.14) for the relative coordinate r; and $-e'\nabla_{r_o}\chi_s(r)$ represents the adiabatic time-dependent shift of momentum, specified in mesoscopic timescale.

In a multiply-connected conductor, $\chi_s(r)$ cannot take a unique value at a given position r. Figure 16.5 sketches a superconducting domain surrounded by a normal phase that is shaded in the figure. In a practical metal, the body may have a void shaded and bordered by Σ, which can be in normal phase or an empty space. Inside the closed curve C_1 the metal is superconducting, whereas C_2 may contain the different phase inside as shown. Such C_1 and C_2 represent simply-connected and multiply-connected spaces, respectively. Equation (16.10) is a formula at an arbitrary local point r, hence in integral form for C_1, we have $\int_{S_{1+}} p_s \cdot dS_{1+} = -\int_{S_{1-}} p_s \cdot dS_{1-}$, where dS_{1+} and dS_{1-} are surface elements above and below the page, respectively, both S_1 and S_2 subtending the curve C_1 in common. For such a simply connected conductor, fluxes of p_s threading in and out of C_1 are equal and opposite, leading to (16.11).

Persisting currents in a superconducting coil last a considerably long time. Evidenced by such persisting currents, the Meissner effect should be re-phrased for a multi-connected body with respect to the trapped flux of magnetic field lines threading the border Σ. Choosing the curve C_2 including Σ as a doubly connected space, two integrals $\oint_{S_2} p_s \cdot dS_2$ and $\oint_\Sigma p_s \cdot d\Sigma$ can be calculated, but the former vanishes because of (16.11), so that only the latter contributes the total integral. In this case, it is convenient to use the magnetic induction vector B, and for the trapped flux inside Σ, we write the induction law as

$$-\frac{\partial}{\partial t}\oint_\Sigma B \cdot dS_\Sigma = \oint_\Sigma (\text{curl } E_s) \cdot dS_\Sigma = \oint_\Sigma E_s \cdot dl_\Sigma,$$

where we have ignored the surface penetration of B, for simplicity, and used the *Stokes theorem* to derive the last expression. Here, S_Σ represents a closed surface subtending the curve Σ, and dl_Σ is the line element along Σ. Using (16.4*b*) to replace E_s by J_s, we obtain

$$\frac{\partial}{\partial t}\left(\oint_{\Sigma} \boldsymbol{B} \cdot \mathrm{d}\boldsymbol{S}_{\Sigma} + \Lambda \oint_{\Sigma} \boldsymbol{J}_{\mathrm{s}} \cdot \mathrm{d}\boldsymbol{l}_{\Sigma}\right) = 0. \tag{16.15a}$$

Defining the total flux $\Phi = \oint_{\Sigma} \boldsymbol{B} \cdot \mathrm{d}\boldsymbol{S}_{\Sigma} + \Lambda \oint_{\Sigma} \boldsymbol{J}_{\mathrm{s}} \cdot \mathrm{d}\boldsymbol{l}_{\Sigma}$ through the curve Σ, (16.15*a*) is written as

$$\frac{\partial \Phi}{\partial t} = 0 \quad \text{or} \quad \Phi = \text{const.} \tag{16.15b}$$

The first surface integral in (16.15*a*) can be replaced by $\int_{\Sigma} (\text{curl } \boldsymbol{A}) \cdot \mathrm{d}\boldsymbol{S}_{\Sigma} = \oint_{\Sigma} \boldsymbol{A} \cdot \mathrm{d}\boldsymbol{l}_{\Sigma}$, and hence the total flux is expressed as

$$\Phi = \oint_{\Sigma} (\boldsymbol{A} + \Lambda \boldsymbol{J}_{\mathrm{s}}) \cdot \mathrm{d}\boldsymbol{l}_{\Sigma}. \tag{16.16}$$

For a simply-connected body, we have shown in (16.10*a*) and (16.10*b*) that the vector $\boldsymbol{A} + \Lambda \boldsymbol{J}_{\mathrm{s}}$ can be expressed as $\nabla \chi_{\mathrm{s}}$, and the London equation (16.8) was obtained by choosing the London gauge to result in $\Phi = 0$. In a multiply-connected body, in contrast $\Phi = \text{const.}$ because of a trapped magnetic flux in Σ, for which the function χ_{s} cannot be uniquely determined, indicating that the supercurrent density $\boldsymbol{J}_{\mathrm{s}}$ returns to any point on the passage repeatedly. Mathematically, the line integral in (16.15*a*) can be restricted to one circulation along the path Σ, where $\psi(\boldsymbol{r}') = \psi(\boldsymbol{r})$ on returning to $\boldsymbol{r}' = \boldsymbol{r}$, accompanying a phase shift $\frac{e'\chi_{\mathrm{s}}}{\hbar} = 2\pi$. Therefore, the London gauge can be selected as $\chi_{\mathrm{s}} = \frac{2\pi\hbar}{e'}$, and in such a superconductor the trapped flux in a hole Σ can be expressed as

$$\Phi = \oint_{\Sigma} \nabla \chi_{\mathrm{s}}(\boldsymbol{r}).\, \mathrm{d}\boldsymbol{l}_{\Sigma} = \frac{2\pi\hbar n_{\mathrm{s}}}{e'} = \chi_{\mathrm{s}} n_{\mathrm{s}}. \tag{16.17}$$

Although classically Φ is determined by $e' = 2e$ or corresponding persistent current $\boldsymbol{j}_{\mathrm{s}}$, the quantized flux Φ is expressed by an integral multiple of $\Phi_{\mathrm{o}} = \frac{2\pi\hbar}{e'}$, namely $\Phi = n_{\mathrm{s}}\Phi_{\mathrm{o}}$ where the unit flux Φ_{o} is called a *fluxoid.* Owing to the trapped flux in voids in practical conductors, supercurrents can exist on their surfaces, even after removing the applied magnetic field. This explains a persisting current observed in a superconducting coil. Practical superconductors are thus signified by such a trapped magnetic flux in their superconducting phases, in addition to Ohm's currents in their normal phases.

16.3 The Ginzburg–Landau equation

In the foregoing we discussed that the superconducting phase is characterized by an *ordered momentum* $\boldsymbol{P}_{\mathrm{s}}(\boldsymbol{r})$, while in a multiply-connected conductor the flux Φ of an applied field is trapped in voids (16.17). Inside the layer, for distributed $\boldsymbol{J}_{\mathrm{s}}(\boldsymbol{r})$ and $\boldsymbol{A}(\boldsymbol{r})$, their spatial correlations should be taken into account to determine the thermodynamic properties.

Considering the wave function $\psi(\boldsymbol{r})$ to represent distributed order variables, the distributed density can be expressed by

$$n_s(\boldsymbol{r}) = \psi^*(\boldsymbol{r})\psi(\boldsymbol{r}), \tag{16.18}$$

representing the number of elemental phase soliton potentials that is equivalent to the number density of Cooper pairs.

Ginzburg and Landau [4] wrote *Gibbs' free energy density* for the superconducting state as

$$g_s = g_n + \frac{1}{2m'}\,|(\boldsymbol{p}_s - e'\boldsymbol{A})\psi|^2 + \alpha\,|\psi|^2 + \frac{1}{2}\beta(|\psi|^2)^2 - \frac{1}{2\mu_o}\boldsymbol{M}^2. \tag{16.19}$$

Here on the right side, the second term represents the *kinetic energy* of the order parameter, the third and fourth are the adiabatic potential as in Landau's theory of a binary system, and the last term is the magnetic field energy density. Interpreted by Landau's expansion in chapter 4, the third and fourth terms represent the adiabatic potential associated with lattice deformation. The Ginzburg–Landau function (16.19) describes a superconducting transition in detail, as it includes the kinetic energy. It is particularly important to determine the role of long-range order in the superconducting domain that coexists with the normal conducting domain.

For thermal equilibrium, the total Gibbs function $G_s = \int_V g_s\,\mathrm{d}V$ should be minimized against arbitrary variations of ψ and $\boldsymbol{A}$. Hence, from the relation $\delta G_s = 0$, we obtain

$$\delta g_s = \left(\frac{\partial g_s}{\partial \psi}\right)_A \delta\psi + \left(\frac{\partial g_s}{\partial \boldsymbol{A}}\right)_\psi \cdot \delta\boldsymbol{A} = 0.$$

Hence from $\left(\frac{\partial g_s}{\partial \psi}\right)_A = 0$ and $\left(\frac{\partial g_s}{\partial A}\right)_\psi = 0$, we can derive the equations

$$\frac{1}{2m'}(-i\hbar\nabla - e'\boldsymbol{A})^2\psi + (\alpha + \beta\,|\psi|^2)\psi = 0, \tag{16.20}$$

and

$$\boldsymbol{J}_s(\boldsymbol{r}) = -\frac{i\hbar e'}{2m'}(\psi^*\nabla\psi - \psi\nabla\psi^*) - \frac{e'^2}{m'}\psi^*\psi\boldsymbol{A}, \tag{16.21}$$

respectively. Equation (16.20) is known as the Ginzburg–Landau equation, and equation (16.21) provides the formula for the supercurrent density.

In the absence of an applied field, i.e. $\boldsymbol{A} = 0$, equation (16.20) is a nonlinear equation of propagation in an adiabatic potential field $\Delta\boldsymbol{U} = \alpha\psi + \beta\,|\psi|^2\psi$. For simplicity, we consider one-dimension wave function as $\psi(x)$ at a given time t, for which (16.20) can be expressed as

$$\frac{\hbar^2}{2m'}\frac{\mathrm{d}^2\psi}{\mathrm{d}x^2} + \alpha\psi + \beta\,|\psi|^2\psi = 0, \tag{16.22}$$

where we consider $\alpha < 0$ and $\beta > 0$ for $T < T_c$. In fact, the differential equation of this type was already discussed, hence the previous results can be used for the present argument. We are in fact interested in such a solution expressed as $\psi = \psi_0 f(\phi)$, where $\phi = kx$ represents the spatial phase of propagation.

We set the boundary conditions $\psi = 0$ at $x = 0$ and $\psi =$ const. at $x = \infty$, signifying the threshold and completed superconducting order, respectively. Near $x = 0$, we can assume ψ is very small, so that (16.19) at the threshold is approximately expressed as

$$\frac{\hbar^2}{2m'}\frac{d^2\psi}{dx^2} + \alpha\psi = 0. \tag{16.23}$$

This is a simple harmonic oscillator equation, whose solution for the lowest energy is given as $\psi \sim \exp i(x/\xi)$, where $\xi = \sqrt{\frac{\hbar^2}{2m'|\alpha|}}$.

For $x \to \infty$, on the other hand, we consider that the non-linear term $\beta\,|\psi|^2\psi$ in (16.22) is significant for a finite amplitude ψ_0. Letting $\left(\frac{d\psi}{dx}\right)_{x\to\infty} = \psi_0$ in the non-linear equation of (16.23), we obtain $\psi_0 = \sqrt{|\alpha|/\beta}$, and the solution of (16.22) is

$$\psi(x) = \sqrt{\frac{|\alpha|}{\beta}}\tanh\frac{x}{\sqrt{2}\,\xi}.$$

Here, the parameter ξ signifies an approximate distance for the density $n_s = |\psi|^2$ to reach a plateau in the direction x, which is a significant measure for the supercurrent in a type 1 superconductor [2]. By definition, ξ depends on $m'|\alpha|$, which is an intrinsic constant of the superconductor, and is not the same as the penetration depth λ of an applied field.

Using $|\psi|^2 = \frac{|\alpha|}{\beta}$ in (16.19), we obtain the relation

$$g_s - g_n = -\frac{\alpha^2}{2\beta},$$

which should be representing an ordered state of the superconductor. Since $\frac{\alpha^2}{2\beta} = \frac{1}{2}\mu_0 H_c^2$, we have

$$H_c = \sqrt{\frac{\alpha^2}{\mu_0\beta}}. \tag{16.24}$$

The corresponding supercurrent density can therefore be expressed as

$$\boldsymbol{J}_s(\boldsymbol{r}) = -\frac{e'^2}{m'}\,|\psi|^2\boldsymbol{A},$$

which is London's equation (16.7*a*), and therefore the penetration depth λ can be calculated as

$$\lambda = \sqrt{\frac{m'\beta}{\mu_o e'^2 \alpha}}. \tag{16.25}$$

On the other hand, equation (16.20) can be written for the transition threshold as

$$\frac{1}{2m'}(-i\hbar\nabla - e'\boldsymbol{A})^2\psi = \alpha\psi.$$

If the magnetic field B is applied parallel to the y-axis, this equation is expressed as

$$-\frac{\hbar}{2m'}\left(\frac{\partial^2}{\partial x^2} + \frac{\partial^2}{\partial z^2}\right)\psi + \frac{1}{2m'}\left(i\hbar\frac{\partial}{\partial y} + e'B\right)^2\psi = \alpha\psi.$$

Setting $\psi = \psi_o(x)\exp i(k_y y + k_z z)$, we obtain

$$\frac{1}{2m'}\left\{-\hbar^2\frac{d^2}{dx^2} + \hbar^2 k_z^2 + (\hbar k_y - e'Bx)^2\right\}\psi_o = \sigma\psi_o,$$

which can be modified as

$$\frac{1}{2m'}\left\{-\hbar^2\frac{d^2}{dx^2} + e'^2B^2x^2 - 2\hbar k_y e'Bx\right\}\psi_o = \left(\alpha - \frac{\hbar^2\left(k_y^2 + k_z^2\right)}{2m'}\right)\psi_o.$$

Using a coordinate transformation $x - \frac{\hbar k_y e'B}{2m'} \to x'$, this equation can be expressed as a harmonic oscillator equation, i.e.,

$$\left(-\frac{\hbar^2}{2m'}\frac{d^2}{dx'^2} - \frac{1}{2}m'\omega_L^2 x'^2\right)\psi_o = \left(\alpha - \frac{\hbar^2 k_z^2}{2m'}\right)\psi_o,$$

where $\omega_L = \frac{e'B}{m'}$. Therefore, the threshold H_{c2} in a type 2 superconductor is determined by the lowest eigenvalue of the above equation, i.e. α, if $k_z = 0$. That is, from the relation $\frac{1}{2}\hbar\omega_L = \alpha$, we can solve this equation for $B = \mu_o H_{c2}$. Combining the relations (16.23), (16.24) and (16.25) for ξ, λ and α against H_c, respectively, we can derive the equation

$$H_{c2} = \sqrt{2}\kappa H_c \quad \text{where} \quad \kappa = \frac{\lambda}{\xi}. \tag{16.26}$$

When $\kappa > \frac{1}{\sqrt{2}}$, we have $H_{c2} > H_c$, characterizing a type 2 superconductor [2]. The ratio κ is thus a significant parameter to determine the type of superconductors; type 1 or 2 depending on the value of κ, either $\kappa < \frac{1}{\sqrt{2}}$ or $\kappa > \frac{1}{\sqrt{2}}$, respectively.

It is interesting that we can write H_{c2} in terms of the quantized flux $\Phi = n_s\Phi_o$, where $\Phi_o = \frac{2\pi\hbar}{e'}$. Combining relations (16.24) and (16.26) with the unit flux Φ_o, we can state that

$$H_{c2} = \frac{2m'\alpha}{e'\hbar} \cdot \frac{e'\Phi_o}{2\pi\hbar} \cdot \frac{\hbar^2}{2m'\xi^2} = \frac{\alpha\Phi_o}{2\pi\xi^2}. \tag{16.27}$$

We note that H_{c2} is equal to the unit flux per area $2\pi\xi^2$ that signifies the ordered area at the superconducting threshold.

16.4 Field theories of superconducting transitions

The field theory of superconducting states of multi-electron system are supported by the soliton theory included in the Cooper pair, so that existing BCS theory does not require any revision. Nevertheless, the theories are statistical and compatible with thermodynamic principles, as summarized in this section.

16.4.1 Bardeen–Cooper–Schrieffer ground states

Writing the BCS Hamiltonian as $\mathcal{H}_{\text{BCS}} = \sum_k \mathcal{H}_k$ where

$$\mathcal{H}_k = \varepsilon_k\left(a_k^\dagger a_k + a_{-k}^\dagger a_{-k}\right) - V\sum_{k'} a_{k'}^\dagger a_{-k'}^\dagger a_{-k} a_k,$$

the operators $a_k^\dagger$ and a_k are quantum-mechanical for electrons, whose time variations are determined by Heisenberg's equations

$$i\hbar\frac{\partial a_k^\dagger}{\partial t} = \left[a_k^\dagger, \mathcal{H}_k\right] \quad \text{and} \quad i\hbar\frac{\partial a_k}{\partial t} = [a_k, \mathcal{H}_k].$$

Using the identity relations $a_k a_k = 0$ and $a_k^\dagger a_k^\dagger = 0$ in these equations, we derive

$$i\hbar\dot{a}_k = \varepsilon_k a_k - a_{-k}^\dagger\left(V\sum_{k'} a_{-k'} a_{k'}\right) \quad \text{and} \quad i\hbar\dot{a}_{-k}^\dagger = -\varepsilon_k a_{-k}^\dagger - a_k\left(V\sum_{k'} a_{k'}^\dagger a_{-k'}^\dagger\right).$$

In the original BCS theory, the perturbation V is considered as a constant, although related with the lattice. Nevertheless, we discuss here the ground states where such perturbation should be constant, so that the following argument is acceptable in spite of the soliton theory.

We can define the quantity $\Delta_k = V\sum_{k'} a_{-k'} a_{k'}$ and its complex conjugate $\Delta_k^* = V\sum_{k'} a_{k'}^\dagger a_{-k'}^\dagger$, indicating interactions between Fröhlich's condensates for $|\varepsilon_k| < \varepsilon_n$. If $|\varepsilon_k| > \varepsilon_n$ on the other hand, we have to take $\Delta_k = \Delta_k^* = 0$ for no electron pairs to be formed. Using Δ_k, these equations of motion are linearized as

$$i\hbar\dot{a}_k = \varepsilon_k a_k - \Delta_k a_{-k}^\dagger \quad \text{and} \quad i\hbar\dot{a}_{-k} = -\varepsilon_k a_{-k}^\dagger - \Delta_k^* a_k. \tag{16.28}$$

We can solve (16.28) for a_k and a_{-k} that are proportional to $\exp\left(-i\frac{\varepsilon_k'}{\hbar}t\right)$, if we can find ε_k' to satisfy the determinant equation

$$\begin{vmatrix} \varepsilon_k' - \varepsilon_k & \Delta_k \\ \Delta_k & \varepsilon_k' + \varepsilon_k \end{vmatrix} = 0 \quad \text{or} \quad \varepsilon'^2_k = \varepsilon_k^2 + \Delta_k\Delta_k^*. \tag{16.29}$$

The real ε_k' represents an eigenvalue for a Cooper pair. Eigen-operators for a Cooper pair can be determined by linear combinations of these one-electron operators a_k and a_{-k}

$$\alpha_k = u_k a_k - v_k a_{-k}^\dagger, \qquad \alpha_{-k} = u_k a_{-k} + v_k a_k^\dagger, \tag{16.30a}$$

$$\alpha_k^\dagger = u_k a_k^\dagger - v_k a_{-k}, \qquad \alpha_{-k}^\dagger = u_k a_{-k}^\dagger + v_k a_k. \tag{16.30b}$$

The reverse relations are

$$a_k = u_k \alpha_k + v_k \alpha_{-k}^\dagger, \qquad a_{-k} = u_k \alpha_{-k} - v_k \alpha_k^\dagger, \tag{16.30c}$$

$$a_k^\dagger = u_k \alpha_k^\dagger + v_k \alpha_{-k}, \qquad a_{-k}^\dagger = u_k \alpha_{-k}^\dagger - v_k \alpha_k. \tag{16.30d}$$

These relations from (16.30*a*) to (16.30*d*) are known as the *Bogoliubov transformation* [5]. The coefficients u_k and v_k are real for the symmetric and anti-symmetric combinations of one-electron operators, respectively, with regard to inversion $\boldsymbol{k} \to -\boldsymbol{k}$, i.e. $u_k = u_{-k}$ and $v_k = -v_{-k}$, normalized as $u_k^2 + v_k^2 = 1$. And, for the operators $\alpha_k^\dagger$ and α_k, we have the following anti-commutator relations.

$$\left[\alpha_k, \alpha_{k'}^\dagger\right]_+ = u_k u_{k'}\left[a_k, a_{k'}^\dagger\right]_+ + v_k v_{k'}\left[a_{-k}^\dagger, a_{-k'}\right]_+ = \delta_{kk'}\left(u_k^2 + v_k^2\right)$$

and

$$[\alpha_k, \alpha_{-k}]_+ = u_k v_k\left[a_k, a_k^\dagger\right]_+ - v_k u_k\left[a_{-k}^\dagger, a_{-k}\right]_+ = u_k v_k - v_k u_k = 0.$$

Differentiating (16.28), we obtain $\varepsilon_k' u_k = \varepsilon_k u_k + \Delta_k v_k$. Combining this result with (16.29), we derive the relation

$$\Delta_k\left(u_k^2 - v_k^2\right) = 2\varepsilon_k u_k v_k.$$

It is noted that this expression is identical to $\tan\theta_k = \frac{\Delta_k}{\varepsilon_k}$ in section 14.4.2, if we write

$$u_k = \cos\frac{\theta_k}{2} \quad \text{and} \quad v_k = \sin\frac{\theta_k}{2}. \tag{16.31}$$

Using the operator α_k, the ground state of the system may be represented by a wave function

$$\alpha_{-k}\alpha_k \Phi_{\text{vac}} = (-v_k)\left(u_k + v_k a_k^\dagger a_{-k}^\dagger\right)\Phi_{\text{vac}}.$$

Although it is not normalized, omitting the factor $-v_k$, we can confirm that

$$\langle\Phi_{\text{vac}}|\left(u_k{}^* + v_k{}^* a_{-k} a_k\right)\left(u_k + v_k a_k^\dagger a_{-k}^\dagger\right)|\Phi_{\text{vac}}\rangle = \left(u_k^2 + v_k^2\right)\langle\Phi_{\text{vac}}|\Phi_{\text{vac}}\rangle,$$

hence the function

$$\Phi_\text{o} = \prod_k \left(u_k + v_k a_k^\dagger a_{-k}^\dagger\right)\Phi_{\text{vac}} \tag{16.32}$$

is considered for the normalized wave function for the ground state. This wave function (16.32) was originally postulated in the BCS theory, implying that Cooper pairs and unpaired condensates are determined by probabilities v_k^2 and u_k^2, respectively. With Bogoliubov's operators creation and annihilation of a Cooper pair can be described conveniently; the normalization $u_k^2 + v_k^2 = 1$ assumed in the BCS theory is consistent with Bogoliubov's transformation.

We can further verify that

$$\alpha_{k'}\Phi_o \propto \alpha_{k'}(\alpha_{-k'}\alpha_{k'})\prod_{k\neq k'} \alpha_{-k}\alpha_k\Phi_{\text{vac}} = 0, \quad \text{for} \quad \alpha_{k'}\alpha_{k'} = 0,$$

signifying annihilation of a pseudo-particle. Also, from the relations

$$\alpha_{k'}^{\dagger}\Phi_o = \left(u_{k'}a_{k'}^{\dagger} - v_{k'}a_{-k'}\right)\left(u_{k'} + v_{k'}a_{k'}^{\dagger}a_{-k'}^{\dagger}\right)\prod_{k\neq k'}\left(u_k + v_k a_k^{\dagger}a_{-k}^{\dagger}\right)$$

and

$$\Phi_o = a_{k'}^{\dagger}\prod\nolimits_{k\neq k'}\left(u_k + v_k a_k^{\dagger}a_{-k}^{\dagger}\right)\Phi_{\text{vac}},$$

where we see that the operator $\alpha_k^{\dagger}$ creates a single particle; $\alpha_k^{\dagger}\alpha_{-k'}^{\dagger}$ is for the pair creation.

The number operator in the $\boldsymbol{k}$-state can be expressed in terms of Bogoliubov's operators as

$$n_k = a_k^{\dagger}a_k = \left(u_k\alpha_k^{\dagger} + v_k\alpha_{-k}\right)\left(u_k\alpha_k + v_k\alpha_{-k}^{\dagger}\right).$$

Using θ_k given by (16.31), u_k and v_k are related to

$$u_k^2 = \cos^2\frac{\theta_k}{2} = \frac{1}{2}\left(1 + \frac{\varepsilon_k}{\varepsilon_k'}\right) \quad \text{and} \quad v_k^2 = \sin^2\frac{\theta_k}{2} = \frac{1}{2}\left(1 - \frac{\varepsilon_k}{\varepsilon_k'}\right). \tag{16.33}$$

allowing for these probabilities to be interpreted in terms of the ratio $\frac{\varepsilon_k}{\varepsilon_k{}'}$.

Finally, the previous result of the energy gap E_g can be verified with Φ_o in (16.32). Namely, the expectation value of the kinetic energy is expressed as

$$\left\langle \Phi_o | a_{k'}^{\dagger}a_{k'} | \Phi_o \right\rangle = \left\langle \Phi_o | v_{k'}^2\alpha_{-k'}\alpha_{-k'}^{\dagger} | \Phi_o \right\rangle = v_{k'}^2,$$

and the potential energy term is

$$-\left\langle \Phi_o | a_{k'}^{\dagger}a_{-k'}^{\dagger}a_{-k''}a_{k''} | \Phi_o \right\rangle = \left\langle \Phi_o | u_{k'}v_{k'}u_{k''}v_{k''}\alpha_{-k'}\alpha_{-k'}^{\dagger}\alpha_{-k''}\alpha_{-k''}^{\dagger} | \Phi_o \right\rangle = u_{k'}v_{k'}u_{k''}v_{k''}.$$

Hence,

$$\langle\Phi_o|\mathcal{H}_{BCS}|\Phi_o\rangle = 2\sum_k \varepsilon_k v_k^2 - V\sum_{k,k'} u_k v_k u_{k'} v_{k'}$$
$$= \sum_k \varepsilon_k(1 - \cos\theta_k) - \frac{V}{4}\sum_{k,k'} \sin\theta_k \sin\theta_{k'}$$
$$= -\sum \varepsilon_k \cos\theta_k - \frac{\Delta^2}{V}.$$

The energy gap is then given as $E_g = \langle\Phi_o|\mathcal{H}_{BCS}|\Phi_o\rangle + \sum_k |\varepsilon_k|$.

16.4.2 Superconducting states at finite temperatures

Thermodynamically, the BCS result can be applied to a finite temperature. At a temperature T the ground state Φ_o should be modified by adiabatic excitations of condensates, as described by Ginzburg–Landau' theory. In the following, we discuss the BCS theory modified in a thermodynamic environment.

The number of quasi-particles of Cooper pairs can be assumed to be temperature dependent, for which we consider the complete set of temperature-dependent states

$$|\Phi_o\rangle,\ \alpha_k^\dagger|\Phi_o\rangle,\ \alpha_k^\dagger\alpha_{k'}^\dagger|\Phi_o\rangle,\ \ldots,$$

forming the basis. Here $|\ldots\rangle$ expresses temperature-dependent *ket*-states.

To study thermodynamic properties of a superconductor, we need to introduce the *statistical average number of quasi-particles* as given by the function

$$f_k = \left\langle \alpha_k^\dagger \alpha_k \right\rangle,$$

which should be evaluated with respect to the soliton theory as discussed in chapter 14.

Assuming statistical independence of excitations, the entropy of the system can be given by

$$S = k_B\sum_k \{f_k \ln f_k + (1 - f_k)\ln(1 - f_k)\}.$$

We re-express the Hamiltonian of interacting pairs $\mathcal{H}_k = \varepsilon_k(a_k^\dagger a_k + a_{-k}^\dagger a_{-k}) - V(a_k^\dagger a_{-k}^\dagger a_{-k} a_k)$ by the Bogoliubov transformation. The kinetic energy term can be converted to

$$\sum_k \left\langle \varepsilon_k\left(u_k\alpha_k^\dagger + v_k\alpha_{-k}\right)\left(u_k\alpha_k + v_k\alpha_{-k}^\dagger\right)\right\rangle = \sum_k \varepsilon_k\left\{v_k^2(1 - f_k) + u_k^2 f_k\right\}, \tag{16.34}$$

in which we noted that $\alpha_k^\dagger\alpha_{-k}^\dagger = 0$. The first term on the right side of (16.34) can be interpreted as representing the condensate system that has no quasi-particle at the state of wave vector $\boldsymbol{k}$, with thermal probability v_k^2. The second term, on the other hand, u_k^2 is the probability for the state to be occupied by quasi-particles. The factors $1 - f_k$ and f_k in these terms are the statistical weights due to average numbers of quasi-particles.

Similarly, the interaction term can be calculated as $\sum_{k,k'} V(\boldsymbol{k}, \boldsymbol{k}'; \boldsymbol{q})(1 - 2f_k)(1 - 2f_{k'})$. It is noted from the definition given in chapter 14 that $V(\boldsymbol{k}, \boldsymbol{k}'; \boldsymbol{q})$ can be factorized $V_k(\boldsymbol{q})V_{-k}(-\boldsymbol{q})$. Therefore, the thermodynamic potential can be expressed as

$$\begin{aligned} g(T, p) &= \langle \mathcal{H}_{\mathrm{BCS}} \rangle - TS \\ &= \sum_k \varepsilon_k \{ u_k^2 f_k + v_k^2 (1 - f_k) \} \\ &\quad - \sum_{k,k'} V_k V_{k'} u_k u_{k'} v_k v_{k'} (1 - 2f_k)(1 - 2f_{k'}) \\ &\quad - \mathrm{k_B} T \sum_k \{ f_k \ln f_k + (1 - f_k) \ln (1 - f_k) \}. \end{aligned} \tag{16.35}$$

We first minimize this Gibbs function with respect to v_k, which is the same procedure as in the zero-temperature case, arriving at expressions similar to (16.24), i.e.

$$u_k^2 = \frac{1}{2}\left(1 + \frac{\varepsilon_k}{\sqrt{\varepsilon_k^2 + \Delta_k(T)^2}}\right) \quad \text{and} \quad v_k^2 = \frac{1}{2}\left(1 - \frac{\varepsilon_k}{\sqrt{\varepsilon_k^2 + \Delta_k(T)^2}}\right),$$

where $\Delta_k(T) = \sum_k V_k u_k v_k (1 - 2f_k)$, representing one-half of the gap at the superconducting transition, which is now verified as temperature-dependent.

Minimizing the Gibbs function (16.35) with respect to f_k, we obtain

$$\mathrm{k_B} T \{ \ln f_k + \ln (1 - f_k) \} + \sqrt{\varepsilon_k^2 + \Delta_k(T)^2} = 0,$$

i.e.

$$f_k = \frac{1}{1 + \exp \dfrac{E_k}{\mathrm{k_B} T}} \quad \text{where } E_k = \sqrt{\varepsilon_k^2 + \Delta_k(T)^2}.$$

Here, the function f_k is the *Fermi–Dirac distribution function* for temperature-dependent E_k of *free single-electrons*, representing Cooper pairs, although consistent to *Anderson's order* f_k *variables with the soliton theory*.

In the absence of H, the superconducting transition at T_c is second-order characterized by no latent heat, signifying that $E_k(T_c) = 0$. On the other hand, in the presence of H, the transition is discontinuous, as signified by a finite energy gap $E_k(T)$ that is temperature-dependent. With the BCS theory, the value of E_k is thus calculable, as shown in figure 14.3, where the calculated result is compared with experimental values from some superconducting metals, showing a reasonable agreement.

Exercises

1. Both regular and super-currents obey the equation of continuity, while characterized by (16.1*a*) and (16.1*b*). Noting that $\boldsymbol{E}$ in these equations are

applied and intrinsic electric fields, respectively, discuss the difference with regard to the Maxwell equations.

2. The BCS theory was worked out based on the hypothesis that Cooper pairs are free particles of fermion statistics. However, it is realized that this assumption is supported by quantized soliton potentials that behave as boson particles. Are these conflicting? Resolve the issue, if you consider it otherwise.
3. The Meissner diamagnetism is undeniable evidence of supercurrents. Does it conflict with the Maxwell theory of electromagnetism? Or, is it a consequence? Discuss this fundamental issue, considering that the soliton mechanism is responsible for superconductivity.

References

[1] Kuper C G 1968 *An Introduction to the Theory of Superconductivity* (Oxford: Clarendon)
[2] Kittel C 1986 *Introduction to Solid State Physics* (New York: Wiley)
[3] Tarnawski Z 1996 *Mol. Phys. Rep.* **15/16** 103
[4] Ginzburg V L and Landau L D 1950 *Zh. Éksp. Teor. Fiz.* **20** 1064
[5] Bogoliubov N N 1947 *J. Phys. Moscow* **11** 23
Bogoliubov N N 1958 *Nuovo Cimento* **X7** 794

Chapter 17

Magnetic crystals

Microscopic magnetic moments $\boldsymbol{\mu}_n$ in a crystal are primarily in Larmor's precession in the intrinsic and applied magnetic fields, where inversion $\mu_z \rightleftarrows -\mu_z$ holds along the effective magnetic axis. Restricted in magnetic symmetry, such $\boldsymbol{\mu}_n$ in precession violate local structure as a function of time, while exhibiting magnetic order in crystals. In this chapter, reviewing the microscopic origin of magnetic moments $\boldsymbol{\mu}_n$, their spatial correlations are discussed with respect to lattice and magnetic symmetries.

Originating from electron exchanges among nearest neighbors, strong spin covalence in short range is responsible for basic magnetic interactions, while long-range interactions are relatively insignificant, and therefore expressed adequately in mean-field approximation. In the spin-half case, their *ferromagnetic* and *antiferromagnetic* phases appear with the binary mechanism, where the magnetic order takes place like structural changes in non-magnetic crystals. However, magnetic transitions are related to soft modes of magnetic displacements, signifying space–time displacements arising from solitons. In this chapter, we review existing magnetic theory with the soliton concept, starting with ionic magnetism particularly for those readers unfamiliar with magnetic crystals.

17.1 Microscopic magnetic moments

Microscopic magnetic moments are intrinsic properties of constituent ions, independent of crystal structure, and are hence detectable by an applied magnetic field. In this section, the origin of ionic magnetic moments is reviewed, prior to discussing magnetic problems in crystals.

Magnetic moments of *transition-* and *rare-earth elements* are characterized by *unpaired electrons* in ionic states. Incompletely filled 3d and 4f inner shells of a magnetic ion are signified by an odd number of electrons that are responsible for magnetism of various types in these crystals.

doi:10.1088/978-0-7503-2572-1ch17

Assuming that electrons are charged mass particles orbiting independently around the nucleus, their energies and angular momentum, ε_i and $\boldsymbol{l}_i$ are conservative quantities. It is significant, however, that magnetic moments can be regarded as dynamically conservative on the basis of *persistent currents* in atomic orbitals. In this case, the total $\sum_i \varepsilon_i$ and $\sum_i \boldsymbol{l}_i$ should be constant, ignoring their mutual interactions. In quantum mechanics, on the other hand, each electron has an intrinsic spin angular momentum $\boldsymbol{s}_i$, so that we need to consider their total spin $\boldsymbol{S} = \sum_i \boldsymbol{s}_i$ as conservative, in addition to total orbital angular momentum $\boldsymbol{L} = \sum_i \boldsymbol{l}_i$.

This assumption, called Russel–Saunders coupling, is primarily an appropriate approach to specify ionic states. In this scheme, $\boldsymbol{L}$ and $\boldsymbol{S}$ are quantized along an arbitrary direction z in space, so that their steady z-components, M_L and M_S, are specified by $+L$, $L-1$, ... , $-L$ and $+S$, $S-1$, ... , $-S$, respectively.

Further, the total angular momentum $\boldsymbol{J} = \boldsymbol{L} + \boldsymbol{S}$ is conserved in the absence of an applied magnetic field, where the component J_z along the z-direction can take discrete values $M_J = L + S$, $L + S - 1$, ..., $|L - S|$. Specified by J, the ionic energies are determined by the principal quantum number n and J, the total energy $\sum_i \varepsilon_i$ is M_J-fold degenerate, which is referred to as the *J-multiplet*.

Primarily independent of $\boldsymbol{S}$, the orbital angular momentum $\boldsymbol{L}$ interacts with the spin momentum $\boldsymbol{S}$ *magnetically*. Considering classical definitions for $\boldsymbol{L} = m(\boldsymbol{r} \times \boldsymbol{V})$ and $\boldsymbol{j} = e\boldsymbol{v}$,[note1] where e and m are charge and mass of an electron, Biot–Savart's law states that the orbital current $\boldsymbol{j}$ generates a magnetic field proportional to $\left\langle \frac{1}{r^3} \right\rangle (\boldsymbol{r} \times \boldsymbol{j}) = \frac{e}{m} \left\langle \frac{1}{r^3} \right\rangle \boldsymbol{L}$ at the center of the closed orbit. Accordingly, the magnetic interaction energy with the spin momentum $\boldsymbol{S}$ can be expressed as

$$\mathcal{H}_{LS} = \lambda \boldsymbol{L} \cdot \boldsymbol{S}, \tag{17.1}$$

which is called the *spin–orbit coupling* energy. Although the constant λ is difficult to evaluate with sufficient accuracy, experimental values from atomic spectra can be utilized in (17.1) as adequate in most applications.

In crystals, the ionic state is perturbed by the anisotropic crystal field, where electronic orbits cannot be considered to be spherical as in isolated ions. In other words, an ion is deformed by the lattice potential in such a way that the ground state is signified by the expectation value $\langle \boldsymbol{L} \rangle = 0$. Using the definition $\boldsymbol{L} = i\hbar(\boldsymbol{r}) \times \boldsymbol{\nabla}$, we can verify that the volume integrals

$$\langle \boldsymbol{L} \rangle_\mathrm{o} = \int_V \psi_\mathrm{o} \boldsymbol{L} \psi_\mathrm{o} \, \mathrm{d}^3\boldsymbol{r} = \int_V \psi_\mathrm{o}^* \boldsymbol{L} \psi_\mathrm{o}^* \, \mathrm{d}^3\boldsymbol{r} = -\left(\int_V \psi_\mathrm{o} \boldsymbol{L} \psi_\mathrm{o} \, \mathrm{d}^3\boldsymbol{r} \right)^* = -\langle \boldsymbol{L} \rangle_\mathrm{o}^*, \tag{17.2}$$

[note1] The orbital current $\boldsymbol{j} = e\boldsymbol{v}$ is a persistent current, similar to superconducting current discussed in chapters 14, 15 and 16. However, the orbital current is induced by changing orbital momentum $m\boldsymbol{v}$ of electrons in the Coulomb potential of nucleus that is different from moving soliton potential of superconducting carriers.

for the ground state ψ_0, so that $\boldsymbol{L}$ is considered as insignificant in crystals as verified by $\langle \boldsymbol{L} \rangle_o = 0$, which is referred to as *orbital quenching*.

For multiplet states signified by $\boldsymbol{J} = \boldsymbol{L} + \boldsymbol{S}$, the spin–orbit coupling $\mathcal{H}_{LS}$ is considered as a perturbation. Since $\boldsymbol{J}$ is a constant in this case, we write the equation $\frac{\mathrm{d}}{\mathrm{d}t}(\boldsymbol{L} + \boldsymbol{S}) = 0$, hence

$$\frac{\mathrm{d}\boldsymbol{L}}{\mathrm{d}t} = -\frac{\mathrm{d}\boldsymbol{S}}{\mathrm{d}t}. \tag{17.3}$$

Considering $\mathcal{H}_{LS}$ for perturbing the motion described by (17.3), we can write

$$\frac{\mathrm{d}\boldsymbol{L}}{\mathrm{d}t} = (\lambda \boldsymbol{S} \times \boldsymbol{L}) \quad \text{and} \quad \frac{\mathrm{d}\boldsymbol{S}}{\mathrm{d}t} = (\lambda \boldsymbol{L} \times \boldsymbol{S})$$

where the right-hand sides represent *torques* on $\boldsymbol{L}$ and $\boldsymbol{S}$ exerted by magnetic fields $\lambda\boldsymbol{S}$ and $\lambda\boldsymbol{L}$, respectively. Consequently, equations of motion for $\boldsymbol{L}$ and $\boldsymbol{S}$ can be expressed as

$$\frac{\mathrm{d}\boldsymbol{L}}{\mathrm{d}t} = \lambda(\boldsymbol{J} \times \boldsymbol{L}) \quad \text{and} \quad \frac{\mathrm{d}\boldsymbol{S}}{\mathrm{d}t} = \lambda(\boldsymbol{J} \times \boldsymbol{S}), \tag{17.4}$$

indicating that $\boldsymbol{L}$ and $\boldsymbol{S}$ are in *precession* around $\boldsymbol{J}$, as illustrated in figure 17.1.

The ground state of the ion is characterized by the smallest component $M_J = |L - S|$, which is perturbed by $\mathcal{H}_{LS}$ for $\lambda > 0$ to obtain the lower energy. The perturbed state can therefore be visualized by such a parameter a as given by the relations

$$\boldsymbol{L} = a\boldsymbol{J} \quad \text{and} \quad \boldsymbol{S} = (1 - a)\boldsymbol{J}. \tag{17.5a}$$

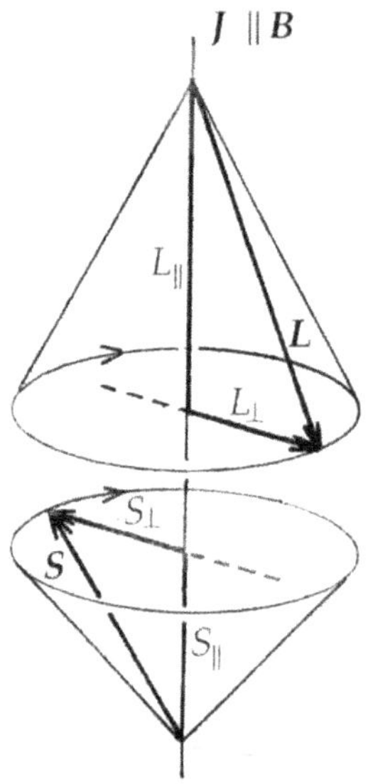

Figure 17.1. Larmor precession of $\boldsymbol{L}$ and $\boldsymbol{S}$ around $\boldsymbol{J}$.

Regarding the total momentum $\boldsymbol{L} + \boldsymbol{S} = \boldsymbol{J}$, the factor a indicates a degree of orbital quenching by $\mathcal{H}_{LS}$. Using quantum relations $\boldsymbol{J}^2 = \hbar^2 J(J+1)$ and $\boldsymbol{L}^2 = \hbar^2 L(L+1)$ with the relation $\boldsymbol{J} \cdot \boldsymbol{L} = a\boldsymbol{J}^2 = \boldsymbol{L}^2 + \boldsymbol{S} \cdot \boldsymbol{L}$, we obtain

$$a = \frac{J(J+1) + L(L+1) - S(S+1)}{2J(J+1)}. \tag{17.5b}$$

It is noted that vectors $\boldsymbol{J}$, $\boldsymbol{L}$ and $\boldsymbol{S}$ can usually be defined by either angular momenta or *quantum numbers*, but (17.5*b*) remains the same, because the factor $\hbar^2$ cancels out in any case.

Using the latter definition, the magnetic moment of an ion in the ground state can be expressed as

$$\boldsymbol{\mu} = -\gamma(\boldsymbol{L} + 2\boldsymbol{S}) = -\gamma(2 - a)\boldsymbol{J} = -g_J\beta\boldsymbol{J}, \tag{17.6a}$$

where

$$g_J = \frac{3J(J+1) - L(L+1) + S(S+1)}{2J(J+1)}, \tag{17.6b}$$

and $\beta = \frac{e\hbar}{2mc} = 0.927 \times 10^{-20}$ emu is *Bohr's magneton*. The expression (17.6*b*) is known as the *Landé factor*.

To determine the total spin S in an incomplete shell specified by L, electrons must obey *Pauli's exclusion principle*; only one electron can occupy each single electronic state. Consider a 3d shell signified by principal quantum numbers $n = 3$ and the orbital $L = 2$, for example, we can accommodate up to $2(2L + 1) = 10$ electrons. Accordingly, the number of electrons in the incomplete 3d-shell of transition elements can be either one of the number n in the range $1 \leqslant n \leqslant 9$. Such an accommodation number specified by n should be determined by empirical *Hund's rule* in atomic spectroscopy for arranging electrons at maximum $M_S = \sum_{\text{shell}} m_s$. Using this rule, we have $M_S = \pm\frac{1}{2}$ for odd n, and $M_S = 0$ for even n. Therefore, for $\lambda > 0$, $n = 1, 2, 3, 4$ correspond to the lowest spin–orbit coupling, whereas for $\lambda < 0$, $n = 6, 7, 8, 9$. The latter case can be re-phrased by *holes* $n' = 10 - n = 4, 2, 2, 1$ to be consistent with $\lambda > 0$. If $n = 5$, in particular, the lowest energy state is ${}^6S$; hence $\mathcal{H}_{LS} = 0$, signifying $L = 0$ and $S = \frac{1}{2}$.

Table 17.1 lists multiplet states of transition-group ions and experimental values of λ. It is noted that 3d electrons of transition ions orbit closely to the outer shell, so that their orbital momentum is strongly quenched in a crystal field. On the other hand, the 4f shell of rare-earth ions is relatively deep inside ions, so that the quenching of $\boldsymbol{L}$ is weak and hence insignificant. Table 17.2 shows multiplet states of rare-earth elements.

17.2 Brillouin's formula

In magnetic crystals, magnetic ions at lattice sites cannot be independent of each other, interacting magnetically. However, if ignoring interactions and applying a

Table 17.1. Spin–orbit coupling constants[a].

Ion	Ti^{3+}	V^{3+}	Cr^{3+}	Mn^{3+}	V^{2+}	Cr^{2+}	Fe^{2+}	Co^{2+}	Ni^{2+}	Cu^{2+}
d-Shell	d^1	d^2	d^3	d^4	d^3	d^4	d^6	d^7	d^8	d^9
J-State	^2D	^3F	^4F	^5D	^4F	^5D	^5D	^4F	^3F	^2D
$\lambda(cm^{-1})$	154	104	87	85	55	57	−100	−180	−335	−828

[a] From [4].

Table 17.2. Magnetic 4*f*-shells in rare-earth ions.

Ion	f^n	Ground State	p[a]
La^{3+}	f^0	^1S	Diamagnetic
Ce^{3+}	f^1	$^2F_{5/2}$	2.5
Pr^{3+}	f^2	$^3H_4$	3.6
Nd^{3+}	f^3	$^4I_{9/4}$	3.8
Pm^{3+}	f^4	$^5I_4$	–
Sm^{3+}	f^5	$^6H_{5.2}$	1.5
Eu^{3+}	f^6	$^7F_0$	3.6
Gd^{3+}	f^7	$^8S_{7/2}$	7.9
Tb^{3+}	f^8	$^7F_6$	9.7
Dy^{3+}	f^9	$^6H_{15/2}$	10.5
Ho^{3+}	f^{10}	$^5I_8$	10.5
Er^{3+}	f^{11}	$^4I_{15/2}$	9.4
Tm^{3+}	f^{12}	$^3H_6$	7.2
Yb^{3+}	f^{13}	$^2F_{7/2}$	4.5
Lu^{3+}	f^{14}	^1S	Diamagnetic

[a] $p = g_J\sqrt{J(J+1)}$ is an observed parameter defined by $\chi = N\beta^2 \boldsymbol{p}^2/3k_B T$, and referred to as the effective magneton. Data from [4].

uniform magnetic field $\boldsymbol{B}$ to the crystal, we can hypothetically deal with their *paramagnetism* in a crystal. In this section, we discuss *Brillouin's formula* of paramagnetic susceptibility, which is nevertheless useful for evaluating magnetized properties of crystals.

In an applied magnetic field $\boldsymbol{B}$, a magnetic ion characterized by the angular momentum $\boldsymbol{J}$ has a Zeeman energy

$$\varepsilon(J, M_J) = -\boldsymbol{\mu}_n \cdot \boldsymbol{B} = -g_J\beta B M_J, \quad \text{where} \quad M_J = J, J-1, \ldots, -J. \tag{17.7}$$

The magnetization of a crystal determined as $M(B, T) = N\langle\boldsymbol{\mu}_n\rangle$, where N is the number of $\boldsymbol{\mu}_n$, can be calculated statistically as

$$M(B, T) = \frac{N}{Z}\sum_{-J}^{+J}(-g_J\beta M_J)\exp\frac{g_J\beta M_J B}{k_B T}$$

where $Z = \sum_{-J}^{+J} \exp \frac{g_J \beta M_J B}{k_B T}$ is the partition function. Writing $x = \frac{g_J \beta B}{k_B T}$ for convenience,

$$Z = \sum_{-J}^{J} \exp x M_J = \exp(-xJ) + \cdots + \exp xJ$$

$$= \frac{\exp(-xJ) - \exp x(J+1)}{1 - \exp x} = \frac{\sinh(J + \frac{1}{2})}{\sinh \frac{1}{2}x},$$

therefore

$$M(B, T) = N k_B T \frac{\partial Z}{\partial B} = N k_B T \frac{\partial \ln Z}{\partial x} \frac{\partial x}{\partial B}$$

$$= g_J \beta \left\{ \frac{\left(J + \frac{1}{2}\right) \cosh \left(J + \frac{1}{2}\right) x}{\sinh \left(J + \frac{1}{2}\right) x} - \frac{\cosh \frac{1}{2} x}{2 \sinh \frac{1}{2} x} \right\}.$$

This is re-expressed as

$$M(B, T) = g_J \beta J \mathcal{B}_J(x),$$

where

$$\mathcal{B}_J(x) = \frac{2J+1}{2J} \coth \frac{(2J+1)x}{2J} - \frac{1}{2J} \coth \frac{x}{2J} \tag{17.8}$$

is known as Brillouin's function.

Using (15.6), we can verify the Curie's law of paramagnetic susceptibility, namely

$$\chi = \lim_{x \to 0} \frac{M(B, T)}{B} = \lim_{B \to \infty} N g_J \beta J \mathcal{B}_J \left(\frac{g_J \beta J B}{k_B T} \right) = \frac{N g_J{}^2 \beta^2 J(J+1)}{3 k_B T} = \frac{C}{T},$$

where Curie's constant is given by $C = \frac{N g_J{}^2 \beta^2 J(J+1)}{3 k_B}$. Curie's law applies only to crystals at elevated temperatures, because interactions of magnetic ions are ignored for paramagnetism.

17.3 Spin–spin exchange correlations

Magnetic moments of constituent ions are primarily independent of the lattice, where the magnetic and the crystal symmetries coexist. Inversion symmetry $\boldsymbol{\mu}_n \to -\boldsymbol{\mu}_n$ holds by reversing its direction of an applied field, i.e. $\boldsymbol{B} \to -\boldsymbol{B}$. Characterized by fast motion at elevated temperatures, the average $\langle \boldsymbol{\mu}_n \rangle$ vanishes in the paramagnetic phase. However, if the motion is slowed down on lowering temperature, these $\boldsymbol{\mu}_n$ can be observed like classical vectors. We consider an

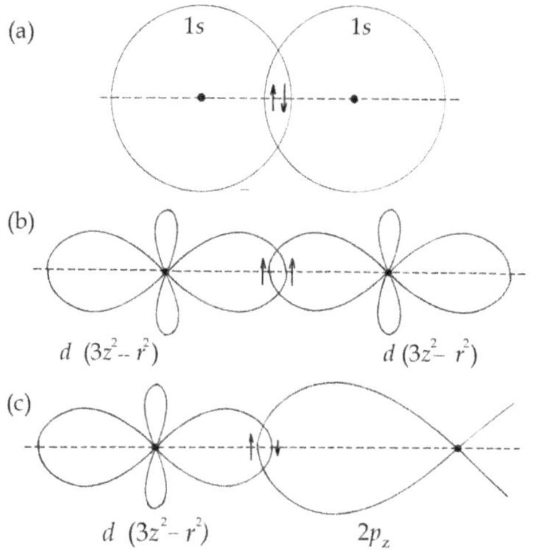

Figure 17.2. Electron exchange between two ions. (a) Two 1s orbitals in H_2. (b) Two $3d(3z^2 - r^2)$ orbitals of Fe^{3+} ions. (c) 3d orbital of a magnetic ion and p orbital of a negative ion.

adiabatic potential ΔU_n at the site n that emerges in the direction of magnetization to drive $\boldsymbol{\mu}_n$ at and below the critical temperature T_c, similar to binary pseudospins. Owing to the relation $\boldsymbol{\mu}_n = g_S\beta \boldsymbol{S}_n$, inversion symmetry $\boldsymbol{S}_n \to -\boldsymbol{S}_n$ holds by reversing $\boldsymbol{B}$, while both $\boldsymbol{\mu}_n$ and $\boldsymbol{S}_n$ are in *precession* around the direction of $\boldsymbol{B}$. Microscopically, magnetic moments should be in relative motion with respect to the ions, while mutual interactions are attributed to *electron exchange* between $\boldsymbol{\mu}_n$ and $\boldsymbol{\mu}_{n'}$.

In early physics, mutual magnetic dipole–dipole interactions were considered as too weak for magnetic ordering, but Heisenberg proposed that electron exchange between ions in short distance can be responsible for ferromagnetic interactions, referring to its correct order of magnitude. Following his idea, the exchange interaction is considered for two Fe^{3+} ions in iron and similar cases in other magnetic crystals, similar to a covalently bonded hydrogen molecule H_2.

Figures 17.2(a) and (b) show two 1s orbitals in H_2 and two adjacent orbitals of d $(3z^2 - r^2)$ in an iron crystal, respectively, showing that the magnitude of overlapped wave functions are essential for magnetic interactions.

Denoting interacting atoms by A and B, the wave functions of two electrons can be written as $\varphi_{A,B}(\boldsymbol{r}_i)\chi_{A,B}(s_{i,j})$, where $i = 1, 2$ and $j = +\frac{1}{2}, -\frac{1}{2}$. The total wave function can be expressed in determinant form, to satisfy Pauli's principle, namely

$$\Psi \propto \begin{vmatrix} \varphi_A(\boldsymbol{r}_1)\chi_A(s_1) & \varphi_B(\boldsymbol{r}_1)\chi_B(s_1) \\ \varphi_A(\boldsymbol{r}_2)\chi_A(s_2) & \varphi_B(\boldsymbol{r}_2)\chi_B(s_2) \end{vmatrix},$$

indicating that the sign of Ψ changes by exchanging electrons between $\boldsymbol{r}_1$ and $\boldsymbol{r}_2$. Also noted is that combined spin functions of $\chi_A\left(\frac{1}{2}\right)\chi_A\left(\frac{1}{2}\right)$, and $\chi_B\left(-\frac{1}{2}\right)\chi_B\left(-\frac{1}{2}\right)$, signifying the total spin +1 and −1, respectively, which are denoted by α and β in the following discussion.

Considering four states $\varphi_{\alpha\alpha}$, $\varphi_{\alpha\beta}$, $\varphi_{\beta\alpha}$ and $\varphi_{\beta\beta}$ as the basis functions, we calculate the matrix of Hamiltonian $\mathcal{H}$ for two electrons, including their kinetic energies and all Coulomb potentials. The diagonal element $\mathcal{H}_{\alpha\alpha}$, for instance, is written as

$$\begin{aligned}\mathcal{H}_{\alpha\alpha} &= \sum_{s_1,s_2} \iint \varphi_{\alpha\alpha}{}^{*}\mathcal{H}\varphi_{\alpha\alpha}\, \mathrm{d}^3\boldsymbol{r}_1\, \mathrm{d}^3\boldsymbol{r}_2 \\ &= \iint \varphi_{\mathrm{A}}{}^{*}(\boldsymbol{r}_1)\varphi_{\mathrm{B}}{}^{*}(\boldsymbol{r}_2)\mathcal{H}\varphi_{\mathrm{A}}(\boldsymbol{r}_1)\varphi_{\mathrm{B}}(\boldsymbol{r}_2)\mathrm{d}^3\boldsymbol{r}_1\, \mathrm{d}^3\boldsymbol{r}_2 \\ &\quad - \iint \varphi_{\mathrm{A}}{}^{*}(\boldsymbol{r}_1)\varphi_{\mathrm{B}}{}^{*}(\boldsymbol{r}_2)\mathcal{H}\varphi_{\mathrm{B}}(\boldsymbol{r}_1)\varphi_{\mathrm{A}}(\boldsymbol{r}_2)\mathrm{d}^3\boldsymbol{r}_1\, \mathrm{d}^3\boldsymbol{r}_2,\end{aligned}$$

where φ_{A} and φ_{B} are orthogonal functions for $\mathcal{H}_{\alpha\alpha}$, etc. Writing these terms in $\mathcal{H}_{\alpha\alpha}$ as K_{AB} and $-J_{\mathrm{AB}}$, the interaction $\mathcal{H}(\mathrm{A, B})$ can be expressed by a determinant

$$\mathcal{H}(\mathrm{A, B}) = \begin{vmatrix} K_{\mathrm{AB}} - J_{\mathrm{AB}} & 0 & 0 & 0 \\ 0 & K_{\mathrm{AB}} & -J_{\mathrm{AB}} & 0 \\ 0 & -J_{\mathrm{AB}} & K_{\mathrm{AB}} & 0 \\ 0 & 0 & 0 & K_{\mathrm{AB}} - J_{\mathrm{AB}} \end{vmatrix}. \tag{17.9a}$$

Here J_{AB} is referred to as the *exchange integral*. Transforming this determinant to a diagonal form, we obtain two eigenvalues $\varepsilon_3 = K_{\mathrm{AB}} - J_{\mathrm{AB}}$ and $\varepsilon_1 = K_{\mathrm{AB}} + J_{\mathrm{AB}}$, whose eigenfunctions are

$$\Psi_{\alpha\alpha},\quad \frac{1}{\sqrt{2}}(\Psi_{\alpha\beta} + \Psi_{\beta\alpha}),\quad \Psi_{\beta\beta} \quad \text{for} \quad s_1 + s_2 = 1$$

and

$$\frac{1}{\sqrt{2}}(\Psi_{\alpha\beta} - \Psi_{\beta\alpha}) \qquad \text{for} \quad s_1 + s_2 = 0, \tag{17.9b}$$

representing parallel and antiparallel spins, respectively. Referring to the total spin $\boldsymbol{S} = s_1 + s_2 = 1$ and 0, the first three functions and the second function in (17.9*b*) are referred to as *triplet* and *singlet* states, respectively, whose energy gap is given by $2|J_{\mathrm{AB}}|$.

The exchange integral can be calculated as

$$J_{\mathrm{AB}} = \iint \varphi_{\mathrm{A}}{}^{*}(\boldsymbol{r}_1)\varphi_{\mathrm{B}}(\boldsymbol{r}_2)\frac{e^2}{|\boldsymbol{r}_1 - \boldsymbol{r}_2|}\varphi_{\mathrm{B}}(\boldsymbol{r}_1)\varphi_{\mathrm{A}}(\boldsymbol{r}_2)\mathrm{d}^3\boldsymbol{r}_1\, \mathrm{d}^3\boldsymbol{r}_2, \tag{17.10}$$

which is numerically evaluated as positive for s-orbitals in H_2, and negative for p-orbitals in O_2. The sign of J_{AB} determines if the ground state is singlet or triplet. For the ferromagnetic state of iron crystals, Heisenberg's theory predicted that the exchange integral estimated in figure 17.2(b) is positive for parallel spins, giving rise to ferroelectric spin arrangement in crystals.

Exchanging *unpaired* electrons between neighboring magnetic ions binds them covalently together with parallel spins, whereas negative J_{AB} gives rise to antiparallel spins. Such an exchange can occur indirectly in other compounds via an ion in between. Crystals of perovskite KMF_3, where the intervening ion M can be either one of Mn^{2+}, Fe^{2+}, Co^{2+}, Ni^{2+}, exhibit *antiferromagnetism.* These crystals show antiferromagnetic phases below the critical temperature T_N, called the Neél temperature; figure 17.3 illustrates the arrangement of magnetic spins in these crystals. In $KMnF_3$ for example, below $T_N = 88$ K, unpaired electrons are exchanged through $Mn^{2+}\cdots F^{-}\cdots Mn^{2+}$, as shown in figure 17.2(c). In this case, the integral J_{AB} was considered as negative for antiferromagnetic spin arrangement.

Spin–spin correlations associated with exchanging unpaired electrons can be indicated by $s_{Az}s_{Bz} = \pm\frac{1}{4}$, referring to the A–B-axis as z, for the two-spin states to be triplet and singlet. Nonetheless, this relation can be generalized to the scalar product $\boldsymbol{s}_A \cdot \boldsymbol{s}_B = \pm\frac{1}{4}$ of two classical vectors $\boldsymbol{s}_A$ and $\boldsymbol{s}_B$, which is valid if both spins are in rapid precession, so $\langle s_{Ax}s_{Bx} + s_{Ay}s_{By}\rangle = 0$ in the first order. Therefore, the exchange integral J_{AB} can be replaced conveniently by $\frac{1}{2}(1 \pm 4\boldsymbol{s}_A \cdot \boldsymbol{s}_B)J_{AB}$, for which $\boldsymbol{s}_A \cdot \boldsymbol{s}_B = \frac{1}{4}$ and $-\frac{1}{4}$ specify triplet and singlet states, respectively. Hence, the correlation energy can be written as

$$\mathcal{H}_{ex} = -2J_{AB}\boldsymbol{s}_A \cdot \boldsymbol{s}_B, \tag{17.11}$$

excluding the constant term. Except for the factor 2, this formula (17.11) for spin–spin correlations is in identical form to the previous one for the non-magnetic correlation of pseudospins. Note from (17.8), the integral J_{AB} is a function of distance $|\boldsymbol{r}_A - \boldsymbol{r}_B|$ and highly directional in crystal space via an intervening ion, if any, between magnetic ions. The latter indirect exchange is called *superexchange.* For thermodynamics, the sign of J_{AB} in any case is significant to determine the spin-spin coupling to lower energy by $\mathcal{H}_{ex}$.

17.4 Collective propagation of Larmor's precession

There is an *easy axis* of magnetization in a magnetized crystal, as shown for iron in figure 17.4. In this context, the easy direction can be identified as the axis for electron

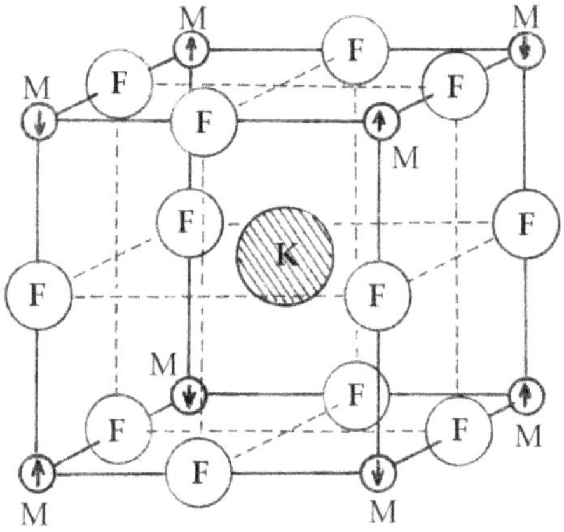

Figure 17.3. Structure of KMF_3: M represents Mn, Fe, Co and Ni.

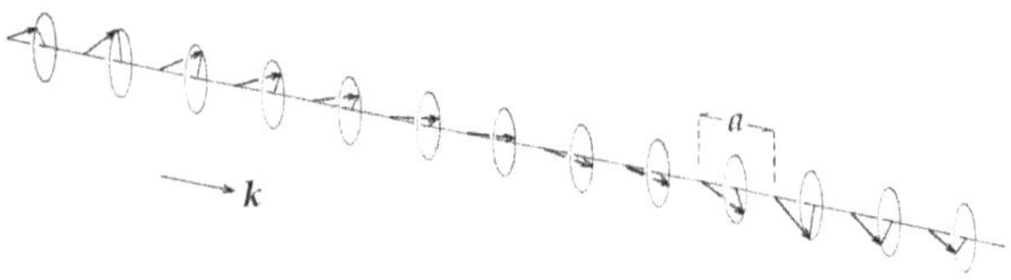

Figure 17.4. One-dimensional array of spins in Larmor precession in an internal field $\boldsymbol{B}_{\text{int}} \| \boldsymbol{k}$.

exchanges in the lattice, which were actually substantiated by neutron diffraction experiments.

In the axis for electron exchanges, collective motion of $\boldsymbol{\mu}_n$ occurs as described by Fourier series, showing their propagation in the periodic lattice. In the condensate model, it is a valid assumption that such a propagation is driven by an adiabatic potential that synchronizes collective $\boldsymbol{\mu}_n$ in this direction. Considering a microscopic moment $\boldsymbol{\mu}_n$ as displacements from the site n, the collective $\boldsymbol{\mu}_n$ should be in-phase with sinusoidal *magnetic lattice displacements* $\boldsymbol{u}_{M,n}$. Disregarding uncertainties in the transition region, such a collective $\boldsymbol{\mu}_n$ mode can be written as

$$\boldsymbol{\mu}_n = \boldsymbol{\mu}_o \exp i(\boldsymbol{q} \cdot \boldsymbol{r}_n - \omega t_n),$$

or directly by the transformed Fourier amplitude

$$\boldsymbol{\mu}_o = \boldsymbol{\mu}_n \exp\{-i(\boldsymbol{q} \cdot \boldsymbol{r}_n - \omega t_n)\}.$$

Considering space–time uncertainties $\Delta \boldsymbol{r}_n$ and Δt_n at sites n, it is convenient to express these modes by the phase variables $\phi_n = \boldsymbol{k} \cdot \boldsymbol{r}_n - \omega t_n$ with phase uncertainties $\Delta\phi_n = \boldsymbol{k} \cdot \Delta \boldsymbol{r}_n - \omega \Delta t_n$.

In the presence of a strong internal magnetic field $\boldsymbol{B}_{\text{int}}$, the magnetic moments $\boldsymbol{\mu}_n$ are in precession around its direction of $\boldsymbol{B}_{\text{int}}$, where the angles of precession are synchronized with lattice translation, as sketched in figure 17.4. Hence, the collective mode of $\boldsymbol{\mu} = \sum \boldsymbol{\mu}_n$ is composed of longitudinal $\mu_{\|}$ and transverse $\mu_{\perp}$ with respect to the direction of $\boldsymbol{B}_{\text{int}}$, both propagating at the same phase $\phi = \boldsymbol{k} \cdot \boldsymbol{r} - \omega_{\text{int}} t$, where $\omega_{\text{int}} = \gamma B_{\text{int}}$ is the Larmor frequency in B_{int} that is considered as the consequence of soliton potential proportional to $\text{sech}^2\,\phi$.

Writing these moments as classical vectors $\boldsymbol{\mu}_m = \mu \boldsymbol{e}_m$ and $\boldsymbol{\mu}_n = \mu \boldsymbol{e}_n$, where $|\boldsymbol{e}_m| = |\boldsymbol{e}_n| = 1$ for unit vectors, the short-range interaction energy at the site n is given by

$$\mathcal{H}_n = -\sum_m 2J_{mn} \boldsymbol{\mu}_m \cdot \boldsymbol{\mu}_n = -\mu^2 \sum_m 2J_{mn} \boldsymbol{e}_m \cdot \boldsymbol{e}_n, \tag{17.12}$$

where J_{mn} is an exchange integral between m and n sites.

Assuming μ to be a constant, these *real* vectors can be written as

$$\boldsymbol{e}_m = \boldsymbol{e}_{+q} \exp i\phi_m + \boldsymbol{e}_{-q} \exp(-i\phi_m) \quad \text{and} \quad \boldsymbol{e}_n = \boldsymbol{e}_{+q} \exp i\phi_n + \boldsymbol{e}_{-q} \exp(-i\phi_n),$$

where $\phi_m = \boldsymbol{q} \cdot \boldsymbol{r}_m - \omega t_m$ and $\phi_n = \boldsymbol{q} \cdot \boldsymbol{r}_n - \omega t_n$ are phases at sites m and n, respectively. In chapter 5, we showed that the average of $\mathcal{H}_n$ over the timescale of observation t_o is given by

$$\langle \mathcal{H}_n \rangle_t = -2\mu^2 \Gamma_t \boldsymbol{e}_{+q} \cdot \boldsymbol{e}_{-q} J(\boldsymbol{q}), \tag{17.13a}$$

where

$$J(\boldsymbol{q}) = \sum_m J_{mn} \exp i\boldsymbol{q} \cdot (\boldsymbol{r}_m - \boldsymbol{r}_n) \tag{17.13b}$$

and

$$\Gamma_t = \frac{1}{2t_0} \int_{-t_0}^{+t_0} \exp\{-i(t_m - t_n)\} \mathrm{d}(t_m - t_n) = \frac{\sin \omega t_0}{\omega t_0}. \tag{17.13c}$$

Equations (17.13*a*) and (17.13*b*) represent such a cluster centered at $\boldsymbol{\mu}_n$ with a limited number of interacting neighbors, which can be considered as a *seed* for magnetic condensation. For $J(\boldsymbol{q})$, however, we consider electron exchanges only with nearest neighbors to keep local symmetry unchanged, as next-nearest neighbor interactions J' disrupt local symmetry.

In the presence of a magnetic field $\boldsymbol{B}$, we consider that the magnetic moments $\boldsymbol{\mu}_n$ are in propagation mode. Minimizing $J(\boldsymbol{q})$ with respect to $\boldsymbol{q}$, we can determine the axis of propagation $\boldsymbol{q}$ for $\boldsymbol{B} = 0$, which should be the same axis for minimum strains in the lattice. If $\boldsymbol{B} \neq 0$, on the other hand, the collective motion is described in terms of phases ϕ_n; all $\boldsymbol{\mu}_n$ waves are synchronized in-phase to acquire stability in the crystal. This process for $\Delta\phi_n \to 0$ can be observed as thermal relaxation for $\boldsymbol{B} \| \boldsymbol{q}$. Further, such a mode of $\boldsymbol{\mu}_n$ below T_c is commensurate with the lattice, as signified by either $\boldsymbol{q} = 0$ or $\boldsymbol{q} = \boldsymbol{G}/2$, corresponding to *ferromagnetic* or *antiferromagnetic* arrangements of $\boldsymbol{\mu}_n$.

17.5 Magnetic Weiss field

Such magnetic correlations as (17.10) are considered for the effective intrinsic field $\boldsymbol{B}_{\text{int}}$ at site n that was defined by

$$\mathcal{H}_n = -\boldsymbol{\mu}_n \cdot \boldsymbol{B}_{\text{int}} \quad \text{where } \boldsymbol{B}_{\text{int}} = \sum_m 2J_{mn}\boldsymbol{\mu}_m,$$

representing the Weiss field. Postulated originally by Weiss, $\boldsymbol{B}_{\text{int}} \propto \boldsymbol{M}$, where $\boldsymbol{M}$ is the macroscopic magnetization, we can assume $\boldsymbol{B}_{\text{int}} = -2J\langle \boldsymbol{\mu}_m \rangle$ in the mean-field approximation, where $J = \langle J_{mn} \rangle$, for $\boldsymbol{k} = 0$ or $\boldsymbol{G}/2$. In the field-theoretical approximation, on the other hand, $\boldsymbol{B}_{\text{int}}$ is determined by the soliton potential proportional to $\operatorname{sech}^2 \phi$, where $\operatorname{sech} \phi$ at $\phi = 0$ determines magnetic domain boundaries. Accordingly, in this case we can write $\boldsymbol{B}_{\text{int}} = -\lambda \boldsymbol{M}(\phi)$, allowing us to discuss a time-dependent Weiss field. For thermodynamic applications, $\boldsymbol{M}(\phi)$ must be renormalized for virtual spatial uniformity, hence we must consider that magnetized crystals should always be in *ellipsoidal* shape.

Assuming such a uniform $\boldsymbol{M}$ in a given crystal, the Weiss formula can be expressed as $\boldsymbol{M} = \chi_0(\boldsymbol{B} + \lambda \boldsymbol{M})$, where χ_0 is a paramagnetic susceptibility for the Curie law $\chi_0 = C/T$. Thus, we obtain the equation $M = \frac{C}{T}(B + \lambda M)$, hence

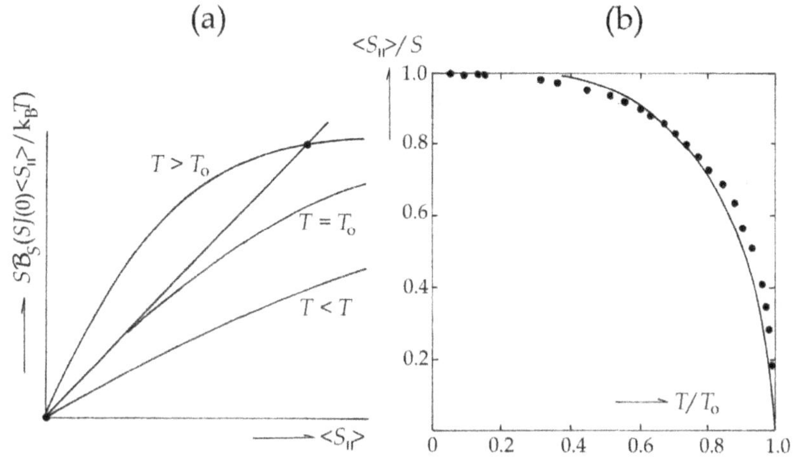

Figure 17.5. Brillouin plots. (a) $SB_S\{SJ(0)\langle S_{||}\rangle/k_BT\}$ versus $\langle S_{||}\rangle$. (b) $\langle S_{||}\rangle/S$ versus T/T_0.

$$\chi = \frac{M}{B} = \frac{C}{T - T_0} \qquad \text{where} \qquad T_0 = C\lambda. \tag{17.14}$$

This relation is the Curie–Weiss law indicating a singularity at $\chi \to \infty$, i.e. $T \to T_0$, called the *Curie temperature*.

Considering $\boldsymbol{B}_{int}$ as if applied externally, the Brillouin formula can be written as

$$M(B, T) = g\beta\langle S_{||}\rangle = g\beta S\mathcal{B}_S\left(\frac{g\beta SB_{int}}{k_BT}\right) \qquad \text{where} \qquad B_{int} = g\beta 2J\langle S_{||}\rangle.$$

Therefore,

$$\langle S_{||}\rangle = S\mathcal{B}_S\left(\frac{2g^2\beta^2 SJ\langle S_{||}\rangle}{k_BT}\right). \tag{17.15}$$

For a small value of $g\beta SB_{int}/k_BT$ in the vicinity of critical temperature T_c, the Brillouin function can be approximated as

$$\mathcal{B}_S \cong \frac{S+1}{3S}\frac{g\beta SB_{int}}{k_BT} = \frac{g^2\beta^2 S(S+1)}{3k_B}\frac{2J}{T}\frac{\langle S_{||}\rangle}{S}.$$

Denoting $T_0 = \frac{2g^2\beta^2 JS(S+1)}{3k_B}$, we obtain the relation $\frac{\langle S_{||}\rangle}{S} = \frac{T_0}{T}\frac{\langle S_{||}\rangle}{S}$ from (17.15), which implies that $\langle S_{||}\rangle = S$ if $T = T_0$. If the spin–spin interaction in (17.14) is limited to the shortest distance, this is consistent with assuming uniform $\boldsymbol{M}$.

Equation (17.16) can be solved numerically for $\langle S_{||}\rangle/S$, in a similar manner to that discussed in section 0.6 in Bragg–Williams' theory. Figure 17.5(a) shows a graphical sketch, where the curve of $\mathcal{B}_S\left(\ldots, \frac{\langle S_{||}\rangle}{S}\right)$ versus $\frac{\langle S_{||}\rangle}{S}$ is crossed by the straight line of the 45°-slope at $T = T_0$. If $T < T_0$, we have a non-zero solution $\langle S_{||}\rangle \neq 0$, whereas

$\langle S_{\|} \rangle = 0$, if $T > T_0$. Therefore, the temperature $T = T_0$ in this analysis should be the critical temperature.

Figure 17.5(b) shows another plot of $\frac{\langle S_{\|} \rangle}{S}$ versus $\frac{T}{T_0}$, where the numerical curve calculated for $S = \frac{1}{2}$ fits reasonably well to the experimental data for Ni. However, calculated curves for larger values of S, as $S \to \infty$, deviate substantially from $S = \frac{1}{2}$, so that the latter curve, supported experimentally, implies that the nearest-neighbor correlations are overwhelming in an iron-group magnet. In this context, spin–spin correlations in the long range are considered as insignificant, at least in Ni.

Of further note from the above analysis with the Brillouin function is that for small values $\frac{\langle S_{\|} \rangle}{S}$, the slope of Brillouin function against $\frac{T}{T_0}$ for Ni is almost vertical in the vicinity of T_c for $S = 1/2$, as seen from figure 17.6(b). In contrast, the calculated curve for $S \to \infty$ looks simliar to the graph of an equation $y = \tanh y$ that represents a typical binary ordering. The difference between these curves suggests that an internal Weiss field exists in the latter case, but is absent or insignificantly small in the former case for Ni and Fe. A large S can logically be attributed to virtually all spins as $\boldsymbol{S} = \sum_n s_n$, on the other hand there is no concern about the cluster size in the Brillouin theory. Hence, a difference may be attributed to long-range electronic spin–spin correlations among s_n constituting $\boldsymbol{S}$, which, however, is insignificant to consider in magnetic crystals, as evidenced by the Brillouin analysis.

It is an interesting comparison with ferroelectric crystals, where Cochran considered only short-range interactions counteracting with polarizations on surfaces for his model discussed in chapter 7, as there is no significant polarization density at intermediate distances, if dielectric crystals are strongly polarized. Nevertheless,

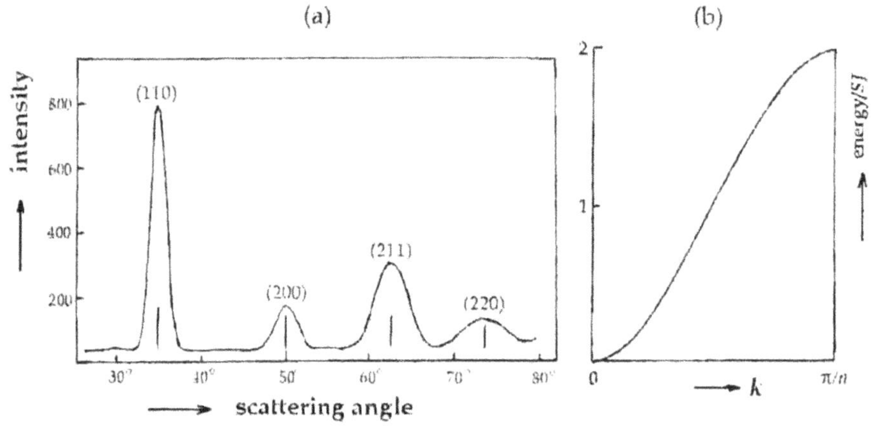

Figure 17.6. (a) A neutron diffraction spectrum from Fe. (b) Magnon energy versus k^2 of ferromagnet $MnPt_3$. Reproduced with permission from [1].

such an interpretation is not realistic, unless substantiated in the timescale of experiments.

In any case, in a given magnetic crystal, the Weiss field of correlated magnetic spins $\boldsymbol{S}_n$ of magnetic ions is realistic to deal with magnetic phase transitions.

17.6 Spin waves

It is noted that the magnetic spin correlation (17.11) is invariant for transformation in ferromagnetic crystals along the easy direction, where the Weiss field $\boldsymbol{B}_{\text{int}}$ is characterized by inversion symmetry. The magnetic field drives spins $\boldsymbol{S}_n$ in Larmor's precession, whose collective motion is driven by an adiabatic potential $\boldsymbol{B}_{\text{int}}$ synchronized with the applied field $\boldsymbol{B}_\text{o}$. However, dominated by nearest-neighbor interactions in magnetized crystals, the non-linear effect is regarded usually as negligible for spin waves of $\boldsymbol{S}_n$. Such a wave is commensurate at $\boldsymbol{q} = 0$ and $\boldsymbol{q} = \pm \boldsymbol{G}/2$, exhibiting ordered spin arrangements, as verified by neutron diffraction experiments.

Spin waves should dominate magnetic crystals at temperatures for $T < T_\text{o}$, for which the short-range interactions (17.12) can be expressed as $-2J\boldsymbol{S}_n.\ (\boldsymbol{S}_{n-1} + \boldsymbol{S}_{n+1})$ with J for the nearest neighbors, and hence the Weiss field is expressed as

$$\boldsymbol{B}_{\text{int}\,n} = -\frac{2J}{g\beta}(\boldsymbol{S}_{n-1} + \boldsymbol{S}_{n+1}). \tag{17.16}$$

The spin angular momentum is determined by the torque $\boldsymbol{\mu}_n \times \boldsymbol{B}_{\text{int}\,n}$, so that we can write the equation of motion as

$$\frac{\mathrm{d}S_n}{\mathrm{d}t} = 2J(\boldsymbol{S}_n \times \boldsymbol{S}_{n-1} + \boldsymbol{S}_n \times \boldsymbol{S}_{n+1}).$$

By considering small $|\boldsymbol{S}_n|$, this equation can be linearized as

$$\begin{aligned} S_{n,x} &= 2JS(2S_{n,y} - S_{n-1,y} - S_{n+1,y}),\\ S_{n,y} &= 2JS(2S_{n,x} - S_{n-1,x} - S_{n+1,x}) \end{aligned} \tag{17.17}$$

and

$$S_{n,z} = S\ (\text{const.}),$$

where x and y are transverse directions to propagation along the z-axis, noting that (17.17) gives the basic nearest-neighbor equations for $\boldsymbol{S}$ in Toda's lattice. Letting $z = ka$, where a is the distance between nearest neighbors, we look for the traveling-wave solutions

$$S_{n,x} = u \exp i(nka - \omega t) \quad \text{and} \quad S_{n,y} = v \exp i(nka - \omega t),$$

where u and v are constants. Therefore,

$$-i\omega u = 4JS(1 - \cos ka)v \quad \text{and} \quad -i\omega v = -4JS(1 - \cos ka)u,$$

from which we obtain the dispersion relation

$$\omega = 4JS(1 - \cos ka) \approx 2JSa^2k^2 \qquad \text{for} \qquad ka \ll 1. \tag{17.18}$$

The dispersion relation (17.18) of such spin waves was actually confirmed with neutron inelastic scattering experiments, showing a significant dispersion relation $\omega = \omega(\boldsymbol{q})$ resulting from neutron inelastic scatterings for wave number change $\Delta\boldsymbol{K} = \pm\boldsymbol{q}$ along specific symmetrical axes $\boldsymbol{q}$ signified by anisotropic J. Figure 17.6(a) shows a neutron diffraction pattern from iron crystals, measured with varying scattering angle, where crystal planes of ordered spins are identified in strong intensities. The dispersion relation (17.18) was verified by scanning scattering angle in the vicinity of $\Delta\boldsymbol{K} = \pm\boldsymbol{q}$. Figure 17.6(b) is an example of experimental spin-wave dispersion, showing a reasonable agreement with (17.18).

It is interesting to note that the short-range interaction as expressed by (17.17) suggests Toda's exponential lattice, where distant magnetic spins are all correlated with Toda's potentials theoretically. Therefore, we can consider that the spin wave constitutes *displacement wave* in the soliton theory, although physically no significantly new outcome can be predicted. In any case, it is logical to consider the spin $\boldsymbol{s}_n$ as a classical displacement vector.

17.7 Magnetic anisotropy

Calculating with the formula (17.13*b*) of section 17.4, magnetic spins are clustered via electron exchange $J(\boldsymbol{q})$ for minimum correlations in a specific direction determined by $\nabla_k J(\boldsymbol{k}) = 0$. Determined by this equation, the vector $\boldsymbol{q}$ should be parallel to a symmetry axis for *easy magnetization*. Figure 17.7 shows magnetization curves of a single crystal of Fe, where the easy axis is the [100] direction. On the other hand, for $\boldsymbol{B}\|[111]$ the magnetization versus $\boldsymbol{B}$ curve is at a slower rate to reach saturation with increasing $\boldsymbol{B}$ than $\boldsymbol{B}\|$ [100].

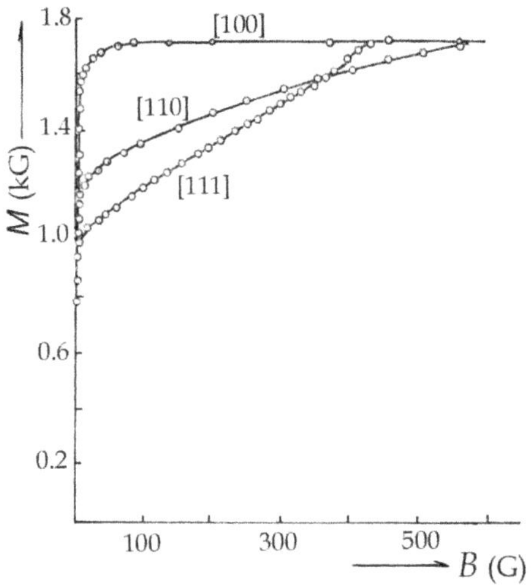

Figure 17.7. Magnetization curves for Fe. [100] is the easy magnetization axis. Reproduced with permission from [1].

A magnetized crystal can be strained, depending on the direction of applied field $\boldsymbol{B}$. In figure 17.8, the graphic area under a magnetization curve represents a free energy for magnetization, that is

$$\Delta F = -\int_0^B \boldsymbol{M} \cdot \mathrm{d}\boldsymbol{B}. \tag{17.19}$$

The process for saturated magnetization $\boldsymbol{M}$ is quasi-static in practice, during which the lattice is strained with increasing $\boldsymbol{B}$; this phenomenon is called *magnetostriction*.

The strain energy in the lattice can be expressed in terms of small displacements $\boldsymbol{u}_n$ as

$$\Delta U = \sum_{n,ij} \boldsymbol{u}_{n,i} K_{n,ij}^{(1)} u_{n,j} + \sum_{n,ij} u_{n,i}^2 K_{n,ij}^{(2)} u_{n,j}^2 + \cdots, \tag{17.20a}$$

where indexes i and j represent lattice points. On the other hand, with $\boldsymbol{B}$ applied in a direction off the easy axis, the magnetic correlation energy among $\boldsymbol{\mu}_n$ can be expressed by

$$\Delta U_{\mathrm{mag}} = \sum_{n,ij} \mu_{n,i} K_{n,\,ij}^{(1)} \mu_{n,j} + \sum_{n,ij} \mu_{n,\,i}^2 K_{n,\,ij}^{(2)} \mu_{n,\,j}^2 + \cdots, \tag{17.20b}$$

which is related to (17.20a) by the thermodynamic process

$$\Delta U + \Delta U_{\mathrm{mag}} \rightarrow \text{minimum}, \tag{17.21}$$

owing to Born–Huang's principle; the total strain energies should be minimized under the equilibrium condition with the surroundings. If $\boldsymbol{B} = 0$, the magnetic tensor $K_{n,ij}$ in (17.20b) should be symmetric and conformal with the strain tensor $K_{n,ij}$ in (17.20a), whereas for $\boldsymbol{B} \neq 0$, inversion symmetry of $\boldsymbol{\mu}_n$ should primarily be established with regard to $\boldsymbol{B}$, so that K_{ij} and K_{ij} are not conformal in general. In

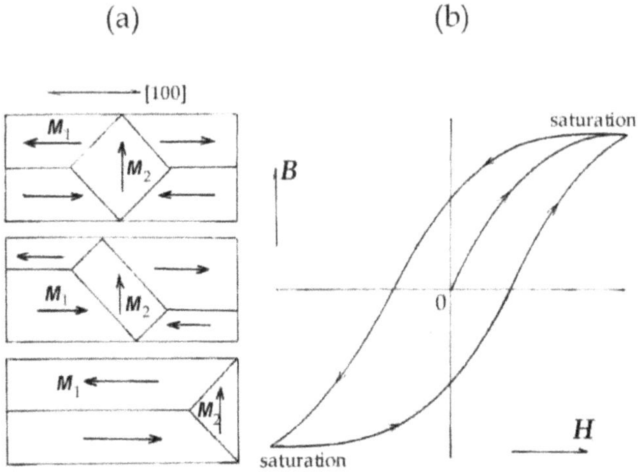

Figure 17.8. (a) Domain patterns in ferromagnetic iron. (b) Hysteresis in a magnetization process. Reproduced with permission from [1].

this case, the tensor $K_{n,ij}$ in (17.20*b*) is not symmetrical in each domain, i.e. $K_{n,ij} \neq K_{n,ji}$. Writing

$$K_{n,ij} = \frac{1}{2}(K_{n,ij} + K_{n,ji}) + \frac{1}{2}(K_{n,ij} - K_{n,ji}),$$

with respect to the magnetic spin inversion axis determined by $\boldsymbol{q}$ in the magnetic lattice, the tensor $K_{n,ij}$ is obviously a symmetrical tensor if $K_{n,ij} = K_{n,ji}$; but is antisymmetric if $K_{n,ij} = -K_{n,ji}$. Dzialoshinsky proposed that the spin–spin correlations can be characterized by off-diagonal elements $J_{mn} = -J_{nm}$ for the latter case, which is, however, consistent with inversion $\boldsymbol{q} \rightleftarrows -\boldsymbol{q}$ of the *magnetic displacement* $\boldsymbol{u}_M \propto \mu$ between these symmetric and antisymmetric potentials, placing the latter at the P mode for lower lattice energy.

Spin–spin correlations can be evaluated by applying an external field in an easy direction of magnetization, where the Weiss field can be defined by evaluating spin–spin correlations dominated in the short range. In an arbitrary direction of $\boldsymbol{B}$, however, we need to include additional interactions between $\Delta \boldsymbol{U}$ and $\Delta \boldsymbol{U}_{\mathrm{mag}}$. Accordingly, we express the spin–spin interaction as

$$\mathcal{H}_{SS} = -2\sum_{m.n} J_{mn}\boldsymbol{S}_m \cdot \boldsymbol{S}_n + \sum_{m.n}\left(\sum_{i.j} S_{m,i}K_{mn}S_{n.j}\right). \tag{17.22a}$$

If $K_{mn} = K_{nm}$, the quantity in the brackets can be expressed as $-\sum_n \boldsymbol{B}_{\mathrm{aniso}} \cdot \boldsymbol{S}_n$, where $\boldsymbol{B}_{\mathrm{aniso}} = \langle \sum_{m,i} S_{m,i}K_{mn} \rangle$ is another internal field due to anisotropic spin arrangements in the mean-field approximation. Combining with the magnetic Weiss field $\boldsymbol{B}_{\mathrm{int}} = 2\langle \sum_m J_{mn}\boldsymbol{S}_m \rangle$, $\mathcal{H}_{SS}$ in (17.22*a*) can be interpreted as

$$\mathcal{H}_{SS} = -\sum_n (\boldsymbol{B}_{\mathrm{int}} + \boldsymbol{B}_{\mathrm{aniso}}) \cdot \boldsymbol{S}_n. \tag{17.22b}$$

When $\boldsymbol{B}$ is parallel to the easy axis, we have $\boldsymbol{B}_{\mathrm{int}} \| \boldsymbol{B}_{\mathrm{aniso}}$, otherwise $\boldsymbol{B}_{\mathrm{int}} \|$(easy axis) and $\boldsymbol{B}_{\mathrm{aniso}} \| \boldsymbol{B}$. As will be shown in the following sections, it is important to consider both $\boldsymbol{B}_{\mathrm{aniso}}$ and $\boldsymbol{B}_{\mathrm{int}}$ together for dealing with the Zeeman behavior of magnetization in magnetic crystals.

On the other hand, for the antisymmetric tensor $K_{mn} = -K_{nm}$, the spin–spin correlation energy can be calculated as

$$\begin{aligned}\mathcal{H}_{mn} &= \sum_{m.n}\sum_{i,j} S_{m,i}K_{mn}S_{n,j} = \sum_{m.n}\sum_{i,j}\frac{1}{2}(S_{m,i}K_{mn}S_{n,j} + S_{n,j}K_{nm}S_{m,i}) \\ &= \frac{1}{2}\sum_{m.n}\sum_{i,j} K_{mn}(S_{m,i}S_{n,j} - S_{m,j}S_{n,i}).\end{aligned}$$

Defining a vector $\boldsymbol{D} = \left(\frac{K_{mn}}{2}, \frac{K_{mn}}{2}, \frac{K_{mn}}{2}\right)$, this expression implies a new type of spin–spin couplings, which can be written as

$$\mathcal{H}_{mn} = \boldsymbol{D} \cdot (\boldsymbol{S}_m \times \boldsymbol{S}_n), \tag{17.23}$$

which is known as Dzialoshinsky–Moriya's interaction [2, 3]. Although not always significant in magnetic crystals, a coupling such as (17.23) is required for unusual magnetic crystals called *antisymmetric magnets*, such as α Fe_2O_3(α-hematite). Nevertheless, we shall not pursue antisymmetric crystals, as such couplings exist only in rare examples.

The external work by $\boldsymbol{B}_o$ on a magnetic crystal is expressed by $-\boldsymbol{M} \cdot \boldsymbol{B}_o$, for which we can consider a torque $\boldsymbol{T}_o$ for rotating the vector $\boldsymbol{M}$, namely $\boldsymbol{T}_o = \boldsymbol{M} \times \boldsymbol{B}_o$, which strains the lattice structure. Nonetheless, a magnetized crystal is strained by $\boldsymbol{T}_o$ or $\boldsymbol{M}$, so that the traditional Weiss field $\boldsymbol{B}_{int}$ in strained crystals should be revised as

$$\boldsymbol{B}_{int} = \lambda \boldsymbol{M} + \boldsymbol{B}_{aniso}, \quad \text{where} \quad -\boldsymbol{M} \cdot \boldsymbol{B}_{aniso} = \Delta U_{mag}. \tag{17.24}$$

Here, $\boldsymbol{B}_{aniso}$ is the effective field due to ΔU_{mag}, and $\lambda \boldsymbol{M}$ represents the Weiss field for $\boldsymbol{B}_o = 0$.

The ferromagnetic phase shows a complicated domain pattern below T_c. As illustrated in figure 17.8, movable domains by $\boldsymbol{B}_o$ are not simply binary, but there are 45°-walls between magnetized domains in perpendicular directions, i.e. $\boldsymbol{M}_1 \perp \boldsymbol{M}_2$; in addition to ordinary 180°-walls between oppositely magnetized domains, i.e. $\boldsymbol{M}_1 \| -\boldsymbol{M}_1$. Inside domain walls, the lattice should be significantly strained by rotating magnetization, hence signified by matrix elements $K^{(1)}_{n,ij}$ and the related potential $\Delta U^{(1)}_{mag}$ determined by the local symmetry of a sample crystal that was presumably not prepared homogeneouly. Despite strains, domain walls are movable by a field $\boldsymbol{B}$ in modest strength, pushing perpendicular magnetization towards surfaces. Such domain patterns can in principle be explained by phase solitons associated with $\boldsymbol{\mu}_n$ that have singularities with C_8-symmetry in a cubic Fe lattice, while the energy transfer to the lattice is elastically determined by the surface/volume ratio in domain structure.

17.8 Antiferromagnetic and ferrimagnetic states

Insulating oxides, fluorides and sulfates of transition elements exhibit a variety of *antiferromagnetic* and *ferrimagnetic* states, depending on the nature of electron exchange mechanisms in crystals. Neutron scattering experiments show diffraction patterns of crystal planes of spins below critical temperatures, providing evidence for ordered spins in two sublattices. Figure 17.9 shows the unit structure in MnF_2 crystals, indicating that Mn^{2+} ions at the body-center positions and those at the corners form sublattices of spins oriented in different directions. These spin sublattices, A and B, are magnetically specified by vectors $\boldsymbol{M}_A$ and $\boldsymbol{M}_B$, respectively, which cannot be independent, but interacting via electron exchange at nearest distances. We consider Weiss fields $\boldsymbol{B}^{(A)}_{int}$ and $\boldsymbol{B}^{(B)}_{int}$ in sublattices A and B, which are expressed as

$$\boldsymbol{B}^{(A)}_{int} = \lambda \boldsymbol{M}_A + \lambda' \boldsymbol{M}_B \quad \text{and} \quad \boldsymbol{B}^{(B)}_{int} = \lambda \boldsymbol{M}_B + \lambda' \boldsymbol{M}_A.$$

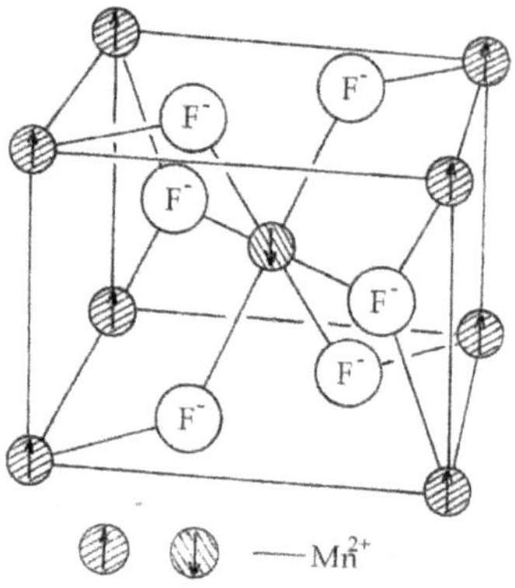

Figure 17.9. Unit structure in antiferromagnetic MnF_2.

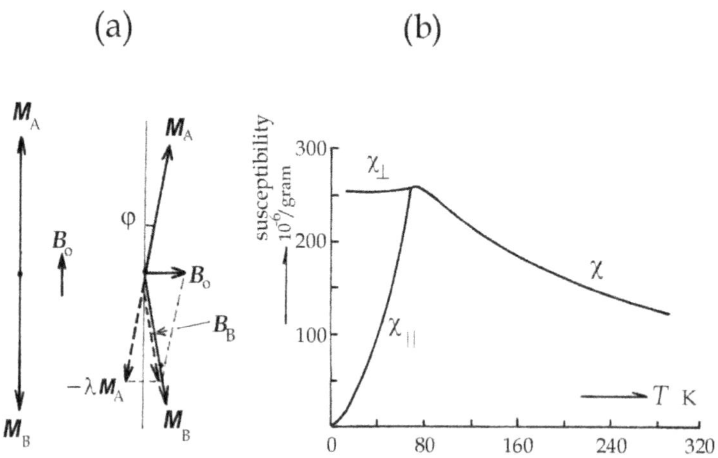

Figure 17.10. (a) A model for calculating $\chi_{||}$ and $\chi_\perp$. (b) Magnetic susceptibility versus T of antiferromagnetic MnF_2.

In the presence of a weak external field $\boldsymbol{B}_o$, *sublattice magnetizations* $\boldsymbol{M}_A$ and $\boldsymbol{M}_B$ become parallel to $\boldsymbol{B}_o - \lambda \boldsymbol{M}_B$ and $\boldsymbol{B}_o - \lambda \boldsymbol{M}_A$ in equilibrium, respectively. Figure 17.10(a) illustrates two cases of $\boldsymbol{B}_o || \boldsymbol{M}_A - \boldsymbol{M}_B$ and $\boldsymbol{B}_o \perp \boldsymbol{M}_A - \boldsymbol{M}_B$.

In the perpendicular case, we can see the relation

$$2\lambda|\boldsymbol{M}_{A,B}|\sin\varphi = |\boldsymbol{B}_o|,$$

or the susceptibility is given by

$$\chi_\perp = \frac{2|\boldsymbol{M}_{A\cdot B}|\sin\varphi}{|\boldsymbol{B}_o|} = \frac{1}{\lambda - \lambda'} \tag{17.25a}$$

where

$$\lambda = \frac{2}{Ng^2\beta^2}\left\{J\left(\frac{\boldsymbol{G}}{2}\right) - J(0)\right\}.$$

On the other hand, in the parallel case,

$$M_A = \frac{Ng\beta S}{2}\mathcal{B}_S\left\{\frac{g\beta S(B_o - \lambda M_B - \lambda' M_A)}{k_B T}\right\} \quad \text{and}$$

$$M_B = \frac{Ng\beta S}{2}\mathcal{B}_S\left\{\frac{g\beta S(B_o - \lambda M_A - \lambda' M_B)}{k_B T}\right\},$$

from which we calculate

$$\chi_{\parallel} = \lim_{B_o \to 0} \frac{M_A - M_B}{B_o}. \tag{17.25b}$$

From this, it is clear that for $T \to 0$, $M_A = M_B$ and hence $\chi_{\parallel} = 0$, while we should have $\chi_{\parallel} = \chi_{\perp}$ at the critical point, as numerically illustrated in figure 17.10(b).

17.9 Fluctuations in ferromagnetic and antiferromagnetic states

In ferromagnetic and antiferromagnetic states, the magnetization vector $\boldsymbol{M}$ is in thermodynamic equilibrium with the crystal lattice. As when in equilibrium with the strain energy, the fluctuations in $\boldsymbol{M}$ are not of random character, but anisotropic with respect to symmetry axes. Such fluctuation modes can be studied by *magnetic resonance*, where an external magnetic field $\boldsymbol{B}_o$ is applied on the sample crystal. Conventional magnetic resonance is an experimental method at $\boldsymbol{G} = 0$, but here the resonance at $|\boldsymbol{G}| = \frac{1}{2}$ can be detected by neutron inelastic scatterings in magnetic crystals with an applied field $\boldsymbol{B}_o$; nonetheless, we also call such experiments magnetic resonance.

In a magnetized crystal strained by an applied field $\boldsymbol{B}_o$, the magnetic strain potential ΔU_{mag} can be represented by $\boldsymbol{B}_{aniso}$, as defined by (17.24). Hence, when $\boldsymbol{B}_o$ is in parallel to a symmetry axis, we have $\boldsymbol{B}_o \| \boldsymbol{B}_{aniso}$, around which $\boldsymbol{M}$ is in Larmor's precession. In scattering spectra with $\boldsymbol{B}_o$, a magnetic resonance can take place at a frequency determined by $\boldsymbol{B}_o$ plus $\boldsymbol{B}_{aniso}$.

17.9.1 Ferromagnetic resonance

A ferromagnetic sample crystal needs to be prepared in ellipsoidal form in order to characterize it by a uniform magnetization, whereas an antiferromagnetic crystal has no macroscopic magnetization as a whole. Typical experimental arrangements for a ferromagnetic and antiferromagnetic resonance are illustrated in figures 17.11(a) and (b), respectively.

Inside of a ferromagnetic crystal in ellipsoidal shape, we consider a uniform $\boldsymbol{B}$ that is expressed by components $B_i = B_o - N_i\boldsymbol{M}_i$, where $i = x, y, z$; we assume $\boldsymbol{B}_o \|$ z-axis, and N_x, N_y and N_z are *demagnetization factors* of an ellipsoid. Such a demagnetizing field $(-N_x\boldsymbol{M}_x, -N_y\boldsymbol{M}_y, -N_z\boldsymbol{M}_z)$ can be regarded as an *internal long-range field* in magnetized crystals. We apply then an oscillating field $\boldsymbol{B}_1 = (\boldsymbol{B}_1)_o \exp i\omega t$ in the x–y-plane to observe magnetic resonance absorption of $\boldsymbol{B}_1$ radiation. In this case, by the torque $\boldsymbol{M} \times \boldsymbol{B}_1$, the perpendicular components M_x and M_y are rotated, as described in the following. We write

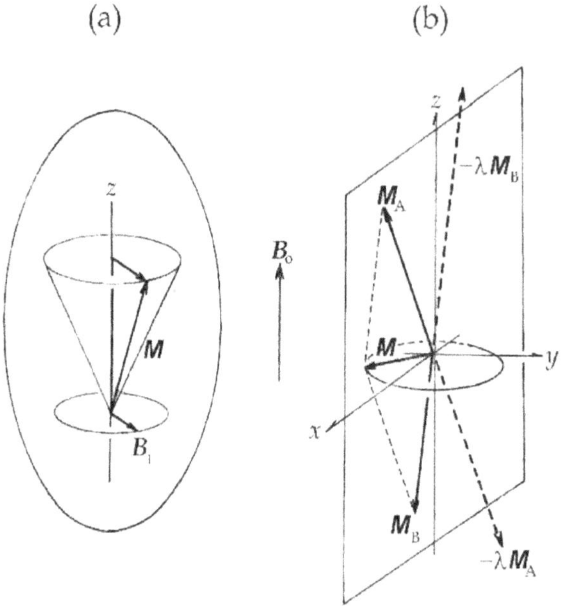

Figure 17.11. (a) Larmor's precession in a ferromagnetic crystal in ellipsoidal shape. (b) Larmor's precession in an antiferromagnetic crystal.

$$B_x = B_0 - N_x M_x,\ B_y = B_0 - N_y M_y \quad \text{and} \quad B_z = B_0 - N_z M_z;$$

hence the equations of motion of $\boldsymbol{M}$ are

$$\begin{aligned} \dot{M}_x &= \gamma(M_y B_z - M_z B_y) = \gamma\{B_0 + (N_y - N_z)M_z\}M_y, \\ \dot{M}_y &= \gamma(M_z B_x - M_x B_z) = -\gamma\{B_0 - (N_z - N_x)M_z\}M_x \end{aligned} \tag{17.26}$$

and

$$\dot{M}_z = 0,$$

where the constant γ represents $g\beta/\hbar$ of an atomic magnetic moment. From the last equation, we have $M_z = M$, which is a constant of time. The equations in (17.26) have a steady-state solution, if

$$\begin{vmatrix} i\omega & \gamma\{B_0 + (N_y - N_z)M\} \\ -\gamma\{B_0 - (N_z - N_x)M\} & i\omega \end{vmatrix} = 0. \tag{17.27}$$

Therefore, the ferromagnetic resonance frequency in the applied field B_0 can be obtained from (17.27), by solving

$$\omega_0^2 = \gamma^2\{B_0 + (N_y - N_z)M\}\{B_0 + (N_x - N_z)M\}. \tag{17.28}$$

For a spherical sample, $N_x = N_y = N_z$, so that

$$\omega_0 = \gamma B_0.$$

If B_0 is applied in perpendicular and parallel directions to a flat-plate specimen, we have $N_x = N_y = 0$, $N_z = 1$, and $N_x = N_z = 0$, $N_y = 1$, respectively, for which the resonance frequencies are obtained as

$$\omega_0 = \gamma(B_0 - \boldsymbol{M}) \quad \text{and} \quad \omega_0 = \gamma\sqrt{B_0(B_0 + \boldsymbol{M})}.$$

Values of γ are normally expressed as $g\beta$, where experimental values of g are reported as 2.10, 2.18 and 2.21 for metallic specimens of Fe, Co and Ni, respectively. These ω_0 are called ferromagnetic resonance frequencies, which may be regarded as of a *soft mode*, expressing Larmor's frequencies of $\boldsymbol{M}$ in the effective field $\boldsymbol{B} = \boldsymbol{B}_0 - (N)\boldsymbol{M}$.

17.9.2 Antiferromagnetic resonance

While ferromagnetic resonance experiments are performed with external fields $\boldsymbol{B}_0$ and $\boldsymbol{B}_1$, the antiferromagnetic resonance is studied by neutron inelastic scatterings at Brillouin-zone boundaries. Nevertheless, theoretically we can interpret magnetic fluctuations by analogy with ferromagnetic resonance experiments. As illustrated in figure 17.6(b), magnetic fluctuations $\Delta\omega$ can be displayed in intensity spectra around the center ω_0.

Consider an antiferromagnetic crystal that is characterized by two sublattice magnetizations $\boldsymbol{M}_A$ and $\boldsymbol{M}_B$. Although primarily antiparallel near the transition point, these are strongly perturbed by internal exchange fields $-\lambda\boldsymbol{M}_B$ and $-\lambda\boldsymbol{M}_A$ plus the anisotropic fields $\boldsymbol{B}_{aniso}$ and $-\boldsymbol{B}_{aniso}$, respectively. The anisotropic field is taken in the direction z of easy magnetization axis, representing lattice strains that are responsible for magnetic fluctuations in the motion of $\boldsymbol{M}_A$ and $\boldsymbol{M}_B$. In the absence of external fields, $\boldsymbol{M}_A$ and $\boldsymbol{M}_B$ are in precession around the effective fields $\pm\boldsymbol{B}_{aniso}$. Hence, we can write

$$\boldsymbol{B}_A = -\lambda\boldsymbol{M}_B + \boldsymbol{B}_{aniso} \quad \text{and} \quad \boldsymbol{B}_B = -\lambda\boldsymbol{M}_A - \boldsymbol{B}_{aniso},$$

as shown in figure 17.11(b). Assuming that $M_{Az} = +M$ and $M_{Bz} = -M$, equations of motion for the x- and y-components are

$$\begin{aligned} \dot{M}_{Ax} &= \gamma\{M_{Ay}(\lambda M + B_{aniso}) - M(-\lambda M_{By})\} \\ \dot{M}_{Ay} &= \gamma\{M(-\lambda M_{Bx}) - M_{Ax}(\lambda M + B_{aniso})\} \end{aligned} \tag{17.29}$$

and

$$\begin{aligned} \dot{M}_{Bx} &= \gamma\{M_{By}(-\lambda M - B_{aniso}) - (-M)(-\lambda M_{Ay})\} \\ \dot{M}_{By} &= \gamma\{(-M)(-\lambda M_{Ax}) - M_{Bx}(-\lambda M - B_{aniso})\}. \end{aligned}$$

Writing $M_{A+} = M_{Ax} + iM_{Ay}$; $M_{B+} = M_{Bx} + iM_{By}$, and assuming that all of these M rotate as proportional effectively to $\exp(-i\omega_s t)$, where ω_s is the frequency determined by the timescale of precession. Hence, the above equations become

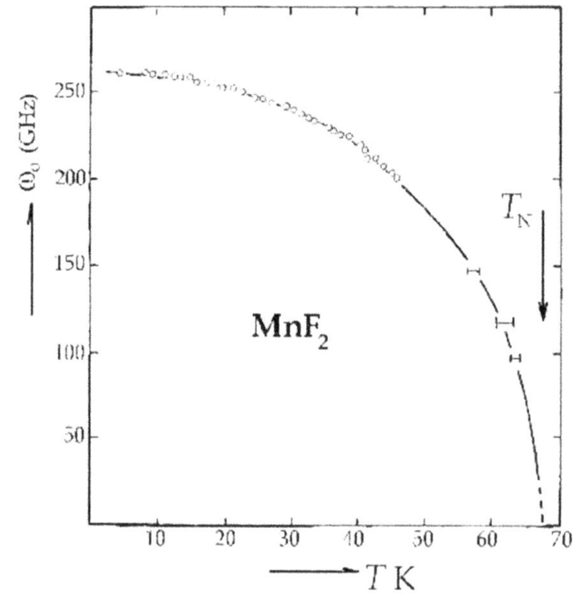

Figure 17.12. Antiferromagnetic resonance frequency versus temperature for MnF_2.

$$-i\omega_s M_{A+} = i\gamma\{M_{A+}(B_{aniso} + B_{ex}) + M_{B+}B_{ex}\}$$
$$-i\omega_s M_{B+} = i\gamma\{M_{B+}(B_{aniso} + B_{ex}) + M_{A+}B_{ex}\},$$

where $B_{ex} = \lambda M$. These equations have a solution, if

$$\begin{vmatrix} \gamma(B_{aniso} + B_{ex}) - \omega_s & \gamma B_{ex} \\ \gamma B_{ex} & \gamma(B_{aniso} + B_{ex}) + \omega_s \end{vmatrix} = 0.$$

Accordingly, the resonance frequency is given by

$$\omega_{so}^2 = \gamma^2 B_{aniso}(B_{aniso} + 2B_{ex}). \tag{17.30}$$

However, the condition $B_{ex} > B_{aniso}$ is common among antiferromagnets studied, so that

$$\omega_{so} \approx \gamma\sqrt{B_{aniso}\, B_{ex}}$$

is an adequate formula to identify the fluctuation mode. For a typical antiferromagnet MnF_2, estimated values were $B_{ex} = 540$ kG and $B_{aniso} = 8.8$ kG at 0 K, for which the resonant frequency ω_{so} was predicted as 280 GHz, while experimentally the resonance was found at 261 GHz. In fact, ω_{so} was temperature-dependent in MnF_2, as shown in figure 17.12, implying that such magnetic fluctuations represent an *adiabatic soft mode with a diminishing frequency* toward $T_N = 67$K.

In the foregoing theory of antiferromagnetic resonance, Equations (17.29) and (17.30) are obtained from displacements u_M in equations of motion, describing the time variation in periodic magnetic fields. Hence, the Weiss postulate $B_{ex} = \lambda \boldsymbol{M}(\phi)$ is interpreted with magnetic lattice displacements as $B_{ex} = -\nabla u_M \propto \nabla u_L$, where

$u_M \propto -u_L$. The displacement $(\Delta \boldsymbol{r})_M$ is driven by the soliton potential energy proportional to u_M^2, to which the temperature-dependent B_{ex} is related.

In this section, the ferromagnetic resonance formula is written with an applied field $\boldsymbol{B}_1 \exp i\omega t$, whereas in antiferromagnetic resonance it is driven with $(u_M)_0 \exp i\omega_s t$, where the frequency ω_s is determined by the timescale of the displacement $(\Delta \boldsymbol{r})_M$. Nevertheless, both resonances are analyzed by analogy of conventional magnetic resonance experiments with external $\boldsymbol{B}_1 \exp i\omega t$ (see chapter 6).

References

[1] Kittel C 1976 *Introduction to Solid State Physics* 5th edn (New York: Wiley)
[2] Moriya T 1963 *Magnetism I* p (New York: Academic), 85
[3] Kanamori J 1963 *Magnetism I* p (New York: Academic), 128
[4] Kanamori J 1968 *Magnetism* (Tokyo: Baifukan) (in Japanese)

Chapter 18

Crystalline polymers and liquid crystals

We planned to discuss the phase relation in crystalline polymer and liquid crystals with respect to the soliton theory. Subject to experimental results available in literature, however, we encountered the fact that most studies on these systems were not carried out in the thermodynamic environment, so it is not appropriate to discuss theoretical matters with *optically observed results*. This is because optical observations do not necessarily provide the same conclusion on the structure determined in thermodynamic experiments. Although it is the matter of the time-scale of observations, the results cannot be evaluated with regard to the solution theory, so we could discuss the matter, but referring only to optical measurements on domain structure that was studied with *polarized light*.

Although the soliton theory established for thermodynamic experiments is not applicable to optical observation in principle, some useful ideas might emerge from exploring the problem. Nevertheless, some computational studies designed for the former were found to be useful to gain insight for interpreting the latter results at least qualitatively.

18.1 Transversal correlations in crystalline polymers

While the soliton theory in crystals deals primarily with propagation of collective order variables, we emphasized the significance of transversal correlations among them from experimental point of view. Following Hopfinger *et al* [1], we discuss in this section that such transversal interactions are evident from computational studies of crystalline polymers.

18.1.1 Polyvinylidene fluoride

Polymerized vinylidene fluorides, known as polyvinylidene fluoride or PVDF, are a technologically important plastic material, regarding piezoelectric and pyroelectric applications among others.

doi:10.1088/978-0-7503-2572-1ch18

PVDF has a glass transition temperature of about −35 °C, and is normally 50%–60% crystalline. To fabricate piezoelectric plastics, PVDF is mechanically stretched to orient the polymer molecules, and then *poled* under applied tension. The polymer exists in three phases, α, β and γ, with respect to *trans* (T) and *gauche* (G) configurations of the unit $-CH_2-CF_2-$ in chain structure, namely TGTG′, TTTT and TTTGTTTG′, respectively. Thus, poled PVDF is ferroelectric, exhibiting piezoelectric and pyroelectric properties.

The β-phase polymer is produced by stretching film of α-polymer. Probably, the backbone structure of chained molecules is *tortuous* in α-phase, whereas *planar zigzag* in β-phase, so that a longitudinal stress can make the transition from α to β easily. We may consider the β-PVDF is in a stable crystalline phase. Figure 18.1 is a sketch of molecular configuration in β-phase of PVDF, showing orthorhombic arrangement in the crystalline phase.

18.1.2 Numerical evidence of transverse correlations in β-PVDF

The stable structure of PVDF polymer in β-phase is illustrated in figure 18.2(b), showing a long straight zig-zag chain of –C–C–C–C– in planar structure. Hopfinger *et al* [1] assumed that repeated monomer units $-CH_2-CF_2-$ execute torsional fluctuations along the chain, which are not only correlated longitudinally, but also in transversal directions. For a poling process of β-PVDF polymer, the results of their computational studies indicated clear evidence for deformed structure in nonlinear propagation.

Denoting the torsional angle of a unit monomer by ϕ_n, where n is its site, the total correlation energy can be expressed by

$$\mathcal{H} = \sum_n \left\{ \frac{1}{2} I \dot{\phi}_n^2 + K(\phi_n - \phi_{n\pm1})^2 + J_1 \cos \phi_n - J_2 \cos 2\phi_n \right\},$$

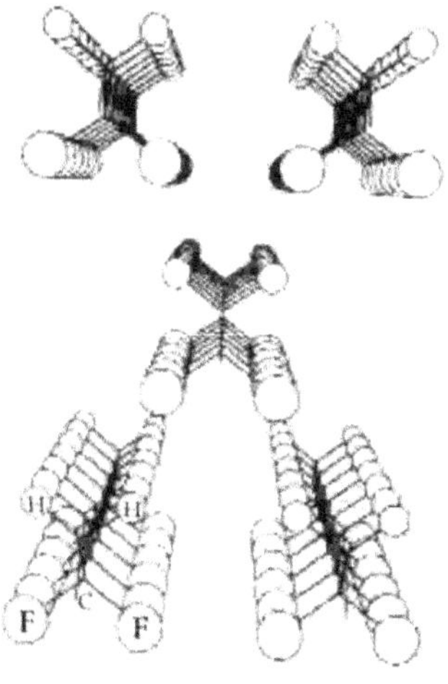

Figure 18.1. Structure of crystalline β-PVDF.

where I and K are the moment of inertia and spring constant, respectively; the terms of J_1 and J_2 are added to deal with remote correlations. Here, it is realized that transversal correlations are taken into consideration in such a classical Hamiltonian $\mathcal{H}$.

For a poling process, the equation of motion is written as

$$I\ddot{\phi}_n = -J_1 \cos \phi_n - 2J_2 \cos 2\phi_2 + K(\phi_{n+1} - 2\phi_n + \phi_{n-1}) - 2I\kappa\dot{\phi}_n + F_n(t),$$

where κ is a damping constant, and $F_n(t)$ represent an external force due to *poling field*.

In the continuum approximation, applicable at low temperature, the speed of propagation v was calculated for polymer of thickness x_0, as

$$\frac{x_0}{a} = \sqrt{\frac{K}{4J_2 + \left(J_1^2/I\kappa\right)}} \quad \text{and} \quad \frac{v}{a} = \sqrt{\frac{K}{I + \left(4J_2I^2/J_1^2\right)}},$$

where a is the repeat unit in a polymer molecule. Numerical solutions of the equation of motion were displayed as a computer-generated movie-film. A sample frame is shown in figure 18.2(a), where the variable ϕ is characterized by twisted propagation of polymer chain. This computer simulation thus verified that transverse correlations always exist with classical vector order variables. Figure 18.2(b) shows the structure of a static polymer to compare with twisted structure that *can be interpreted in terms of the cnoidal potential with Kac's theory in chapter 11*. However, for the correct analysis, their results should further be studied under varying external force $\Delta F_n(t)$ to perform the correct analysis.

Exhibiting typical nonlinearity, the poling force $F_n(t)$ in the above analysis can be characterized in soliton theory as proportional to d(sech ϕ)/dϕ, as it is essentially a function of the phase ϕ due to inevitable transversal correlations, sharing the same conclusion with the soliton theory. This is also a consistent view on the *ϕ-particle model* proposed by Rice [2] on CDW systems. Therefore, such a computational result can be considered for supporting the phase soliton theory for transverse correlations between chain molecules.

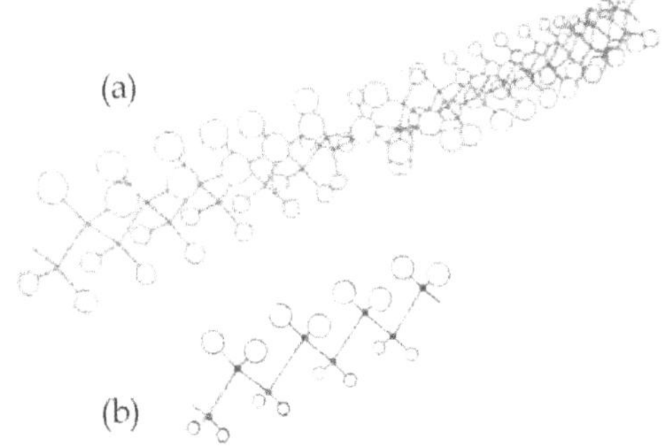

Figure 18.2. (a) A computer-simulated structural propagation in β-PVDF. (b) A schematic view of –$(CH_2–CF_2)_n$– chain.

Considering the computer simulation as illustrated by figure 18.1, the twisted propagation of angles $\pm\phi$ are evident from figure 18.2(a), which can be attributed to transversal correlations between molecular chains. Further referring to figure 18.2(b), unknown C_m-screw symmetry may appear along the molecular axis in a different crystalline phase of β-PVDF, where there may be a domain structure with respect to the variable ϕ, permitting us to identify the origin of fluctuations in specific thermodynamic setting.

18.2 Liquid crystals

Experimental studies on liquid crystals have not been carried out logically under thermodynamic principles with respect to today's literature that are conveniently observed results by *optical observations under polarizing light*. A typical observation of liquid crystals is made through parallel glass plates, as illustrated by figure 18.3(a) and (b), where the external pressure p is not controllable while the temperature T can be maintained as constant. Technically, the experiments at constant p were a technically difficult task in practice. In this section, we therefore can comment only on physics of liquid crystals in brief qualitative manner, assessing the validity of soliton theory.

18.2.1 Lattice-like structure of liquid crystals and the correlation energy between parallel layers

Many large organic molecules in rod- and ring-shapes in solvent liquid are known as coagulated, forming a lattice structure constructed by their correlation energy. Thermodynamically, they are embedded in solvents as heat reservoir, exhibiting the domain pattern characterized with a polarizing microscope. In this sense, the experimental situation of liquid crystals is similar to polymer crystals discussed in section 12.8.

18.2.2 Onsager's order variables

For a rod-like molecule, Onsager proposed that its orientational parameter can be used for an order variable for ordering processes in liquid crystals [3]. However, it is required for the order variable to be defined in more accurate detail than in conventional binary order, in order for soliton theory to deal correctly with transition phenomena.

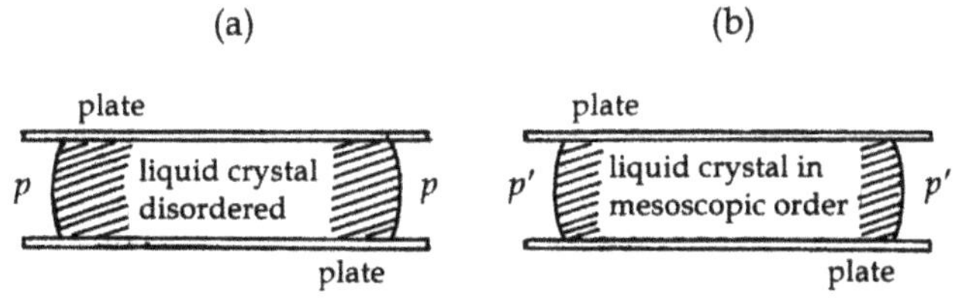

Figure 18.3. Experimental setups: (a) for a disordered phase, and (b) an ordered phase in liquid crystals.

Large molecules in liquid crystals are generally in flexible shape, although their overall structure appears as if rigid, allowing Onsager to see them as rod-shaped objects. However, as considered with the Born–Huang principle, they can be represented as rigid after *phasing* as proposed by Landau; otherwise, the Bragg–Williams ordering mechanism can hardly be suggested for the rigid model. Figure 18.3(b) shows a molecular arrangement for typical domain structure in liquid crystals, whereas a helical stack of periodically ordered layers is recognized characterized by the period $-\boldsymbol{G}/2 \leqslant \boldsymbol{q} \leqslant \boldsymbol{G}/2$ in lattice structure. Theoretically, it is expected that a periodic pattern of *shifting and twisting* $\mathrm{sech}^2\,\phi$-peaks, is like that shown in figure 18.2(b) for a crystalline polymer under intense polarized light. However, no such experimental studies have actually been reported for discussion in the present literature.

Such a model of order variable can be easily acceptable for rod-shaped molecules, while for ring-shaped molecules a vector perpendicular to the ring can be only adequate, thereby treating these complicated cases in common with the soliton theory. However, to confirm the model, it is necessary to study the critical region.

18.2.3 Optical observation of the mesoscopic structure

Assuming that the light beam of observation is polarized in parallel to a symmetry axis, the domain pattern characterized by $\boldsymbol{q} \rightleftarrows -\boldsymbol{q}$ and $\Delta n = 0$ can be studied optically for isothermal changes. Experimental devices, as illustrated in figure 18.3, are properly designed to view domain profiles for isothermal changes specified by ΔT at $\Delta p = 0$. On the other hand, the *pseudopotential* determines adiabatic changes in topological correlation characterized by $\boldsymbol{q} \rightleftarrows -\boldsymbol{q}$ and $\Delta n \neq 0$, as illustrated in figure 18.4(a and b), which should, however, be analyzed under *varying external pressure* $\Delta p_o \exp i\bar{\omega}t$ *at constant temperature.*

Most domain patterns in liquid crystals under various geometrical and thermal conditions in the surroundings are described in the book by de Gennes and Prost [4], where the observed patterns are discussed with no reference to the soliton theory, but with respect to room temperature under atmospheric pressure. All examples they showed in their book are considered as consistent in thermodynamic interpretation, however, requiring results of experimental confirmations. We consider that a large heat capacity of solvent is responsible for entropy production in liquid crystals, which needs to be specified.

Experimentally, more serious questions should be focused on the timescale of measurements. Considering the timescales of inversion $\boldsymbol{q} \rightleftarrows -\boldsymbol{q}$ in mesoscopic states that are comparable with *visible light* for optical observation, *the adiabatic variation cannot be resolved in principle*, making interpretable results different from thermodynamic investigations.

De Gennes and Prost discussed many important cases in their book [4], as previously discussed for a typical case in polymer molecule in computer-simulated twist for β-PVDF.

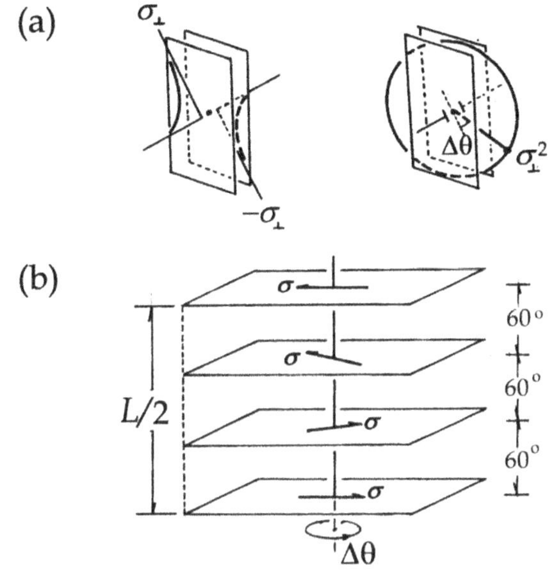

Figure 18.4. A typical model for a helical domain structure in liquid crystals. The periodic pattern should by analyzed for layer domain structure either by a specifically designed optical detector or under variable applied pressure.

18.2.4 Static distortions in liquid crystals

According to the soliton theory, those domain patterns are observed optically consequent on binary transitions in liquid crystals within given boundary conditions. Such patterns clearly show the elastic properties of a solvent material, attributing to a change of the Gibbs' potential along a specific axis of binary order variable $\boldsymbol{\sigma} = n_s\boldsymbol{\sigma}_o$, where n_s and $\boldsymbol{\sigma}_o$ are the soliton number and unit vector, respectively.

Considering special derivatives of $\boldsymbol{\sigma}_o$, a strain tensor of rank two can be defined as

$$e_{ij} = \frac{1}{2}(\partial\sigma_{oj}/\partial i - \partial\sigma_{oi}/\partial j),$$

we obtain that

$$e_{zz} = 0, \quad e_{zx} = \frac{1}{2}(\text{curl}\,\boldsymbol{\sigma}_o)_y, \quad e_{zy} = -\frac{1}{2}(\text{curl}\,\boldsymbol{\sigma}_o)_x \text{ and } \text{div}\,\boldsymbol{\sigma}_o = e_{xx} + e_{yy};$$

with which the Gibbs function can be expressed as

$$G(\boldsymbol{\sigma}_o) = \frac{1}{2}K_1(\text{div}\,\boldsymbol{\sigma}_o)^2 + \frac{1}{2}K_2(\boldsymbol{\sigma}_o\cdot\text{curl}\,\boldsymbol{\sigma}_o)^2 + \frac{1}{2}K_3(\boldsymbol{\sigma}_o\times\text{curl}\,\boldsymbol{\sigma}_o)^2.$$

The first term of K_1 in this expression can be used for an *isothermal process*, because of the relation div $\boldsymbol{\sigma}_o \neq 0$ for $\Delta V \neq 0$, otherwise insignificant $\Delta V = 0$. In contrast, the other terms of K_2 and K_3 are essential for an *adiabatic process.*

That was considered as the basic equation of continuum theory for *nematic phases* surrounded by given boundaries under (p, T) conditions. In this expression, the first term of K_1 can be ignored for an isothermal process, which otherwise is significant for an adiabatic process, due to div $\boldsymbol{\sigma}_o = 0$ or $\neq 0$, respectively.

There are a large number of interesting cases in liquid crystals for thermodynamic analysis, which are mostly discussed in the book by de Gennes and Prost [4], and all are assumed to be consistent with the soliton theory. Therefore, we shall discontinue our arguments at this stage, leaving all details to their discussions in [4].

Exercises

1. Discuss that the optically observable $\langle \Delta n \rangle_{\text{opt}}$ may not be resolved with respect to observing time-scale t_o of the practical measurement.
2. Discuss your experiments designed to observe isothermal and adiabatic processes in liquid crystals. Expecting to encounter technical difficulties with your apparatus, discuss practical problems in detail to see if soluble within available technology.

References

[1] Hopfinger A J, Lewanski A J, Sluckin T J and Taylor P L 1978 Solitary wave propagation as a model for poling in PVDF *Solitons and Condensed Matter Physics* ed A R Bishop and T Schneider (Berlin: Springer)

[2] Rice M J 1978 *Charge Density Wave Systems: The ϕ-Particle Model in Soliton and Condensed Matter Physics* ed A R Bishop and T Schneider (Berlin: Springer)

[3] Savin A V 2001 Topological solitons in crystalline poly-tetrafluoroethylene *Polym. Sci. Ser.* A **43** 860

[4] de Gennes P G and Prost J 1993 *The Physics of Liquid Crystals* 2nd edn (Oxford: Clarendon)

IOP Publishing

Solitons in Crystalline Processes (2nd Edition)
Irreversible thermodynamics of structural phase transitions and superconductivity
Minoru Fujimoto

Concluding remarks

Phase inversion of pseudospins in collective mode along an easy axis is considered as the basic mechanism for structural transitions, leading to mesoscopic domain structure in crystals. Nonlinear pseudospins are collectively driven in phase with the Weiss adiabatic potential that emerges at the critical temperature consequent on mutual correlations in crystals in finite size. All of this has been confirmed with existing experimental results.

Entropy production during transitions constitutes a basic process for mesoscopic disorder, whereby the energy dissipation occurs in the lattice due to inelastic scatterings of random phonons, which is identical to the Born–Huang principle proposed in their theoretical work on lattice dynamics, subjecting the process in the hierarchy of boson statistics of solitons.

Nonlinear pseudospin waves, detected in crystals as soft modes, are composed of longitudinal and transversal components. The former describe propagation through crystalline media, while the latter occur as related to mutual correlations among neighboring modes. With modulated amplitude and phase, such nonlinear waves can be attributed to a complex soliton potential that is responsible for structural changes between mesoscopic domains specified by symmetry group and pseudo-symmetry. Known as the Weiss field, the soliton potential leads to separation of a crystal with respect to the surroundings, where the entropy production exhibits a change in the soliton density, along with isothermal and adiabatic anomalies.

Such binary transitions in crystals are described in terms of phase solitons, where the critical anomalies arise from singularities of soliton waves interacting with phonons by inelastic scatterings. In entropy production processes, initial fluctuations are inevitable in isothermal changes, while transversal phase fluctuations are precursory for domain structure to attain new equilibrium conditions; all can logically be analyzed algebraically by the Toda theorem. Unlike random phonons, the soliton mobility is restricted in space–time and by pseudosymmetry in the lattice. Nevertheless, referring to collective modes, order variable is by no means identical to the soliton variable itself, but should be determined in individual phase.

Representing nonlinear displacements, so-called soft modes exist in irreversible processes as characterized by temperature-dependent frequencies in isothermal processes, and pressure-dependent in adiabatic processes as well. For piezoelectric and elastomagnetic pseudospins, soft modes signify mesoscopic structural changes in dielectric and elastic lattices.

Interpreting Fröhlich's electron–lattice interaction by the soliton theory, the superconductivity can take place in association with solitons, where the classical Biot–Savart law can relate the Cooper pair with the Meissner effect observed bellow T_c. Characterized by charge–current continuity theorem, the soliton-potential energy in momentum space is responsible for superconducting charge–current relation in superconducting materials of all types, where a symmetric pair of two charged solitons displays the nonlinear process by means of Cooper pairs.

In contrast, magnetic systems are dominated by short-range spin correlations, so that the nonlinearity is not particularly significant, where the mean-field approach is sufficiently adequate for most magnetic phenomena.

Although complex mathematically, the soliton theory can be abstracted for irreversible processes in equilibrium crystals to be composed by stepwise isothermal and adiabatic processes in sequence from one conservative state to another, as if repeating the Carnot cycle. Restricted by structural symmetry, the soliton wave in two components is essential in hyperbolic variations under equilibrium conditions, where the transversal component can be responsible for interactions with surroundings. Guided by a Klein–Gordon equation, the Korteweg–deVries equation predicts the Weiss field, whereby mesoscopic domain phases can be algebraically analyzed with respect to transversal correlations. Corresponding to longitudinal correlations, the soliton theory is essential for propagation, while the transversal component yields domain structure; both signified by the soliton density in discrete structure.

Regarding structural phase changes, significant findings are that critical transition points occur either in elemental *isothermal* or *adiabatic* processes initiated by either clusters or nonlinear displacements, respectively, as substantiated particularly by high-pressure experiments of superconductivity. Although signified by structural difference, all superconducting crystals seem to have a common mechanism for charged particles in a conduction band to modulate the host lattice by site-ionization, as proposed originally by Fröhlich. All in all, principles of irreversible thermodynamics are established for crystalline processes with quantized solitons by new lattice symmetry.

Experimentally however, the adiabatic processes need to be studied under variable external pressure, although prohibited in some cases by the present technology.

Sharing basic thoughts with Professor de Gennes on liquid crystals, it is my pleasure to quote the following:

'Well, we have seen each other', said the unicorn 'If you believe in me, I'll believe in you.
Is that a bargain?'
'Yes, if you like.' said Alice.

IOP Publishing

Solitons in Crystalline Processes (2nd Edition)
Irreversible thermodynamics of structural phase transitions and superconductivity
Minoru Fujimoto

Appendix A

Hyperbolic and elliptic functions

A.1 Hyperbolic functions

Definitions:

$$\sinh x = \frac{e^x - e^{-x}}{2}, \quad \cosh x = \frac{e^x + e^{-x}}{2}, \quad \tanh x = \frac{\sinh x}{\cosh x} = \frac{e^x - e^{-x}}{e^x + e^{-x}}$$
$$\operatorname{csch} x = \frac{1}{\sinh x}, \quad \operatorname{sech} x = \frac{1}{\cosh x}, \quad \coth x = \frac{1}{\tanh x} = \frac{\cosh x}{\sinh x}$$
$$\sinh(-x) = -\sinh x, \quad \cosh(-x) = \cosh x, \quad \tanh(-x) = -\tanh x$$

Relations:

$$\cosh^2 x - \sinh^2 x = 1$$
$$\operatorname{sech}^2 x = 1 - \tanh^2 x$$
$$\operatorname{csch}^2 x = \coth^2 x - 1$$

Derivatives:

$$\frac{\mathrm{d} \sinh x}{\mathrm{d}x} = \cosh x, \quad \frac{\mathrm{d} \cosh x}{\mathrm{d}x} = \sinh x$$
$$\frac{\mathrm{d} \tanh x}{\mathrm{d}x} = \operatorname{sech}^2 x = \frac{1}{\cosh^2 x}, \quad \frac{\mathrm{d} \coth x}{\mathrm{d}x} = -\operatorname{csch}^2 x = \frac{-1}{\sinh^2 x}$$
$$\frac{\mathrm{d} \operatorname{sech} x}{\mathrm{d}x} = -\tanh x \operatorname{sech} x, \quad \frac{\mathrm{d} \operatorname{csch} x}{\mathrm{d}x} = -\coth x \operatorname{csch} x$$

doi:10.1088/978-0-7503-2572-1ch20

A.2 Elliptic integrals

Elliptic integrals of the first kind:

$$F(\kappa, \varphi) = \int_0^{\varphi} \frac{d\varphi}{\sqrt{1 - \kappa^2 \sin^2 \varphi}} = \int_0^{\sin \varphi} \frac{dz}{\sqrt{(1 - z^2)(1 - \kappa^2 z^2)}},$$

and the complete elliptic integral is

$$F\left(\kappa, \frac{\pi}{2}\right) = \int_0^{\frac{\pi}{2}} \frac{d\varphi}{\sqrt{1 - \kappa^2 \sin^2 \varphi}} = \int_0^1 \frac{dz}{\sqrt{(1 - z^2)(1 - \kappa^2 z^2)}} = K(\kappa).$$

Elliptic integral of the second kind:

$$E(\kappa, \varphi) = \int_0^{\varphi} \sqrt{1 - \kappa^2 \sin^2 \varphi}\, d\varphi = \int_0^{\sin \varphi} \sqrt{\frac{1 - \kappa^2 z^2}{1 - z^2}}\, dz,$$

and the complete elliptic integral is

$$E\left(\kappa, \frac{\pi}{2}\right) = \int_0^{\frac{\pi}{2}} \sqrt{1 - \kappa^2 \sin^2 \varphi}\, d\varphi = \int_o^1 \sqrt{\frac{1 - \kappa^2 z^2}{1 - z^2}}\, dz = E(\kappa),$$

$$= \int_0^{K(\kappa)} dn^2(u, \kappa) du.$$

$dn(u, \kappa)$ is Jacobi's dn-function.

Defining $\kappa' = \sqrt{1 - \kappa^2}$, $K(\kappa') = K'$ and $E(\kappa') = E'$, we have the relation

$$EK' + E'K - KK' = \frac{\pi}{2} \quad \text{(Legendre's relation.)}$$

A.3 Jacobi's elliptic function

From the elliptic integral $u(x) = \int_0^x \frac{dz}{\sqrt{(1 - z^2)(1 - \kappa^2 z^2)}}$, we write the reverse function as

$$z = \text{sn}\, u = \text{sn}(u, \kappa),$$

which is Jacobi's sn-function. Considering the relations with trigonometric and hyperbolic functions, we also define the corresponding cn- and dn-functions by

$$\text{cn}^2\, u = 1 - \text{sn}^2\, u \quad \text{and} \quad \text{dn}^2\, u = 1 - \kappa^2 \text{sn}^2\, u.$$

In the limit of $\kappa \to 0$,

$$\text{sn}\, u \to \sin u, \quad \text{cn}\, u \to \cos u \quad \text{and} \quad \text{dn}\, u \to 1.$$

On the other hand, if $\kappa \to 1$, we have

$$\text{sn}\, u \to \tanh u, \quad \text{cn}\, u \to \text{sech}\, u \quad \text{and} \quad \text{dn}\, u \to \text{sech}\, u.$$

Differential formula:

$$\frac{\mathrm{d}\,\mathrm{sn}\,u}{\mathrm{d}u} = \mathrm{cn}\,u\ \mathrm{dn}\,u, \qquad \frac{\mathrm{d}\,\mathrm{cn}\,u}{\mathrm{d}u} = -\mathrm{sn}\,u\,\mathrm{dn}\,u, \qquad \frac{\mathrm{d}\,\mathrm{dn}\,u}{\mathrm{d}u} = -\,\kappa^2\,\mathrm{sn}\,u\,\mathrm{cn}\,u.$$

Expansion formula:

$$\mathrm{sn}\,u = u - \frac{1+\kappa^2}{3!}u^3 + \frac{1+14\kappa+\kappa^4}{5!}u^5 + \cdots,$$

$$\mathrm{cn}\,u = 1 - \frac{1}{2}u^2 + \frac{1+4\kappa^2}{4!}u^4 - \frac{1+44\kappa^2+16\kappa^4}{6!}u^6 + \cdots,$$

$$\mathrm{dn}\,u = 1 - \frac{\kappa^2}{2}u^2 + \frac{(4+\kappa^2)\kappa^2}{4!}u^4 - \frac{(16+44\kappa^2+\kappa^4)\kappa^2}{6!}u^6 + \cdots$$

Reference books

Elliptic functions and integrals are not quite familiar mathematics with physicists and engineers today. I myself learned them from Professor Morikazu Toda's book *Introduction to Elliptic Functions* in Japanese (Nippyo, Tokyo, 2001), where I found the following references in the available literature.

Bowman F 1961 *Introduction to Elliptic Functions with Applications* (New York: Dover)

Greenhill A G 1959 *The Applications of Elliptic Functions* (New York: Dover)

Hancock H 1958 *Lecture on the Theory of Elliptic Functions* (New York: Dover)

Cayley A 1961 *An Elementary Treatise on Elliptic Functions* (New York: Dover)

E T Whittaker and G N Watson 1958 *A Course of Modern Analysis* (Cambridge: Cambridge University Press)

Milne-Thomson L M 1950 *Jacobian Elliptic Function Tables* (New York: Dover)

Online references

http://en.wikipedia.org/wiki/Hyperbolic_function
http://en.wikipedia.org/wiki/Elliptic_function
http://en.wikipedia.org/wiki/Elliptic_integral

www.ingramcontent.com/pod-product-compliance
Lightning Source LLC
Chambersburg PA
CBHW082132210326
41599CB00031B/5955
9780750325714